Wiley Survival Guide in Global Telecommunications

BROADBAND ACCESS, OPTICAL COMPONENTS AND NETWORKS, AND CRYPTOGRAPHY

WILEY SURVIVAL GUIDES IN ENGINEERING AND SCIENCE

Emmanuel Desurvire, Editor

Wiley Survival Guide in Global Telecommunications: Signaling Principles, Network Protocols, and Wireless Systems *Emmanuel Desurvire*

Wiley Survival Guide in Global Telecommunications: Broadband Access, Optical Components and Networks, and Cryptography
Emmanuel Desurvire

Probability and Random Processes: Handy Reference
Venkatarama Krishnan

Wiley Survival Guide in Global Telecommunications

BROADBAND ACCESS, OPTICAL COMPONENTS AND NETWORKS, AND CRYPTOGRAPHY

Emmanuel Desurvire

A John Wiley & Sons, Inc., Publication

Copyright © 2004 by John Wiley & Sons, Inc. All rights reserved.

Published by John Wiley & Sons, Inc., Hoboken, New Jersey.
Published simultaneously in Canada.

No part of this publication may be reproduced, stored in a retrieval system, or transmitted in any form or by any means, electronic, mechanical, photocopying, recording, scanning, or otherwise, except as permitted under Section 107 or 108 of the 1976 United States Copyright Act, without either the prior written permission of the Publisher, or authorization through payment of the appropriate per-copy fee to the Copyright Clearance Center, Inc., 222 Rosewood Drive, Danvers, MA 01923, 978-750-8400, fax 978-646-8600, or on the web at www.copyright.com. Requests to the Publisher for permission should be addressed to the Permission Department, John Wiley & Sons, Inc., 111 River Street, Hoboken, NJ 07030, (201) 748-6011, fax (201) 748-6008.

Limit of Liability/Disclaimer of Warranty: While the publisher and author have used their best efforts in preparing this book, they make no representations or warranties with respect to the accuracy or completeness of the contents of this book and specifically disclaim any implied warranties of merchantability or fitness for a particular purpose. No warranty may be created or extended by sales representatives or written sales materials. The advice and strategies contained herein may not be suitable for your situation. You should consult with a professional where appropriate. Neither the publisher nor author shall be liable for any loss of profit or any other commercial damages, including but not limited to special, incidental, consequential, or other damages.

For general information on our other products and services please contact our Customer Care Department within the U.S. at 877-762-2974, outside the U.S. at 317-572-3993 or fax 317-572-4002.

Wiley also publishes its books in a variety of electronic formats. Some content that appears in print, however, may not be available in electronic format.

Library of Congress Cataloging-in-Publication Data:

Desurvire, Emmanuel, 1955-
  Wiley survival guide in global telecommunications: Broadband Access, Optical
  Components and Networks, and Cryptography / Emmanuel Desurvire.
      p. cm.
  Includes bibliographical references and index.
  ISBN 0-471-67520-2 (cloth)
  1. Wireless communication systems. 2. Cryptography. I. Title.

TK5103.2.D49 2004
621.382--dc22

2004041198

Printed in the United States of America

10 9 8 7 6 5 4 3 2 1

Contents

Foreword, ix
Preface, xiii
Acronyms, xix

Introduction: The Network Cloud, 1

CHAPTER 1
Broadband Wireline Access, 11

1.1 From Telephone and Cable-TV Networks to the Internet, 11

1.2 Digital Subscriber Line (DSL), 14
    1.2.1 xDSL Flavors, 17
    1.2.2 DSL Modulation and Coding Schemes, 22
    1.2.3 DSL-signal Physical Impairments, 31
    1.2.4 DSL Transfer Modes and Framing, 34
    1.2.5 xDSL Services, 50
    1.2.6 Voice Over DSL (VoDSL), 51
    1.2.7 Incumbent and Competitive Local-Exchange Carriers (ILEC/CLEC), 52

1.3 Fiber in the Loop (FITL), 55
    1.3.1 Passive Optical Networks (PON), TDM vs. WDM, 58
    1.3.2 ATM and Ethernet Framing (APON/EPON) and Gigabit-PON (GPON), 66
    1.3.3 Optical Access, 75
    1.3.4 Hybrid Fiber-Coaxial (HFC) Systems, 76

1.4 Home Networking, 82
    1.4.1 The Evolution of Intelligent Home Appliances, 82
    1.4.2 An Internet Intrusion, 84
    1.4.3 Wiring the House Network, 86
    1.4.4 House-Network Control, 88
        My Vocabulary, 89

CHAPTER 2
Optical Fiber Communications, Components and Networks, 91

2.1 Optical Communications: from Wireless to Wireline, 91
    2.1.1 The Early Times of Wireless Optical Communications, 92

2.1.2 The Conceptual Grounds of Wireline Optical Communications, 94

2.2 Basic Physics of Light-Wave Signals, 96
    2.2.1 Light's Electromagnetic Nature, 97
    2.2.2 The Speed of Light, 98
    2.2.3 Light Reflection, Refraction and Polarization, 100
    2.2.4 Classical and Quantum Natures of Light, 110
    2.2.5 Optical Amplification, 116
    2.2.6 Light and Photo-Current Generation in Semiconductors, 121
    2.2.7 Lasers and Coherent Light Generation, 123

2.3 Optical Waveguides, 132
    2.3.1 Ray Propagation in Index Layers and Waveguiding, 133
    2.3.2 Mirror Waveguide, 137
    2.3.3 Dielectric Waveguide, 138
    2.3.4 Glass Fiber Waveguides, 149
    2.3.5 Fiber Loss and Dispersion, 156
    2.3.6 Single-Mode Transmission-Fiber Types and Dispersion Compensation, 162
    2.3.7 Polarization-Mode Dispersion (PMD), 168
    2.3.8 Fiber Nonlinearities, 174

2.4 Passive Optical Components, 194
    2.4.1 Connectors, Couplers, Splitters/Combiners, Multiplexers/Demultiplexers, and Polarization-Based Devices, 196
    2.4.2 Optical Filters, 217
    2.4.3 Compensation and Power Equalization, 228
    2.4.4 Optical Fiber Amplifiers, 234

2.5 Active Optical Components, 249
    2.5.1 Laser Sources and Transmitters, 249
    2.5.2 Photodetectors and Receivers, 262
    2.5.3 Photonic Switching and Optical Cross-Connects, 271
    2.5.4 All-Optical Signal Regeneration, 285

2.6 WDM Networks, 292
    2.6.1 Wavelength Standards and Digital Hierarchy, 293
    2.6.2 Point-to-Point WDM Transport, 296
    2.6.3 WDM Network Topology and Wavelength Connectivity, 304
    2.6.4 Network Protection and Virtual-Topology Design, 317
    2.6.5 Network Evolution and Convergence, 329
        Exercises, 335
        My Vocabulary, 339

CHAPTER 3
Cryptography and Communications Security, 345

3.1 Message Encryption, Decryption and Cryptanalysis, 345
    3.1.1 Mono- and Multi-Alphabetic Encryption, 348
    3.1.2 Frequency Analysis, 352
    3.1.3 Other Classical Ciphers, 363

3.2 Modern Cryptography Algorithms, 382
    3.2.1 Encryption With Binary Numbers, 382
    3.2.2 Double-Key Encryption, 389
    3.2.3 Modular-Algebra Basics, 391
    3.2.4 Cryptography Without Key Exchange, 396
    3.2.5 Public-Key Cryptography and RSA, 398

3.3 Communications Security and Applications, 407
    3.3.1 Data Encryption Standard (DES), 407
    3.3.2 Advanced Encryption Standard (AES), 418
    3.3.3 Other Encryption Algorithms and Standards, 427
    3.3.4 Digital Signature and Authentication, 432
    3.3.5 Network and Internet Security, 441
    3.3.6 Current and Futuristic Applications of Cryptography, 451

3.4 Quantum Cryptography, 461
    3.4.1 Photons, Polarization States and Measurements, 461
    3.4.2 Quantum Key Distribution (QKD), 468
        My Vocabulary, 475

Solutions to Exercises, 479
Bibliography, 501
Index, 507

Foreword

This book is for the curious.... Most of us open the refrigerator, take out a bottle of milk and never wonder how it stays cold. Or we may notice the sun rise or set without ever needing a wider explanation of the daily celestial phenomenon that governs our lives. If this description fits you, you probably need read no further; you are excused. But if you do wonder about the technological and scientific marvels we encounter every day, this *Survival Guide in Global Telecommunications* may provide considerable personal satisfaction and, depending on your professional or academic status, can offer practical benefits as well.

The book (in two independent volumes) is addressed to a diverse spectrum of readers, who will find it accessible at different levels. The format is designed for the nontechnical reader who is curious about the telephone and the Internet. It is also designed for the nontechnical reader who encounters technology in his professional life, for example: at home, setting up an office computer; in government, legislating or regulating telecom technology; in law, litigating telecom-related patents or contracts; in finance, investing in high-tech companies. In addition, university students will find the book a handy resource. The book is also addressed to the technically savvy reader who wants to quickly get up to speed in new areas of telecommunications that arise due to rapid career evolution or the prospect of future job movement.

A large impediment to entering an unfamiliar technical field is the jargon that has been developed by the practitioners over many years as a form of shorthand. The author, Emmanuel Desurvire, is careful to define each new term in language familiar to a beginner. He also uses these terms repeatedly in context, such that their meanings become intuitively familiar. Another impediment to learning about a particular subject in an ordinary text is the need to wade through the introductory chapters in order to comprehend the few pages of interest. Each chapter in the *Survival Guide* is constructed to be self-contained. Thus, the curious reader can jump right in and read on until she is satisfied or snowed.

I have enjoyed 50 years as a research engineer and physicist, mainly studying optical fiber telecommunications at AT&T Bell Labs (at one time the premier telephone research lab in the world). Much of my enjoyment came from understanding how things work. I also got a great deal of pleasure from making novel contributions of my own. I have no idea where these drives come from, but like the enjoyment of food or friendship, they are real. Although I retired from Bell Labs in 1996, I still found this book personally fascinating for several reasons.

On leafing through the Contents, I found topics relevant to my current interests as a consultant and visiting professor. Some topics, I had known from decades earlier but had forgotten their underpinnings; others, I had never really understood or found a satisfactory explanation. In each case, the discussion in the text was clear and correct, and was presented in an effective tutorial fashion, starting with basics and adding complexity.

My first job after retiring from Bell Labs was as a Congressional Fellow on the professional staff of the U.S. Congress. One assignment was in the Congressional Research Service (CRS) of the Library of Congress whose mission is to provide Congress with background information on the myriad aspects of its legislative and oversight duties. In 1996, the Internet became a commercial-, rather than a government-managed, service. Questions of Internet regulation, taxation, security, governance, encryption and wire-tapping were just popping up. I was asked to prepare an Issue Brief describing the Internet for the staffs of representatives, senators, and committees. It is important to know that of the 535 members of Congress, only a handful have any technical training and their overburdened staffs are largely 20- or 30-somethings trained in liberal arts or political science (which is not a science). Thus, the CRS Issue Briefs are designed to give them a quick tutorial on anticipated technical subjects. After a month or so, I presented my 20-page draft to my boss. She said the limit was 6 pages and I couldn't use technical words like *digital*. It was a struggle, but she did manage to boil my explanation down to 6 pages, although there was no sensible way to avoid *digital*. Today, of course, Congress does know the word "*digital*" at some level. However, I dare say that it would benefit from the nice discussions and examples presented in the *Survival Guide*.

Another of my post-retirement jobs has been as expert witness in patent cases involving telecom issues. Most patent lawyers have some technical training but cannot know every field they are likely to encounter, which is why they hire an expert. The expert may consult with lawyers to help prepare their case, and may also be called upon to testify in deposition or in court to provide technical opinions. Communicating technical ideas in simple language to attorneys can be difficult, but it is a piece-of-cake compared with presenting a technical argument to a jury of high school graduates in the brief time allowed by the judge. Clearly, the attorneys would benefit from reading the explanations offered in *Survival Guide*, and might even want to cite it in their briefs.

Up until about 10 or 15 years ago, the telecom business was managed, designed, and operated by experienced engineers and specialists. The telecom bubble has created numerous job opportunities for an army of self-styled consultants, CEOs, CTOs, CFOs, marketing advisors, strategic planners, and other emerging professionals, whose technical backgrounds are often very weak. As demonstrated in the examples above, these people will find the *Survival Guide* an invaluable tool in getting their jobs done.

The two volumes of Emmanuel Desurvire's *Survival Guide* are a remarkable achievement in my opinion. First, I have wrestled with presenting technical ideas to the uninitiated in class, Congress, and court and know that it requires careful planning, ingenuity, and patience. Desurvire has created a novel instructional instrument for the purpose. Second, I have known Emmanuel since 1986, when he was an exceptionally talented researcher in my department at Bell Labs. In 1990, he became a professor at Columbia University and, in 1993, he joined Alcatel, a global telecom supplier, in France where he was a researcher and, later, a predevelopment project manager. With this background, he has realized an unprecedented feat: he has covered the entire telecom field through these two books, in order to explain to both young and mature minds how things work, he has succeeded in presenting complex ideas in an entertaining style with a minimum use of mathematics.

Thanks to its effective format and pleasing teaching style, I believe that the first two pioneering volumes of the *Survival Guide Series in Global Telecommunications* will find a place on a diversity of reference shelves around the world. Indeed, I expect that it will serve as the template for sequels on other topics.

IVAN P. KAMINOW

Holmdel, New Jersey
June 2004

Preface

These days, the words telecommunications and telecom are often identified with the concept of new technologies, which bears various subliminal meanings. These concern the enthusiastic hopes and illusions of our budding *Information Age*, the promise of a rapidly deployed global-communications culture, of instant person-to-person connectivity, of better and faster interaction between two parties anywhere in the world. Overall, the new and inescapable technology, which reaches everyone at home or at work, is the **Internet**, the panacea of all private or professional communications needs. *What* is being communicated is normally more important than *how* it is communicated. But, paradoxically, communication means and its bandwidth performance (the effect of being instantaneous) remain the user's top concern. What is important is to get a maximum of information from which to select. These new telecom technologies attempt to catch up to this growing perception and need.

The rapid evolution of telecom networks and related broadband access services now seems to differentiate humankind into five basic categories, in growing order of population:

- The rare telecom generalists, who conceptually grasp the full picture, from close familiarity with some of the technologies to accurate market analysis;
- The more diversified experts who deploy, integrate and manage telecom systems, the global and local, incumbent or competitive operators and the service providers;
- The greater number of individuals, who are technical contributors, scientists and engineers, who concentrate their top-level expertise on microscopic aspects of network sublayers, software applications and hardware circuit/board design;
- The majority of end users, and consumers, professionals or private, who are essentially unaware of telecom technologies' features and the intricacies of their complex, pyramidal integration, but are skilled in using their commodities;
- The populations in developed and under-developed countries, who rarely or never use any of these technologies, because of the lack of any telecom infrastructure, or their belonging to an older generation underclass situation.

The penultimate, privileged category conceive of "telecom" as being a mere *commodity*, just like transportation. They do not need to have a background in railway engineering or air-traffic control, or to acquire conductor/pilot licenses in order to take trains or planes. Similarly, the telecom generalists of the first category, do not need either an engineering background or an international scientific reputation to employ their skills (though some rare individuals may enjoy both!). Generalists may not even know anything technical at all, but they master the global concepts, the latest developments in programs and trends, to the point of sounding like top technology experts. At this level, considerations of technology integration, evolution and limits, market projections and investment opportunities are all taken into account, which defines an expertise field of its own. To such a category also belongs the rarer academic generalist, who can provide a six-hour tutorial on any aspect of telecom to any audience level, sometimes with talent. But generalists also make mistakes, since their views may be obscure, confused or biased, and their sources may be the same press in which other generalists express themselves, forming a closed informationless loop. The intelligent ones often visit the labs, talk to the engineers, visit the troops. They don't look at top management as a promoting machine where "yes, sir or madam" remains the master keyword. They like to share their knowledge and views with broad audiences, having understood that the more information you freely give, especially to the young, the more you may acquire.

The third and intermediate population, that is, the telecom scientists and engineers, are people whose over-specialization and heavy-duty agenda keep them away from any global picture whatsoever. As top experts and sometimes prime contributors to new technologies, they enjoy the creativity, the inventiveness, the challenges and success of their projects and products. They may even experience the awe of the engineering beauty of their realizations, which keeps their vocation alive and makes them work hard, including at weekends. From any reasonable viewpoint, such a population is key to any future development and innovation in telecom. Yet we are talking of an endangered species. With the current evolution, there are fewer and fewer opportunities to make a career in telecom based on science or engineering. The result is an accelerated over-specialization which is merely project- or product-oriented. As technologies grow in complexity, it is important that instead of over-specializing, experts do cultivate a general background in the whole field. By default, the risk is of rapidly becoming obsolete as a key specialist, should the associated products be phased out, owing to new market orientations or radical evolutions.

Yet, in order to meet the market and be "on time" in product development cycles, experts have very little leisure to expand their background. Because of this, most of them do not (or are physically unable to) pay attention to technologies that are too remote from their area of responsibility. Their immediate boss will say that they are not paid for self-education and growth, but for project effectiveness. The same situation applies to innovation: innovation yes why not, but only within the project framework. Upper management experiences the same constraints, from a broader expertise standpoint but with fewer career risks. Mobility encourages the most inspired to climb the ladder anyway, leaving their expert teams behind. In big corporations with many intermediate decision levels and a transverse structure, the result is a loss of potential, if not a lack of technical vision and reactivity. Moreover, issues of experts' "employability," "retraining" and technical "career management" become increasingly acute. The palliative solution is that employees in the technical field, at any operational level, should be obliged constantly to broaden their expertise, beyond their field of immediate responsibility. The corporation must believe that it is worth investing in this human potential. Even if people may leave to start their own businesses or join more creative teams (as in the normal cause of events), younger generations will be attracted to the potential, so that the net flow and benefits are globally positive. In any case, telecom engineering specialists are at high risk if they become over-specialized, and trust a big corporation, or a small business, to provide them lasting job security or possibly make them wealthy.

With the development of telecom as a commodity, the above concerns and needs are real. It would be useless to attempt to solve the issue by the force of a single "retraining" handbook, however big or small, written by all-knowing and all-mighty generalists. What a book could provide, however, is a rapid introduction. This would build mental bridges towards new domains, integrate or reactivate pieces of prior knowledge, bring familiarity with new concepts (not simply silly "acronyms"!), stimulate curiosity and thirst for exploring new directions. Our "survival guide" approach may represent what is needed to escape the risk of over-specialization and competency stagnation, to reach out towards new technical horizons. It is also addressed to scientists and engineers, working outside telecom, or young people just graduating, who may consider joining the telecom adventure. Where to start in this thick jungle and where it leads one, are key problems to all. We will come back to this later.

End users and customers have it best of all. Operators are competing to lower their prices and increase service quality. The telecom commodity has a price, but no more than that. There is a long way to go to fiber-to-the-home, broadband services on demand, and multimedia fixed/mobile communications, although progress is now irresistible. Once service providers have achieved the right account balance between capital investments, operating costs and service revenues, a new world of telecom will start on a fresh basis, to the great benefit of the aforementioned categories.

We are left with the last category, the people who, for differing reasons of geography and social status, cannot have access (yet) to the new telecom commodity. For them, the picture may also change. Submarine fiber-optic cables have been extensively deployed over the last decade, connecting the most remote locations to the Internet, which is complemented at local or continental scale by GEO satellite coverage. Wireless infrastructures may also compensate for the failed development, degradation, or obsolescence of old telephone systems. With appropriate help, these countries will be able to catch up rapidly with the Internet, including in schools and universities, and not just in business. In both advanced and less advanced developing countries, telecom technologies create the so-called *digital divide*. Should this neologism help society to propose new solutions, it would then be useful. Underclasses may not need e-mail and Internet browsing so much as the reassurance that their children will enjoy equal training and aware in the new telecom technologies for equal education, vocation, and job/career potential. Teachers and professors may also use the Internet to facilitate their communications, develop on-line education programs, and learn new effective ways to educate and train their classes.

And what about this book? It is **not** a pocket guide, **not** a guided tour, **not** a crash course, **not** an illustrated glossary of acronyms, and **not** a kind of Bible reference. The ambition is not to provide a layer of varnish on otherwise superficial understanding. Rather, it is to explain how things work and open up conceptual horizons. If the reader goes through the whole of this book, s/he is on track towards a better career realization, and why not, moving in the direction of the top two aforementioned levels. The idea is to disperse the feeling of being overwhelmed by telecom complexity, and maybe more important, **not** being afraid of it. Even better would be to find it exciting and challenging. Telecom is a beautiful science. It does not have that reputation, but it really is, and this is what this book attempts to convey. Telecom is full of clever and fun stuff, provided one pays due attention and takes due interest.

A move towards a higher-level of insight or into a deeper level of understanding requires some effort, concentration and acquisition. This book does not provide **ad-hoc** recipes as a substitute for true expertise and training. Rather, it guides the reader directly into the depth (and again, fun!) of the issues, without the unnecessary burden of cross-references, historical descriptions, and heavy-duty mathematics.

Concerning mathematical formalism, we made it a self-imposed rule that all engineering and physics should be explained exclusively with elementary operations and functions $(+, -, :, \times, \%, \text{sine}, \log, \exp, \text{etc.})$. These functions are available from any scientific pocket calculator for instant verification. Top-level generalists only recognize $(+, -, :, \times, \%)$, which makes ultimate sense for business but these are low-level, if not worthless, knowledge references to technology specialists. Our challenge has been to explain waves and signal processing with reasonably simple mathematics, such as sine waves, which even high-school students are able to understand. We have dared to strip away from the telecom field any derivatives and integrals, which proves that one could learn more without such an advanced background.

We have included a wealth of easy and practical exercises the purpose of which is to illustrate the concepts, to correctly assess the magnitude of the effects involved (from 10^{-12} in optics to 10^{100} in cryptography) and to become familiarized with standard units. We certainly believe (and we are not the first) that no engineering can coexist with fuzzy concepts, mistaken magnitudes and mishandled standard units. We also believe that engineering experience is not fully integrated unless one is able to address practical applications without gross mistakes. The path to becoming a new telecom expert in a new telecom field is narrow, but we believe that asking oneself the right questions, and developing a true interest for the real "stuff," with curiosity and sustained effort, is probably the safest way to go.

In most books, normal practice has it that the author describes the contents in the introduction: what the reader will find, and where. Here, we shall be innovative. This book contains different fields of knowledge that are identified by the chapter titles. Any chapter can be addressed separately, with little or no cross-reference. Should the reader be interested first in *broadband access, cryptography, WDM technologies,* or *home networking,* there is practically no need to go through any preceding chapters. At the end of each chapter, we have provided a summary of the main keywords and acronyms. We call these lists **My Vocabulary**. These summaries are not subindexes. They just represent a list of concepts that are associated with the field covered in the corresponding chapter. If the reader goes through the list at first, it looks like a headache. But, after reading the chapter, the list should be clarified nearly "crystal clear." If a few items still do not make sense, this could mean that the concept was not acquired or remains difficult. Another application of these **My Vocabulary** lists is to use the items as search keywords in the Internet. The inexperienced reader might be surprised by how much knowledge and clarification can be gained by such self-guiding exploration. There are many academic and professional sites from which tutorials, white papers and other hotline news can be freely consulted and downloaded. Our Bibliography, far from being exhaustive, constitutes an indicative start which includes real Web-site "jewels," but we won't say which ones, the purpose here being not to distribute points or compile a "hot list." Besides, we trust that the reader knows how to use Web links and rapidly generate his/her personalized URL database.

Could one ever become an expert in *all* fields of telecom? Obviously, the answer is no, unless we redefine what expertise means. As a matter of fact, the telecom concept has become sufficiently elusive that we may dare to innovate. This is because telecom's expertise is no longer a matter of scientific/technologic specialization, let alone being a heroic contributor of devices, systems, standards or service/software applications. Rather, it is a matter of integrating all technology aspects into one coherent viewpoint, being aware not so much of the picky details as of the great essentials. Since after the telecom deregulation, these essentials concern more economic than scientific issues, the engineer has been overtaken by the "marketeer," to be taken over in turn by the "client" and the "investor."

Another factor contributing to the demise of engineering expertise is the increasing role of software and the widespread view according to which every technology should follow "Moore's law," to become a mere and cheaper commodity. In this book, we show at least how Moore's law could be revised for wavelength division multiplexing (WDM). Based upon this deceptive faith of self-fulfilling Moore's laws in every direction and field, which "innovation seminars" have a tendency to overdo, new engineers should not tarry on the basic rules of telecom, but instead view things globally, with an emphasis on specialized market directions and new opportunities thereof. In big corporations, the sum of all these decaying factors has led the pure engineering/scientist role to a very low level of esteem. Such a fate would be acceptable if telecom networks did not rest upon 30 years of pure engineering science. The public now regards software and applications as "technology," confusing the *service* with the *physical or hardware channel*. Telecom technology can hardly be reduced to software and services, but the physical layer, without which there would be no telecom, can hardly claim to represent the whole. The network-layer model, where the physical layer has been conventionally put at the bottom, is one more repelling factor for engineering vocations! But such a perspective has changed with the developments of Internet on WDM and, possibly, quantum communications. In telecom, physics and engineering may always keep the upper hand as a faithful servant, regardless of the business implications or plans, a feature that has been verified at least with the radio, the transistor, the laser, the optical fiber, and semiconductor chips.

Can engineers absorb so much complexity as to master all "conceptual layers" of telecom networks? The practical and radical answer is no. This is why this book is published under a series called "Survival Guides." Let's face it: during the busy (and sometimes exhausting) course of our professional and family lives there is so little time we can dedicate for our own education, that we can't hope to master everything, for example music, if it is not our bread-earning job. But we surely can understand better how things work, with limited time investment, for our own benefit and potential career orientations. This may be one definition of the scientific spirit: inquire, then ask the right questions and find the answers, and possibly dig further if not satisfied.

Our sincere hope is that this book will help engineers and scientists to "survive" by catching up to, and rising above the fear of inadequacy, and the mistrust of new technologies and standards. All the above through minimal conceptual effort. The reward is to get a thrilling sense of understanding some of the important things that every aspect of telecom touches upon, being able to make the right decisions, to train or teach, and be better prepared for the telecom future.

EMMANUEL DESURVIRE

May 2004

Acronyms

A3,/A5/A8 Authentication/ encryption/confidentiality algorithms (GSM)
A5/n Version n of A5 (GSM)
AALn AAL cell type (ATM)
AC Alternative current
ADM Add-drop multiplexing
ADSL Asymmetric digital subscriber line
AES Advanced encryption standard
AES-n AES with n-bit key ($n = 128$, 192, or 256)
AEX AS-bearer extension byte (ADSL)
AGC Automatic gain control
AM Amplitude modulation
ANSI American National Standards Institute
AOR All-optical regeneration or regenerator
AOTF Acousto-optic tunable filter
APC Automatic power control
APD Avalanche photodiode
APD-FET Hybrid photoreceiver
APON ATM passive optical network
APS Automatic protection switching
AR Antireflection (coating)
As Arsenic
AS Authentication server
AS0-AS3 Downstream bearer channels (ADSL)
ASAM ATM subscriber access multiplexer
ASCII American standard code for information exchange
ASE Amplified spontaneous emission
ASK Amplitude shift keying
ASON Automatically-switched optical network
ASP Application service provider
ATM Asynchronous transfer mode
ATM-DSLAM DSLAM for ATM access
ATU ADSL termination/transceiver unit
ATU-C ADSL termination/transceiver unit (central office)
ATU-R ADSL termination/transceiver unit (remote)
AWG American wire gauge
AWG Arrayed-waveguide grating
BB84 Bennett and Brassard 1984 quantum-cryptography algorithm
BBOR Black-box optical regenerator
BER Bit error rate
BIP Bit interleaved parity [field] (APON)
BLSR Bidirectional line-switched ring,
BLSR/n BLSR with n = 2 or 4 fibers
BOL Beginning of life (system)
BPON Broadband PON
BPSR Bidirectional path-switched ring
BPSR/n BPSR with n = 2 or 4 fibers
BSHR Bidirectional self-healing ring, with 2 or 4 fibers
C × D Capacity-distance figure of merit
CA Certification authority
CAP Carrierless amplitude-phase (modulation)
CATV Common-antenna (or community-access) television
CBC Chain block coding
CCA Common cryptographic architecture
CD Compact disk

xix

CD-ROM Compact disk, read-only memory
CDMA Code-division multiple access
cdma2000 3G CDMA system family
CFB Cipher feed-back (DES)
CGM Cross-gain modulation
CLEC Competitive local-exchange carrier
CO Central office
COBRA Commutation optique binaire rapide (rapid binary optical commutation)
CP Customer premises
CPE Customer premises equipment
CPM Cross-phase modulation (=XPM)
CPU Central processing unit
CR Clock recovery
CR-LDP Constraint-based routing label distribution protocol (MPLS)
CRC Cyclic redundancy check (code)
CSBH Channel-substrate buried heterostructure (laser)
CWDM Coarse WDM
dB Decibel
DBA Dynamic bandwidth allocation (PON)
dBm Decibel-milliwatt
DBR Distributed Bragg-reflector (laser)
DCF Dispersion-compensating fiber
DCPBH Double-channel planar buried heterostructure (laser)
DCS Digital cross-connect
DD Direct detection
DDM Direction-division multiplexing
DDoS Distributed DoS
DEA Data encryption algorithm (ANSI)
DEA-1 Data encryption algorithm (ISO)
DES Data encryption standard
DFB Distributed feedback (laser)
DFF Dispersion-flattened fiber
DFT Discrete Fourier transform

DGD Differential group delay
DGEF Dynamic GEF
DH Double heterostructure (laser)
DM Dispersion management
DM Dispersion-managed (soliton, system)
DMT Discrete multi-tone (coding)
DMUX Demultiplexing/demultiplexer
DoS Denial of service
DSA Digital signature algorithm
DSF Dispersion-shifted fiber
DSL Digital subscriber line
DSLAM DSL access multiplexer
DSS Digital signature standard
DSSS Direct-sequence spread-spectrum
DVD Digital video disk
DW Digital wrapper
DWDM Dense wavelength-division multiplexing
DWMT Discrete-wavelet multi-tone (modulation)
DXC Digital cross-connect
Dy Dysprosium (element)
E/O Electro-optic (effect, conversion)
EA Electro-absorption (effect)
EADM Electrical add-drop multiplexing
EAM Electro-absorption modulator
EBCDIC Extended binary coded decimal interchange code
EC External-cavity
EC-DBR External-cavity DBR (laser)
ECB Electronic code book (DES)
ECC Error-correction coding / error-correcting code
ECL EC laser
EDFA Erbium-doped fiber amplifier
EFM Ethernet in the first mile (PON)
ELC Explicit label control (GMPLS)
ELSR Edge label-switching router (MPLS)
EM Electromagnetic [wave, field]
EMI Electro-magnetic interference
EOL End of life (system)
EPON Ethernet PON

Er Erbium (element)
ER Extinction ratio
ESI Equivalent step-index (fiber)
ESP Encapsulating security payload (IP)
ETSI European Telecommunications Standards Institute
EU European Union
eV Electron-volt
EXC Electrical cross-connect
FA-OH Frame-alignment overhead (DW)
FAX Facsimile (machine)
FBG Fiber Bragg grating
FDD Frequency-division duplexing
FDD-DMT Frequency-division duplexed DMT
FDDI Fiber distributed data interface
FDM Frequency-division multiplexing
FDMA Frequency-division multiple access
FEC Forward error correction
FET Field-effect transistor
FEXT Far-end cross-talk
FFT Fast Fourier transform
FITB Fiber into the building
FITL Fiber in the loop
FM Frequency modulation
FP Fabry-Pérot (cavity or interferometer, filter)
FS-VDSL Full-service VDSL
FSAN Full-service access networks
FSC Fiber-switching capable (GMPLS)
FSC-LSP Fiber-switching capable LSP (GMPLS)
FSK Frequency shift keying
FSO Free-space optics
FSR Free spectral range
FTL Fourier-transform limited (pulse)
FTP File transfer protocol (TCP/IP)
FTTB Fiber to the building/business
FTTC Fiber to the curb
FTTCa(b) Fiber to the cabinet
FTTH Fiber to the home
FTTN Fiber to the neighborhood/ network/node
FTTO Fiber to the office
FTTP Fiber to the pole/premise
FTTU Fiber to the user
FTTx Generic name for fiber-loop access
FWM Four wave mixing
g.hsdsl (Single/double-pair) high-speed DSL
Ga Gallium
GaAs Gallium-arsenide
gcd Greatest common divider
Ge Germanium
GEO Geosynchronous Earth orbit
GEF Gain-equalizing filter
GFF Gain-flattening filter
GFP Generic framing procedure (GPON)
GigE Gigabit Ethernet
GI-POF Graded-index POF
GMPLS Generalized multiprotocol label switching
GNFS General number field sieve
GOF Glass optical fibers
GPON Gigabit PON
GPS Global positioning system
g.shdsl Single-pair high-speed DSL
GSM Global system for mobile telecommunications
GVD Group-velocity dispersion
H/E Head-end (HFC)
HDSL High bit-rate DSL
HDSL2 HDSL with one copper-wire pair, or second-generation HDSL
HDSL4 HDSL with two copper-wire pairs
HDTV High-definition TV
He-Ne Helium-neon (laser)
HFC Hybrid fiber coaxial system
HLR Home location register (GSM)
HMAC Hash message authentification code
HOH HDSL overhead
HR High-reflection (coating)

HTTP Hypertext file transfer protocol (internet)
HTTPS HTTP over SSL
HTU HDSL terminal unit
HTU-C HDSL terminal unit (central office)
HTU-R HDSL terminal unit (remote user end)
IAD Integrated access device
IC Integrated circuit
ICI Intercarrier interference
ICV Integrity check value (IPsec)
ID Identification document
IDEA International data encryption algorithm
IDSL ISDN-DSL
IEC International Engineering Consortium
IEEE Institute of Electrical and Electronics Engineers
IETF Internet Engineering Task Force
IF Intermediate frequency
IFBG In-fiber Bragg grating
IFFT Inverse FFT
IKE Internet key exchange
ILM Intregrated laser modulator
IM Intensity modulation
IM-DD Intensity modulation and direct detection
In Indium
InGaAsP Indium-gallium-arsenide-phosphide
InP Indium-phosphide
IP Internet protocol
IP-DSLAM DSLAM for Internet access
IPES Improved proposed encryption standard
IP-PON Internet-protocol PON
IPsec IP security protocol
IPv4, IPv6 IP addressing versions 4 or 6
IR Infrared
IrDA Infrared Data Association
IRS Internal Revenue Service (U.S.)
ISAKMP Internet Security Association key management protocol
ISD Information spectral density
ISDN Integrated services digital network
ISI Intersymbol interference
ISO Internations Standards Organization
ISP Internet service provider
ITRS International Technology Roadmap for Semiconductors
ITU International Telecommunications Union
ITU-R ITU for radio systems
ITU-T ITU for telecommunications systems
IXC Interexchange carrier
LAN Local area network
LASER Light amplification by stimulated emission of radiation
LCD Liquid-crystal display
LCF Laser control field (APON)
LD Laser diode
LDP Label distribution protocol (MPLS)
LEA Large effective-area (fiber)
LED Light-emitting diode
LEPA-HDSL Local-exchange primary access for HDSL (ISDN)
LER Label edge router (MPLS)
LEX LS-bearer extension byte (ADSL)
LF Low frequency
LH Long haul
LHS Left-hand side
LMP Link-management protocol (GMPLS)
LO Local oscillator
LOS Line of signt (radio)
LS0-LS2 Duplex bearer channels (ADSL)
LSC Lambda-switching capable (GMPLS)
LSC-LSP Lambda-switching capable LSP (GMPLS)
LSP Label-switched path (MPLS)
LSR Label-switched router (MPLS)

LY Light-year
M12 ... Multiplexing DS1 tributaries into DS2, etc.
MAN Metropolitan area network
MASER Microwave amplification by stimulated emission of radiation
MD2, MD4, MD5 Message-digest algoritms 2, 4 or 5
MDU Multidweller unit (VDSL)
MEMS Micro-electromechanical system
MF Medium frequency
MFD Mode-field diameter
MH Medium haul
MIC Message integrity check (PEM)
MM Multimode (fiber)
MMS Multimedia messaging service
M-PAM M-ary PAM (format)
MPLS Multiprotocol label switching
MPλS Multiprotocol lambda switching (MPLS)
MPOF Microstructured polymer optical fiber
M-PSK M-ary PSK (format)
MQAM M-ary QAM (format)
MQW Multiquantum-well (laser)
MS-DPRING Multiplexed-section dedicated protection ring
MSOH Multiplexer section overhead (field)
MS-SPRING Multiplexed-section shared protection ring
MTU Maximum transmission unit
MTU Multitenant unit (VDSL)
MUX Multiplexing/multiplexer
MxU MTU or MDU
MZ (or MZI) Mach-Zehnder (interferometer)
N-bit CFB CFB encoding over N-bit blocks
NA Numerical aperture
Nd Neodymium (element)
NEXT Near-end cross-talk
NF Noise figure
NFS Number field sieve

NIST National Institute of Standards and Technologies
NRZ Nonreturn to zero
NRZI Nonreturn to zero inverted
NSFNET National Science Foundation network (USA)
NTR Network timing reference (DSL)
NVoD Near video-on-demand
NZDSF Non-zero dispersion-shifted fiber
NZDSF NZDSF with positive or negative dispersion at 1.55μm wavelength
O/E Opto-electronic (modulation, conversion, regeneration)
O/E/O O/E followed by E/O (modulation, conversion, regeneration)
O/O All-optical (regeneration)
OADM Optical add-drop multiplexing
OAM Operations, administration and maintenance
OAM&P OAM and provisioning
OAN Optical access network
OAS Optical amplifier section (SDH/SONET)
OC Optical circuit (SONET)
OC Optical channel (SDH/SONET)
OCC Optical connection controller (ASON)
OCH or OCh Optical channel (SDH/SONET), same as OC
OCH-OH Optical channel overhead (DW)
OCH-PE Optical channel payload envelope (DW)
OCH-SPRING Optical channel shared protection ring
OD Outside diameter (fiber)
ODU Optical channel data unit (DW)
ODU-OH ODU overhead (DW)
OFB Output feedback mode (DES)
OFDM Orthogonal FDM
OH Overhead
OKG Optical quantum generator
OLT Optical line terminal (PON)

OMS Optical multiplex section (SDH/SONET)
ONT Optical network terminal/termination (PON)
ONU Optical network unit (VDSL, PON)
OOK On-off keying
OPA Optical preamplifier/preamplification
OPS Optical packet switching
OPU Optical channel payload unit (DW)
OPU-OH OPU overhead (DW)
OSI Open systems interconnection (model)
OTS Optical transmission section (SDH/SONET)
OTU Optical channel transport unit (DW)
OTU-OH OTU overhead (DW)
OXC Optical cross-connect
P Phosphorous
PAM Pulse amplitude modulation
PBS Polarization beam-splitter
PC Personal computer
PCD Polarization-dependent chromatic dispersion
PCE Power conversion efficiency
PDF Probability density function
PDG Polarization-dependent gain
PDL Polarization-dependent loss
PEM Privacy-enhanced mail
PGP Pretty good privacy
PIN Personal identification number
PIN or p-i-n p-i-n hetero-junction (photodiode)
pin-FET PIN/FET integrated photoreceiver
PKC Public-key cryptography
PKCS#1 PKC standard #1
PLL Phase locked loop
PLOAM Physical-layer OAM [cell] (APON)
PM Phase modulation
PM Polarization-maintaining (fiber/device)
PMD Polarization-mode dispersion
PMF Polarization-maintaining fiber
PMMA Polymethyl-metacrylate (fiber)
POF Polymer optical fiber
POH Path overhead (field)
PON Passive optical networks
POP Point of presence
POTS Plain old telephony service
ppm Part per millions
Pr Praseodymium (element)
PRBS Pseudo-random bit sequence
PSC Packet-switching capable (GMPLS)
PSC-LSP Packet-switching capable LSP (GMPLS)
PSK Phase shift keying
PSP Principal states of polarization
PSTN Public switched telephone network
PTT Post, telegraph and telephone
P2MP Point to multipoint (network link)
P2P Point to point (network link)
QAM Quadrature amplitude modulation
QED Quantum electrodynamics
QKD Quantum key distribution
QoS Quality of service
QPSK Quadriphase PSK
RADSL Rate-adaptative DSL
RAN Residential access network
RC4 Rivest cipher 4 (WLAN, WAP, SSL)
RDF Reverse-dispersion fiber
RE Rare earth (element)
RF Radio frequency
RFA Raman fiber amplifier
RHS Right-hand side
RIPEM Riordan PEM
rms Random mean-square (value)
ROM Read-only memory (IC)
RS Reed-Solomon (code)
RSA Rivest-Shamir-Adleman (cryptography standard)
RSOH Regenerator section overhead (field)

RSVP-TE Traffic engineering resource reservation protocol (MPLS)
RXCF Receiver control field (APON)
RZ Return to zero
S-CDMA Synchronous CDMA (HFC)
S-HTTP HTTP over SSL or secure HTTP (=HTTPS)
SA Saturable absorber/absorption
SA Security association Ipsec
SAN Storage-area network
SBS Stimulated Brillouin scattering
SCG Server gated cryptography
SDH Synchronous digital hierarchy
SDM Space-division multiplexing
SDMT Synchronized DMT
SDSL Symmetric DSL
SELD Surface-emitting laser diode
SGC Server gated cryptography protocol
SH Short haul
SHA Secure hash algorithm
SHA-1 Secure hash algorithm 1
SHG Second-harmonic generation
Si Silicium
SI-POF Step-index POF
SI-SRS Self-induced SRS
SIM System/subscriber identification module (GSM)
SKEME Secure key exchange mechanism
SKIP Simple Key Management Scheme for Internet Protocols
SLD Semiconductor laser diode
SM Single-mode (fiber)
SM Synchronous modulation (3R regeneration)
SMF Single-mode fiber
SNR Signal to noise ratio
SOA Semiconductor optical amplifier
SOA-MZ SOA-based Mach-Zehnder (interferometer)
SOH Section overhead (field)
SOHO Small office/home office
SONET Synchronous optical network
SOP State of polarization

SPD Single-photon detector
SPI Security parameter index (IPsec)
SPI Security parameter index (IP)
SPM Self-phase modulation
SPS Single-photon source
SPS Standard positioning service
SRS Stimulated Raman scattering
SSFS Soliton self-frequency shift
SSL Secure socket layer
SSL Vn SSL version n
SSLR Shared static link restoration
SSPR Shared static path restoration
STM Synchronous transfer mode (DSL)
STT Secure transactions technology
SWP Spatial walk-off polarizer
TBP Time-bandwidth product
TCP Transmission control protocol
TDD Time-division-duplexing
TDD-DMT Time-division-duplexed DMT
TDEA Triple data encryption
TDM Time-division multiplexing
TDM-LSP TDM-switching capable LSP (GMPLS)
TDMA Time-division multiple access
TE Transverse electric
TE Traffic engineering
TEM Transverse electromagnetic (wave)
TEM$_{lm}$ Transverse electromagnetic mode
teraKIPS Terabit.km/s (capacity-distance growth law)
TFF Thin-film filter
THG Third-harmonic generation
THP Tominson-Harashima precoding (CAP)
TIS/PEM Trusted information systems PEM
TLS Transport layer security
Tm Thulium (element)
TM Transverse magnetic
ToS Theft of service
TV Television
UDP User datagram protocol (TCP/IP)

ULH Ultra long-haul
ULSR Unidirectional line-switched ring
ULSR/n ULSR with n = 2 or 4 fibers
UMTS Universal mobile telecommunications system
UPSR Unidirectional path-switched ring
UPSR/n UPSR with n = 2 or 4 fibers
USHR Unidirectional self-healing ring
UV Ultraviolet
VC Virtual container (SDH)
VCO Voltage-controlled oscillator
VCSEL Vertical-cavity surface-emitting laser
VDSL Very-high bit-rate (or very high-speed) DSL
VLR Visitor location register (GSM)
VLSI Very large scale integration
VoATM Voice over ATM
VoD Video on demand
VoDSL Voice over DSL
VoFR Voice over frame relay
VoHFC Voice over HFC
VoIP Voice over internet (protocol)
VoMSDN Voice over multiservice data networks
VoP Voice over packet
VTD Virtual topology design
VTD-MILP Mixed-integer linear program for VTD
WAN Wide-area network
WDM Wavelength division multiplexing
WDSL Wireless DSL
WECA Wireless Ethernet Compatibility Alliance
WEP Wired equivalent privacy (WLAN)
WGR Waveguide grating router
WIC Wavelength-interchanging cross-connect
Wi-Fi Wireless Fidelity (trademark for 802.11b)
Wi-Fi5 Wireless Fidelity (trademark for 802.11a)
WLAN Wireless LAN
WSC Wavelength-selective coupler
WSC Wavelength-switching cross-connect
WXC Wavelength cross-connect
WWDM Wide WDM (PON)
WWW World Wide Web
X.509 ISO protocol for network authentication
xDSL Generic name for DSL family
XGM Cross-gain modulation (=CGM)
XOR Exclusive OR
XPM Cross-phase modulation (=CPM)
XT Cross-talk
Xtalk Cross-talk
Yb Ytterbium (element)
YIG Yttrium-iron garnet
3DES Triple DES
3R Signal regeneration by signal Repowering, Retiming and Reshaping
4-PAM Quaternary PAM (modulation/coding format)

Introduction
The Network Cloud

During the year 2000, I went through an unusual experience at the company's cafeteria. People were busy walking around, picking up food and drink from the booths. As I was doing the same, my attention was suddenly caught by a strange scene. A young Asian man was walking alone, erratically, eyes rolling and unfocused, waving one hand in the air with the other in his back pocket, engaged in an animated conversation with himself. In such a situation, one immediately feels some unease, as if witnessing a madman who has lost control in a public place. But he did not really look like a madman neither did he seem drunk, and the scene and place and looks were not right. What could HE ever be doing?! Getting a bit closer, it took me a few moments to realize that a near-invisible, minute wire was connecting his mouth, ear and something in his pocket. He was engaged in a hands-free, wireless telephone conversation! The weird thing is that he was having this conversation in the midst of this busy crowd all too hungry to pay any attention, too concentrated himself to feel self-conscious. Apparently, this was an important call, maybe with a "big" customer. Lesson of the day: the telecom network has become so all-pervasive and user-flexible that we hardly even realize it, even as trained engineers. You can be called upon by your customers, boss, colleagues, friend, spouse, anywhere and anytime, especially if you choose to carry with you this network terminal, as light as a pen or a credit card. The signal quality is such that you don't need to isolate yourself from the noise, and you can be reached deep inside any building, whatever you are doing, from anywhere in the world.

This is where we start. Let's imagine that the person at the other end of the line is driving a car in some country a few time zones away, or seated on a flight to Paris, or at his/her office desk. Once established, the connection between the two persons is seamless, instantaneous and as immaterial as a cloud of nothingness, as networks are often pictured and as illustrated in Figure I.1. Nothingness means that practically any communication means could be activated from within this transparent or immaterial cloud, as illustrated in Figure I.2. At the entry edge of the cloud, intelligent "boxes" pick up the incoming signals and convert them into something meaningless for human beings, but wholly meaningful and intelligent for the network: *bits*, short for "binary digits." The signal bits are aggregated into various packets, frames,

---

Wiley Survival Guide in Global Telecommunications: Broadband Access, Optical Components and Networks, and Cryptography, by E. Desurvire
ISBN 0-471-67520-2   © 2004 John Wiley & Sons, Inc.

2 Introduction

FIGURE I.1 A telephone connection through the "network cloud."

streams or combinations thereof, packaged or encapsulated with label information (headers, control bits, trailers), compressed at higher speeds (bit-rate conversion), mixed with other incoming signals (multiplexing), until they reach the core of the cloud, referred to as the "backbone" for the physical infrastructure, or the "core" for the network hierarchy level. Such a backbone transports the resulting

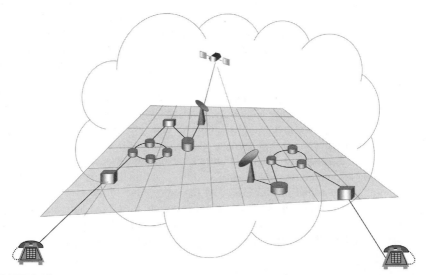

FIGURE I.2 Detail of the "network cloud," showing access/gateway stations (cubes), local sub-networks (rings), wireless edge network and radio-satellite connection, forming a highly complex yet virtually seamless and quasi-instant communication medium.

information (of which a substantial fraction is for network use only) anywhere in the world, via terrestrial or sub-marine cables or satellites, depending upon the physical distance to cover through the immaterial cloud, to reach the other edge. What is said about this phone call also applies to intercomputer or data communications, whether or not initiated, activated and controlled by actual human beings.

This description points to the tremendous complexity of any global communication process and the corresponding network operation, regardless of the type of signal to be exchanged (namely: voice, video or data). The first characteristic of the network is that its existence and reality rests upon a multiplicity of old and new *hardware and software technologies*. The technologies include telephone wires, microwave links, coaxial cables, optical fibers, satellites, digital switches, mainframe computers, hybrid electronics and integrated circuits and all their interconnections.

The second characteristic is *network intelligence*. The network always *knows* where to send your signal message. It makes sure that it will reach its destination through the most secure and fastest route, regardless of the underlying technology and transmission protocol available on the different path segments. This is why so many bit-processing stages and conversions occur through the cloud, which enable not only the precise routing function for the message, but its management with millions of simultaneous other network users. The only way to achieve such an apparent miracle is to structure the network cloud into several hierarchical layers, as shown in Figure I.3. Schematically, these network layers are the *access*, the "*metro*" (for metropolitan), and the *core*. Within the metro layer, one may also distinguish the *metro edge* (interface with the access) and the *metro core*

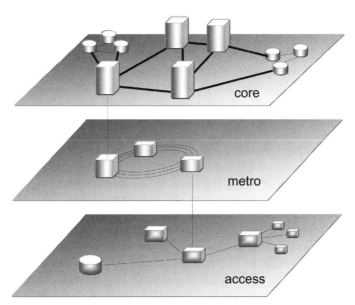

FIGURE I.3 Network segmentation in three functional layers (bottom to top): access, metropolitan and core.

(central functions, traffic aggregation or deaggregation, and interface with the core). The meaning of the different boxes shown in the figure is not important at this introduction stage.

Each layer can be operated nearly independently, with decentralized management and control, and by several owner agents (operators) and service providers (vendors). Each layer has its own ingrained technology favorites, according to the *legacy* (the system's deployment history and heritage), the evolving needs and the technology progress. Clearly, the core requires the most powerful technologies, that is, those providing the highest *bandwidth*, or bit-per-second transmission capability. *Lightwave technologies*, also referred to as *Optics*, or *Photonics* at component, system and backbone network-layer levels, meet such a need. At the bottom end, the access layer, the type of *user* and *user bandwidth* needs determine which technology is best suitable, from microwave to radio to electrical or even optical. Users can be either *fixed* or *mobile*. Mobile users can also *roam* the network, for instance using the same cellular telephone or laptop computer in different places and even different countries.

Schematically, there exist two global categories of network user, namely *residential* (home, personal, private, communities, associations, institutions, government..) or *business* (office, corporation, enterprise, retailing, agencies..). But the difference between the two may not be so clear cut. The possibility of *telecommuting* (more accurately termed *teleworking*) turns home into office, and residential block into small business center. To each user category corresponds a broad, ever-progressing variety of communications *services*, of which the ancestral *telephone*, whether fixed or mobile, is the most common and routinely used. But voice services are not limited to mere "phone calls": there are also voice mail, toll-free numbers, automatic call back and caller identification, among several other possibilities.

In the most developed countries, *cable-TV* and *Internet* have become indispensable services to tens or hundreds of millions of private users, while in the least developed, a single telephone line could be shared by hundreds of inhabitants. The Internet has come into the network picture as a revolution, both as a telecom-technology driver and a serendipitous service offer. Most people erroneously conceive the Internet, the "World Wide Web," to be the above-described network cloud. Yet, the concept is partly true, since the cloud is anything you want to put in it, as we have seen. By itself, the Internet also creates its own network, hence the spider-like web denomination. But in reality, the Internet is a new way of packaging transmission over the old and traditional network. It may borrow the routes of our *plain old telephone service* (POTS) and find a path through any network channel it may find appropriate. We come here to this notion that signals can be transmitted not only though various physical media (atmosphere, outer space, electrical wire, optical fiber), but also under different intelligent arrangements. The rules governing such arrangements are referred to as *protocols*, as are the way governments receive and treat their visitors, in the right order of importance, function and precedence. Protocols should not be confused with *standards*, which are the rules that define their specifications. The standards may differ from one continent to another (e.g., United States, Europe, Asia), while the resulting protocols may or may not be mutually compatible at different traffic-hierarchy levels. A basic example is

SONET (for North America) and SDH (for Europe, Asia and other). These two international standards happen to be compatible at some traffic concentration levels. But even then, their framing protocols remain wholly different.

The analogy between communications and transportation networks is often helpful for making one-to-one comparisons. Indeed, cars, trucks, trains, airplanes or boats can be viewed as representing as many different transportation "protocols" for people and goods, each following their own traffic rules and priorities (the standards). Protocols may or may not be compatible. Cars or trucks can be put into trains or boats to go faster, farther or cheaper, immediately recovering their autonomy at the other end of the route. Air travelers may take the subway to reach the airport and a bus shuttle or taxi or car rental to reach their final destination. Each time they switch protocol and adopt the local standard, which is one of the fun experiences of travelling (ever tried the Tokyo subway for the first time during rush hour?). Likewise, voice can be transported through the *Internet protocol* (IP), as referred to by the **VoIP** acronym. In our network cloud, the same signal payload my swap protocols several times. The network elements receiving and sending signals must also be "multi-protocol", meaning that they recognize several protocol types as they come in (of course they have been designed to handle this at this network point). One important function of network elements is protocol conversion, usually at the transition point between two network layers (e.g., core to edge, edge to metro, etc., and the reverse). Another function is *traffic aggregation* (upstream) or *deaggregation* (downstream), which may involve both protocol conversion and bandwidth conversion to higher/lower bit-rates within a given protocol. Finally, protocols may evolve within a given standard. For instance, the Internet is moving from IPv4 to IPv6. Contrary to general belief, it is not just a new addressing system, but a powerful way to use the standard, which for instance may consist in encapsulating IP packets into other IP packets after encrypting them, a secure Internet over the Internet. Standards may also evolve, but not necessarily with *backward compatibility*. For instance, the cryptography standard DES will be changed into a radically different advanced version, AES. Yet this will not prevent one from still using DES, the point being that computers recognize which standard is used. As a matter of fact, telecom could be entirely described through a suite of standards and protocols, with their complex history and intricacies. Hence a need for some abstraction and focus. At the opposite end, telecom could be viewed as entirely physical, a mesh of wires, cables, fibers, radio links and terminals handling different flows of data. Such a perspective is also entirely correct. Would not it be nice to have a deep appreciation of both perspectives? This is the kind of reconciliation that this book is attempting.

Communication networks are thus characterized by multilayered, multifunctional, and multistandard complexity. It is quite difficult to apprehend such a global picture at once, because of the inevitable confusion one often makes between the different applications, services, protocols and technologies. The *Internet* is the best example of this intertwined conception. When we "log on" the Internet, we use a computer application which connects to a service (a server with its web pages and search engines), which in response generates IP packets to feed our PC. But all these different stages remain invisible, albeit spurious losses of connection and exasperating downloading times occasionally remind us of the *physical reality* of the network. A failed connection could also be explained by the child who played with the telephone plug in the basement! Those experiences of wire-and-plug fixing also remind us of the benefits of basic engineering, and that the Internet is not just abstract software.

To the nonspecialist, and also the professional, another factor of confusion is introduced by *marketing*. Indeed, telecom marketing has this tendency to name technologies after a product brand, and not the reverse. As a result, the same technology can bear multiple brand names, making it absolutely unclear what it is about and how it actually works. Market considerations could re-name the same technologies, for example what you see here is *not* a car. The truth is, that telecom is not rocket science with dependable engineering reference and designation. The combined scientific and engineering aspects of telecom fully deserve the *hi-tech* label, as compared to other innovative but more trivial markets. Yet, "telecom technologies" do not come down to mere science and engineering. They are deeply rooted in evolving services and customer applications. This is to the point that version X.Y of this cellular phone is promoted as a "new technology," because it is compliant to a new standard and has extra command buttons or display features. This is where (we believe) most people with scientific and engineering training, especially those with the highest education, often find it hard to rationalize and see clearly into the telecom mess. For such people, things make sense only if there is a fundamental starting point, a logic track, a rational chain of deductions and conclusions based upon common principles and definitions, and possibly historical factors. The development of telecom technologies hardly follows any scientific logic (we would like the opposite) because it is plugged in to irrational market fluctuations and instant needs. But telecom has its own fundamentals and scientific rationale, and this is what this book attempts to summarize and convey. Telecom also has its own history based upon market, legacy, product success, service evolution, and technology innovations. There is probably no other field in modern society in which anyone has something sensible to say, from the mere user to the advanced specialist, while at the same time being unaware of the global picture.

Writing a book that would fully describe telecom, from technologies to standards and service applications could be viewed as a *risky* if not a pretentious endeavor. Why? Because telecom, unlike applied and exact sciences, is market-sensitive, which involves inputs from many independent contributors, participants, competitors, investors and overall, market strategies. For this reason alone, telecom is *not* a science. Rather, it is an evolving body of knowledge, with subjective and incomplete perceptions. The tragedy of the so-called "Internet bubble" showed the limitations of market analysis and predictions, which would be a joke if not for the loss of jobs and the waste of capital. Any telecom handbook, no matter how thick and how many contributors, would never be able to present the field in a decisively final and satisfactory way to all these participants. As a matter of fact, it would have to be rewritten every six months, or even more frequently, each time a new standard or service appears on the market.

Being aware that in telecom no statement can ever be final, and most important of all, that market is "just" one aspect of the field, we may then try to rationalize and consolidate its body of knowledge. The task apparently looks immense and hopeless. But we can reduce it to a minimum, making an inventory of all technology fields one after another, and extract what seems to represent the core concepts. A first difficulty, but easy to overcome, is *subjectivity*. Can technologies be dissociated from their brands or mother companies (has everything been ever invented in *that* specific place)? Are not technologies evolved from the works of several independent teams or inventors? Could one avoid going through a lengthy account of historical background and cross-references? The answer to all these questions is an emphatic yes, as long as the goal is to explain how thinks work, and not to pretend that the solutions came overnight to people and teams "skilled in the art." A second level of difficulty is the language barrier, or more precisely, the technology *jargon*. This jargon takes the form of more or less meaningful alphanumerical scripts or *acronyms*, such as 802.11b, RC-4, RS(255,231), or EDFA. The first is a leading wireless-LAN standard, the second a popular encryption algorithm, the third a frequently used error-correction code and the last a ubiquitous optical fiber amplifier. Could one make sense of and memorize all of these? The answer is a partial yes, should the *concept* always precede the acronym. An expert should not be reduced to a living glossary of acronyms, but he or she should at least be able to tell what acronyms mean in practical terms, and point to their immediate equivalents or derivatives. One may forget what GSM or UMTS mean but not that they are cellular-telephone standards, and be able to tell the difference without having to borrow presentation slides from a colleague. Finally, the supreme impediment to understanding telecom is *mathematical formalism*. Some experts seem to have the talent to render things obscure and deterring, or even worse, to convey the feeling that there is no other way to grasp a telecom concept than to suffer intense maths derivations, with a toolbox of integrals, non-linear equations and perturbation methods! Although respectable and essential, these branches of knowledge do not need to be introduced to telecom (after all, $E = mc^2$ may be sufficient for an introduction to relativity). On the other hand, formalism is required when it comes to measuring performance and understanding system limits. This is what market and investors need the most from engineers. As often in physics, the answer should be simple and easy to grasp (regardless of its possible academic complexity). Imagine you must explain to your CEO a new discovery, without advanced notice and only two slides. Good engineering and gaining management support also require indispensable communications skills!

For all of the above reasons, the author prepared this book according to the following principles. No use shall be made of:

> *Formulae involving differential/integral calculus;*
>
> *Mathematical derivations other than using $\pm$, $\times/:$, % and elementary functions that can be pressed on a pocket calculator;*
>
> *Quotations of $/Euro market figures, unless meaningful as a stand-alone or historical reference;*
>
> *Comparisons of competing technologies;*
>
> *Predictions concerning market trends;*

Quotations of companies or products, unless generic or historic (e.g., Bell Labs transistor);

The author neither does claim to:

Convey any "insider" viewpoint, implying any form of higher-source knowledge or company bias;
Be 100% accurate and complete in describing current or future standards;
Suggest market trends for any technology;
Provide reference grounds for business decision making.

This book, which is entirely based upon publicly available information, intends only to be a helping tour guide, either as an introduction to new fields of expertise or a stimulation towards new vocations or professional evolution, or just for checking out one's knowledge. Hence the series' brand name, "Survival Guide," destined for a new practice we could call telecom bushwalking.

Before entering a new telecom field jungle, a first check up might be to test one's *vocabulary*. We prefer this term to "jargon," because the spirit is to communicate, not to annoy or conceal. At the end of each chapter, we have introduced a **My Vocabulary** list. The reader will find below such a list to show the principle right from this basic introduction. We may suggest this approach as a new self-teaching or class-teaching method, being aware that this may represent yet another case of reinventing the wheel. The individual reader may cross out some words from the list, but only those for which the meaning and significance are absolutely clear. He or she may even distinguish the "reasonably" clear ones. Finally, the ones that he or she would tend to avoid should be circled. These are the most important, because they point to *knowledge gaps*, or even worse, *cognitive* knowledge gaps. For class use, the teacher may ask the students to volunteer a few words and acronyms on a piece of paper. The collection could then be pinned on separate boards by different subgroups. Each subgroup would then have to figure out a good conceptual definition and present the results to the class. If need be, a 15-mn Internet search, right on the premises, may be allowed. A class vote for the team that made the best contribution, with some token reward, might conclude the session. Try this in your company as well. The management will appreciate the absence of consulting fees and travel expenses for a very high return in employee training and team motivation.

Finally, the main chapters in this book, which concerns optical communications components and networks, ends with *exercises*. These exercises, which are not "problems," are recommended as a means for the reader to prove that a given concept has been understood. There is no shame in looking directly at the solution if one is definitely stuck or does not even understand the question. Comprehensive and extensively detailed "Solutions to the Exercises" can be found at the end of the book and are included as another tool for the teacher to use. The class participants, equipped with pocket calculators (or working without, as computing wizards) have only 10 mn to find the solution. The class looks at the results and goes together through the proper demonstration. It might be important to look at the reasons why some went wrong. Orders of magnitude? Algebra? Wrong numbers? Wrong reasoning? This could be yet another way to prepare good engineers. Mouse clicks are no substitute to this indispensable form of intelligence and skill.

We hope that this book which covers *broadband wireline access*, *optical communications components* and *networks* and *cryptography*, will be useful to the novice student and the skilled expert as well. Note: The other fields of *signaling principles*, *network protocols* and *wireless systems*, are covered in a second, independent volume of this series. Telecom is a nice field to be in or to work for. Like music, it has a background and rules of its own, and the more one is familiar with them, the more one is in a position to appreciate, to contribute and to innovate.

a MY VOCABULARY

Can you briefly explain each of these words or acronyms in the telecom network context?

Access	Edge	Microwave	Residential
Acronym	Header	Multiplexing	Routing
Aggregation	Internet	Network (layer)	Signal
Backbone	IP	Packet	Standard
Bandwidth	Layer	Payload	Transport
Bits	Legacy	Photonics	Voice
Business	Lightwave	POTS	VoIP
Core	Metro	Protocol	

CHAPTER 1

Broadband Wireline Access

This chapter covers the issues of broadband wireline access, from the digital subscriber line to the concepts of fiber-to-the-loop and fiber-to-the home and their networking/protocol implications.

■ 1.1 FROM TELEPHONE AND CABLE-TV NETWORKS TO THE INTERNET

Before the emergence of the *Internet*, one could conceive of "networks" as being separated into two distinct and unrelated service categories: those made for *two-way communications*, and those made for *one-way broadcasting*. As we shall describe, the corresponding traffic patterns and characteristics were utterly different, as well as the way these networks services could be accessed.

The first category includes *public-switched telephone networks* (PSTN), also dubbed *plain old telephony service* (POTS). The PSTN was initially conceived as a means to deliver basic voice-telephony services over entire countries, continents and across the oceans; it was designed and optimized to establish and switch *voice channels or circuits*, which were first analog then digital. Except for Mother's day or New Year's Eve, the PSTN traffic flow followed regular and predictable patterns, with peaks and troughs during day and night hours, respectively, and characterized by relatively limited connection times (as regulated by telephone bills). On the other hand, *local-area networks* (LAN) were conceived for computer communications on local private/public premises for business, administration or university uses. The LANs were optimized to transport and switch *data packets* having random initiation times and variable payload lengths. This is where the term "access" came from, meaning the possibility for a machine to send data to another when the network is available or the destination machine not already in use. Because the corresponding traffic patterns are of a burst-prone and

---

Wiley Survival Guide in Global Telecommunications: Broadband Access, Optical Components and Networks, and Cryptography, by E. Desurvire
ISBN 0-471-67520-2 © 2004 John Wiley & Sons, Inc.

unpredictable nature, specific LAN protocols were developed to handle priorities, solve access contentions and more or less equitably regulate the flow between the different users (see Chapter 2, Section 2.5 of Desurvire, 2004). Both PSTN and LAN are essentially wired networks, except when using GEO satellites as relaying interfaces to expand coverage (see Chapter 4, Section 4.3 of Desurvire, 2004). With the recent advent of *cellular telephony* and *wireless LAN* (see Chapter 4, Sections 4.2 and 4.4 of Desurvire, 2004), access to both network types is no longer dependent upon wires and wire plugs, but contrary to popular belief, their core-network infrastructures are still made of hard wires. The only exception to this are *satellite constellations* with intersatellite service and on-board switching capabilities (see Chapter 4, Section 4.3 of Desurvire, 2004), but their development and pervasiveness has remained limited so far.

The second network category concerns one-way/broadcast systems, which were primarily developed for radio and TV services. The service mostly concerned home and fixed-appliance reception, until the apparition of lighter "portable" radio receivers, based upon transistors to replace vacuum tubes, to be installed in cars or hand-carried (because of their greater weight and sizes, portable TVs followed a different history of customer usage). Access to broadcast networks only required cheap radio-antenna appliances and was mostly free, the case for TV depending upon national laws, as well as channel/contents ownership. New classes of pay-per-channel service for broadcast systems were then introduced through *cable TV* (or "cable", for short) and *satellite radio/TV* (or "satellite") systems, with both analog- or digital-quality features. The concept of *common-antenna television* (CATV) came into the picture to simplify the cable network and share the resource of local cable/satellite reception appliances. Such services also introduced the concept of "on demand" contents delivery. Demands from the end-users, for instance concerning video (VoD), would generally be placed through the PSTN, thus establishing a two-way loop between the customer and the cable/satellite-TV operator. In this loop, the *uplink channel* is associated with the demand (limited both in time and bandwidth), and the *downlink channel* with the service contents (with longer time and greater bandwidth). To anticipate the next section, this service feature is that of an *asymmetric subscriber line*, for which the upstream and the downstream traffic are not assigned the same bandwidths. The existence of "on-demand" services in a given geographical area defines a *local subscriber loop* (or *local loop*, for short), which is one of the topics developed in this chapter. Loops may are generally wired with high-bandwidth cables, but radio-access was also developed for deeper or more economical urban penetration. In this case, services have proudly called themselves "wireless cable," which may sound odd or oxymoronic to the engineer, but makes sense if "cable" means "service access."

The unpredicted emergence and irresistible growth of the *Internet* has dramatically upset the above, orderly network picture, as segregated between broadcast- and telephone-service systems. To simplify and avoid any complex explanation, the Internet signal protocol was designed as a new and convenient way to exchange data (text and pictures) between computers while using the POTS/PSTN, from ordinary telephone plugs. The necessary interface between the computer and the plug is the *modem* (for MOdulator/DEModulator), a device which makes the digital-to-analog signal conversion and the reverse. The modem then dials a certain phone number which reaches a special server called an *Internet gateway*. The analog signal is back-converted to digital and internet packets, called *datagrams* (see Chapter 3, Section 3.3 of Desurvire, 2004) are forwarded to this destination, with a similar reverse process at the end, all the way to the receiving computer. Computers can also download Internet data from remote servers in the form of text and picture files, called *web pages*. An important feature is that all these Internet connections are made at the price of a local phone call (connection to the gateway), regardless of how far away the remote computer or internet server may be located. As a most attractive feature, and long before the way it looks now, Internet was *first and foremost* a free service to be connected to, bringing value-added contents (Web pages) and novel personal communication means (electronic messaging, or *e-mail*). An extensive description of current Internet uses and associated jargon is made in Desurvire, 2004, Chapter 3, Section 3.3.

Telephone operators were once accustomed to predictable and placid voice traffic patterns, with limited user-connection times. But the development of the Internet brought an unprecedented type of traffic, essentially burst-prone with high peak-to-peak variations at any time of day and night! Moreover, Internet users would remain "*on-line*" for extended times, thus threatening to saturate the circuit-switching resource of entire urban areas, if not occasionally paralyze the local phone system by pre-empting all available lines. Since then, the picture has immensely evolved. Another key factor in this overall change is the 1996/1998 (US/EU) deregulation, which dismantled the telephone-system monopolies of *incumbent local-exchange carriers* (ILECS), opened the telephone market to new entrants called *competitive local-exchange carriers* (CLECS). Furthermore, CATV operators were allowed to carry voice-telephony on their networks, and likewise for ILEC/CLEC operators with video programming. Such a context led to the appearance of *internet service providers* (ISP), covering a wide range of functions from local gateway access, support of internet-services (e.g., Web browsing and e-shopping), and local-loop services such as VoD and multimedia (MM) communications. The picture is even more complex with the recent development of wireless access networks, fixed and mobile, especially concerning 3G cellular telephony. Network bandwidth limitations and bottlenecks were rapidly removed with the implementation of optical technologies (Chapter 2) all the way to metropolitan areas and in some case, to the business and residential levels. Furthermore, optical technologies have made it possible to link together the major continental networks through a mesh of broad-band undersea cables, making possible the implementation of the Internet as a true *World-Wide Web* (www).

For the above reasons, *access* is no longer associated with a specific type of network technology (fixed or mobile, wired or wireless, telephone or data), but is rather understood as some general commodity which can provide various types of voice, data and video services. While the Internet immediately appeared to be the *Holy Grail* of modern telecommunications, it is important to keep in mind that it is essentially a *transport protocol* (namely TCP/IP, see Desurvire, 2004, Chapter 3, Section 3.3). Such a protocol conveys higher-level and more sophisticated applications, on one hand, and is also carried in turn by lower-level protocols such as ATM, Ethernet or SDH/SONET, for instance (see Desurvire, 2004). For this same reason, the notion of local access is not confined any longer to connecting to either CATV or the Internet, but rather to services in general, preferably of the "*broadband* (BB)" type. Another important feature is that broadband access is becoming progressively available anywhere (urban, suburban and country areas) and at affordable prices. This is also becoming true (or at least as a realistic hope) is most developing countries, sometimes through satellite-mediated technologies which compensate for the absence of a fully developed wired infrastructure.

This chapter is dedicated to broadband *wireline* BB access, which rests upon "wired" technologies for the access links. The links can be indifferently made of copper-pair telephone wires, coaxial cables or fiber-optic links, with a wide range of capacity and distance performance. Broadband access is also possible through *wireless* technologies such as LMDS and 3G–4G, but it remains safe to emphasize that wires (especially optical fibers) carry more bandwidth than radio channels, setting aside any consideration of network deployment and installation costs which puts BB wireline access at an advantage. As a matter of fact, wireline access has been revolutionized by the technology called *asymmetric digital subscriber line* (ADSL) and its derivatives, referred to as the *xDSL* family. As further described in the next section, xDSL made it possible to better exploit the bandwidth of ordinary twisted-pair copper telephone wires, thus connecting individual homes and businesses into the core of a new class of BB services.

■ 1.2 DIGITAL SUBSCRIBER LINE (DSL)

The principle of *digital subscriber lines* (DSL) was introduced in 1989–1990, as a new means for the telephone operators to provide high-bandwidth services (e.g., VoD) through their telephone access networks, and as a competitive solution with respect to cable-modem/cable-TV distribution. Since then, the DSL service offer has evolved towards *high-speed Internet access*, with an "always-on" connection at a flat monthly fee. Moreover, such a service can also be implemented without interrupting the normal use of the telephone line. This option means that one could browse the Internet while sending or receiving calls through the same telephone line, as opposed to have to chose between one service or the other, or having to install two telephone lines for each use. The fact that DSL can be implemented by just plugging an inexpensive modem device between the computer and the home/office telephone plug, with "same-day" service activation, is another factor of high appeal. On the high-end service side, DSL implementation may still require the equivalent of two telephone lines and more complex terminals, as further explained below. In this case, the service and customer investment cost are substantially higher. To complete the picture, BB-DSL services can also be carried over optical fibers for the sake of extended capacity, access range, and signal-quality performance. Such an evolution is further described in the next section of this chapter, which is dedicated to *fiber in the loop* (FITL) and *fiber to the "x"* (FTTx) access services.

The DSL technology thus makes possible the bidirectional transport of high-speed data over ordinary twisted-pair copper telephone wires. The bit rates are generally different for the uplink and the downlink streams (the latter being higher), hence the name *asymmetric* DSL, or *ADSL*; or they can be made to match for full duplex capability (e.g., interactive multimedia), which is referred to as *symmetric DSL* or *SDSL*. Since telephone wires have intrinsic bandwidth limitations, the higher the bit rate, the shorter the maximum transmission distance for a specified bit-error rate. Namely, this distance defines an access zone between the customer premises (modem and telephone plug) and the network interface. Such an interface is called a DSL *access multiplexer* or *DSLAM*, as described in the following text. The different classes of upstream/downstream bit rates and related access distances have led to a flurry of new service appellations, which are grouped under the generic *xDSL* family name. Despite what the acronym DSL suggests, the modulation format for the high-speed signals is analog, not digital (more accurately, it corresponds to a *discrete, multitone [DMT] coding* scheme and other variants, see Section 1.2.1). In this section, we shall describe the characteristics of the different xDSL services and also briefly detail some of the main DSL coding principles. For clarity, we will use the term ADSL when referring to the technology in general.

Figure 1.1 shows the basic layout of ADSL access. The uplink channel reads from left to right, as follows. In the customer premises (CP), the computer is connected to the system through a DSL modem, which does the digital-to-analog/DSL format conversion. The telephone and modem lines are combined together into the twisted-pair telephone wire through what DSL terminology calls a

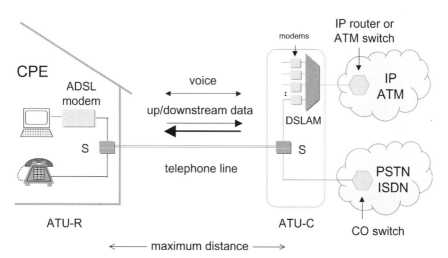

FIGURE 1.1 Basic layout of asymmetric digital subscriber line (ADSL) access, showing the remote terminal unit (ATU-R) and the central-office terminal unit (ATU-C) with twisted-pair telephone link (CPE = customer premises equipment, S = splitter, DSLAM = DSL access multiplexer, CO = central office, ISP = internet service provider, ATM = asynchronous transfer mode, PSTN = public-switched telephone network, ISDN = integrated-services digital network).

"splitter." In fact, this splitter is a frequency multiplexer, where voice occupies the smaller, low-frequency part of the spectrum and data the broader, high-frequency part. Such a CP equipment forms altogether the ADSL *remote termination/transceiver unit*, which is called *ATU-R*. At the other line end, the voice and data signals reach a unit called ATU-C (for ATU central-office, or CO). In the ATU-C, the voice and data signals are separated through the same frequency-splitter device as previously described. The data signal is forwarded to the closest ATM switch or IP router via the aforementioned interface called *DSLAM* (pronounce *"dee-slam"*). The first function of the DSLAM is to be a two-way modem, i.e., a device which converts analog DSL signals into digital data (uplink data traffic), and the reverse (downlink data traffic). On the other branch, the voice signal is sent towards the closest PSTN switch. Such an architecture thus makes it possible to offload the intensive data traffic from the PSTN and to make optimal use of the two network interfaces. The term DSLAM is sometimes also used to designate the whole ATU-C function as a terminal or access point to DSL services for multiple users. As the figure indeed suggests, the second function of the DSLAM is to combine different DSL input lines into a single output data multiplex. This output can be carried over extended distances by means of optical fibers, in order to reach a remote ISP or ATM network gateway. Figure 1.1 should also be read from right to left, describing the downlink channel. In the downlink channel, the input Internet or ATM signal from the packet network is converted into a DSL signal through the DSLAM. The DSL data signal and the return-voice signal from the PSTN are then *combined* in the ATU-C, carried over by the copper-wire telephone line, and finally *split* towards the computer and the telephone units, respectively. Therefore, the telephone line carries a two-way traffic containing both voice and data.

In the configuration shown in Figure 1.1, the ATU-C appears to be separate from the CO. In fact, the ATU-C is generally integrated into the CO premises as a terminal, with separate feeders for voice and data switching and multiple DSL input/output lines. As further described below, high-speed DSL applications (such as HDSL or SHDSL) require two copper lines (4 wires) rather than one, and more rarely up to three (6 wires). In this case, the network topology shown in Figure 1.1 remains basically unchanged, the link between ATU-R and ATU-C becoming a 4-wire (6-wire) line, with the CP-modem and DSLAM capacities being upgraded accordingly.

In very-high-speed DSL (VDSL), the maximum distances allowed for DSL-signal transmission over telephone lines are reduced from less than a mile to 1,000 feet, which is generally too short to reach a neighboring CO. In this case, the traffic is split at the maximum-distance point, as illustrated in Figure 1.2. The analog voice traffic, if any, follows its normal telephone-wire route, but the high-speed data portion is fed from an *optical network unit* (ONU). The ONU transports the data from the CO to the customer premises (CP) via a fiber-optic link, which can be several kilometers long. In VDSL, the CP are referred to as *multitenant* (MTU) and *multidweller* (MDU) *units*, which concern business and residential uses, respectively. Some also use the term MxU in order to designate the two CP types.

1.2 Digital Subscriber Line (DSL)

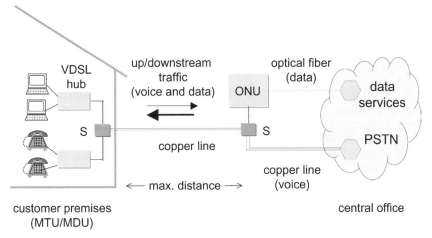

FIGURE 1.2 Basic layout of very high bit-rate DSL (VDSL) access, showing the remote customer premises (MTU/MDU = multitenant/dweller unit) and the central office. At some intermediary distance from the customer premises, the traffic is split between analog voice (copper wire) and high-speed digital data (optical fiber), via a splitter (S) and an optical network unit (ONU).

There is no unique design for DSLAM. The ones that are used for ATM network access are called *ATM-DSLAM*, or alternatively, *ASAM* (ATM *subscriber access multiplexer*), while those used for the Internet are called *IP-DSLAM*. But they are also qualified by their different types of xDSL service class, of voice/data signal capacities and protocols, and of number of input/output DSL ports they are made to handle. A more general appellation for DSLAM is *IAD*, for *integrated access device*. For instance, an IAD can be a 4-8-16 port device with T1 TDM (1.544 Mbit/s or 24 multiplexed voice circuits at 64 kbit/s), T1 ATM (T1 voice over ATM) and ADSL/VDSL/g.shdsl multiple compatibilities. Sophisticated versions of IAD also feature PBX/centrex functions, built-in IP routers, IP-address assignment and firewalls. It is clear then that the DSLAM should not be viewed as uniquely confined to central offices, but rather as an IAD gateway to be placed, anywhere that is best suitable, from the CP to CO locations or in between.

1.2.1 xDSL Flavors

In this subsection, we consider the different service classes (also referred to as "*DSL flavors*", or "*DSL technology suite*") to be found in the xDSL family. Such classes can be characterized by five key parameters:

- *Upstream data rate*,
- *Downstream data rate*,
- *Maximum access distance, or access range*,

- *Ability to combine voice telephony*, or
- *Number of required copper-wire pairs (one or two)*.

The exact specifications in data rates and distances are dependent upon physical parameters such as the copper-wire *gauge* and *diameter* (see further subsection below), the DSL coding scheme used, the corresponding signal-quality reference (BER and SNR margin) and the type of error-correction algorithm. There are also trade-offs between performance and cost, which differentiate products within the same xDSL service class. The main xDSL classes and indicative specifications, as now standardized, are summarized in Table 1.1, and can be listed as follows:

- **ADSL *(asymmetric digital subscriber line)*.** Exists both in long-range (5.5-km) and short-range (3.6-km) options; the corresponding downstream capacities are 1.7 Mbit/s to 7.1 Mbit/s, respectively; the upstream capacities are about ten times smaller, namely 176 to 640 kbit/s, respectively; as a way to visualize what these capacities represent, they are equivalent to what could be obtained by time-multiplexing together 25 to 100 telephone circuits for the downlink or 3 to 10 telephone circuits for the uplink (assuming 64-kbit/s baseband rates); note that the access range is between 3.5 and 5.5 km (2 to 3.5 miles), which potentially covers sub-urban areas of 38 or 95 km^2, or approximately 20,000 or 100,000 residential users.

- **HDSL *(high bit-rate DSL)*.** The downstream bit rates are 1.5 Mbit/s or 2.0 Mbit/s, which is near the maximum ADSL capability (1.7 Mbit/s); a major difference with ADSL is that the upstream rate is made to match the downstream rate, which represents about 2–3 times the maximum ADSL upstream rate; the bi-directional and "symmetric" bit rates, more accurately defined as 1.544 Mbit/s or 2.048 Mbit/s, do correspond exactly to the T1 and E1 payloads used in the protocol SDH (*synchronous digital hierarchy*), see Chapters 2 and 3, Desurvire (2004), Tables 2.2 and 3.1, respectively; the matching performance for the upstream/downstream links is obtained at the cost of using two copper pairs (4 wires). The second HDSL generation, HDSL2, requires only one pair (one thus refers to the first and the second "generations" as HDSL4 and HDSL2, respectively); in either case, the access range is limited to about 12,000 feet or 3.5 km (2.2 miles); such a distance is twice to four times the haul of ordinary T1 or E1 telephone trunks between two repeaters (3,000 to 6,000 feet), which is advantageous for extending the local telephone distribution to businesses and other premises needing such grouped capacity (leased lines), although the reliability may not be the same as with the usual T1/E1 TDM approach; with HDSL2, the link requirement is only a single copper-wire pair (as opposed to 4 pairs in repeated E1/T1 links; the connection fee can thus be significantly reduced; note that other possible and less conventional implementations of HDSL (not listed in the table) use from one to three wires.

- **VDSL *(very-high data rate [or very high bit ratespeed] DSL)*.** Represents a third and ultimate grade in the xDSL system family; downstream capacities range from 13 Mbit/s (range of 1.4 km/0.8 mile) to a whopping 55 Mbit/s (range of 300 m or 1,000 feet); because the access range is so limited and distance-sensitive, VDSL implementation requires the ATU-C or DSLAM terminals to be installed

TABLE 1.1 Different service classes of the xDSL family, showing upstream and downstream data rates, access range, ability to carry telephone voice, and number of copper-wire pair requirements

xDSL Service	Upstream Data Rate	Downstream Data Rate	Access Range (feet/kilometers)	Voice	Number of Pairs
ADSL	176 kbit/s	1.7 Mbit/s	18,000/5.5	Yes	1
	640 kbit/s	7.1 Mbit/s	12,000/3.6		
HDSL, or HDSL4 (HDSL2)	1.544 Mbit/s	1.544 Mbit/s	12,000/3.6	Yes	2
	2.048 Mbit/s	2.048 Mbit/s	12,000/3.6	Yes	2
	(1.54 Mbit/s)	(1.544 Mbit/s)	(12,000/3.6)	(Yes)	(1)
VDSL	640 kbit/s	13 Mbit/s	4,500/1.37	Yes	1
	3.2 Mbit/s	26 Mbit/s	3,000/0.910		
	6.4 Mbit/s	51.8–55.2 Mbit/s	1,000/0.300		
SDSL	1.1 Mbit/s	1.1 Mbit/s	24,000/7.3	No	1
	2.312 Mbit/s	2.312 Mbit/s	9,000/3		
RADSL	128 kbit/s	640 kbit/s	21,300/6.5	Yes	1
	176 kbit/s	1.54 Mbit/s	18,000/5.5		
	1 Mbit/s	2.56 Mbit/s	12,000/3.6		
IDSL	128–144 kbit/s	128–144 kbit/s	18,000/5.5	No	1
DSL Lite, or G.lite	384 kbit/s to 500 kbit/s	1 Mbit/s to 1.5 Mbit/s	18,000/5.5	Yes	1
G.shdsl, or SHDSL	192 kbit/s	192 kbit/s	40,000/12.2	Yes	1 (2.3 Mbit/s)
	2.3 Mbit/s	2.3 Mbit/s	6,500/2.0		2 (4.6 Mbit/s)
	4.62 Mbit/s	4.62 Mbit/s	6,500/2.0		

relatively close to the residential/business areas in some cabinet called an *optical network unit* (ONU); the rest of the distance between CO and ONU is covered by an optical-fiber link; this also permits the multiplexing of data for servicing large numbers of users (see also FITL and FFTx, in next section); the different VDSL service classes are described in the xDSL framing subsection.

- **SDSL (symmetric DSL).** As the name indicates, offers the same capacity for the uplink and the downlink; traffic symmetry is however a feature found in other xDSL classes like HDSL, for instance, which also makes SDSL a generic term; the best SDSL performance corresponds to capacities up to 1.1 Mbit/s over 24,000 feet or 7.3 km (4.5 miles) and 2.312 Mbit/s over 9,000 feet or 3 km; the rationale for SDSL, in fact confusingly referring to as HDSL, is to offer an economical replacement solution for E1 (2.048 Mbit/s) and T1 (1.544 Mbit/s) telephony; as previously mentioned for HDSL the advantage over conventional E1/T1 trunks is two-fold: a reduced number of wire pairs and an extended distance between repeaters.

- **RADSL (rate-adaptative DSL, pronounced "rad-zel").** Variant of ADSL where the capacity is adapted according to changing line conditions in terms of outside cabling and length.

- **IDSL (integrated digital network service [ISDN] DSL).** Made for direct, bidirectional inter-operability with 128–144 kbit/s ISDN, using the same modems; in the so-called *BRI* service class of ISDN, the capacity of 128 kbit/s represents two 64-bit/s digital telephone circuits, while 144 kbit/s corresponds to the overhead complement, see Vol. 1, Chapter 2, Section 2.2; with normal ISDN, the BRI actually requires a 5-wire link, as opposed to 2 wires in IDSL; thus IDSL can be deployed by the telephone operator to simplify the distribution of cables and wiring in densely populated business or residential premises; another feature of IDSL, which provides added-value from the classical ISDN is that it is an "always-on" service, which is well adapted to Internet use.

- **DSL Lite, also known as Universal DSL and as splitterless DSL.** Derived from the standard ITU-T (*International Telecommunications Union, Telecommunication systems*) called *G.lite*, which is a cost-effective version of ADSL that removes the need for splitters in the CO and the CP (implementation however requires micro-splitters for the different subbands); compared to standard ADSL (see Table 1.1), the DSL Lite performance lies in between for the upstream (384 kbit/s vs. 176–640 kbit/s) and about 1.5 times under the minimal capacity offer (1 Mbit/s v. 1.7 Mbit/s) for the downstream; the downstream figure is however almost as good as HDSL (1.54 Mbit/s), while the distance (as for long-range ADSL) is 1.5 times longer (18,000 feet vs. 12,000 feet), corresponding to 5.5 km or 3.5 miles;

- **G.shdsl (singledouble-pair, high-speed DSL), also called SHDSL.** Is defined as a "symmetric" DSL service with three capacity offers of 192 kbit/s, 2.3 Mbit/s and 4.6 Mbit/s; the last figure requires a two-pair line; the distances covered range from 40,000 to 6,500 feet; at equal capacities, the g.hsdsl/SHDSL approach is supposed to offer longer distances than previous SDSL and is meant to replace it.

This description shows that it is not easy to compare the relative merits of the different xDSL service classes, as they are so redundant in either capacity or access range, in addition to covering different options under the same name. The plot in Figure 1.3 puts them into a clearer perspective, from the highest capacities and shortest ranges (VDSL at 55 Mbit/s over 300 m) to the lowest capacities over the longest ranges (G.shdsl at 192 kbit/s over 12 km).

One may wonder why the above bit rates take so many different (and uncommon) values, and about the rules that define the capacity scales for upstream and downstream traffic options. This issue is addressed in a following subsection concerned with *DSL framing*. As will be described, both upstream and downstream ADSL traffic is divided into subchannels, or *bearers*, which scale by multiples of 32 kbit/s according to different options. Other DSL applications such as HDSL use frames but the subchannels are T1/E1 circuits. In all cases, the framing includes various overhead fields for the purposes of synchronization, payload delimitation and error correction, which leads to various possibilities of uncommon bit-rate values, as will be explained.

For the DSL end-user, a prevalent figure of merit is the *time* it takes to download data files (Table 1.2). To visualize how the different xDSL capacities compare, consider the download times for a hefty 10-Mbyte file (80 Mbit) and a standard CD-ROM (700 Mbytes). With a standard 56 kbit/s voice-channel modem, the downloads require about 24 minutes for the file and 28 hours (!) for the CD. With the best DSL offer (VDSL at 55 Mbit/s), these times are changed into 1.45

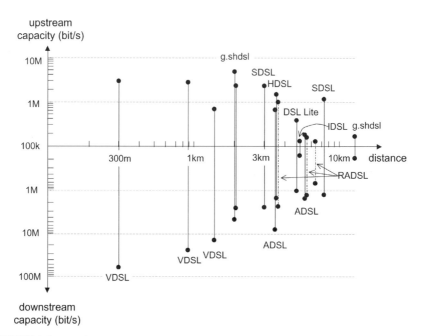

FIGURE 1.3 Maximum upstream and downstream capacities and corresponding access range of the xDSL family, featuring different options for ADSL, HDSL, VDSL, SDSL, RADSL, IDSL, DSL Lite and G.shdsl (Note: for clarity, IDSL and DSL Lite were placed at a distance somewhat shorter than their actual 5.5-km range). See also data in Table 1.1.

TABLE 1.2 Download times corresponding to 10-Mbyte and 700-Mbyte (full CD-ROM) data files, according to different DSL modem classes

Modem Class	10-Mbyte File Download Time	700-Mbyte File Download Time
56 kbit/s voice	23 mn 48 s	27 h 46 mn
128 kbit/s ISDN	10 mn 24 s	12 h 8 mn
1.54 Mbit/s HDSL	52 s	1 h 36 mn
1.7 Mbit/s ADSL	47 s	54 mn 54 s
7.1 Mbit/s ADSL	11 s	13 mn 9 s
55 Mbit/s VDSL	1.45 s	1 mn 42 s

seconds and less than 2 minutes, respectively, which illustrates huge differences for practical purposes. In between, the standard ADSL service (1.7–7.1 Mbit/s) is in the 1 mn–10 s ballpark for the 10-Mbyte file and 1 h–10 mn for the CD, which represent reasonably acceptable download times.

1.2.2 DSL Modulation and Coding Schemes

Different signal modulation techniques for DSL, referred to as *DSL line codes*, have been developed in order to meet three main criteria: (a) efficient use of bandwidth; (b) combined optimization of transmission speed and distance; and (c) robustness to intrinsic noise and external interference from nearby wire pairs. The last issue concerning physical impairments and their causes is addressed in the next subsection. Here, we shall focus on the criteria of *bandwidth efficiency* and show that there are different line code options to meet the bit-rate × distance requirements of each xDSL system. It is clear that the choice of a given line code for a given xDSL system is dictated by other criteria as well, namely standardization and its history, chip technologies, robustness to imperfections and overall considerations of complexity and cost. A coding solution that works best in the laboratory under ideal or optimized operating conditions may not be suitable in the real environment, or just be too complicated to implement in practice. Despite this fact, some of the standard DSL line-coding approaches are quite sophisticated. Two important conceptual aspects of DSL coding to be distinguished from each other are (a) the arrangement of input/output bits into *frames* (DSL framing), and (b) the packaging of the frame bits into "analog" symbols (several bits per symbol) prior to transmission and their restitution after transmission. The different DSL framing standards and user-options are described in a later subsection. Here, we shall focus on the symbol-coding techniques.

There are two classes of DSL line codes: *single-carrier* or *multicarrier*. In the fist case, the information contained in the symbol occupies the entire channel bandwidth (although such occupation may not be uniform, as we shall see). In the second, the channel bandwidth is sliced into many subchannels which carry only a portion of the symbol information. We consider first the simpler single-carrier line codes, then describe the multiple-carrier case.

As previously mentioned, xDSL modulation formats are analog, meaning that they use analog symbol waveforms to encode digital information. Symbols can encode groups of two bits or more. Such a D/A encoding operation requires *multilevel* (or *M-ary*) *amplitude* and *phase modulation* (AM/PM) of a single carrier. Different schemes are described in Desurvire (2004), Chapter 1, Section 1.2. The description shows that M-ary amplitude modulation alone is called *M-PAM* (for M-ary pulse amplitude modulation) and phase modulation alone is called *M-PSK* (for M-ary phase-shift keying). In particular, 2-PSK and 2-PAM are identical (changing the amplitude by ± 1 being equivalent to modulating the phase by $\pm \pi$). The specific case of *4-PAM*, also called *2B1Q* for "two-bits, one quaternary," is of interest because each of the 4 resulting symbols carries two bits (or 1 *quat*). While the 2B1Q line coding is used in DSL applications, it is not as versatile and powerful as other M-ary coding approaches based upon the simultaneous use of AM and PM, which we describe next.

Two leading single-carrier, quaternary line codes for DSL are called *quadrature amplitude modulation* (QAM) and *carrierless amplitude and phase modulation* (CAP). It is first worth going into the details of QAM in order to get a clearer view of how the two modulation types actually work, from transmitter to receiver.

The principle of QAM is described in Desurvire (2004), Chapter 1, Section 1.2. To summarize, QAM is a modulation scheme where two carrier quadratures at frequency ω, namely $\sin(\omega t)$ and $\cos(\omega t)$, are combined with discrete-level amplitudes a_m and b_m, respectively, for a duration T_S called the symbol period (symbol rate $= 1/T_S$). The resulting time-dependent waveform is written:

$$f(t) = a_m \sin(\omega t) + b_m \cos(\omega t) \tag{1.1}$$

The cosine and sine parts of $f(t)$ are usually referred to as the *in-phase* and *quadrature components*, respectively. Each of the amplitudes can take $\sqrt{M}$ possible values ($M = 2^k$ with $k =$ integer), corresponding to M different symbols. The plots of the different symbol possibilities in the (a_m, b_m) plane is referred to as a *constellation*. An example of a constellation with $M = 16$ is provided in Desurvire (2004), Chapter 1, Figure 1.13. Such a format is referred to as *16-ary QAM*, and more generally, *M-ary QAM* (MQAM). Since each symbol represents $k = \log_2 M$ bits, the bit rate B is k times the symbol rate, or $B = k/T_S$. Assuming that the symbol spectral width is exactly $\Delta f = 1/T_S$, the *spectral efficiency*, as expressed in units of bit/s/Hz, is equal to $B/\Delta f = k = \log_2 M$. Thus 64-ary QAM ($64 = 2^6$) has a theoretical spectral efficiency of 6 bit/s/Hz. Current technology makes it possible to achieve *256-QAM*, corresponding to 8 bit/s/Hz.

Consider next the principle of coding and decoding from transmitting to receiving ends, as illustrated in Figure 1.4. The function of the QAM transmitter is to encode incoming bit sequences into analog-waveform symbols, defined by $f(t)$ for the duration T_S. The first function is to determine the amplitudes (a_m, b_m) from the incoming bit sequence. Taking for instance the example of 16-ary QAM in Figure 1.13, any group of 4 bits is mapped into a unique point in the constellation

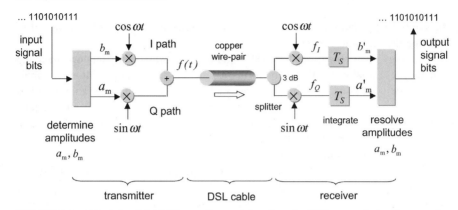

FIGURE 1.4 Principle of QAM line coding and decoding for DSL transmission (see text for description).

(for instance, 0000, 0010, 0110 and 0100 correspond to points located in the four outer corners of the constellation). Each symbol point is therefore uniquely defined by the two coordinates (a_m, b_m), to which is associated a set of discrete voltages [$V(a_m)$, $V(b_m)$]. Since there is a one-to-one correspondence between coordinates and voltages, we shall just use (a_m, b_m) to describe either one. The signal paths belonging to a_m are called the "Q-path" and that belonging to b_m is the "I path." In another part of the transmitter, a cw signal generated from an RF source at frequency ω (called oscillator) is split into cosine and sine components (this is realized by a 3-dB RF splitter, not shown in Figure 1.4, which introduces a $\pi/2$ relative phase shift in one output branch). The two voltages (a_m, b_m) are then applied to modulate the quadrature (sine) and in-phase (cosine) components, over the Q-path and I-path, respectively. The modulated signals are finally combined into the waveform $V(t) \propto f(t)$, which is applied to the copper-wire pair. The signal is thus transmitted through the cable up to the receiving end. At this point, the signal waveform is equally split into two branches. Within each branch, the signals are amplitude-modulated by $\cos(\omega t)$ and $\sin(\omega t)$, respectively. The resulting signals thus take the respective forms:

$$\begin{cases} f_I(t) = [a_m \sin(\omega t) + b_m \cos(\omega t)] \cos(\omega t) = a_m \sin(\omega t) \cos(\omega t) + b_m \cos^2(\omega t) \\ f_Q(t) = [a_m \sin(\omega t) + b_m \cos(\omega t)] \sin(\omega t) = a_m \sin^2(\omega t) + b_m \cos(\omega t) \sin(\omega t) \end{cases}$$
(1.2)

The next function of the receiver (Figure 1.4) is to *integrate* the above signals over the symbol period T_S. To "integrate" a signal $f(t)$ means summing up the values $f(0), f(\tau), f(2\tau), \ldots, f(N\tau = T_S)$, taken at small time intervals $\tau = T_S/N$. Looking at equation (1.1), we see that the terms in $\cos^2$ and $\sin^2$ are always positive numbers. It can be shown that their integration is equal to $T_S/2$. On the other hand, the terms in $\cos \times \sin$ oscillate between $+1$ and -1. If the symbol period T_S is sufficiently long compared to the RF oscillation period $T = 2\pi/\omega$, the number and absolute amplitudes of positive and negative terms are the same, leading to a zero sum or integral. By analogy with vectors, one says that the sine and cosine carriers are *orthogonal*, yielding a zero cross-product integral, i.e., $\cos \times \sin \to 0$ (arrow meaning time integration). The end-result is that the integration of in-phase and quadrature signals yields nonzero voltages of $b_m T_S/2$ and $a_m T_S/2$, respectively, which are proportional to the original constellation amplitudes (a_m, b_m)!

But in real systems, the transmission line introduces signal distortion and noise (see next subsection). The result of the integration could be called (a'_m, b'_m), which represents a pair of random numbers of mean (a_m, b_m) and deviation $[\sigma(a_m), \sigma(b_m)]$. As shown in Figure 1.4, it is the final task of the receiver to resolve the random values (a'_m, b'_m) into the discrete set (a_m, b_m) and match the result to the corresponding constellation point and bit sequence. Because of the uncertainty (however small), the constellation point can be mistaken with a neighbor, resulting in errors in the resolved bit sequence. The errors translate into a *bit-error-rate* (BER), which is one of the standard measures of system transmission quality. The BER can be reduced by precoding the initial bit sequence to be transmitted with *forward-error correction* (FEC). As described in Desurvire (2004), Chapter 1, Section 1.6, FEC is based upon *Reed–Solomon* (RS) and cyclic-*redundancy check* (CRC) algorithms, which allow flexible degrees of bit-error correction through the use of redundancy bits, at the expense of actual payload information.

The second single-carrier DSL modulation technique, *CAP*, is very similar to QAM. Current technology makes it possible to achieve *256-CAP*, corresponding to 8 bit/s/Hz. One therefore uses the same constellation encoding/mapping principle, resulting in in-phase and quadrature signal components being generated at the transmitter and resolved at the receiver. But such I/Q components are not obtained by modulating the $[\cos(\omega t), \sin(\omega t)]$ signals with the amplitudes (a_m, b_m) as in QAM. Instead, one uses *pulses* with temporal shape $p(t)$ and amplitudes (a_m, b_m), leading to the signal pair $[a_m p(t), b_m p(t)]$. The pulses are then passed through digital filters called *Hilbert (-transform) pairs*. At the filter outputs, the resulting signals are $[a_m h(t), b_m \hat{h}(t)]$, where $h(t), \hat{h}(t)$ are characteristic filter transfer functions in the time domain (hence the concept of "carrierless" amplitude modulation). The two signals are then combined to give the waveform

$$f(t) = a_m h(t) + b_m \hat{h}(t) \qquad (1.3)$$

which is input to the DSL cable. The CAP receiver works on the same reverse principle as for QAM. The amplitudes (a_m, b_m) are retrieved by passing the output signal through a similar Hilbert pair of adaptive filters, each filter blocking/passing the component of the opposite/same transfer function. This is based upon the property that the time-integral of $h(t)\hat{h}(t)$ is zero, while that of $h^2(t), \hat{h}^2(t)$ is non-zero. The final function of the receiver is to resolve the actually received amplitudes (a'_m, b'_m) into the single constellation point (a_m, b_m), and finally, convert the result into the corresponding bit sequence. As with QAM, pre-coding with FEC (RS, CRC) algorithms makes it possible to maximize the BER.

Compared to QAM, CAP requires more digital circuitry. However, it can be shown that the two formats are mathematically equivalent if the QAM symbols

26 *Broadband Wireline Access*

use the same pulse shape function $p(t)$ as in CAP, which amounts to rotating the constellation about the plane origin. But as we have seen, pulse-shaping is not required in QAM, which makes its implementation more straightforward. The two approaches are in fact very different when considering error correction, namely how it is applied in the frame blocks (see subsection below) and upstream/downstream channels. The bit sequences to be transmitted in QAM or CAP formats are also scrambled, interleaved and precoded according to different standard algorithms, which are beyond this book's scope to describe. For advanced reference, the CAP transceivers additionally use a precoding scheme called *Tomlinson–Harashima precoding*, or THP, to alleviate the effect of *intersymbol interference* (ISI). The spectrum allocations of QAM and CAP are different, as illustrated in Figure 1.5 in the case of ADSL. It is seen that both use the low-frequency portion of the spectrum for the upstream channel and the high-frequency portion for the downstream channel, but with different start and stop frequencies. The stop frequencies correspond to different symbol and bit rates. Two classes are defined for QAM, which correspond to the same start-frequencies (34 kHz upstream and 148 kHz downstream) and different stop-frequencies of 134/189 kHz (upstream) and 1,104/1,366 kHz (downstream), respectively. By convention, the corresponding symbol rates, as expressed in *kilo-bauds* (10^3 symbols/s) are 20/40/84/100/120/136 kBauds (upstream), and 40/126/160/252/336/504/806/1008 kBaud (downstream). Different bandwidth-use and baud-rate options are also defined for CAP, which are explicitly indicated in the figure.

As elementary as they may appear, the "single-carrier" line codings that are QAM and CAP are progressively abandoned by most LEC in profit of more complex approaches, as we shall describe. One of the intuitive reason for this evolution is that in QAM/CAP, the information is evenly distributed across the upstream or downstream band spectra. But the noise, signal interference and

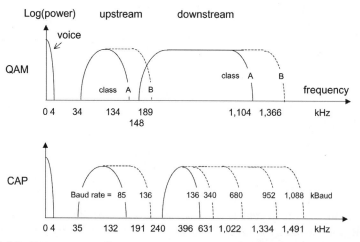

FIGURE 1.5 Power spectra allocated to ADSL with QAM or CAP line codings, according to different Baud rates or classes, showing upstream- and downstream-channel partition, in addition to the common 0–4 kHz voice channel.

distortion impairments of the line are frequency-dependent, being usually higher at high frequencies, with local peaks or resonances, and are time-varying. The signal-to-noise ratio (SNR) corresponding to each band is further degraded by such frequency-dependent impairments. While the SNR can be improved by means of dynamic pre-/post-equalization schemes, the filtering functions and digital circuits are quite complex to implement. This is the reason why *multicarrier modulation* was introduced.

The leading multicarrier modulation format in DSL is known as *discrete multi-tone* (DMT), also called sometimes *orthogonal frequency-division multiplexing*, or OFDM. The rationale for DMT is to improve the line performance by splitting the information bits into different sub-bands. Conceptually, this is equivalent to carrying information at low bit rates in many independent subchannels, as opposed to using a single, high-bit-rate channel. But one can also distribute the bit rate in a non-uniform fashion in order to get some advantage. Indeed, it is possible to allocate higher bit rates to the subbands (frequency "bins") that exhibit the highest SNR, and lower bit rates to the subbands that are in the opposite case. The bin bit/symbol rate is thus maximized where the SNR is the highest, which minimizes the BER. Conveniently, one can also "avoid" the frequency regions where deleterious noise/interference resonances are found, leaving the corresponding subbands empty. Finally, it is possible to dynamically adapt the bit-rate/subband allocation according to time-varying conditions or variable system parameters/imperfections.

The principle of DMT is the same as in the previously described QAM case. The difference is that the bit-information is split into many parallel blocks to be encoded at different frequency tones, also called *subcarriers*. As we have seen, the frequency subbands (or subchannels) are also called *bins*. Figure 1.6 shows a conceptual view of a DMT transmitter. It is made of N parallel QAM encoders at tone/bin frequencies ω_k. The resulting QAM signals, $f_k(t, \omega_k) = a_{mk} \sin(\omega_k t) + b_{mk} \cos(\omega_k t)$ are then summed into the overall waveform $\sum f_k(t)$. While the DMT transmitter representation shown in the figure is conceptually correct, it is not used as such in real systems, especially for considerations of receiver complexity. Indeed, one could retrieve the constellation amplitudes (a_{mk}, b_{mk}) through the same *orthogonality* principle as previously described for single-carrier QAM, based upon the additional property that multi-frequency carriers are orthogonal in the sense that:

$$X = [a_{mk} \sin(\omega_k t) + b_{mk} \cos(\omega_k t)] \times \sin(\omega_l t) \to 0 \text{ or } a_{mk} T_S/2$$
$$Y = [a_{mk} \sin(\omega_k t) + b_{mk} \cos(\omega_k t)] \times \cos(\omega_l t) \to 0 \text{ or } b_{mk} T_S/2$$
(1.4)

In the above expression, the arrow means time-integration over the symbol period T_S, and the nonzero result corresponds to the matching-frequency case, i.e., $k = l$. But implementing N parallel integrator/receiver modules of this type would be fastidious. A more practical approach consists in using *discrete Fourier transform* (DFT). The *Fourier transform* is a mathematical operation which allows one to convert a signal from the time-domain into the frequency-domain and the reverse (*inverse Fourier transform*). Digital circuits can perform this operation through an algorithm called *fast Fourier transform* or FFT. The QAM/bin signals $f_k(t, \omega_k)$ may then be viewed as representing the individual frequency components of a

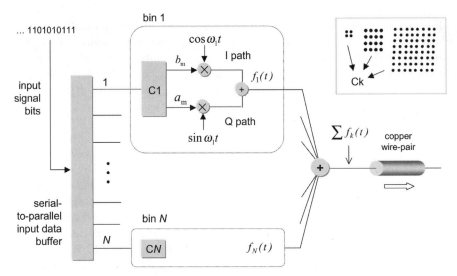

FIGURE 1.6 Conceptual view of a discrete multi-tone modulation (DMT) transmitter, showing constellation allocation for each encoder and corresponding frequency bin at tone frequency f_k. The inset on top right shows different optional allocations of M-QAM constellations to each individual encoder k.

time-domain signal $\sum f_k(t)$. But instead of effecting this direct summation into an analog signal, one can apply the *inverse FFT* (or IFFT) to convert the same time-domain signal into a digital form. From the receiver end, the advantage is that the reverse operation (FFT, which is the reverse of IFFT) can be implemented to retrieve the individual QAM/bin signals. As Figure 1.7 illustrates, the realistic DMT system is thus made of the following element sequence, from transmitter to receiver:

- Serial-to-parallel conversion via input data buffer (N output ports);
- DMT subchannel encoder (N bins);
- IFFT module (N parallel input ports [frequency bins], one serial output port [time-domain signal]);
- digital-to-analog converter;
- transmission line;
- analog-to-digital converter;
- FFT (one serial input port [time-domain signal], N parallel output ports [frequency bins]);
- DMT subchannel decoder and constellation resolution (amplitudes → quats);
- Parallel-to-serial conversion via output data buffer.

To complete the above picture, a trick is to insert an overhead *cyclic prefix* at the beginning of the IFFT block to be transmitted. Such a prefix lengthens the bit sequence by an amount which is sufficient to suppress intersymbol/carrier

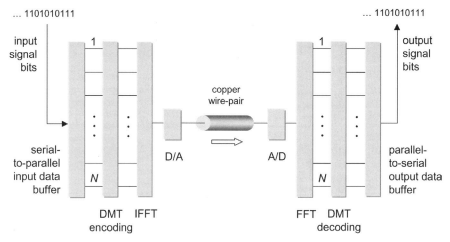

FIGURE 1.7 Realistic implementation of DMT system, from transmitter to line to receiver, with digital fast Fourier transform (FFT) and inverse (IFFT) operations (A/D and D/A = analog to digital conversion and reverse).

interference (ISI/ICI), the explanation of which being beyond the scope of our description. At the other end of the line, this "dummy" prefix is removed prior to the FFT operation.

For efficient bandwidth use, each bin encoder may use a different M-ary constellation. The higher-order or denser constellations, which have the highest bit-per-symbol rates, are allocated to bins which exhibit the highest SNR. Once the SNR spectrum is fully characterized (under initial signal loading conditions), a bandwidth/bin allocation table can be established. Figure 1.8 shows how the bandwidth/bin is allocated to both upstream and downstream channels according to the measured SNR criteria. Note that regions where the SNR happens to be severely degraded (e.g., transmission gap due to frequency-resonant tapping loss, isolated peak of AM radio-station interference, high-frequency cross-talk background) are left unused. The allocation can also be dynamic, according to changing system conditions. The bandwidth/bin loading thus ensures optimum SNR conditions for each subchannel, and optimum bit-rate with lowest BER for each of the subchannels as well.

The prevailing standard decomposes the 0–1,104-kHz DSL spectrum into 256 bins, numbered 0 to 256. This number is not arbitrary, because standard FFT circuits are usually based on 512 sampling points, corresponding to two analog-coordinates per constellation point. The bin spectral width is thus $1,104/256 = 4.3125$ kHz, which provides a sufficient *guard band* to separate the modulated tones. For instance, a 256-QAM represents 8-bit/s/Hz spectral efficiency, corresponding to a potential capacity of $4\text{ kHz} \times 8\text{ bit/s/Hz} = 32\text{ kbit/s}$ with 0.3 kHz as guard-band. To recall, 32 kbit/s is for DSL the standard bit-rate granularity. Therefore, the 4.3125-kHz bin interval conveniently fits all possible bit-loading combinations under all options from 4-QAM to 256-QAM. By convention, bins 0 and 256 are

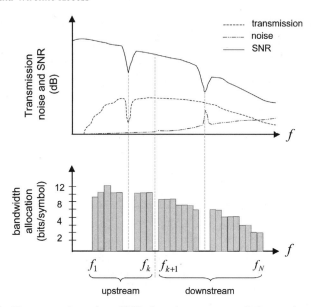

FIGURE 1.8 Frequency-dependent SNR, based upon transmission and noise spectrum of DSL transmission line (top), and corresponding optimum DMT bandwidth allocation (bit/symbol) to the different upstream/downstream frequency bins (bottom), avoiding degraded-SNR regions.

never used to carry data. Bin 0 gets the DC current present in the loop, while bin 256, which contains the 1,104-kHz/un-modulated carrier is used as *Nyquist frequency* reference: the Nyquist sampling rate is therefore 2,208 kHz. Bins 1–6 (up to 25 kHz) are used for usptream voice. Note that this represents a greater bandwidth allocation for PSTN voice services compared to the older, 4-kHz DSL standards. Digital-voice based on ISDN (see Desurvire, 2004, Chapter 2, Section 2.1), as more simply coded through 2B1Q or higher-level M-PAM, can be introduced through bins 18–28 for upstream. Overlooking ISDN channel allocation options, the split between upstream and downstream channels is normally made at bin 32. In practice, 7–32 are reserved for upstream and 33–250 for downstream. Bins above 250 are generally left unused because of their higher loss and lower SNR (hence low capacity potential). Bin 64 (tone 276 kHz) is reserved as a *pilot tone*, which serves a frequency reference for bit re-assembly at the receiver end. As previously described the bins can be loaded with variable bit rates, depending upon their associated SNR. Recall that in DSL, the bit-rate granularity is 32 kbit/s. However, the individual bins can be loaded with lower bit rates equal to integer fractions of 32 kbit/s, the choice of constellation density (bit/symbol) being left as a second degree of optimization (i.e., 4-, 16-, 64-, 128-, and 256-QAM). For instance, using 128-QAM (7 bit/symbol) with 16 kbit/s would result in a symbol date of $16/7 = 2.28$ kBaud (or kilo-symbol/s). This example illustrates that only low-speed electronics are necessary to perform the various FFT operations. Note that the bit rate can also be *adaptive*, meaning adapted to variable or evolving traffic needs or patterns. Even more sophisticated signal-transform algorithms can be implemented. Although we shall not discuss them here, it is useful to mention (for future advanced reference) the *Discrete-wavelet multitone* (DWMT) transform. The principle is similar to QAM, except that orthogonal "wavelet" functions are used to carry the bin coordinates, instead of the sine/cosine quadratures. One of the advantages of DWMT over DMT is a shorter signal-processing time.

These descriptions exemplify the importance of DMT for efficient DSL transmission and bandwidth management. Because of the multiplicity of line-coding options already implemented in commercial systems (QAM, CAP, DMT), it is not clear at present which one should win. As we have seen QAM/CAP is based upon broadband, single-tone signaling, corresponding to short symbol pulses. In contrast, DMT is based upon narrowband, multitone signaling, corresponding to long symbol pulses. Thus QAM/CAP is sensitive to time-dependent noise (bursts) but is relatively immune to frequency-dependent noise (AM-radio interference and cross-talk), while the contrary holds for DMT. It is clear, however, that unlike QAM/CAP, the DMT line code offers high flexibility in adapting to and countering any type of noise impairment. In addition, digital signal processing (FFT) has become relatively inexpensive and easy to implement, which lifts the argument of complexity and cost against DMT. Finally, the discrete 256-bin allocation simplifies *bandwidth management*, in particular between the upstream and downstream channels, and *interoperability* between different existing commercial systems. Another feature of DMT is the ability to remove painstaking bandwidth-splitting issues between upstream and downstream channels and achieve full-duplex, fully symmetric operation. This ability is of great interest for SDSL (and especially VDSL) systems. One way to achieve symmetric operation is to *interleave* bins corresponding to upstream and downstream channels, e.g. even-bins for upstream (↑), odd-bins for downstream (↓). Such a bin-alternating (↑↓↑↓ ...) configuration is known as the *"zipper"* approach, or *frequency-division-duplexed DMT* (FDD-DMT). For its implementation, two requirements are (1) the time and frequency synchronization of both DSL terminals, and (2) the addition of a suffix (in addition to the prefix) to the IFFT block for compensating line propagation delays. It is worth mentioning another variant of DMT, which is called *time-division-duplexed DMT* (TDD-DMT) or *synchronized DMT* (SDMT) for short. The idea is that upstream and downstream data are sent through the line in an alternating way, like in a *ping-pong* game. Such a "ping-pong" approach makes it possible to transform a highly asymmetrical line into a fully symmetrical one, since any amount of available bandwidth can be fully and equally used from both ends, while taking turns. Another advantage of time duplexing is that parts of the DSL modems can be used for both transmission and reception, which simplifies the equipment and reduces costs.

1.2.3 DSL-Signal Physical Impairments

As we have seen, DSL signals are made of a superposition of discrete frequency bands or tones, which range from 20 kHz to 1.1 MHz. The signal integrity is affected by several kinds of transmission impairments, which are of physical and of electromagnetic origins, respectively.

The first physical impairment is *transmission loss*. This loss first depends upon the *wire gauge*, according to the ANSI (*American National Standards Institute*) definitions, which are referred to as 24–26 AWG (*American wire gauge*). There is a one-to-one correspondence between the AWG number and the wire diameter, the higher number corresponding to the smaller wire diameter and hence, the greater intrinsic resistance. In the United States, telephone trunks are typically based upon copper-wires gauges of either 24 AWG (0.511 mm or 0.0201″) or 26 AWG (0.404 mm or 0.0159″). The characteristic line resistances of these standard wires, as expressed in Ohms per 1,000 feet (304.8 m) are 25.67 Ω/1,000 feet and 41.02 Ω/1,000 feet. Recall that, *resistance* only concerns continuous or zero-frequency signals, while *impedance* concerns amplitude-modulated signals. In the second case, the wire impedance is its resistance plus a frequency-dependent contribution. Since the signal power loss is proportional to the wire impedance, it is therefore sensible that long-distance telephone trunks use the lowest-resistance gauge (24 AWG), while locally the gauge can be made higher (26 AWG) for considerations of shorter distances and economy. Indeed, a near 20% ($x = 0.404/0.511$) diameter reduction between the two wire types corresponds to 40% (x^2) in the wire volume and mass, which makes economical sense recalling that copper belongs to the family of precious metals.

As in most communication systems, the channel loss is characterized by an attenuation coefficient expressed in units of *decibels per kilometer* (dB/km). The wire impedance being frequency-dependent, so is the attenuation coefficient. With twisted-pair channels, the attenuation coefficient rapidly increases with frequency, as illustrated by the data shown in Table 1.3. Consider frequency tones near 1 MHz and a 24-AWG channel. From the data in Table 1.3, it is seen that a 300-m line has an attenuation of 20 dB/km × 0.3 km = 6 dB, representing a transmission coefficient of $10^{-6/10} = 0.25$ (only 25% of the power is transmitted). With a 1-km line, the attenuation is 20 dB/km × 1 km = 20 dB, representing a transmission coefficient of $10^{-20/10} = 0.01$ (only 1% of the power is transmitted). If the line length is extended to 2 km, the dB attenuation is twice as high, representing a transmission of 0.01%, or 1/10,000. In contrast, tones near 100 kHz have an attenuation of 3 dB/km, which corresponds to 300-m, 1-km, and 2-km line losses of 3 dB/km × 0.3 km = 0.9 dB (81%), 3 dB/km × 1 km = 3 dB (50%), and 3 dB/km × 2 km = 6 dB (25%),

TABLE 1.3 Frequency dependence of attenuation coefficient of twisted-pair channels, as corresponding to two types of ANSI wire standards, 24 AWG and 26 AWG

Frequency (MHz)	24 AWG (dB/km)	26 AWG (dB/km)
0.1	≈3	≈4
1	20	25
5	47.5	60
10	67	85

AWG = American wire gauge

respectively. In the first case (300 m), the low-frequency (LF) and high-frequency (HF) components of the spectrum have 81% and 25% transmissions, respectively. In the last case (2 km), the transmissions are 25% (LF) and 0.01% (HF). This shows that line loss introduces a strong power imbalance between the LF/HF components, acting like a low-pass filter. In the time domain, this frequency-dependent filtering corresponds to both amplitude and phase distortion for the DSL signal. The bit-word value associated with the DSL symbol is therefore modified, resulting in bit-error-rate (BER) increase after signal demodulation. To alleviate this effect and reduce the BER, corrective *frequency-equalization* can be used at the receiver level to compensate the power differences between HF and LF components. Another source of amplitude/phase distortion comes from the fact that the frequency tones travel at different velocities (an effect referred to as *spectral dispersion*). Dispersion thus introduces time delays between LF and HF tones, which must be compensated at the receiver. Equalization therefore alleviates both effects of attenuation and dispersion. One refers to *adaptive equalization* as the ability to optimize the received signal under time-varying or locally different line conditions. Adaptive equalization between LF and HF components is possible only within some power dynamic range and noise-background conditions.

A second source of transmission limitations is *signal cross-talk* (Xtalk or XT, for short) which represents the parasitic power leakage of one channel into another. As the name indicates, the term "crosstalk" comes from early analog-switched telephony. It corresponds to the effect of hearing a weak background conversation while being on the telephone. Nowadays such an occurrence is rare, because of the PSTN conversion to digital switching and processing technologies. More generally, crosstalk between different subchannel carriers (case of DMT transmission) acts as a parasitic noise source which causes *intersymbol interference* (ISI), also called *intercarrier interference* (ICI). Since twisted-pairs are insulated, their mutual power leakage is due to both electromagnetic (EM) and capacitance couplings. In the first case, the metallic wires act as mutually coupled antennas; in the second, the wire closeness allows for HF currents to leak because of cross-capacitance effects. Another source of crosstalk is moisture and water, which in old cables with damaged or inadequate insulation cause continuous currents to leak between pairs. One distinguishes the *near-end crosstalk* (NEXT) and the *far-end crosstalk* (FEXT), which correspond to couplings from close and remote sources, respectively, as illustrated in Figure 1.9. Since telephone cables can

FIGURE 1.9 Effects of near-end crosstalk (NEXT) and far-end crosstalk (FEXT) in twisted-wire pair transmission channels, with power leakage between pairs illustrated by the dashed arrows. In this example, transmitter A causes NEXT on receiver B and FEXT on receiver B′. Likewise, transmitter B causes NEXT on receiver A and FEXT on receiver A′.

contain several wire pairs, the effects of NEXT and FEXT increase somewhat in proportion to the number of these pairs. If all pairs carry the same xDSL service type (or flavor, e.g., ADSL exclusively), one refers to the crosstalks as being *self-NEXT* and *self-FEXT*, respectively. If the wire pairs carry different xDSL flavors (e.g., ADSL and VDSL), the corresponding crosstalks are called *foreign-NEXT* and *foreign-FEXT*. Another source of crosstalk noise is the coupling between the DSL cable and other EM sources such as amateur radio systems and AM radio broadcasting, whose power spectrum may overlap with the 100 kHz–1.1 MHz range of DSL. This effect is referred to as *alien Xtalk*, or more generally, *radio-frequency interference* (RFI). The effect of RFI is a noise peak in the RF spectrum, which results in local SNR degradation, as previously illustrated in Figure 1.8. Another source of noise in DSL lines is *signal echo*. As the name indicates, the echo is a reflection of the signal which is fed back to the transmitter. The effect is due to electrical-impedance mismatch between line elements. Such a mismatch can concern two spliced wires having different gauges/diameters, a line with a connector, a corroded splice or un-terminated taps. Regardless of the DSL coding type, the signal reflection coefficient is given by $T = |Z - Z'|/(Z + Z')$, where the two impedances Z, Z' (as expressed in *ohms*, symbol Ω) are frequency-dependent. The echo powers of HF and LF tones are therefore also different. Symmetric or duplex DSL systems, like HDSL, include *echo-cancellation* circuits. Once the line impedance has been characterized, it is possible to calculate the echo power spectrum. The echo-cancellation circuit is then able to subtract, in real time, the "calculated" echo signals from the actually received signals.

1.2.4 DSL Transfer Modes and Framing

A DSL link represents more than a mere copper wire transporting bits on top of analog voice, even if such a picture is strictly correct. As with other physical or Layer-1 transport protocols (e.g., SDH/SONET), the DSL bits are *framed,* according to certain rules and payload substructure options, in a fixed-length format. In this case, the frame substructure is that of ATM (*asynchronous transfer mode*). The fact that the frames contain ATM cells does not exclude the payload being made of other-than-ATM data, such as IP (or IP-over-ATM), as discussed below. A second option is to use the fixed-size frame as a way to transport a mere bit stream. In this case, the frame's substructure and any inside bits are not meaningful to the DSL equipment. Such a mode of operation is referred to as the *(bit-) synchronous transfer mode* or STM. The third and last option is to transmit frames having variable lengths, which can accommodate variable-length payloads such as IP datagrams, for instance. In this case, one refers to the *packet mode*, which can be seen as different from STM, since the traffic is structured and the DSL equipment is aware of its packet nature. It is also different from ATM since the frames have variable lengths. Thus ATM, STM and packet mode are the three standard ways to operate DSL. The STM operation does not provide any payload-multiplexing feature, unlike the other two. Concerning IP payloads, this gives two alternatives: Either IP is transported over framed-ATM, or it is transported directly in the (framed) packet mode. Whether "IP-over-DSL" should or should not use ATM as a mediating subchannel structure, remains an open debate, which we shall address later on.

Since the DSL traffic is bidirectional and most generally asymmetric, one should also distinguish between upstream and downstream channels, and within these channels, those which are "one-way" (simplex) and "to-way" (duplex). In ADSL, these channels are called *bearers*. The term channel/bearer should not be interpreted in terms of traffic lines, recalling that there is only one trunk, the copper-wire pair. Rather, the channel/bearer represents a dedicated information time slot (namely a sequence of specific bytes) inside the ADSL frame, within which a unique traffic is transported either upstream or downstream. In short, bearers form a set of transparent data-stream options which users may choose from. Not all xDSL types use bearer channels. In HDSL, for instance, the subchannels are the T1/E1 circuits. A detailed description of all the xDSL standards and options would be fastidious and possibly confusing. In this subsection, we will only consider three representative cases ADSL, HDSL and VDSL, which together provide a key background reference for approaching the diversified and complex xDSL world.

We shall then begin with a description of ADSL. The standard ADSL frame includes seven bearers, as illustrated in Figure 1.10. Four bearers are exclusively dedicated to downstream traffic (simplex) and the three others are for either upstream or downstream (duplex) traffic. The four downstream/simplex bearers are called AS0 to AS3, and the three duplex bearers are called LS0 to LS2. Not knowing the reason for such names being chosen, bear it in mind that "A" always concerns the downstream value, like the "A" in ADSL. One also refers to the two bearer families as the *AS* and *LS subchannels*. As the figure also shows, there are two separate channels. One is for the timing of voice (downstream signal), and is called *network timing reference,* or NTR. The other is a bi-directional

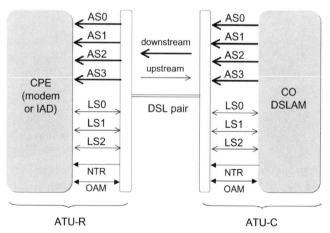

FIGURE 1.10 Bearer channel definitions for uni-directional and bidirectional upstream/downstream DSL traffic. See Figure 1.1 captions for acronym definitions (NTR = network timing reference, OAM = operations, administration and maintenance).

channel for the general purposes of *operation, administration and maintenance* (OAM).

We consider first the *ATM* mode of ADSL operation. In this case, one only uses AS0 and LS0 as the dedicated subchannels for downstream and upstream traffic, respectively. One refers to the subchannel pair as a *port*, or Port0. As an option, AS1 and LS1 can also be used, making a second port called Port1. This simplified grouping by ports is illustrated in Figure 1.11. Since ATM can provision any type of payload and multiplexed traffic (voice, data and video), there is no *a-priori* reason to have a second port. The rationale is in fact to separate IP-only traffic from the other types, which simplifies the ATM traffic management. Consistently, if IP carries voice or video, only Port 0 is required. The same is true if the traffic is only ATM-based. As described in Chapter 3, Section 3.2 of Desurvire (2004), ATM cells are of three possible type: user (voice, data, video, IP), control (setting up or clearing channel connections), and management (monitoring, configuring and signaling of network elements). Thus management at ATM layer 2 level is made through the AS and LS subchannels. This must not be confused with the DSL management through the OAM channel, which concerns the physical layer 1.

The different payload bit-rate options have been made to correspond to the ATM standards described in Chapter 3 of Desurvire (2004). As previously mentioned, the DSL bearer-channel bit-rates go by multiples of 32 kbit/s. Four ATM-ADSL (simplex) transport classes have thus been defined according to the following AS0 (or AS1) channel bit rates:

- Class 1: 1.760 Mbit/s (55 × 32 kbit/s)
- Class 2: 3.488 Mbit/s (109 × 32 kbit/s)
- Class 3: 5.216 Mbit/s (163 × 32 kbit/s)
- Class 4: 6.944 Mbit/s (217 × 32 kbit/s)

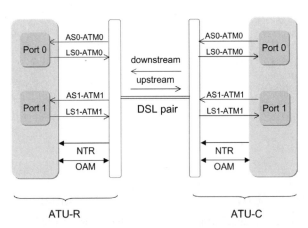

FIGURE 1.11 DSL bearers in framed-ATM operation, showing subchannel grouping by ports (Port 0 and optional Port1) and respective upstream/downstream traffic directions.

In the duplex or bidirectional case, two possible bit rates have been defined for the AS0-LS0 (or AS1-LS1) bearers:

- Class 1: 448 kbit/s (14 × 32 kbit/s)
- Class 2: 672 kbit/s (21 × 32 kbit/s)

Such bit rate definitions do not make too much sense when directly compared to the standard rates used in the AAL (*ATM adaptation layer*) service classes of ATM. The simplex and duplex rates are made to be compatible with the AAL1 (constant bit-rate) and the AAL5 (variable bit-rate) service classes. One can also distinguish two possible transfer modes, namely a "*fast*" one which has minimal latency (adapted to interactive video services) and an "*interleaved*" one which has higher latency by minimal bit-error-rates (adapted to data transport). As will be described a few paragraphs below, the "fast" and "interleaved" channels correspond to two possible payload segments, or buffers, of the ADSL frame. Hence ADSL either has "dual latency" as these two buffers are used indifferently, or single latency if only one is used. The latency options can be different for upstream and downstream. Three *latency classes* can thus be envisaged for ATM over ADSL services: (a) single latency in either direction; (b) single latency upstream, dual latency downstream; and (c) dual latency for both upstream and downstream.

We consider next the more complex *STM* mode of ADSL operation. By convention, the STM subchannel data rate must be an integer-multiple of either of these two options:

1. 1.536 Mbit/s = 48 × 32 kbit/s (transport classes 1 to 4)
2. 2.048 Mbit/s = 64 × 32 kbit/s (transport classes "2M-1" to "2M-3")

These two options correspond to *transport classes* labeled 1–4 in the first case and (oddly so) labeled 2M-1 to 2M-3 in the second. One recognizes 2.048 Mbit/s to be the value used by the E1 standard for digital telephony in Europe and more generally, outside the United States (32 circuits of 64-kbit/s). Although such an occurrence is not coincidental, the idea is not to make DSL yet another means for voice-channel multiplexing. Rather it is about reusing the existing E1 clock rates and their multiples. Table 1.4 shows the rules defining the various *transport class* possibilities based upon (a) or (b) options, which concern the AS0–AS2 bearer channels (simplex traffic). The trade-off is either to maximize the bit rate in a given ASx bearer (as determined by the constraints of DSL range) or to maximize the number of ASx channels (as also limited by range constraints).

Within the previous ADSL classes, and as an option, duplex traffic can also be supported when using the LS bearer channels (Figure 1.10). In all transport classes the bearer LS0 is exclusively used to carry a control signal, and is for this reason called the *C channel*. In classes 4 and 2M-3, the control signal is at 16 kbit/s, while in other classes it is at 64 kbit/s. In all cases, the C channel concerns both simplex (AS1–AS3) and duplex (LS1–LS2) communications.

We consider next the framing structure of ADSL. It is based upon individual frames, whose appended sequence of 68 units (numbered 0 to 67) forms a *superframe*. The ADSL frame structure itself is independent of the coding scheme (e.g.,

TABLE 1.4 Simplex ADSL downstream transport classes and corresponding bearer-channel (AS0–AS3) bit rates, as based upon base-rate options of 1.536 Mbit/s (referred to as Classes 1 to 4) and 2.048 Mbit/s (referred to as classes 2M-1 to 2M-3)

Base rate	Bearer	Class 1				Class 2				Class 3			Class 4
1.536 Mbit/s	AS0	4	3	2	1	3	2	1	1	2	1	1	
	AS1		1	2	1		1	2	1		1	1	
	AS2			1	1			1	1				
	AS3				1				1				1
	Total simplex capacity	6.144 Mbit/s				4.608 Mbit/s				3.072 Mbit/s			1.536 Mbit/s

Base rate	Bearer	2M-1				2M-2			2M-3
2.048 Mbit/s	AS0	3	2	1	1	2	1	1	1
	AS1		1	1	1		1	1	
	AS2				1				
	Total simplex capacity	6.144 Mbit/s				4.096 Mbit/s			2.048 Mbit/s

The numbers shown correspond to integer multiples of the base rates. Values that are underlined correspond to a default-configuration setting.

DMT or CAP). The codes just carry bits through multi-level amplitude/phase analog symbols, as described previously. A group of 8 such bits forms a *byte*, and the byte is the basic structural unit of a frame. Each frame carries both AS and LS bits, which illustrates the fact that the so-called AS/LS channel/bearers just represent dedicated byte-fields or payload "containers" within each ADSL frame. The byte-sizes corresponding to the AS0–AS3 and LS0–LS3 fields will be detailed after having described the frame and superframe structure and functions. The frame length is 250 µs, corresponding to an emission rate of 4,000 times per second, or 4 kHz. This also corresponds to the symbol emission rate of DMT/CAP data. Consistently, the superframe length is 17ms (68 × 250 µs). An extra frame flag is also inserted between superframes, which defines their boundaries and ensures superframe synchronization. The actually transmitted symbol rate is therefore (69/68) × 4 kHz. But for user data purposes, the meaningful rate is 4 kHz. Since DSL is a point-to-point link with only two definite ATU ends, there is no need for specific addressing or frame identification schemes, unlike in other framed protocols such as SDH/SONET, frame-relay or Ethernet, for instance.

The generic ADSL frame is made of two regions called *buffers*, as illustrated in Figure 1.12. The two buffers are preceded by a "*fast byte*", whose varied overhead functions are described further below. The first and the second buffers are called *fast-data buffer* and *interleave(d)-data buffer*, respectively (one also refers to them as "*fast frame*" and "*interframe*"). The rationale for the subdivision of payload into buffers is to provide different transport qualities according to the type of data. The category of "fast" data corresponds to *delay sensitive* but *noise tolerant* payloads. Fast-data tolerance to noise is provided by *forward error correction* or FEC (see Chapter 1, Section 1.6 of Desurvire, 2004), which is provisioned by a

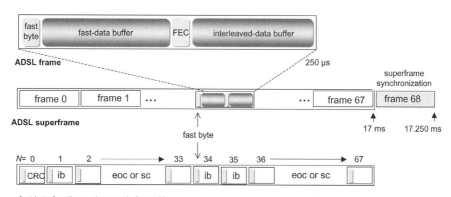

FIGURE 1.12 Top: ADSL generic frame structure, showing fast-byte and FEC fields with fast-data and interleaved-data buffers. Middle: ADSL superframe made of 68 ADSL frames. Bottom: fast-byte function attribution corresponding to the frame order number (CRC = cyclic redundancy check, ib = indicator bits, eoc = embedded-operation channel bits, sc = synchronization control bits).

dedicated field located at the end of the buffer. In contrast, the category of "interleaved" data corresponds to unprotected (no FEC) and delay-insensitive traffic. In order to increase its tolerance to noise, however, the bits are scrambled according to some interleaving patterns. Payload attribution between fast-data and interleaved-data buffer categories is left to the choice of the DSL operator. Obviously, digital-voice and video represent delay-sensitive traffic, and should therefore be carried by the first buffer. In contrast, internet web pages and file downloads are usually carried by the second buffer. In this case, the occasional loss of information due to noise can be compensated by re-sending the IP contents (via the TCP control protocol), which causes only minor delays.

As previously stated and shown in Figure 1.12, the ADSL frame always starts with a *"fast byte"* which precedes the fast-data field. This fast byte serves different possible purposes, depending upon the location of the ADSL frame within the superframe. The first frame's fast byte serves the purpose of error correction (*cyclic redundancy check* or CRC) for the superframe. A second overhead function is that of *"indicator bits* (ib)" (fast byte of frames 1, 34–35) for link management. Among other management functions, these bits are used by the ATU (C or R) to indicate problems of signal loss/fading or defective frames received from the opposite direction or ATU. They are also used to indicate whether FEC is activated, as concerns both buffer types, but for the purpose of protecting the sequence of frames (which is different from FEC in the frame's fast-data field). Two other overhead functions in the fast-data byte are that of an *embedded operation channel* (eoc), containing "configuration" bits, and of *synchronization control* (sc), containing "synchronization" bits. As shown in Figure 1.9, the "eoc" and "sc" bits only concern the fast bytes of frames 2–33 and 36–67. To identify whether these fast bytes are of the "eoc" or the "sc" type, the first bit in the byte is set to either 1 or 0, respectively (in the 7-0 bit notation, this corresponds to xxxxxxx1 and xxxxxxx0, respectively). The "sc" bits are used for the synchronization control of the ASx/LSx bearer channels (see following text). Within each ADSL frame, buffer bits, except the preceding fast byte, are shuffled or scrambled. This makes it possible to avoid accidental synchronization failures for both the frame and the superframe.

We consider now the details of buffer payloads. The issue is to assign the upstream and downstream traffic from ATU-R and ATU-C to the different bearers (ASx [downstream only] and LSx [two-way]), and distribute these bearers into the two data buffers (fast or interleaved). Both buffers can potentially receive any of the ASx and LSx bearers, and these bearers can be attributed different byte sizes, including zero when the bearer is not used. This byte-size allocation between different bearers is different for each transport class, namely Classes 1–4 (1.536 Mbit/s base rate) and Classes 2M-1 to 2M-3 (2.048 Mbit/s base rate), as shown in Table 1.4 and further described later.

Figure 1.13 shows the detailed substructure of both fast-data (F) and interleaved-data (I) buffers. First, it can be seen that both buffers present similar features. Both are filled in according to the precise bearer sequence AS0, AS1, AS2, AS3, LS0, LS1 and LS2. To recall, the bearer byte-sizes are optional (according to the transport class) and the size is zero when the bearer is not used. We observe that

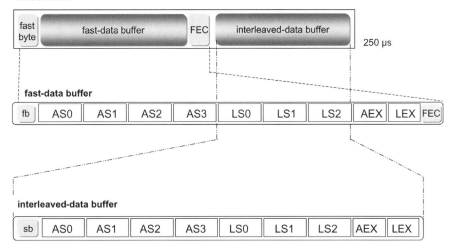

FIGURE 1.13 Detailed substructure of fast-data and interleaved-data buffers in ADSL frame. Fields AS0–AS3 and LS0–LS2 correspond to the bearer channels (fb = fast byte, sb = synchronization byte, AEX = ASx extension byte, LEX = LSx extension byte, FEC = forward error correction).

the interleaved-data buffer has a synchronization byte (sb), which plays a role similar to that of the fast byte in the fast-data buffer. At the end of the AS0 ... LS2 sequences are two optional 1-byte fields called AEX and LEX, for AS/LS "extension bytes". The field AEX is set to 1 byte ($A_{F,I} = 1$) if any ASx bearer field is used in the buffer, and is 0 byte otherwise ($A_{F,I} = 0$). The field LEX is 1 byte if any ASx or LSx bearer field is used in the buffer, and is 0 byte otherwise ($L_{F,I} = 1$ or 0). To simplify the description, the purpose of AEX and LEX fields is to ensure synchronization when either or both ASx or LSx bearer channels are not synchronous with the timing base of the receiving modem. If either one or both are synchronous, the corresponding AEX/LEX functions are deactivated. For the fast-data buffer, such information is provided by the "sc" fast-data byte (set to xxxxxxx0). For the interleaved-data buffer, it is provided by the synchronization byte "sb" (set to xxxxxxx0). Without getting into the full structure details of the "sc" and "sb" bytes, suffice it to say that in both cases, the bits 3–2 (xxYYx0) and 7–6 (YYxxxxx0) designate which bearer channel ASx or LSx should be specifically used for synchronization of the AS or LS bearers, respectively.

Since there are several possible ways to fill up the two ADSL-frame buffers according to the above overhead and bearers structure, some selection rules must be introduced. Call $B_F(ASx)$, $B_F(LSx)$, $B_I(ASx)$ and $B_I(LSx)$ the number of bytes contained in the channels ASx and LSx in either fast- or interleaved-data buffer. Distinguish the LS0 channel, normally used only for synchronization, by calling its payload $C_{F,I}(LS0)$ instead of $B_{F,I}(LS0)$. The total data payloads in each of the buffers (B_F, B_I), as expressed in number of bytes, can thus be expressed as the following sums:

$$B_F = \sum_{x=0}^{3} B_F(ASx) + \sum_{x=1}^{2} B_F(LSx) + C_F(LS0)$$

$$B_I = \sum_{x=0}^{3} B_I(ASx) + \sum_{x=1}^{2} B_F(LSx) + C_I(LS0)$$

(1.5)

For downstream traffic, the selection rule is that bearer data ASx or LSx can be only assigned to either the fast or the interleaved buffer types. Thus, if $B_F(ASx) \neq 0$ (allocation of AS bearer "x" to fast-data buffer), then we have $B_I(ASx) = 0$. The reverse applies when ASx is allocated to the interleaved-data buffer ($B_F(ASx) = 0$, $B_I(ASx) \neq 0$). Consistently, the same rule applies to the downstream traffic carried by LSx bearers. A special case is the bearer LS0, which is used as a control channel at 16 kbit/s (transport classes 4 and 2M-3) or at 64 kbit/s (other classes), as seen in Table 1.5. In the 16-kbit/s case, the control channel is in fact not LS0 but the LEX byte, as set to the value 11111111 = 255. At a 4-kHz frame rate (250 μs fixed length), this corresponds to a signaling rate of 8 bits × (4,000/sec) = 32 kbit/s. (An oddity to overlook is that such a control channel rate is in fact 16 kbit/s, because it is carried by only half of the frames). Since in this case no information is conveyed, one sets by convention $C_{F,I}(LS0) = 0$ in the equation (1.5). With the other transport classes (Table 1.5), LS0 consists in a 2-bytes code-word in the interleaved buffer, which yields a 2 × 8-bits × (4,000/sec) = 64-kbit/s control

TABLE 1.5 Optional ADSL duplex bearer-channel (LS1, LS2) bit rates, corresponding to the transport Classes 1–4 and 2M-1–2M-3 shown in Table 1.4

Bearer	Class 1		Class 2 or 3		Class 4
LS0 (LEX)	64 kbit/s		64 kbit/s		(16 kbit/s)
LS1	160 kbit/s		160 kbit/s		160 kbit/s
LS2	384 kbit/s	576 kbit/s		384 kbit/s	
Total duplex capacity	608 kbit/s	640 kbit/s	224 kbit/s	448 kbit/s	176 kbit/s

Bearer	2M-1		2M-2		2M-3
LS0 (LEX)	64 kbit/s		64 kbit/s		(16 kbit/s)
LS1	160 kbit/s		160 kbit/s		160 kbit/s
LS2	384 kbit/s	576 kbit/s		384 kbit/s	
Total duplex capacity	608 kbit/s	640 kbit/s	224 kbit/s	448 kbit/s	176 kbit/s

The channel LS0 is used for control, except in the 16-kbit/s case where it is carried by the LEX byte. Values that are underlined correspond to a default-configuration setting (see Table 1.7).

channel. Excluding FEC overhead, the total number of bytes $K_{F,I}$ carried by fast- and interleaved-data buffers is thus given by:

$$K_F = 1 + B_F + A_F + L_F$$
$$K_I = 1 + B_I + A_I + L_I \quad (1.6)$$

where the "1" corresponds to the synchronization byte preceding the buffers' payloads. Since ASx or LSx bytes are either in the fast or interleaved buffer and the ADSL frame is emitted at a 4-kHz rate (user-data level), the corresponding payload rate is 8 bits × 4 kHz = 32 kbit/s. We can thus make the correspondence between number of bytes and transport-class base rates (m = integer):

1. m × 1.536 Mbit/s = m × 48 × 32 kbit/s ↔ m × 48 bytes (transport classes 1 [m = 4] to 4 [m = 1])
2. m × 2.048 Mbit/s = m × 64 × 32 kbit/s ↔ m × 64 bytes (transport classes "2M-1" [m = 96] to "2M-3" [m = 32])

As previously mentioned, the ASx/LSx allocation between fast- and interleaved-data buffers is arbitrary and depends upon the type of traffic (basically delay or not delay sensitive, respectively). The number of bytes carried by each bearer ($B_{F,I}(ASx)$, $B_{F,I}(LSx)$) is also optional within each transport class, as previously shown in Tables 1.4 and 1.5. Yet, it is possible to define a *default-setting* configuration which applies in the absence of any other specifications. Such a configuration corresponds to the ASx/LSx bit rates underlined in the two tables. The specific buffer allocation and number of bytes $B_{F,I}(ASx)$, $B_{F,I}(LSx)$ are shown in Table 1.6. Note that this byte allocation between fast- and interleaved-data buffers represents a high-level protocol. On the physical layer, the individual bits within each frame are transported according to a different group selection and ordering, which is the result of pre-coding, scrambling and other FEC implementation. In the case of discrete multi-tone (DMT) coding, the individual bits can be assigned to different frequency tones, or bins, by groups of variable sizes or symbol efficiency (e.g., 2 to 8 bits/symbol), regardless of their belonging to a specific frame field or buffer. As explained in an earlier subsection, such a flexible bit/symbol/bin allocation makes it possible to alleviate frequency-dependent SNR degradation and optimize the throughput capacity with minimum BER.

We consider next the case of HDSL. As we have seen in an earlier subsection, HDSL offers a symmetric traffic based upon the standard bit rates called T1/DS1 (1.544 Mbit/s) and E1 (2.048 Mbit/s). Thus HDSL can be seen as a special link interface between the operator's central office (CO) and the remote customer premises (CP), except that the capacities are 24–32 times greater than an ordinary 64-kbit/s telephone line. An important difference with the telephone system is that HDSL-T1/E1 is used as a *leased private line*. This means that on the CO side the data are not going through a PSTN switch but a special device called *digital (access) cross-connect*, or DCS/DACS. The CO and remote CP termination units for HDSL are called *HTU-C* and *HTU-R*, respectively. With HDSL, there are no such devices as DSLAM (DSL access multiplexer), but the HTU-C is generally able to concentrate several HDSL links, each one being made of 1, 2 or 3 parallel

TABLE 1.6 Default allocation for channel-bearers AS*x*/LS*x* in fast-data (*F*) and interleaved-data (*I*) buffers of ADSL frame, and corresponding number of bytes (B_F, B_I) for 1.536-Mbit/s and 2.048-Mbit/s base rates and different transport classes

Base rate	Bearer	Class 1		Class 2		Class 3		Class 4	
		B_F	B_I	B_F	B_I	B_F	B_I	B_F	B_I
1.536 Mbit/s	AS0				96		96		48		48
	AS1		96		48		48		
	AS2								
	AS3								
	LS0 (*LEX)		2		2		2		(*1)
	LS1	5						5	
	LS2	12		12		12			

Base rate	Bearer	Class 2M-1		Class 2M-2		Class 2M-3	
		B_F	B_I	B_F	B_I	B_F	B_I
2.048 Mbit/s	AS0		64		64		64
	AS1		64		64		
	AS2		64				
	AS3						
	LS0 (*LEX)		2		2		(*1)
	LS1	5				5	
	LS2	12		12			

The corresponding simplex (downstream) and duplex capacities are shown in Tables 1.4 and 1.5. Note (*): In classes 4 and 2M-3, the control channel is not LS0 but the LEX byte, as set to $11111111 = 255$.

wire pairs. When used for servicing ISDN (see Desurvire, 2004; Section 2.1), the two-pair terminal unit in the CO is called *local exchange primary access HDSL*, or LEPA-HDSL. The LEPA-HDSL is thus connected to an ISDN switch.

As shown in Table 1.1, the HDSL traffic is generally carried by two duplex copper-wire pairs (the data payloads being 772 kbit/s or 1,024 kbit/s per pair). A second generation, HDSL2, uses only one wire-pair for carrying the same duplex capacities (this results in the earlier, two-pair version being called HDSL4). Another approach, which was meant to reduce the wire data rate and thus increase distance, used three duplex pairs (784 kbit/s). Because of the progress in digital electronic speed and the extra wire cost, this last approach has been abandoned. Note that in HDSL, the analog coding scheme used is exclusively 2B1Q, although CAP is an experimental possibility (see previous subsection).

We detail first the HDSL (or HDSL4) framing structure for both T1 and E1, which is illustrated in Figure 1.14. As the figure shows, the 6-ms frame is divided into 2 groups, 4 units and 48 blocks by sets of 12. The frame begins with a 14-bit synchronization field and ends with a 26-bit stuffing field. Each set of 12 blocks contains a 2-bit HDSL overhead (HOH) field, which separates one set from the

1.2 Digital Subscriber Line (DSL)

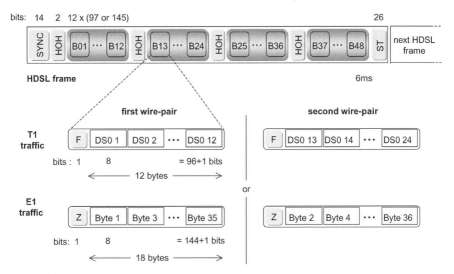

FIGURE 1.14 Framing structure of high-bit-rate DSL (HDSL) for transport of T1 or E1 over two wire-pairs (SYNC/F/Z = synchronization bits, HOH = HDSL overhead, ST = stuffing bits, DS0 = 64-kbit/s channel).

next in the frame. The synchronization and first HOH fields thus form a 16-bit (2-byte) synchronization overhead at the beginning of the frame. The 48 blocks are named B01 to B48 (B01–B12 for the first set, B13–B24 for the second, etc.).

The figure also shows the detail of the Bxx blocks. How is the T1 or E1 traffic being distributed within these blocks? Recall from Chapter 2 in Vol. 1 that T1 represents the multiplexed aggregation of 24 channels at 64 kbit/s, called DS0, while E1 represents the same with 32 channels. A small modification is that HDSL uses 36 channels instead of 32 for E1 transport, the four extra channels being used for compatibility purposes with SDH. Since we have two wire pairs, the traffic must be split and transported in parallel. Thus T1 channels DS01 to DS012 are carried by one pair, and DS013 to DS024 are carried by the other. For E1, the same applies but the 36 channels are split according to even (first pair) and odd (second pair) channel numbers. The figure thus shows how a single Bxx block is loaded with the corresponding bytes, each sequence of 12 (T1) or 18 (E1) bytes being preceded by a 1-bit synchronization field. The Bxx block length is therefore either $12 \times 8 + 1 = 97$ bits (T1) or $18 \times 8 + 1 = 145$ bits (T1). Since there are 48 such blocks in each 6-ms frame, the block rate is 8 kHz, and the bit rate associated with any of the channel bytes is 8 bits $\times$ 8 kHz = 64 kbit/s. Considering next the aggregate bit rates:

- **Case of T1.** There are 12 channels per wire-pair, so the pair capacity is 12×64 kbit/s = 768 kbit/s, corresponding to a total (two-pair) traffic payload of 2×768 kbit/s = 1.536 Mbit/s. This rate is smaller that the 1.544 Mbit/s rate used in T1 telephony, because it does not include the overhead used in the plesiosynchronous hierarchy (PDH) for bit-stuffing in the "12×" multiplexing operation. If one includes the full overhead, we get the bit rate per pair:

$$B = \{14\,\text{bits (sync)} + 4[2\,\text{bits (HOH)} + 12 \times 97\,\text{bits (B}xx)] + 26\,(\text{stuff})\}/6\,\text{ms}$$
$$= 784\,\text{kbit/s}$$

which corresponds to $2 \times 784\,\text{kbit/s} = 1.568\,\text{Mbit/s}$ as the full HDSL-T1 traffic rate. The difference between 1.568 Mbit/s and 1.536 kbit/s thus corresponds to an overhead of 32 kbit/s, or 16 kbit/s overhead per pair.

• **Case of E1.** If there were only 16 channels per wire-pair, the pair capacity would be $16 \times 64\,\text{kbit/s} = 1{,}024\,\text{kbit/s}$, corresponding to a total traffic of $2 \times 1{,}024\,\text{kbit/s} = 2{,}048\,\text{kbit/s}$, which is the standard E1 rate. But there are 18 channels per pair, so the pair capacity is $18 \times 64\,\text{kbit/s} = 1{,}152\,\text{kbit/s}$, corresponding to a total traffic of $2 \times 1{,}152\,\text{kbit/s} = 2{,}304\,\text{kbit/s}$. If one includes the full overhead, we get the bit rate per pair:

$$B = \{14\,\text{bits (sync)} + 4[2\,\text{bits (HOH)} + 12 \times 145\,\text{bits (B}xx)] + 26\,(\text{stuff})\}/6\,\text{ms}$$
$$= 1.168\,\text{Mbit/s}$$

which corresponds to $2 \times 1.168\,\text{Mbit/s} = 2.336\,\text{Mbit/s}$ as the full HDSL-E1 traffic rate. The difference between 2.336 Mbit/s and 2,304 kbit/s thus corresponds to an overhead of 32 kbit/s, or 16 kbit/s overhead per pair, as in the T1 case.

The previous text addressed two-pair configurations. With HDSL-E1, it is also possible to use a 1-pair (HDSL2) or even 3-pair configuration. In the HDSL2 case, the 36 (E1) bytes are carried in a single Bxx block, instead of being split into two subgroups of 18 bytes in the two-pair case. Apart from this difference, the frame format is the same as in Figure 1.14. In the three-pair case, the 36 bytes are divided into three groups (one for each pair). Each group contains bytes numbered as 1-4-7..., 2-5-8..., and 3,6-9..., respectively. Apart from this difference, the frame format is also the same as in Figure 1.14.

Note that with HDSL (T1 or E1), the traffic is bidirectional or duplex, so the above description applies to both transport directions.

Consider finally the case of VDSL. As previously described and shown in Table 1.1, VDSL is a short-range (300-m), high-data rate (52–55-Mbit/s downstream) asymmetric transmission channel. The range can however reach 1,000 m (1,500 m) for lower downstream rates of 25 Mbit/s (15 Mbit/s), still remaining about one order of magnitude above HDSL. In these lower bit rates, symmetric operation is possible (see following text). The basic VDSL layout was described in Section 1.2. As the corresponding Figure 1.2 showed, the reduced-range handicap is compensated by the use of optical fibers which convey the data from the CO to a local *optical network unit* (ONU). With VDSL, one leaves the world of all-copper-wire DSL loops and enters the domains of *fiber in the loop* (FITL), and *passive optical networks* (PON), which provide broadband access via optical fiber cables (see Section 1.3). However, FITL and PON are based upon all-fiber approaches which exclude electrical wires, while VDSL represents a hybrid version combining a short-range DSL loop and a long-range optical loop.

As in other DSL approaches the VDSL features have been defined by several standardization bodies (ITU-T, ETSI, ANSI), but also specifically through several other groups called *FSAN* for *full-service access networks*, *FS-VDSL* for *full-service VDSL*, *DAVIC* for *digital audio-video Council*, *VDSL Alliance* and *VDSL Coalition* (see Web sites in reference list). The VDSL service classes correspond to two categories: *asymmetric* for the highest downstream speeds, and *symmetric* otherwise, with a broad variety of bit-rate/distance capabilities. To make things a bit more complicated, ANSI and ETSI came up with different definitions which have practically no one-to-one correspondence. Table 1.7 lists these classes altogether in order of decreasing downstream capacity, according to their respective ENSI/ANSI names. The corresponding ranges shown are only indicative of best theoretical or observed performance. As expected, the range increases as the downstream capacity

TABLE 1.7 Very-high-data rate (VDSL) service classes as defined by ETSI and ANSI standards, showing upstream/downstream capacities and corresponding optimum reaches

type	VDSL service class ETSI		downstream capacity (Mbit/s)	upstream capacity (Mbit/s)	Reach feet	Reach km
asymmetric		ANSI AS	52 38.2	6.4 4.3	984	0.300
		AM	26 19	3.2 2.3	3,280	1.000
	A4		23.2	4.1	3,280	1.000
	A3		14.5	3.1	4,430	1.350
		AL	13 6.5 6.5	1.6 1.6 0.8	4,920 6,560 6,560	1.500 2.000 2.000
	A2		8.6	2.048 (E1)	5,575	1.700
	A1		6.4	2.048 (E1)	5,905	1.800
symmetric		SS	34 26 19	34 26 19	984	0.300
	S5		28.3	28.3	984	0.300
	S4		23.1	23.1	1,310	0.400
	S3		14.4	14.4	2,790	0.850
		SM	13	13	3,280	1.000
	S2		8.6	8.6	4,265	1.300
		SL	6.5 4.3 2.3	6.5 4.3 2.3	4,920	1.500
	S1		6.4	6.4	4,756	1.450

A = asymetric, S = symmetric, AS = asymmetric short, AM = asymmetric medium, AL = asymmetric long, SS = symmetric short, SM = symmetric medium, SL = symmetric long

decreases. In the asymmetric case, for instance, reducing this capacity by 50% (52 Mbit/s to 26 Mbit/s) makes it possible to more than triple the range (300 m to 1,000 m). In the symmetric case, the downstream capacity must be reduced by about 60% (34 Mbit/s to 13 Mbit/s) to obtain the same range increase. Since a fiber is used to link the CO to the ONU, one may wonder whether the "range" actually matters. In fact, it all depends upon the existing copper-wire plant. If this plant is already in place and fully developed, ONU cabinets need only to be installed within VDSL reach (e.g., 300 m) of each home/building in the service area, which minimizes fiber-deployment cost. However, ONU installation is expensive, especially in urban or suburban environments where space must be acquired or leased. It is also expensive to overlay a fiber network on top of an older copper plant. On the other hand, if the copper plant is under-developed or non-existent (new residential, industrial or developing areas), the fiber can be deployed all the way to the buildings or the individual homes (see *FTTx*, Section 1.3). In such a case, some economy might be realized, even at higher fiber-cable cost, because the network is of a single type, without duplication of trunk lines, with optimization of the distribution path, and overall, there are no ONU cabinets to install.

The transport of VDSL data relies upon three operation modes, which are similar to that of ADSL, namely ATM, STM and packet mode. Since a VDSL link has two segments, one copper-based the other optical fibre-based, the modes can be different in each segment, or the same from end to end. Table 1.8 summarizes the five possible transport-mode combinations. Since ATM provides a well-established quality of service (AAL classes) and traffic flexibility (bit rate and contents type, including IP data) it makes sense to use it as the unique end-to-end transport protocol, rather than mixing it with other modes or the ONU–CP segment. Alternatively, one could argue that it is more sensible to use the packet mode, which allows the transport of frames with variable-length payloads (such as, IP datagrams). Such an issue is further discussed at the end of the subsection. Overall, the synchronous transfer mode (STM) remains the simplest approach, that of a mere "bit pipe," where the VDSL equipment is unaware of the traffic contents.

TABLE 1.8 Possible VDSL transport-mode combinations between central office (CO) and optical network unit (ONU) and between ONU and customer premises (CP), to chose from ATM (asynchronous transport mode), STM (synchronous transport mode) and packet mode

CO–ONU Link	ONU–CP Link
STM	
Packet mode	
ATM	STM
ATM	Packet mode
ATM	

Its advantage is not only to simplify the ONU and CPE transceivers, but also to offer complete flexibility in the choice of protocol and framing format, as best adapted to the type of traffic or payload transported. Finally, the possibility of combining different transport modes in the access loop is a desirable feature in view of making best use of ONU–CP legacy equipment, xDSL migration (e.g., ADSL to VDSL) and de-coupled CO–CP upgrades. Note also that VDSL can be run in conjunction with ADSL, which further increases the number of transport options and migration scenarios.

As previously stated, VDSL represents an ultimate stage in the DSL family, offering one order of magnitude increase in capacity. Since bandwidth and capacity are closely related, VDSL must also occupy ten times more bandwidth than conventional DSL. Indeed, the VDSL spectrum extends itself up to 10–12 MHz, while (from Section 1.2.1) the DSL spectrum is typically limited to 1.1 MHz (ADSL) and 2 MHz (HDSL). A major problem of such a VDSL spectrum allocation is its possible interference with existing radio bands (e.g., AM broadcast, safety services, amateur radio). This interference is due to the fact that aerial copper wires act both as receiving and emitting antennas, unlike buried wires which are less exposed or can even be shielded. The VDSL spectral allocation must therefore avoid certain frequency bands, which vary widely according to national and regional regulations.

What are the possible criteria to chose between *symmetric* and *asymmetric* VDSL? As far as symmetry is concerned, it should mostly apply to business use, for the purposes of interactive video-conferencing and multimedia networking. Symmetric capacities of 6 to 8 Mbit/s (S1–S2 classes) are then sufficient. Private users are more likely to need asymmetric VDSL for both VoD and fast-Internet access purposes, maybe with higher downstream bandwidth in order to receive more than a single video channel (i.e., 15–18 Mbit/s for 2–3 channels). Apart from the Internet, two applications are home entertainment (an elaborate form of cable TV) and teleworking (working from home). The Section 1.2.5 provides further details of such service possibilities.

We conclude this subsection by briefly addressing the issue of whether ATM or IP are the best suited for broadband access through xDSL. Conveying either ATM or IP payloads through modem links into the local loop, while bypassing the PSTN, represents a natural and clean evolution step for packet networks. The debate of ATM versus IP over DSL is irrelevant if the loop is only viewed as a means of access to either network type, depending upon the customer use or business needs. On the other hand, the debate is relevant from the operator's perspective, which seeks to make the best and cleanest use of its core-network resource and offer the services of both highest demand and quality. As a reminder, ATM offers several advantages in terms of service-class options (AAL1 to AAL5), reliability (low bit-error- and packet-loss rates, low outage probability, delay tolerance and control), traffic flexibility (variable bit/cell rate and application types, end-to-end timing relation), support of multiple parallel sessions (independent virtual channels), and overall, a high level of network control and management. However, such a high dependability is at the expense of substantial traffic overhead (recalling that every 48 payload bytes require 5 to 7 overhead bytes, and that cells are also used for management and signaling). This overhead comes in addition to that required to transport

TCP/IP in the cells. Also, ATM is vulnerable to issues of network congestion and link-failures. This is unlike the Internet, which is intrinsically self-healing in such events, due to its intrinsic routing flexibility. The development of IP-based protocols also comes with new service offers competing side-to-side with ATM, for instance concerning voice (VoIP versus VoATM, see Sections 1.2.5 and 1.2.6). The stance of using ATM or IP over DSL is naturally different for incumbent and competitive carriers (see Section 1.2.7). The first are naturally prone to using ATM based upon the aforementioned service/management arguments and reasons of core-network functionality/legacy, while the second favor IP, or other framing types such as Ethernet to encapsulate IP, as an alternative for a more direct access to the Internet POP. Independently of these two independent network views, the key point is that if the quality of ISP services some day matches or surpasses that of ATM, the mediation of ATM in the DSL loop will not be necessary or make sense anymore. For the time being, ATM is still firmly rooted as representing a dependable and practical implementation solution.

1.2.5 xDSL Services

A key feature of xDSL is that it is an "always on" service. There is not need to dial-up a phone number in order to access the Internet. As paradoxical as it may sound, this is an advantage for telephone operators, since the corresponding traffic is in fact by-passed by the splitter/DSLAM and thus represents no load for the PSTN. From the user side, there is no dial-up time involved and access is direct with near-instant Web-page browsing, hence the popular name of "fast Internet" to designate a DSL and its diversified modem capabilities. This is in contrast with the previous generation of dial-up modems (56 kbit/s) which use a voice channel and congest the PSTN, not so much because of the associated traffic as because of the relatively long Internet connection times and their completely random occurrences. The DSL technology has thus freed the POTS from the initial problem where *internauts* would congest the entire telephone system, when not virtually stealing all the phone circuits in the local CO.

Asymmetric DSL (ADSL, VDSL, RADSL) is intrinsically well adapted to applications where only the downstream capacity matters, such as in standard Internet use (e.g., Web browsing, downloading freeware or music) and in *video on demand* (VoD), for instance. On the other hand, symmetric DSL (SDSL, HDSL, IDSL, SHDSL) is appropriate for small and medium business applications. While IDSL supports classic 144 kbit/s voice-and-data ISDN (see Chapter 2, Section 2.1 of Desurvire, 2004), the other symmetric DSL flavors allow two-way traffic at bit rates as substantial as 2.3–4.6 Mbit/s, which are adapted to fast intra- or inter-LAN trunking. More generally, any DSL service (symmetric or asymmetric) makes remote connections possible to LAN or intranets (restricted-access internet for professional use). This opens new perspectives for *"teleworking"* (working from one's home with professional network tools, working while on travel or away from the office). The odd-sounding word of *"telecommuting"* is also equivalently used as a way to promote the benefits of avoiding highway traffic and commuting stress. A majority of users may belong to a wide variety of "small office/home office" (SOHO) workers, who depend upon fast Internet/intranet access or LAN databases in the exercise of their profession. With bidirectional capacities of 400 kbit/s, or more, one can use the applications of *"teleconferencing"* and *"video conferencing,"* in which a group of users from different business locations can hold meetings through real-time voice and video (high-resolution requiring over one Mbit). *Multimedia* (MM) conferencing combines voice, video and computer applications (graphics, pictures, spreadsheets, presentation slides, data files, e-mail, Web pages, forums, etc.) to enable users to share and discuss working materials. Two main applications concern *healthcare* (e.g., consultation of medical databases and records, telediagnosis) and *education* (e.g., students learning from remote geographical locations or from home, and professional training, these being called *"distance learning"*). With the combination of DSL for BB access and the Internet for rapid or real-time content delivery, many other applications and services can be envisaged for the end-users, of which VoD would only represent the very basics. In the next two subsections, we shall consider more specifically the DSL applications concerning the transport of voice, ATM and IP, from the network operator's perspective.

1.2.6 Voice Over DSL (VoDSL)

Paradoxically, one of the most appealing service applications of DSL is *voice* (VoDSL). This is both for economical and practical network-integration reasons. For network operators, whether "incumbent" or "competitive" (see Section 1.2.7), VoDSL represents a key alternative to the age-old *circuit-switched* telephone service, the classical PSTN or POTS. Most generally, VoDSL is one out of many possibilities for transporting voice traffic within the global family of *voice-over-packet* (VoP) services. This VoP family also includes *VoIP*, for *voice-over-internet*, *VoATM* for *voice-over-ATM*, and *VoFR* for *voice over frame relay*. It is clear that voice channels *per se* would not deserve so much consideration and processing sophistication if it were not for the direct impact and economy in network upgrade/scalability and for the flexibility in bandwidth use. As we have seen in this section, (H)DSL offers economic advantage for E1/T1 transport, due to practical system simplifications and reduced trunking cost. In this case, VoDSL is a reality which appears "seamless" to the business/residential user, as only representing a mere, lower-cost telephone service option. As previously mentioned, the xDSL network layout makes it possible to by-pass the PSTN and thus avoid its congestion by Internet traffic. One could reverse the argument and consider that VoDSL is also a way for incumbent operators to free up the PSTN facilities and expand voice services into packet-switched networks (such as frame-relay, ATM or Internet). The choice of network type for supporting VoDSL is mainly dictated by considerations of *quality of service* (QoS), for which ATM has the longest history (see ATM service classes in Chapter 3, Section 3.2 of Desurvire, 2004 and Section 1.2.7). Other network types have their own intrinsic advantages, from ubiquity and routing flexibility (IP), to ease of implementation (frame relay). Developing standards for handling of VoDSL through IP are referred to by the name *VoMSDN* for *voice over multiservice data networks*. Mixed architectures making inter-operable the resources of IP, ATM and POTS networks for VoDSL, are currently under development.

As a bottom line for the business/private telephone users, VoDSL could possibly mean lower bills for normal use, or reduced bills for new forms of intensive use. In addition, the number of telephone extensions that can be attributed to a given region is no longer limited by the voice-network legacy and CO switch sizes; rather, it becomes virtually unlimited, at least within the addressing capabilities of the data networks. Users do not have to pay for multiple telephone-line subscriptions. A single DSL connection suffices for both voice and data connection needs, while the home/office IAD can be equipped with 4 to 16 independent voice ports! From the operator's business viewpoint, traditional voice still provides the large majority of net income, notwithstanding the rapid growth of data traffic. Therefore, it is sensible for carriers to invest in enhanced voice services as a means to fund and promote the development of future data and multimedia services.

As an "always on" telephone line, VoDSL offers several new possibilities when compared to the age-old dial-up telephone. The promise of a superior digital-sound quality may first change the way we appreciate the usual telephone. But the VoDSL offer goes much beyond this. The IAD modem, which interfaces the CP and the network (DSLAM), is indeed capable of several intelligent functions, such as handling phone calls from dial-up to hang up with virtual personal assistants, personalized call waiting, prioritized call-back, caller identification, multi-line management (PBX), multiple-party calls, audio conferencing with data connection, video-telephony and advanced voice/video messaging (referred to in mobile as MMS for *multi-media messaging services*). These future VoDSL services could be seen as just the evolution of the earlier PBX and ISDN services, but with greater bandwidth allocation and access to private users at affordable costs. They are about the same as already offered, or promised, by wireless 3G–4G mobile (see Chapter 4, Section 4.2 of Desurvire, 2004), except that the access is "wired." This makes VoDSL essentially destined to office or residential/home use. However, one can connect wirelessly to a DSL access point by various means of *short-range radio links*, which have now become pervasive in computer and phone appliances (see WI-FI technologies, Chapter 4, of Desurvire, 2004). Users can thus also use VoDSL while on the move, commuting, on business travel, or on vacations, provided local DSL access points are available on the way.

1.2.7 Incumbent and Competitive Local-Exchange Carriers (ILEC/CLEC)

It is beyond the scope of this book to venture into the world of *telecom operators*, also known as *"carriers"* or *"telcos"*. Such a complex world can be approached in three possible ways. The simplest one is to analyze its *background history*, its well-regulated beginnings, evolutionary features and milestones, for which the United States provide a key reference model. A more difficult but more common approach is to consider its current *economy paradigm*, which is essentially deregulated, global and driven by complex, inter-dependent market factors and actors. An even more difficult approach is the inventory of the *technologies*, from the Internet

1.2 Digital Subscriber Line (DSL) 53

to mobile communications to broadband access, and more. It is more difficult because the "technology" concept is rooted into a multiplicity of interleaved notions ranging from engineering, applications, products, services, market, and overall "innovation." While this book represents an introduction to the telecom technologies from the engineering viewpoint, we have touched upon the operator's perspective in many places (see Chapter 2), simply because there is no other way to understand technology rationale, evolution, and mutual convergence. This chapter on "broadband wireline access" provides the opportunity to further deepen such a perspective. Additional background can be found in the Bibliography.

The early world of public telecommunications has been characterized by the monopoly by national governments of the ownership, operation and exploitation of their telephone network. Such a monopoly still prevails in most countries, even in today's global economy, just like paper mail and other communications. The telephone service is then provided by a national *PTT* (*post, telegraph and telephone*), or from more recent terminology, an *incumbent* operator. A key advantage of this monopolistic situation is that there is only one entity in charge of developing the network and assessing future technology needs. In that stage of history, the PTTs used their steady revenues to finance their own research laboratories to invent the "future" of telecommunications, with emphasis on the long term, on science and discovery, and separated from direct business concerns. Such an internal knowledge made them educated in the choice of technologies and product to upgrade their national infrastructure. In some countries, especially the less developed, this monopolistic position also had severe drawbacks: either the revenues were insufficient to invest in expanding and upgrading the network, or they were redirected into other government utilization. Another drawback was that PTT monopoly could stifle technology innovation and be slow in developing low-cost or competitive services. For this reason, large countries allowed several carriers to coexist and compete for price and quality on long-distance communication services. These were called *interexchange carriers* (IXC), since they ensured a service continuation between *local exchange carriers* (LEC or LXC). In this system, the LEC were only in charge of local calls, with the ownership and exploitation of a small regional network built from a mesh of central offices (CO) and short-haul trunks. The combined LEC and IXC provided what has been known as the *public-switched telephone network* (PSTN) or *"plain" old telephone service/system* (POTS). Any of the IXC's network access locations is called a *point of presence*, or POP. More generally, a POP is a site location where a long-distance carrier or service provider can be physically accessed by any local operator.

The other face of the network coin was occupied by the *cable operators*. The service was essentially for local radio and TV distribution. To reach suburban and remote areas, this network also used point-to-point and satellite radio links, providing economical, common-antenna television (CATV) distribution. The business separation between cable TV (video contents) and telephone (voice contents) operators was complete until the emergence of fiber-optics in the 1980s. Cable operators could also install long-haul trunks and sell them to the XEC. The XEC would then be able to directly link together local and metropolitan areas, by-passing the LEC with cheaper connection fees. Because of their market coexistence with LEC, these new operators were called *competitive access providers* or CAP. With the systematic development of this network backbone, the CAP naturally turned into *competitive local-exchange carriers* (CLEC, pronounce "see-LEC"), a category of business and ownership distinct from the traditional, *incumbent LEC* or ILEC. Apart from local and long-distance voice services, two key aspects concerned with CLEC business are the transport of *data* and the direct access to the *Internet*. Thus CLEC would provide seamless data connectivity from *local-area* (LAN) to *metropolitan-area* (MAN) and *wide-area* (WAN) network levels. They would also appear as *internet service providers* (ISP). The ILEC and CLEC family enlarged itself with DLEC (data LEC, pronounce "dee-LEC") for private/business LAN services, and BLEC (building LEC, pronounce "bee-LEC") for campuses, hotels, hospitals and other premises such as business centers and marinas, for instance. The rationale for DLEC and BLEC services is to be a one-vendor, one-bill service offer for small and medium businesses, including all possible features plus network privacy and protection. To complete the picture, the OLEC (optical LEC) are dedicated to installing and managing fiber-optic connections between business offices, buildings or LANs to some optical network POP.

The development of CLEC backbone networks brought several benefits. First, it made it possible to decongest the ILEC/PSTN from popular internet traffic and multiply the number of Internet users. Second, it made it possible to extend the range of data-transport services for business customers, with the emergence of MAN, WAN and *private virtual networks* (PVN). Third, it offered to ILEC new opportunities to reconfigure their networks and access interfaces. But overall, the huge copper-wire plant owned by the ILEC represented a capital asset to be re-discovered by CLEC and other local-loop service providers, thanks to the xDSL techniques described earlier in this chapter.

In the beginning, CLECs would have to lease the ILEC telephone lines in order to implement xDSL access (the xDSL using only the high-frequency band of the wires, without affecting the voice band). For ILEC, another rationale for leasing their lines was that their analog-voice business could be threatened in the future by the introduction from CLEC of inexpensive voice-over-internet (VoIP) services. But new regulations in the US came in by the end of 1999 to oblige the ILEC to freely open their telephone loops, trunks and other switching resources to the CLEC/DLEC, under the form of an "unbundled access" offer. From then on, they had to work together to create innovative, single-source broadband (BB) services for voice, video, data and affordable high-speed Internet access. While the two teams would share the costs of this network expansion and operating expenses, they would also reap new business benefits and revenues. To escape competition in some areas, another possibility for the ILEC was to establish a local subsidiary CLEC and, so to speak, share the network resource with themselves. The effect of the unbundling regulation has boosted the BB access market, both in terms of service-offer diversity and technology performance, as summarized by the xDSL concept.

As seen in previous subsections, xDSL is a bridging technology, via ordinary telephone wires, between home/business customer premises (CP) and "the network cloud." Such a network can be anything from ATM, Internet or PSTN, or all of the above. In the operator's central office, or somewhere in-between, the upstream voice-and-data traffic is split. The voice part is directed to the PSTN, while DSL access multiplexers (DSLAM) combine the data inputs from different lines and send the aggregate to the corresponding network POP. Thus both PSTN and data networks are overlaid without congesting each other, while the end-user is connected through a single telephone line.

The other technology in competition for BB access is based upon *optical fibers*. Its general service denomination is *fiber into the loop* (FITL), with specific *fiber-to-the-X* (FTTx) range definitions, as described in the next section. In high-speed DSL applications for which the access range is limited to a few hundred meters, the DSLAM must be brought closer to the CP, if not placed directly on the CP. In such a case, optical fibers are used to transport the aggregate traffic between the DSLAM and the data-network POP. This shows that optical fibers and copper wires are in fact both competitive and complementary technologies in BB access deployment.

■ 1.3 FIBER IN THE LOOP (FITL)

Section 1.2 described in detail how broadband services can be accessed by using the "ordinary" copper-wire plant which spreads between the customer premises (CP) and the telephone's central office (CO). We have seen that by means of complex DSL coding schemes, these wires are able to handle payload bit rates from about 200 kbit/s to 1 Mbit/s, 10 Mbit/s or 50 Mbit/s, but only over relatively short distances, typically 300 m–1 km for the highest rate to 5–10 km for the lowest rate. These figures define the capacity (C) × distance (D) limits of xDSL. On the other hand, optical fibers have C × D characteristics that are immensely superior to copper wires. The world "immense" means here *several* orders of magnitude (1 THz = 10^6 MHz, 100 km = 10^2 km, gives C × D = 10^8 times that of DSL). In fact, the transmission distance can be increased by another 1 to 2 orders of magnitude by use of in-line *optical amplifiers*, as described in the next chapter, but these applications only concern long-haul transport.

Since bandwidth limitations can be virtually removed by use of optical fibers, why not develop broadband access services exclusively based on optical technologies (trunks and terminals)? The answer is that there is no reason not to, except for bandwidth needs and overall cost comparisons. One may wonder whether private persons or businesses may some day need 1-Gbit/s duplex communications from their premises. The second and main argument concerns installation and equipment costs. While the cost of fiber cables is about twice that of copper wires (20 cents over 10 cents per meter), the fiber-to-fiber connecting or splicing remains relatively expensive (10–25 $/connector, 20 $/splice), compared to practically zero cost with copper wires. Another aspect is deployment costs. In dense urban areas, these deployment costs can be prohibitively high (>100,000 $/500 m), which has led to the introduction of *free-space optics* (FSO), as defined by "wireless" infrared optical-beams with "fiber-like" bandwidth (see Chapter 4, Section 4.4 of Desurvire, 2004). In some cases, however, fiber cables can be introduced through the underground network of ducts and pipes with minimal construction and real estate issues. In suburban and country areas, they can also be hung on telephone poles or buried along the highways, railways, roads and even riverbeds. The installation cost is significantly decreased if the fiber cables are installed at the time of road or building construction, even if intended only for future use, and remaining as "dark fiber." If there are so many solutions for deploying access fiber networks according to geography, resources or obstacles, the extra cost corresponding to installation remains the bottom line, and for equivalent service capacities one may resort to using the copper-based xDSL technologies for this reason alone. A final issue is the cost of terminal equipment, including the means to power the terminal. Indeed, optical fibers only convey signals, while copper wires convey both signal and terminal powering. Such an argument has become somewhat weaker with the current proliferation of home-telephone/fixed-wireless appliances which depend upon an electrical plug and handset batteries to be able just to work and function. As far as the cost of optical technologies in the terminals (lasers and receivers) is concerned, it may not be a serious issue when produced at large scales, just as in the case of CD/DVD players, whose prices have dropped to affordable levels over only a few years.

As we have seen with VDSL, optical fibers are used to bridge the distance between the end-point of the copper link corresponding to maximum DSL range, and the CO. The next conceptual step is to bring the fiber all the way to the CP, referred to as the *fiber in the loop* (FITL) approach. The other acronym *FITB* (fiber into the building) exists but is rarely used. One also uses the prolific terminology of *fiber to the "x"* (FTTx) to designate the family of options where *"x"* can represent different types of terminals or premises:

An individual *end-user* (FTTU);

A *home* (FTTH);

An *office* (FTTO);

A *building* or a *business* (FTTB);

A network-unit *cabinet* (FTTCa, FTTCab);

A telephone *pole* or any undefined *premise* (FTTP);

A residential or city *curb* (FTTC), often identified with FTTCa;

A *neighborhood*, a *network* or a *node* (FTTN).

The underlying concepts behind these acronym definitions are that the fiber trunk may be dedicated to various uses and may have different end-point types to be accessed by either a single user entity (FTTU, FTTH, FTTO) or a group of several users (other FTTx). The definitions have significant overlap. For instance, FTTC, FTTCa and FTTN may just represent FITL for residential use with 10–50 homes; FTTP, FTTB and FTTO may just be the same FITL for business use, servicing 10–50 businesses or offices.

One may view FTTx (or more generally FITL) as the ultimate solution for broadband access, in replacement and upgrade of a previous generation of xDSL based on age-old copper-wire telephone networks. Such a view is only partially true, since not all homes or businesses will ever need all the bandwidth offered by optical networks. Another misconception could be that FITL provide seamless access to "broadband optical networks." As we shall see in subsequent text, FITL only represents the use of optical technologies for local distribution of broadband services to the private end-customer. The expression *deep fiber* is sometimes used to qualify the extent to which FITL penetrates a given geographical area, or expands towards the ultimate network outskirts. The "deep fiber" concept is opposed to "copper" and "coaxial" equivalents. One then speaks of *the last mile* to evoke this relatively short distance remaining to be covered by any broadband networks to ultimately reach every home or building. Yet, this "last mile" notion remains elusive since that distance is already covered in many other ways, by twisted-pair and coaxial cables and many other types of wireless link, which also claim to carry broadband services (e.g., ADSL). It is also network-centric, since for any given customer the service should always be within reach, without any "first mile" to traverse. Finally, identifying broadband access through fibers as representing the ultimate "first mile" or "last mile" is misleading since fibers can convey signals over several tens of miles without need of repeaters. More accurately, the fiber represents a means for broadband access over extended distances, which opens new development and business perspectives for the local loop, and in particular for CLEC operators (see subsection on ILEC and CLEC).

One may thus consider the implementation of optical fibers into the local loop from the networking or LEC-operator standpoint. One then speaks about *passive optical networks* (PON). As the name indicates, a PON is a network segment which only involves "passive" optical elements, this term meaning without electronic interfaces whatsoever. Passive elements essentially include *1:N splitters*, *wavelength multiplexers/demultiplexers* and *optical amplifiers* (see following text). While the acronym does not imply any specific network type, a PON is essentially based upon a *tree topology*, not to be confused with a *star* topology. In such a tree network, the fibers represent the main trunk (spreading over an extended distance from a central node) which at some location splits into different branches (spreading over shorter distances to the endpoints). Such a tree topology is well-adapted to broadband local distribution/broadcast services (point-to-multipoint networking with exclusively downstream traffic). The 64-kbit/s telephone line is then used for upstream traffic, for instance in VoD applications. But PONs are destined to more sophisticated uses, such as multipoint-to-multipoint networking with full duplex traffic over optical fiber. A PON can be a LAN carrying Ethernet or ATM, which in this case is called *EPON* or *APON*, respectively. The PON tree topology is justified by the need to share as much as possible of the resource and cost of the main optical trunk which carries the aggregated traffic. In contrast, an optical star network requires one to deploy from the central node as many independent fiber cables as there are end-user terminals, which is far more costly to implement. Note that other PON topologies such as "ring" and "bus" (see Chapter 2, Section 2.5 of Desurvire, 2004) are possible. But the tree topology so far remains the accepted PON reference. More generally, a PON represents a specific category of *optical network access*, which is briefly described in the next section, but addressed in more details in the next chapter dedicated to optical fiber communications and networks. In this section, we shall focus on PON and their associated FTTx service applications, including *home networking*. To complete the description, we will also cover the case of *hybrid fiber-coaxial* (HFC) distribution networks, which offer economic advantages in comparison to all-optical FTTx.

1.3.1 Passive Optical Networks (PON), TDM vs. WDM

Most generally, the function of a PON is to provide a service to a limited group of remote terminal users by means of optical fiber trunks and splitters, as illustrated in Figure 1.15. The figure shows the most basic PON layout which consists in a central node interface (OLT for optical line terminal) to which the main fiber trunk is connected, a passive power or wavelength splitter (in this last case called "multiplexer"), and the different fiber branches leading to the remote-user premises. If the premises concern a single premise (e.g., FTTH, FTTB), the interface is called optical network terminal/termination (ONT); if there are several subscribers in the premises (e.g., FTTC, FTTN), it is called optical network unit (ONU). Within these premises, the different subscribers are individually connected to the ONU through a

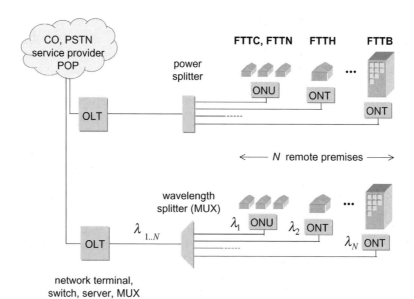

FIGURE 1.15 Basic layout of passive optical network (PON), linking a group of N remote users to a service provider via optical fiber trunks and passive signal splitters; the bottom portion shows the same network with wavelengths dedicated for each terminal (CO = central office, PSTN = public switched telephone network, POP = point of presence of service provider, OLT = optical line terminal, MUX = wavelength multiplexer, ONU = optical network unit, ONT = optical network terminal, FTTC/N/H/B = fiber to the curb/neighborhood/home/building).

coaxial cable or a telephone wire over distances not exceeding 100–300 m. When exclusively considering home or residential premises, one may refer to "*residential access networks*" (RAN), yet another acronym but which makes sense if it is used to distinguish it from the LANs (FTTH/FTTC/FTTN) used for business purposes. With the concepts of teleworking (also dubbed "telecommuting"), the difference between a RAN and a LAN is not this clear. If there are several "tele-workers" in the same residential premises, then we have a LAN, and in the other case a RAN. Since a RAN can also be used for business-networking purposes on a larger scale via the Internet (e.g., remote connection to a Company's intranet server), the notion of "residential" access remains somewhat elusive, just as "local" is in these different acronyms. The more general acronym of *optical access network* (OAN) may be preferable, but we note that it is very rarely used.

In the previous text, the term "splitter" for downstream traffic should also be understood as meaning "combiner" for upstream traffic. Likewise, a "multiplexer" (MUX) is a wavelength combiner in the upstream direction, but it also acts as a "demultiplexer" (DMUX) in the downstream direction. Despite this important conceptual precision, the PON jargon only uses the terms "splitter" and "MUX." As Figure 1.15 indicates, there are two types of PON.

> The first type is based upon *1:N power splitting*, in which all the upstream or downstream signals have the same wavelength. Note that the upstream and downstream wavelengths are usually different (e.g., 1.5 μm downstream, 1.3 μm upstream), which prevents interference between the two traffic types due to possible sources of reflections. Upon crossing the *1:N* splitter/combiner, the signal power is divided by *N*, regardless of the traffic direction. The number of end-users that can be serviced by such a PON is primarily limited by this effect of splitting loss. Because the same wavelength is used in a given traffic direction, end-users must be identified by using different time slots, referred to as *time-division multiplexing* (TDM). Note that this TDM does not correspond to a byte-interleaved approach (as in digital telephony), but is rather based upon packet-interleaving or frame sequencing (ATM or Ethernet, respectively), as discussed in the Section 1.3.2.
>
> The second type of PON is based upon *1:N wavelength multiplexing*, also referred to as *wavelength-division multiplexing* (WDM), as described in detail in Chapter 2. In this WDM case, each terminal user has its own dedicated wavelength (see bottom of Figure 1.15). The wavelength is the same for upstream and downstream. For the end-user, it is like having a permanent circuit connection into the network. The advantage of the WDM approach is that the signal power is not "split." The action of the DMUX/MUX element is similar to that of a prism, which decomposes/recomposes white light into primary rainbow colors (the channel wavelengths). In each of the channels, the signal thus experiences minimal power loss, unlike in the previous, same-wavelength case. Since each terminal uses a different wavelength, the PON implementation is more expensive. Also, the number

of end-users in the WDM-PON is limited by the number of wavelengths that the MUX can actually handle.

In the following, we shall describe the TDM-PON and WDM-PON in further detail.

Consider first the case of TDM-PON, and their associated network topologies. Figure 1.16 illustrates different possible topologies which, in the general case, make use of more than one splitting stages (1:N then 1:N'). The total number of end-users that can access the PON is this NN'. The first topology is called "fiber rich," because the central node is a multiple OLT with many fiber trunks to transport (TDM) aggregate signals. The second topology is called "fiber lean" because a single and main trunk is used to carry the aggregate signals over an extended length, which reduces the amount of fiber in the network. The fiber-rich topology can be justly called a "double star" (because it includes several trees), but the term is also used for the fiber-lean case (even if it is a single tree). Note that for clarity, all the signal paths are shown parallel, but in reality they go in any 360° direction around the central node. The real picture is not so much that of a tree as that of a "snowflake," except that snowflakes have angular symmetry. The reality of a double-star PON looks more like a two-dimensional shrub, whose branches follow broken lines matching the most convenient/economical urban, suburban or country paths.

Since a PON is only progressively deployed from the central hub outwards, multiple 1:N splittings can be introduced according to evolving needs, to service a larger number of areas. To meet these development needs, one of the N splitter branches can thus be reserved as a feeder for future network use, as illustrated in Figure 1.17 with 1:4 splitters. Such an approach causes a loss of PON terminations. Indeed, the concatenation of two 1:4 splitters is seen to provide only 7 terminations instead of 8 (see figure). More generally, N (1:4) splitters provide $3N + 1$ instead

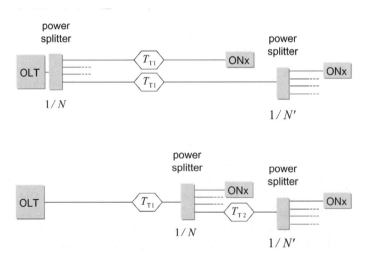

FIGURE 1.16 Fiber-rich (top) and fiber-lean (bottom), double-star PON topologies. The transmission losses (T_T) of different fiber trunk sections are indicated by T_{T1} and T_{T2} (ONx = ONU or ONT).

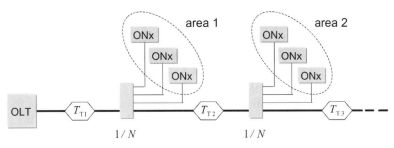

FIGURE 1.17 Progressive deployment of PON over different service areas by using a dedicated branch of the 1:N splitter.

of $4N$ terminations, representing a deployment efficiency of $\frac{3}{4} = 75\%$ for large N. One drawback of this progressive development approach is that the received signal power is different according to the service area, since it decreases as the number of splitting stages increases. For downstream traffic, this may require the use of optical power attenuators in the stages closest to the central hub, and optical amplifiers in the stages that are most remote to the hub. We consider next the issue of net transmission loss from OLT to ONT/ONU, as caused both by power splitting and fiber attenuation.

In the decibel scale (dB), the 1:N power splitting loss is given by $T_S = 10\log_{10}(1/N) = -10\log_{10} N$. If there is a second splitter (1:N') on a given signal path or branch, then the total splitting loss is $T_S = -10\log_{10} N - 10\log_{10} N'$, and so on. Power splitters are typically of the type 1:16 or 1:32, corresponding to splitting losses of -12 dB and -15 dB, respectively. Thus two consecutive splittings from combinations of such devices result in signal losses of -24 dB, -27 dB or -30 dB, respectively (PON with 256, 512, or 1,024 users, respectively). The passive splitter itself has a characteristic *excess loss*, T_E, which is the term used to define its intrinsic or internal loss (typically a dB or less). One may also distinguish the splitter's *insertion loss*, T_I, due to fiber connectors or splices (typically a dB or less for connectors, and 0.5 dB or much less for splices). To simplify, one may include the insertion loss into the excess loss. The second cause of power attenuation is the intrinsic fiber *transmission loss*, T_T. As described in Chapter 2, the transmission loss is characterized by an *attenuation coefficient*, α, expressed in dB/km. Given a fiber trunk of length L (km), the total transmission loss is therefore $T_T = -\alpha L$ (the minus sign is introduced for consistency with previous definition; but all loss types could as well be defined positive). At 1.55 μm wavelength, the attenuation coefficient is typically $\alpha = 0.2$ dB/km. This means that a 10-km or a 50-km trunk has a loss of -2 dB (63% power transmission) or of -10 dB (10% power transmission). This trunk transmission loss is increased by the excess loss from fiber connectors, splices and possible monitoring fiber taps placed along the signal path, which may increase the trunk loss by one dB or more.

The total end-to-end link loss from OLT to ONU/ONT is thus:

$$L(\text{dB}) = \sum T_S(i) + \sum T_T(i) + \sum T_E(i) \qquad (1.7)$$

where all loss sources of given network trunks or components "i" (splitting, transmission and excess) are summed up. Looking at Figure 1.16 (top), and overlooking excess loss, we have for instance $L = -10\log_{10} N + T_{T1}$ for the single-star branch (single splitting), and $L = -10\log_{10} N + T_{T1} - 10\log_{10} N' + T_{T2}$ for the double-star branch (double splitting), respectively. These expressions for signal loss make it possible to calculate the signal *power budget*. This power budget provides the relation between the input (or transmitted) signal power, P_{in}, and the output (or received) signal power, P_{out}, which is written as follows:

$$P_{out}(dBm) = P_{in}(dBm) + L(dB) \quad (1.8)$$

where the powers are expressed in decibel-milliwatt (dBm). Note that the same power budget applies in the upstream direction (P_{in} = signal power transmitted by ONU/ONT, P_{out} = signal power received by OLT). This budget makes it possible to determine the input power required at the transmitting end in order to yield a given output power level at the receiving end. It is possible to alleviate such a power constraint by including *optical amplifiers* along the line. As explained in detail in the next chapter, optical amplifiers are analog devices which boost the optical power level by a characteristic factor G called *gain*, usually expressed in decibels. One can thus achieve the condition $G(dB) + L(dB) = 0$, meaning exact loss compensation by optical amplification, or a power budget identical to zero. Since the distances covered by PON are relatively limited (e.g., ≤ 50 km) compensation of fiber loss is not required as much as compensation of splitting loss, which as we have seen can be significant (-20 dB to -30 dB) in the case of a double star.

How many end-users can a TDM-PON service? The answer depends upon several possible approaches and criteria. First, one can define a fixed number of available time slots and assign them to the end-users according to certain rules. For instance, each user could be attributed a permanent time slot, but this would not be an efficient approach, considering that most users could remain idle for extended periods of time. A better solution is to attribute the time slots according to user requests, service and capacity needs. The slot assignment depends upon the type of protocol used (i.e., packet or frame) and service levels (constant or variable bit rate, delay sensitivity, simplex or duplex transmission, etc.). These issues are solved by classical protocols such as Ethernet (EPON) and ATM (APON). See more about EPON/APON in the next subsection. In theory, the number of users in a TDM-PON should be unlimited. In practice, the limit is defined by the user's maximum *access delay* and minimum *time of network use*. If the central node (OLT) operates at a bandwidth (or bit rate) B and the PON services M out of N users on average, the mean available bandwidth per user is B/M. Thus, a 1-Gbit/s bandwidth could in theory service 1,000 users at 1-Mbit/s net payload rate. But as with the PSTN, network congestion can occur if a number of users larger than the average want to connect at once to the network or stay connected for extended periods of time (e.g., Internet browsing). This shows that the TDM-PON servicing capabilities remain within boundaries, as in any LAN, in spite of the enormous (multiple-THz) bandwidth offered by optical fibers.

1.3 Fiber in the Loop (FITL)

Consider next the case of WDM-PON. As we have seen, each user has its own dedicated wavelength channel, in contrast with TDM-PON where this wavelength is the same for all network users. As an observation, we should note that, unlike wireless or radio technologies, fiber-optics uses the term "wavelength" (λ) instead of "frequency" (f), although there is a one-to-one correspondence ($\lambda = c'/f$, $c' =$ speed of light in the optical fiber; see Chapter 2 of Desurvire 2004). This terminology should not obscure the fact that each wavelength channel occupies a certain amount of frequency space, or bandwidth. In fiber optics, this bandwidth can be expressed either in *nanometers* (1 nm = 10^{-9} m = 10^{-3} μm), which is the wavelength unit, or in *mega/gigahertz* (1 MHz = 10^6 Hz, 1 GHz = 10^9 Hz), which is the corresponding frequency unit in the range of interest. In WDM systems, each channel is thus assigned a wavelength. The set of all wavelengths forms what is called a *wavelength comb*, where wavelengths are spaced from each other by a certain separation $\Delta\lambda$, called *channel spacing*. Referring to wavelengths, such a terminology is somewhat inaccurate, because the wavelengths are in fact not equally spaced, unlike frequencies. See the Exercises in Chapter 2 for a detailed explanation. Overlooking this fact, we can relate the frequency (Δf) and wavelength ($\Delta\lambda$) spacings according to the relation

$$\Delta f = c' \frac{\Delta\lambda}{\lambda^2} \qquad (1.9)$$

where λ is the mean comb wavelength. At 1.55-μm wavelengths and in free space ($c' = 3 \times 10^8$ m/s), we find that a 1-nm wavelength spacing ($\Delta\lambda = 10^{-9}$ m) corresponds to a frequency spacing of $\Delta f = 125$ GHz. At 1.3-μm wavelengths, the same wavelength spacing corresponds to $\Delta f = 177.5$ GHz. When channels are closely spaced, namely with $\Delta f = 50$ GHz down to 25 GHz (1.55-μm wavelength), one refers to *dense WDM* or DWDM. The term "dense" is justified by the fact that the channel spacing is of the same order of magnitude as the bandwidth occupied by the channel (see Chapter 2). Depending upon the bit rate, a minimum spacing must be observed in order to avoid the channel spectra overlapping and interfering in the receiver. Typically, 50–25 GHz and 100 GHz, are safe spacings for bit rates of 2.5–10 Gbit/s, and 40 Gbit/s, respectively. In contrast, when channel spacings significantly greater than 1 nm are used (namely 20 nm), one refers to *coarse WDM* or CWDM. The CWDM approach corresponds to under-using the available fiber bandwidth, or exploiting it only by broad spectral bands (e.g., combining 1.3 μm and 1.5 μm).

How many wavelength channels or end-users, can a PON service? Consider first the fact that the fiber bandwidth is not infinite. Fiber bandwidth is indeed limited by attenuation and dispersion (see Chapter 2). Although optical amplifiers and dispersion-compensation-modules have been developed to alleviate these two limitations, the corresponding device bandwidths are intrinsically finite. For instance, an optical amplifier can be specified to operate with a maximum usable bandwidth of 20 nm, representing 2.5 THz at 1.55-μm wavelength. This limits the number of end-user channels to 100, 50, and 25 for 25 GHz, 50 GHz, and 100 GHz spacings, respectively. If optical amplifiers are not required, the limitation in number of channels is defined by the mere power budget, or the maximum/minimum required signal power at transmitter/receiver ends. The good news is that WDM splitters only have excess and insertion loss, but no splitting loss, unlike in the TDM case. Thus, the received powers are independent of the number of PON users, and are also intrinsically higher (power budgets without $\log N$ contributions).

Another consideration for WDM-PON is that the WDM splitters (MUX) are inherently not cascadable, unlike in the TDM case. Therefore, the number of users is strictly defined by the WDM splitter technology, namely its number of output ports (4, 16, 32, 64,...) and comb spacing (25–50 GHz in DWDM, and 20 nm in CWDM). While the number of channels is thus limited, the net payload bit rate is substantially higher than in the TDM case, since the dedicated wavelength channel can be used at all times for unlimited duration. The drawback of the WDM approach is that it is generally fiber-rich, unless all terminal stations are located in the same vicinity (an unlikely situation!). Therefore, an N-user WDM-PON is made of a common trunk which splits at some location into N fiber feeders. This feature is known as *wavelength routing*, which means that the MUX splitter acts as a passive channel "router". The MUX splitter must then be located as far away from the OLT as possible, in order to optimize the length of the common fiber trunk and minimize the feeders, lengths. An alternative approach makes it possible to reduce the number of feeders. It consists in using a two-stage WDM-splitter arrangement with the two splitters having different characteristics: the first one separating/combining *groups* of WDM channels (CWDM) followed by a second one separating/combining *single* channels (WDM or DWDM). An illustration of this concept is provided in Figure 1.18, in the example of a 32-user PON. The figure shows the fiber-rich approach, which requires 32 fiber feeders, and the fiber-lean approach which requires only 8 fiber feeders. To achieve this economy, the channels have been set into 8 groups of 4, which corresponds to 8 possible service areas. We note that the cascading of WDM splitters does not increase the number of network users, unlike in the TDM case.

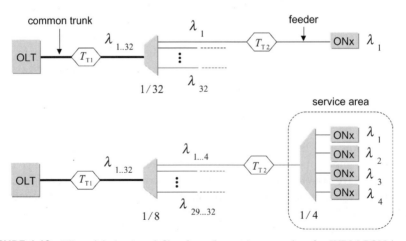

FIGURE 1.18 Fiber-rich (top) and fiber-lean (bottom) approaches for WDM-PON in the example of a 32-user network.

1.3 Fiber in the Loop (FITL)

The above description concerned the WDM-PON downstream channels. How does the upstream or "return" channel work? There are several possibilities. The fanciest one consists in using the same WDM wavelengths for both upstream and downstream. This requires that the PON terminals be equipped with transmitters having a dedicated wavelength (the same as downstream). This approach does not present any technical limitation, since it is already used in the OLT hub, but it remains expensive at an end-user level. A more practical approach consists in using TDM as a shared return channel. This channel can be carried through a separate fiber link. This is sometimes referred to as *direction-division multiplexing* or DDM, an awkward acronym to say that the upstream and downstream channels have different physical paths. The drawback of this solution is that a second TDM-PON must be overlaid to the WDM-PON. A more sensible approach consists in using the WDM-PON fiber infrastructure for the return channel, by means of a single, out-of-band wavelength (e.g., 1.3 μm upstream versus 1.54 μm downstream). This requires the MUX splitter(s) to have the extra capability of combining the two bands. In this case, all end-users are equipped with the same transmitter. The upstream channel resource is then shared through TDM access protocols.

What are the respective pros and cons of TDM and WDM in PON applications? One can summarize these as follows. The least expensive approach is TDM, since there is only one laser wavelength (per direction) and power splitters are substantially cheaper than WDM-MUX. As we have seen, TDM-PON can be progressively deployed according to evolving needs, since one branch of the splitter can be split again to service new groups of users, unlike in the WDM case. The TDM channel cost is shared by all users, unlike in WDM where the channel is user-specific. Thus TDM is ideally suited for broadcasting services, for instance high-definition TV (HDTV) and VoD. But TDM offers no privacy protection, unless channel payloads be crypted by some secure means. The argument that WDM is safer than TDM because channels are not shared is valid only to the extent that optical fiber cables be protected against any form of "eaves-dropping." The disadvantage of TDM is that network upgrades (introducing more terminals) might imply some modifications or changes of the ONT/ONU because of power and frame-rate considerations. As we have seen, another disadvantage is that the access bandwidth is limited by the number of users, since the TDM channel is a shared resource, unlike in the WDM case. Finally, the TDM solution implies a common signal format and protocol (which does not imply that the user bit rate is unique), while in WDM these options remain virtually free (e.g., SDH/SONET, ATM, Ethernet). Since the access protocol is common with TDM, network congestion or packet/cell loss due to spurious "collisions" is possible (e.g. EPON) but this limitation can be alleviated by introducing service classes (e.g. APON) and properly allocating the bandwidth resource. Finally, TDM-PONs are intrinsically limited by splitting loss (dB attenuation in $\log N$), which is absent in WDM-PON. While WDM-PONs have a number of end-users limited by the number of available wavelengths carried by the network, the transmission distances are significantly greater, as only limited by fiber loss. In both cases, however, it is possible to implement optical amplifiers (splitting-loss compensation for TDM, fiber-loss compensation for WDM), but this is not a requirement at all for WDM-PON covering distances as long as 50–100 km. As previously mentioned, WDM-PON is more costly to implement, both from the WDM-splitter (MUX) and user-terminal (ONU) standpoints. As a raw indication of cost, a single-frequency laser is near $1,000, while a MUX is near $500 per wavelength port. With CWDM, these figures should be divided by ten. The overall ONU cost, to be supported by the end user, is estimated to be near $4,000 (or $500 for CWDM). Because the number of wavelengths is limited (16 to 32), the shared "per-user" cost of the ONT is still significant (>$1,000, or $100 for CDWM). Alas, the OLT/ONU costs for either TDM or WDM are not so significant in comparison to the other costs involved in the fiber plant installation. The actual figures depend upon the type of installation, namely *aerial* (telephone poles), or *trench* and *conduit* (existing or new). The very first and the very last solutions correspondingly define the cheapest (>$5,000) to the most expensive (>$30,000). The grand total remains of the order of $5,000 to $35,000 per PON user. For this reason, PON applications may concern businesses and "residential" groups (FTTB, FTTC, FTTN) rather than individual or private homes (FTTH). These figures are however indicative, and as usual prices may drop by one or even two orders of magnitude with massive implementation. But one should keep in mind that PON are not mere "appliances" like home TVs, PCs or DVD players. The optical network deployment costs will always remain relatively high, unlike in the copper-plant case (see the following discussion).

Based upon these considerations, it is clear that WDM-PON are more advantageous than TDM-PON by several respects, albeit more expensive to deploy and thus more expensive to the user. An interesting compromise between the two PON types is to use TDM with a second or additional wavelength. It is like having two shared channels, but with different service applications. For instance, the first wavelength can be used for delay-sensitive data traffic and the second wavelength for video broadcast. The ONT/ONU terminals must then be equipped with dual-wavelength receivers, which may not be significantly more expensive than single-wavelength receivers when using optical and electronic integration technologies. Additional wavelengths can be included for greater differentiation of services. Finally, incremental migration from TDM to WDM is also possible. In this case the (multiwavelength) TDM power splitters are progressively upgraded to achieve wavelength routing for dedicated paths, service areas and individual end customers.

1.3.2 ATM and Ethernet Framing (APON/EPON) and Gigabit-PON (GPON)

The two leading and competing data protocols for TDM-PON, which have been defined by the FSAN and ITU-T standardization groups are ATM (asynchronous transfer mode) and Ethernet. These two protocols, which are described in detail in Chapter 3 of Desurvire (2004) naturally gave birth to dedicated acronyms: *APON* for ATM-PON, *EPON* for Ethernet-PON. Both APON and EPON traffic types are framed, the first case being of fixed-size and the second of variable size. The other acronym BPON (pronounce Bee-PON) was later introduced to designate *any* type of broadband PON, whether APON or EPON, which avoids designating ATM (for instance) as the only protocol solution. In the same spirit, operation of BPON at

1.3 Fiber in the Loop (FITL) 67

Gbit/s rates and above refers to GPON (pronounce Gee-PON), which represents yet another PON standard, as further discussed below. Some refer to these different types of PON standards or protocols as "*PON flavors*" (as in DSL), maybe as a commercial incentive showing that customers may select the technology according to taste!

Consider first the case of APON, which was the first PON type to be standardized. Two configurations are possible, which correspond to symmetric and asymmetric bit rates for upstream and downstream traffic. In the symmetric case, both directions have a bit rate of $B = 155.52$ Mbit/s (corresponding to OC-3 or STM-1 in SONET or SDH hierarchies, see Desurvire, 2004, Chapter 1 and Table 3.1). In the asymmetric case, the downstream traffic has a bit rate of 622.08 Mbit/s ($= 4 \times B$, OC-12/STM-4). The downstream wavelength is 1.490 μm or 1.55 μm (OLT transmitter to ONU receiver) and the upstream wavelength is 1.310 μm (ONU transmitter to OLT receiver). The framing formats corresponding to downstream and upstream traffic are shown in Figure 1.19 and are defined as follows:

Downstream frame: Is made of 56 or 224 cells of equal lengths (53 bytes), the two numbers corresponding to bit rates B or $4B$, respectively; the total downstream frame duration is thus 56 cells $\times$ 53 bytes $\times$ 8 bits/B = 224 cells $\times$ 53 bytes $\times$ 8 bits/$(4 \times B)$ = 152.67 μs; most of the cell contents (54 or 216 cells per frame) are the usual ATM cells, whose format is described in Chapter 3, Section 3.2 of Desurvire (2004); the 2 or 8 remaining ones, which come up every 28th slot, are occupied by PLOAM (physical-layer operations, administration and maintenance) cells, see following description.

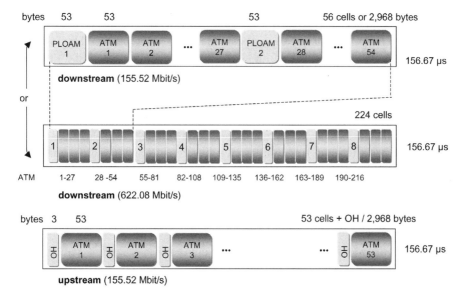

FIGURE 1.19 Framing structure of ATM-PON (APON) corresponding to upstream and downstream channels (PLOAM = physical-layer operation, administration and maintenance, OH = overhead).

Upstream frame: Is made of 53 ATM cells which are preceded by a 3-byte overhead (OH) field; the total upstream frame duration is thus (53 cells × 53 bytes + 53 OH × 3 bytes) × 8 bits/B = 152.67 μs, which is the same as in the downstream case; the 3-byte (24 bits) OH consists in at least 4 bits of guard time, followed by 20 bits (or less) serving as preamble (synchronization and amplitude recovery) and delimiter (start of incoming ATM cell); the lengths and patterns for these three fields are programmed from the OLT end; for control purposes, the upstream frame may occasionally contain PLOAM cells, since they have the same size as the ATM ones (see below).

The 53-byte PLOAM cells in the downstream channel serve several key functions. These functions concern frame *identification* (start of frame, frame version used), *synchronization* (8-kbit/s clock channel), *time-slot granting* (for upstream channel use) and *individual ONU messaging*, and other quality-monitoring functions. Synchronization of the ONU is made by counting the number of bytes transmitted (or groups of 4-bytes in the 622-Mbit/s case) and putting the value in the PLOAM synchronization field. The counter value is reset to zero every 8 kbits or 125 μs. This field thus provides an 8-kHz clock to the remote ONU. Note that this clock rate is the generic one used in SONET/SDH which, in particular, is advantageous from the micro-electronics/IC technology standpoint (see GPON below). The granting and messaging functions occupy the so-called "grant" and "message" fields in the PLOAM cell, which can be more specifically described as follows:

Grant: In APON, there are exactly 53 possible time slots to be granted to the upstream channel; Figure 1.19 shows indeed that there are 53 possible slots for ATM cells in the upstream channel (noting, as an oddity, that 54 ATM cells can be carried the downstream channel); each of the PLOAM cells is thus able to distribute the 53 time slots between the different ONU by coding 27 "grant" fields (26 for the next PLOAM cell in the frame series); these grant fields are labeled GRANT0 to GRANT26; during frame reception, each individual ONU "x" reads its corresponding GRANTx field (starting from the first or the second PLOAM cell), and therefore learns whether it is authorized to transmit upstream cells and in which time slots; the GRANTx field indicates whether the cells to be transmitted by the ONU should be of the ATM (data) or PLOAM (control) type; within the PLOAM cell, several CRC fields are introduced to protect grant fields by groups of seven; if the corresponding CRC results indicates any errors, all grants involved in said 7-fold slot group are discarded by the ONU.

Messaging: The message block in the PLOAM starts with a MESSAGE_PON_ID byte which specifically designates the ONU to which it is addressed; the type of message is then specified by a MESSAGE_ID byte; the message is then contained in a 10-byte message field; the message block ends with a CRC field; if the corresponding CRC results indicate any errors, the message is discarded by the ONU; see following text on message types.

We note that the minimum APON bandwidth provisioning corresponds to the grant of one time slot, representing one ATM cell every 152.67 μs, or a cell rate of 6.55 kHz. Since the ATM cell is 53 bytes with 48/46 bytes of actual payload, the minimum payload bit rate is about 48 bytes $\times$ 8 bits $\times$ 6.55 kHz = 2.515 Mbit/s. If a user was granted all of the 53 time slots (an unlikely event), the maximum available capacity would then be 53 $\times$ 2.515 Mbit/s = 133.3 Mbit/s. These min/max capacities correspond to both upstream and downstream traffic in the 155 Mbit/s case. They must be multiplied four-fold in the 622-Mbit/s downstream option. Therefore, the minimum downstream bandwidth (1 time-slot grant per APON frame) is 4 $\times$ 2.515 Mbit/s = 10.06 Mbit/s. This represents the maximum capacity that a TDM-APON would be able to equally distribute to all users. Note that the upstream bandwidth must be corrected by the small overhead caused by occasional PLOAM cell transmission. This calculation illustrates the fact that despite the "virtually unlimited bandwidth" of optical fibers, the APON does not service more than multiples of 10 Mbit/s as bandwidth available to end users. Such a reality contrasts with the superficial picture that is often attached to FTTx/FITL networks, according to which "optical" bandwidth would be an unlimited commodity.

This description shows that the PLOAM makes it possible for the OLT to centrally manage the PON bandwidth by assigning variable numbers of time-slot grants to the different ONU, while preventing any collision risk (two ONU transmitting in the same slot). While the upstream frames are synchronized, a problem is that they do not have the same arrival times on the power splitter, due to the fact that the fiber feeders between the splitter and the ONU have different lengths. In this case, the different upstream frames would overlap or destructively interfere. To compensate such arrival-time differences, a *ranging* protocol is initiated when a new ONU must be installed. The protocol is implemented by use of PLOAM messages. The principle of ranging is to measure the absolute distance separating the OLT from the ONU. This is done by granting a group of adjacent time slots to the ONU under test and requesting from the ONU a ranging PLOAM cell. The OLT then measures the arrival time which, compared to the ideal time reference provides the information on the delay to be corrected. This information is then conveyed to the ONU. Then the ONU knows at what delayed time its cells should be transmitted, based upon the aforementioned, 8-kHz clock reference. Upon completion of the ranging protocol, all ONU have the same "logical distance" to the splitter, despite differences in their actual physical distances or feeder lengths. Note that a 4-bit guard time is included as overhead in each of the upstream cells to compensate for ranging inaccuracies or time-varying delay conditions due to seasonal temperature changes or any other factors. Finally, the PLOAM includes a last field called BIP (for *bit interleaved parity*). The BIP is used to monitor the *bit-error-rate* (BER) of the downstream traffic.

This synchronization and time-slot granting processes enable the different ONU to transmit individual cells in the right time slots and by the right amounts. All these isolated or groups of upstream cells join together in the splitter to form the overall upstream frame. As previously described, upstream cells are made of a 3-byte OH field followed by a 53-byte ATM or PLOAM payload. As in the downstream case, upstream PLOAM cells serve messaging functions. The cell begins with an identification field (PLOAM format version), followed by message and CRC fields (same structure as in downstream case). The other functions involved are addressed by a *laser control field* (LCF) and a *receiver control field* (RXCF). The first ensures optical power stabilization and the second indicates to the OLT the optimal *decision threshold* for data recovery (see Chapter 2). As in the downstream case, the last field is a BIP field which monitors the upstream channel BER.

As we have seen, PLOAM messages make it possible for the OLT to determine the ONU ranges, which is essential for upstream cell synchronization. But messages can be used for many other useful functions. One concerns ONU *identification*. All ONU eventually receive an individual PON_ID identification number, which is assigned by the OLT, either by periodically interrogating the network or by receiving an ID request from the new ONU. Another function concerns *upstream bandwidth request*. The ONU can send requests to the OLT to indicate the length of its data queues (ATM payload cells waiting to be transmitted). This network feature is referred to as *dynamic bandwidth allocation* (DBA). To complete this in-exhaustive list, PLOAM messages can be used to initiate and validate the process of *encryption security*. Indeed, without encryption, TDM-PONs would not be protected against eavesdropping or hacking, and would have no intrinsic privacy (unlike WDM-PON). This is because the downstream traffic is sent to all ONU, leaving them the task of selecting the payload cells which are destined for them. As described in Chapter 3, encryption is based upon the use of *keys*. These keys are required to transform the payload data into a *cipher* (the unreadable "secret" message) and this cipher back into *plain text* (the original and readable data). To initiate the process, the OLT sends an encryption key request via a downstream PLOAM message to the new ONU to be serviced. The ONU sends the requested key via an upstream PLOAM message, which is then acknowledged by the OLT. This key will then be used by the OLT to encrypt the cells destined for this ONU. A third party having intercepted this key would not be able to decipher anything, because a second, or *private* key is also needed to recover the plain-text, as described in Chapter 3. Thus both OLT and ONU have their own private key to secure their communications in the downstream and upstream channels, respectively. In practice, only 24-bit keys are used, which is considered very weak in the real field of encryption. The solution is to periodically change the keys, just like passwords. The PLOAM request/acknowledgment message protocol is very well adapted to such key updates. Another level of APON security is provided by the use of *passwords* for each ONU. Access to the network is granted when the OLT recognizes the password, which prevents potential "intruders" from masquerading as an ONU, or getting into the APON for malicious hacking purposes.

The APON can be categorized under three standard classes, which define the number of ONU that can be serviced by the OLT hub, and the overall OLT/ONU path loss (min/max ranges), as follows:

- Class A: up to 8 ONU, path loss of 5–20 dB (622 Mbit/s downstream rate only)
- Class B: up to 16 ONU, path loss of 10–25 dB (all bit rate cases)
- Class C: up to 32 ONU, path loss of 15–30 dB (all bit rate cases)

It is seen that, under current standards, the maximum number of users in a given TDM-APON is limited to 32 (class C), with possible increase to 64 in the future. Such limits are dictated by system cost considerations rather than by any technology limitations. In practice, several APON can be overlaid using multifiber cables and multiple independent OLT in the hub, thus increasing the number of ONU that can be serviced in a given geographical area or premises. Such an approach could have a major impact in decreasing the per-user service cost. As we have seen, the maximum downstream bit rate in APON is 622 Mbit/s. Symmetric operation with 622 Mbit/s for both upstream/downstream channels is under study. An upgrade at 2.5-Gbit/s rates is also considered for future developments. See the following discussion regarding comparison between APON, EPON and xDSL.

This description concerned TDM-APON. If WDM is introduced, the picture is very different since every user has its own dedicated channel. Then ATM is just another communication protocol from premises to an ATM core network, and there are no bandwidth limitations other than those imposed by end-user equipment (transceiver) and the number of wavelengths that can actually be serviced. The current standard limits the APON to 16 or 32 wavelengths, corresponding to 200-GHz or 100-GHz spacings, respectively. Because the available bandwidth is not limited by considerations of loss or dispersion (unlike in long-haul, point-to-point transmission; see Chapter 2), the uncertain acronym *wide WDM* (or WWDM) has been introduced. It could be justified if future developments of PONs more fully exploited fiber bandwidth, for instance from 1.2 μm to 1.8 μm (600 nm or 83 THz!), but this would require the development of laser sources covering this spectrum. The fact that PONs are so far limited to the 1.45–1.60-μm spectral range is explained by the convenient reuse of 1.3-μm and 1.5-μm telecommunications technologies destined for long-haul transport (including 1.48-μm optical amplifier laser pumps).

We consider next *Ethernet-based PON* or EPON, which is also referred to as "*Ethernet in the first mile*" or EFM. Note that in this new acronym, the "last mile" for the network operator has now become the "first mile" the end-user customers, but apart from such a new marketing label, there is no conceptual change from any technology considerations. Standardization of EPON is still in process, although the basic operation and network topology remains very similar to that of APON (TDM and WDM). A key difference between EPON and APON, beyond intrinsic access protocols, is that Ethernet frames have variable sizes with variable time slots. The Ethernet frame size can be between 64 and 1,518 bytes (see Desurvire, 2004, Chapter 2, Section 2.5 and Figure 2.23). Another difference is the bit rate. Ethernet standards concern rates of 10 Mbit/s (called 10 Base-T), 100 Mbit/s (called 100 Base-T), 1.25 Gbit/s and 10 Gbit/s, the last two being referred to as *GigE* for *gigabit Ethernet*. Compared to ATM, one advantage of Ethernet is that it concerns a wide majority (90%) of local-area networks (LAN). For the end-user, this implies that there is no need for new terminal equipment other than a shared ONU with multiple ONT feeders. Furthermore, DSL modems with Ethernet interfaces can be readily used in such a network. Another feature of Ethernet is that different LAN can be connected through simple network "bridges" (see Chapter 2). Thus EPON can be easily inter-connected, without need for intermediate gateway switches as would be the case with APON.

The standards concerning EPON (or EFM) concern applications which are point-to-point (P2P for the initiated) and point-to-multipoint (P2MP). Note that EFM also considers copper-plant access solutions, at least for P2P ranges below 1 km and for rates not exceeding 10 Mbit/s. The other EFM classes concern P2P and P2MP solutions over fiber with capacities at and over 1 Gbit/s, and network ranges up to 10 km. It is only the last P2MP solution that deserves the real name of EPON. Under current standards at the time, the number of EPON users (ONU) is limited to 16 or 32, which reflects more a WDM than a TDM protocol signature. Ethernet being a multiple-access and relatively delay-insensitive protocol, there are no definite limits in the number of EPON users, other than those defined by minimum guaranteed bandwidth or maximum network-access time. The framing standard for the upstream channel is still under definition. One possibility would be to assign 1,518-byte time slots fitting the maximum Ethernet frame (e.g., 125 μs or 250 μs), but this would be making very poor use of network bandwidth since most frames are under this size (a typical statistical distribution consists in over 50% of 64-byte frames, 25% of 512-byte frames and the rest in lengths up to 1,518 bytes). Another possibility would be to concatenate the Ethernet frames in order to fill up these slots efficiently, but this would introduce another level of complexity to the protocol. A third solution would be to segment the Ethernet frames in order for them to fit into a unique slot size. This would only require a *segmentation and reassembly* (SAR) process similar to that used in ATM (see Chapter 3, Section 3.2 of Desurvire, 2004). Finally, security concerns must be addressed since there is no such protection mechanism built into Ethernet framing. However, security can also be implemented directly at IP level, for instance though firewalls and encryption, and several other protection/authentication standards already developed for IP networking. These issues have been only recently addressed under the future GPON standard (see below).

Since there are currently two leading PON standards, one may wonder which one is best suited for broadband access under various FTTx scenarios. But as may be clear to the reader, both standards have their own qualities and drawbacks. By definition, ATM is well-suited to delay-sensitive traffic such as in interactive multimedia applications. It also has different service classes between which users may choose in terms of minimum guaranteed bit rate, packet loss, network use, quality of service and so on. The price to pay is the complexity in the traffic segmentation (into cells) and reassembly, including a significant bandwidth overhead (the so-called 10% "cell tax"). Yet the possibility of granting time slots of definite packet size is very advantageous for both bandwidth management and guaranteed user access in the upstream channel. This is in contrast with EPON, where there are no fixed frame sizes and collisions occur (although being detected by the protocol), resulting in occasional frame-transmission failures. Considering gigabit rates, these events may not constitute a limitation at end-user level, even if the Ethernet transmission protocol remains a "best effort" approach. But it is also possible to introduce priorities and shaping into the Ethernet traffic, thus defining classes of bandwidth/service as in ATM. At equal bandwidth-use efficiency and *quality of service* (QoS), therefore, there should be no differences between APON and EPON, meaning that these two transmission protocols should appear *seamless* or *transparent* to the network customer. The key difference is in the *pricing*, the perceived *service reliability*, the settlement of *standards* and the development *history*, all these factors taken together seeming to have given APON an advantageous stance. Yet Ethernet also benefits from a solid reputation and acceptance in the LAN world, with a widespread equipment base. Note that APON can carry Ethernet payloads (by means of the ATM's SAR protocol layer) in a seamless fashion to the Ethernet terminals, but with the 10% ATM cell tax and at lower bit rates (622 Mbit/s downstream, 155 Mbit/s upstream, as we have seen). A solution that is also envisaged is to develop a standard common to APON and EPON, for considerations of inter-operability but also for leaving the option to put payloads into (Ethernet-like) frames or (ATM-like) cells, according to need (fixed- or variable-bandwidth allocation). Beyond natural bias from ATM or Ethernet disciples, the debate still remains open and should evolve according to these new standards.

Since PON (APON and EPON) cover the need for the "last" or "first" access mile, what becomes of DSL? A first consideration is that the xDSL family belongs to a different bit-rate park. As described previously, the highest capacity (VDSL) is 52 Mbit/s for downstream and 32 Mbit/s for symmetric channels. Even in that case, an optical fiber cable with ONU termination is required due to the limited transmission range (300 m). The other xDSL technologies have longer ranges (from 1 km to over 10 km), but the corresponding bit rates are one or two megabit/s at most. For VDSL, a limitation is that the ONU is *not* located in the customer premises (CP), which involves a certain number of additional constraints (resistance to climatic conditions, access, installation and space occupation costs). To alleviate such constraints, a possibility is to bring the fiber all the way to the CP, this representing a "DSL-over-fiber" solution. Note that this "all-fiber" version of DSL is not comparable to another FTTx solution, because it remains a simple P2P link with DSLAM network interface, in contrast to a PON which is a fully managed P2MP network, and is based upon a completely different interface (OLT). Despite its intrinsic bandwidth limitations, the xDSL approach represents an attractive, very economical reuse of the copper plant (the old twisted-pair telephone wires). Independent of bandwidth considerations, the drawback of DSL (when compared to PON) is its sensitivity to electromagnetic and radio-frequency interference (EMI/RFI). Moreover, it does not have the security and protection features of PON. Signals can be tapped along the line (wire or radio means), without the end users being aware of it, unlike with PON. Note that optical signals can also be tapped by different means, but this represents a higher level of difficulty and "investment." For the above reasons, DSL may be seen as representing the first generation of broadband access, to be progressively followed or substituted by PON.

We should note that PON may use frame/cell formats other than ATM or Ethernet, which are only "layer 2" protocols. Thus, Internet-based PON (IP-PON) may appear as a "layer 3" solution. Behind the suggested appeal of the approach lies the fact that networks need layers 1–2 to operate. Furthermore, the tree topology of PON is not adapted to the IP "datagram" format as such. For local network access, IP datagrams must be encapsulated into frames one way or another. Thus an IP-PON may just be an APON or EPON with Internet payloads, but the customer is only aware of the upper service layers (actually TCP and up, see Chapter 3, Section 3.3 of Desurvire, 2004). Since IP payloads are more easily encapsulated into Ethernet frames, the IP-PON represents a key application of EPON, notwithstanding the fact that it does not have the same service offer/quality and flexible-bandwidth access capability of an APON supporting IP traffic.

We consider finally *GPON*, which stand for gigabit-class PON. More than representing mere gigabit/s operation, GPON were defined according to new and demanding service requirements. First GPON must support multiple "client," payloads from TDM (PDH) to SDH/SONET to Ethernet (10 or 100-Base T), ATM, IP and other possible ones. Thus it is a multiservice PON. Second, the physical access range should be 20 km (12 miles), with logical network connectivity under the same protocol of 60 km (35–40 miles). Third, bit rate should be scalable, with both symmetrical and asymmetrical options (622 Mbit/s and 1.25 Gbit/s in the first case, and 2.5 Gbit/s downstream in the second). Finally, GPON should offer enhanced security features in the downstream channel, as previously discussed. A key feature of the emerging (2003) GPON standard is the efficient use of usptream bandwidth, regardless of traffic type, by means of DBA though ONU *pointers*. Such a notion is introduced in the so-called *generic framing procedure* (GFP). The framing format is based upon the same 8-kHz (125-μs) clock rate as in SDH/SONET. Traffic payloads are either *segmented* or *concatenated* in order to fit in the assigned time slots, and thus meet the service-bandwidth requirements with maximum efficiency. The frame structure is however different for downstream and upstream channels, according to the following features:

Downstream: The queued ATM cells ($N \times 53$ bytes) and the GFP section (other client traffic, e.g., TDM or Ethernet) form two distinct payload fields; unlike with APON, the control functions are centralized into a single frame header; this header ensures all functions of synchronization, identification, PLOAM messaging, bandwidth-allocation pointers (DBA), BIP and CRC; the pointers define which ONU is granted time slots and designate the upstream transmission mode (see next item);

Upstream: The frames are also separated into payload and header fields; for payloads, the three possible transmission modes are "ATM-only" (pure ATM-cell payloads), "GFP-only" (fragmented/concatenated, other-than-ATM payloads) and mixed type (including distinct ATM, GFP and PLOAM fields); among other functions, the frame header addresses the different ONU (ID followed by "start" and "stop" time-slot assignments.

Detailed efficiency comparisons while assuming an identical bit rate for APON, EPON and GPON, and assuming a conservative 10% TDM and 90% data have shown that GPON wins with above 90%, while APON and EPON only reach 70% and 50%, respectively. But these numbers may only be taken as an indication rather than an absolute reference, since assumptions are made for the traffic distribution. While GPON thus comes as an additional PON "flavor" with clear advantage in efficient (should one say "smart"?) multitraffic use and bandwidth allocation, price and complexity considerations must also be taken into account. It may turn out that most realistic PON niche applications are exclusively ATM- or Ethernet-based, which does not preclude planning, ahead of time, a larger picture and possible migration to GPON. We also note that GPON is a TDM-PON and therefore does not have the sophisticated features of a P2MP WDM network. But nothing says that WDM could not be another GPON overlay feature. The bottom line issue, to take into account in such a debate, is the service type that most customers will be ultimately requesting. Then it all depends upon the FTTx application. Probably, FTTH and FTTC are concerned by VoD and HDTV entertainment applications, which corresponds to a well-identified, nonchanging traffic type, with known access patterns and limited bandwidth demand (0.5 to 5 Mbit/s per video channel). But there are also requests for high-speed Internet (such as currently serviced by xDSL) and integrated voice services (VoIP). For either small and medium enterprise (SME) or business/corporate applications (FTTB), interactive multimedia and protected data communications are central. Another issue for business is the digital storage of and instant access to company data or archives (*storage area network*, or SAN). By definition, a PON does not necessarily offer any storage-service features different from other network solutions, but secure/protected optical channels (upstream for data loading, downstream for data downloading) represent a vital component for any SAN. For PON operators (ILEC and CLEC) the key challenge is to integrate both service offers (private/business) into a "residential" PON concept, or RAN, by use of a single protocol that would be able to fit all and with appropriate pricing scale. It is likely that the combination of broadband Internet and VoIP (as a "freebie") may in any case represent one of the leading demands.

1.3.3 Optical Access

The definition of *optical access* applies to any all-fiber, multigigabit link between customer premises (e.g., a business LAN) and the greater network infrastructure, as reached through a specific service *point of presence* (POP). The POP can concern *Internet service providers* (ISP), *application service providers* (ASP) or *storage-service providers* (SSP). A local ISP can connect to a larger network via an *Internet service exchange* (ISX) POP, the same way a local operator does via an *interexchange carrier* (IXC) POP, which requires optical access. The peripheral zone covered by the different types of optical access links represents the *network edge*. In the global network picture, this edge also corresponds to an intermediate stage which relates small-size LAN and larger-size *metropolitan* area networks (MAN), also called *metro* networks. As described in the next chapter, MAN/metro networks usually take the form of large (up to 100-km diameter) WDM rings. The previously described PON, which are tree-and-branch optical networks, can be viewed as a specific case of optical access. The applications are however restricted to user data rates below 1 Gbit/s, as limited by TDM or WDM bandwidth allocation. In contrast, optical access more generally refers to links between LAN and MAN having *multigigabit* capacities. The access links, usually under 50-km length, connect to a single MAN node, which acts as a gateway to ISP, ASP, SSP and PSTN. Traffic is based upon Layer 1–4 protocols such as SONET/SDH (Layer 1), Gigabit-Ethernet and ATM

(Layer 2) or TCP/IP (Layers 3–4), with various encapsulation schemes (e.g., IP over ATM, ATM over SDH, etc.) and overall, WDM for bandwidth management. Access networks can also be multipoint, thus forming a network mesh. The advantage of such a mesh configuration is that the traffic can be carried over different possible routes. In the event of node congestion/outage or link failure, the traffic can then borrow an alternative route, which provides uninterrupted service. Another feature of WDM is the ability, according to evolving needs, to upgrade traffic capacity by using additional wavelengths. Thus optical access can be scaled up by tapping into the "virtually infinite" fiber bandwidth reserve. However, such an upgrade is not automatic and free of cost. Indeed, it requires one to install additional wavelength transceivers at both link ends, which may represent a significant investment relative to the overall network cost. As usual, the cost issue (fiber-cable access and $/wavelength) can be alleviated if the resources are shared by several service providers and large groups of end-users.

The specific technologies and features of optical networks are described in Chapter 2.

1.3.4 Hybrid Fiber-Coaxial (HFC) Systems

Another solution for local broadband access, which combines optical and electrical networking, is provided by *hybrid fiber-coaxial* (HFC) systems. Just as the copper-wire plant was extensively deployed for the telephone service, coaxial networks have developed in urban/sub-urban areas for cable TV and other video-on-demand (VoD) services. Since coaxial wires offer an intrinsic bandwidth from a few tens of MHz to the GHz range (according to transmission loss), they are well adapted to broadband-service applications similar to that previously described for xDSL and FTTx, with both downstream and upstream traffic capabilities (in HFC, one also refers to "*forward*" and "*reverse*" traffic directions, respectively). A key difference with xDSL is that coaxial wires can transport high-speed digital signals over distances of several tens of kilometers, the line attenuation being periodically compensated by inexpensive analog/RF amplifiers. Optical fibers are then introduced to further extend this transmission range and carry aggregate traffic, thus linking together different local cable-networks to form a wide-area HFC system. Unlike FTTx, the HFC system is not fiber-intensive, since the local access is 100% coaxial-based. The fiber and associated installation costs are thus shared by a very large number of users, in contrast with FTTx. To access the new HFC services, the end-users only need to get a relatively inexpensive modem and plug it into the house's coaxial terminal (just like ADSL modems with telephone plugs). The device can also include a splitter to separate the data from the analog-TV signal, thus ensuring the continuation of the "cable-TV" service, like a DSL connection does with respect to the existing telephone. Note that HFC modems and DSL modems are not compatible, which indicates that the end user must make a choice between the two.

The generic network layout of an HFC system is shown in Figure 1.20. The topology is that of a tree whose trunk and main branches are made of fiber-optic cable and whose smaller branches are made of coaxial wire (one refers to such a topology as "*tree-and-branch*"). The central HFC *head-end* (H/E) connects to the PSTN and

1.3 Fiber in the Loop (FITL) 77

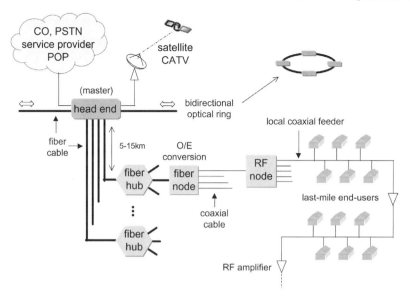

FIGURE 1.20 Generic network layout of hybrid fiber-coaxial (HFC) system.

any other service provider (e.g., ISP), in addition to *common-antenna (or community-access) television* (CATV) services, for instance based on *satellite* (see Chapter 4, Section 4.3 of Desurvire, 2004). The H/E can provision several HFC networks through many independent fiber-optic cables. Each of these cables usually contains several individual fibers that are assigned to carry either upstream or downstream traffic. Because bandwidth is usually asymmetric in HFC systems, the downstream wavelength is 1.5 μm (higher signal-to-noise ratio and WDM-allocation capabilities, see next) while the upstream wavelength is 1.3 μm. The fiber cable may also contain a reserve of "dark fibers" for future network extensions or upgrading the traffic according to new capacity demands and evolving upstream/downstream capacity ratios. The H/E cables terminate on *fiber hubs*, from which the optical traffic is split and transported through smaller cables. In turn, these cables end in *fiber nodes*. In these nodes, the downstream optical signals are converted into radio-frequency (RF) analog/digital data, to be carried by single coaxial wires (also called "coaxial cable" since the wire is coated with one or several layers of protective insulator). Likewise, the upstream RF traffic from coaxial wires is converted into optical signals. The two processes are referred to as *opto-electronic* (O/E) and *electronic-to-optical* (E/O) conversion, respectively (or O/E both ways, for short). The coaxial wires finally reach *RF nodes*. In these nodes, the downstream traffic is split again into *local coaxial feeders*. The RF nodes also act as concentrators for upstream traffic incoming from different coaxial feeders. Each feeder can service 10–100 individual homes, using a bus topology with "T" coaxial connectors. This represents the "last mile" of the HFC system. Note that each T connection corresponds to splitting and excess loss, which sets limits to the number of terminals that can be hooked to the feeder. To extend the feeder range (and also to compensate for splitting/excess loss), in-line

RF amplifiers are used with 300-m to 1.5-km periodicity. The amplification period depends upon the range needed, as well as the number of homes/terminals to be connected along the path, which is dependent upon the type of suburban or urban area, and residential/business infrastructure. Because the signal degrades with the number of cascaded amplifiers, there is a trade-off between the maximum number of home terminals and the maximum feeder range (or terminal distance). Typically, a single RF node can service up to 500 home terminals, while a fiber node can handle several (e.g., 4) RF nodes (i.e., 2,000 homes). Considering that several tens of HFC networks (e.g., 50) can be provisioned by a single H/E, the number of homes that can be so connected is of the order of 100,000. This number is quite impressive when compared to FTTx figures.

As Figure 1.20 also shows, the H/E can be included into a larger-area, *optical ring*, with only one master H/E being connected to the network core and its service POP. Such a ring makes use of several independent wavelengths in order to carry the aggregated traffic from/to all individual H/E nodes. In this configuration, each H/E is assigned a unique wavelength channel, which corresponds to the technique of *wavelength-division multiplexing* (WDM). Note that the ring traffic is usually bidirectional (by means of fiber pairs), which ensures protection against cable breaks or other one-way interruption of traffic (see Desurvire, 2004, Chapter 2, Section 2.5). Since WDM rings are the basis of *metropolitan area networks* (MAN or "metro" for short), it is seen that HFC represents another means of broadband-access into a larger backbone-network picture (see Chapter 2). The maximum distance between the end-user and a ring H/E is of the order of 100 km, which provides an idea of the global HFC coverage (corresponding to a maximum "footprint" of 30,000 km^2).

But WDM is not restricted to aggregating traffic within the HFC optical ring. It can also be used at H/E level to *dynamically* allocate downstream traffic all the way down to the fiber nodes, according to bandwidth requirements and changing needs. Such a feature is illustrated in Figure 1.21, which shows two network layers.

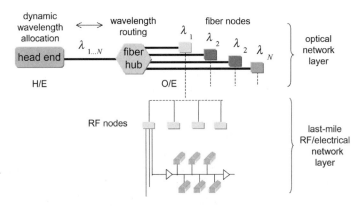

FIGURE 1.21 Optical and RF/electrical network layers in WDM-HFC system, showing dynamic wavelength allocation to fiber nodes.

The optical or WDM layer is concerned with wavelength allocation and routing. Wavelength allocation (sending traffic onto a specific wavelength) is controlled by the H/E while the wavelength routing (steering the wavelength into a specific physical path) is made at fiber hub level. Thus each fiber node receives downstream traffic through a single wavelength carrier, which can change according to the allocation pattern. For instance, the same wavelength can be used for a subgroup of fiber nodes, while the other nodes are attributed different wavelengths. The downstream traffic can thus be managed according to bandwidth needs and usage type, which vary during the course of a single day. For instance, most residential end-users are likely to request TV and video services (VoD) at the end of working days through beyond midnight, while the pattern is naturally different over week-ends and holidays. In contrast, the demand for e-mail and high-speed Internet is high during working/business hours, but this traffic also continues to some extent, from private demand, during evenings and week-ends. Thus dynamic bandwidth allocation through WDM is essential to avoid HFC network congestion and to regulate the flow of variable traffic types and demands. See more on WDM networks and bandwidth management protocols in the next chapter.

Transmission of digital data over the RF-network layer of HFC systems is made through either *analog* or *multilevel (M-ary) digital* coding formats, using *TDMA*, *FDMA* or *CDMA*. The spectral allocation for upstream and downstream channels widely differs according to the dominant service type (e.g., analog-TV/HDTV broadcast versus IP voice/video data) and network-integration history. As a typical reference, the first 50 MHz are generally used for upstream, while the remaining 50-MHz–1-GHz band is for downstream. The actual separation depends upon the country's standards. For instance, the United States and Europe respectively allocate bands of 5–42 MHz and 5–65 MHz for this upstream use. It is important to note here that the correspondence between "Hz" and "bit/s" bandwidth is not uniform and strongly depends upon the coding format. It is indeed possible to transmit several bits in one Hz cycle, which is referred to as the "spectral efficiency" (bit/s per Hz, or bit/s/Hz). For instance, QPSK (*quadriphase PSK*; see Desurvire, 2004) has 1 bit/s/Hz, while M-ary formats such as M-QAM (*quadrature amplitude modulation*) have spectral efficiencies equal to $\log_2 M$. Thus M = 16, 32, 64, 128, or 256 have efficiencies ranging from 4 to 8 bit/s/Hz, respectively.

The 50-MHz (or so) upstream spectrum can be sliced into subchannels spaced by 2.5 MHz (20 channels), for instance. To minimize transmitter cost for the end user, the modulation format is kept relatively simple, for instance QPSK (4-level phase modulation). Each individual QPSK channel having 1-bit/s/Hz efficiency, the overall upstream channel capacity is at maximum 50 Mbit/s (in fact, the net bandwidth per channel is some 20% lower, in order to prevent channel spectral overlap and interference, corresponding here to 40 Mbit/s).

The downstream channel, shared by all users, is usually sliced into 6–8-MHz channels. The signals generated by the H/E can be coded through more sophisticated formats such as 16/32/64/128/256-QAM. Thus, an 8-MHz subchannel band with 256-QAM modulation can carry 50-Mbit/s (<8 MHz × 8 bit/s/Hz = 64 Mbit/s) traffic. Such a capacity can transport about 8–12 high-quality video channels

(4–6 Mbit/s each). The 50-MHz–1-GHz band offers plenty of space to allocate channels according to the traffic type. Typically, two bands of 100–300 MHz and 450–600 MHz are allocated to TV distribution services. The 300–450-MHz band between these TV-bands is available for data services. The upper band of 600–1,000 GHz can be attributed to VoD services. One also refers to *near video-on-demand* (NVoD) for services providing groups of video channels which are not user-specific, which permits simplification of bandwidth allocation for RF nodes. For interactive video/multimedia, the bandwidth must be made symmetrical (meaning equal for upstream and downstream channels). New standards such as S-CDMA (*synchronous CDMA*) were developed to realize symmetrical rates up to 30 Mbit/s, which compares well to the top VDSL service classes (see Section 1.2, Table 1.7). Other provisions are made to assign additional upper-frequency bands to the (confined) 5–50 MHz upstream channel, for instance in the unused 850–1,000 MHz spectral range.

Compared with DSL and PON/FITL, the *"service suite"* (offer) offered by HFC systems is somewhat wider, which is explained by the differences in access range, number of terminals and user profiles, and the contents-provider "cable TV" background. Three differentiating examples are *targeted advertising* (selective multicast), HDTV (multicast) and analog NVoD/VoD (multicast). Another differentiated service offer concerns voice, either as *cable telephony* (as a local-exchange system) or V*oIP telephony* (via the Internet). This is where HFC system operators are in a different position compared with ILEC and CLEC (see definitions and discussion in earlier subsection). Indeed, ILEC must provide "unbundled" access to their copper plant to CLEC in order for the second to deploy new broadband services, including in particular VoIP. In contrast, cable-TV operators are privately owned, and can thus provide Internet access at competitive prices, including VoIP, without needing to share their network resource with (I/C)LEC. Large cable-TV operators also have the option to install classical telephone switches in their HFC and thus become LEC operators as well. These differences should not obscure the fact that VoIP as included into a broadband-IP service-offer remains a common and major business opportunity for all operator types. This is also true for video-telephony and multimedia/video-conferencing applications. In all cases, bandwidth access, service offer/suite and package pricing are the key differentiation factors, and the flavors vary widely according to the end-user need (residential vs. enterprise, and local/geographical combinations thereof).

Because of the relatively high bandwidth provided by existing coaxial networks, and their cost-effective service offers, HFC systems have entered into competition not only with DSL, but also with PON solutions. Yet the HFC solution is not strictly equivalent or transparent from the access standpoint, as we shall describe further. First, HFC access suffers from several shortfalls, which concern issues of *reliability, security/privacy* and *operation/administration/maintenance* (OAM). Reliability not only represents a 99.9% to 99.999% ("3 to 5-nines") guarantee of access, but also the trust that the network will be always available for "life-saving" purposes, such as in the event of a major power failure due to climatic or other major catastrophic events. In this respect, the POTS is the lasting reference model. Security and privacy are issues concerning both operator and end user. Since signals are broadcast through the system, the potential of *theft of service* (ToS) is high, just as in previous cable-TV systems. This contrasts with DSL which is user-specific, and cannot be subject to ToS. The other perceived threat is that of having private information, such as telephone conversations, being broadcast to the network, whether by accident or malicious purpose or simply eavesdropped by third parties (this issue concerns both private and business users). The solution to this threat is *encryption*, as we have described for PON. It is not clear at present how safe and practical HFC encryption is and what protocols may be ultimately used, but this lack of current standards does not indicate any intrinsic limitations, as with many other shared-access networks (e.g., PON, WI-FI). The issue of OAM is central to the HFC business case and market evolution. Without QoS (guaranteed service level/class), protection against accidental outages, efficient bandwidth management, and increasing local reputation, customers may not buy into or may turn away from the HFC service offer, considering that xDSL may have similar services with existing higher QoS.

As previously mentioned, HFC offers a service suite ranging from TV to Internet and *voice*. The possibility for private users to have their telephone expenditure included into a global and single "service bill," is quite attractive, especially considering that most of these users already have mobile telephones, possibly more than one in a given household. How could voice be serviced by HFC? As we have previously seen (see Section 1.2) voice can be carried over packets (VoP), namely over IP (VoIP), ATM (VoATM) or DSL (VoDSL). Thus voice-over-HFC (VoHFC) could be just a transparent definition for VoIP or VoATM, with IP and ATM payloads encapsulated into HFC frames. Because voice is thus packetized, some spurious delays and packet losses may occur, which, with the effect of voice-compression algorithms, could result in the perception of a poor sound quality compared to traditional POTS. But in this matter HFC is on the same footing as DSL or other means of VoP transport, assuming equal bit rates. Another solution is to allocate a 64-kbit/s channel to both upstream and downstream traffic, which can be done under a *synchronous transfer mode* (STM) protocol (see STM description/examples in Section 1.2). An additional solution, which can be overlaid on STM, is referred to as *personal communication services*, or *PCS* (see Chapter 4, Section 4.2 of Desurvire, 2004). Basically, PCS is a microcellular, short-range mobile telephone service operating in the 1.9-GHz radio band. One or several omni-directional antennas can be plugged into the coaxial wires of the feeders. Upon antenna reception, the signals are frequency-converted to fit into the 5–50-MHz upstream channel, and sent to the H/E. The signal is then forwarded by the H/E through the downstream channel into one of the 6–8-MHz bands assigned to the called end-user. If the end-user does not belong to the HFC system, the H/E then forwards the call to the PSTN. The PCS offer is thus competing with that of cellular/mobile services, and it is not clear whether the first presents any advantage over the second, except where the cellular coverage is under-developed or nonexistent, or simply of poor quality.

As previously mentioned, the migration of HFC systems towards *symmetric* bandwidth offer is key to the deployment of interactive, multimedia services, or more generally high-speed data services. New protocols and node configurations have been developed in order to provide circuit-oriented P2P connections.

To increase access speed and improve local connectivity, Ethernet data switches can be implemented at *fiber hub* level (Figure 1.20) instead of at H/E level. In such a configuration, the circuit-oriented data service becomes "nontime-shared," which means higher QoS and security, as in APON and higher efficiency and flexibility as in EPON (see previous subsection). Such an approach, which fully emulates a switched Ethernet service, is transparent to the end-user.

Boosting HFC bandwidth for 1–2-GHz operation is also possible through an alternative modulation format called *wavelet*. Not only are wavelet signals highly resistant to extreme conditions of noise and distortion, but they also have spectral efficiencies as high as 5.5 bit/s/Hz (e.g., 100 Mbit/s in 18 MHz). Because the approach is based upon far more complex modulation algorithms than used in QAM or other formats (although well within current IC-processing capabilities), it may be available commercially within only a few years. The drawback of any new approach that modifies the existing upstream/downstream HFC spectrum allocation is being incompatible and therefore slowly accepted as a replacement solution. On the other hand, new wavelet and QAM formats could be implemented in the upper portion of the coaxial spectrum (850 MHz–2 GHz) without disturbing the operation of current services, which is a sensible strategy.

■ 1.4 HOME NETWORKING

In this section, we shall describe the different elements (technologies, services and applications) which together constitute *home networking*. While this subject may be seen as speculative for futuristic and distant applications, it has already become a reality in most advanced countries. Having an Internet connection and (for instance) a cordless telephone is a first step into this networking world, where wireline and wireless digital technologies are for private home and family use. Home computers with their printer, scanner or external disk drive, and TV/video/stereo centers with their wireless controls, also constitute home networks. Finally, several programmable household functions (e.g., heating, air conditioning) and appliances (e.g., fire or intrusion alarm systems) also exist in every corner of modern houses. The next step of house networking, which concerns the integration of all these different "subnetwork" units into an intelligent whole, corresponds to a more complex technology concept which is far from being fully captured (let alone standardized). In addition, the future house network picture comes with unprecedented services and applications, which should represent many disruptive "ways of life." One may immediately think of video-telephony on giant wall screens, but this could be just another continuation of 3G mobile services being now introduced. As we shall attempt to describe here, futuristic house networking is a disruption rather than a continuation of what we may already know.

1.4.1 The Evolution of Intelligent Home Appliances

The current home communication network of our early 2000s is very different from that known by our grand-parents and even parents. By the mid-1900s, homes were only equipped with a single wired telephone line, a tube-radio set and, somewhat later, an antenna-based television. The early times of home telephony, where calls were personally mediated by a local switching operator, are described in Chapter 2, Section 2.1 of Desurvire (2004). The revolution introduced by microelectronics and microprocessors in the 1970s led to the popular development of "personal" or home computers (PC) in the 1980s. The early models only had alphanumeric monitors and did not have any connection to the outside world. *Facsimile* (FAX) machines then entered the home mainly for teleworking purposes (home office), or as another means of instantly communicating handwritten messages/letters to friends and relatives without having to use stamps/envelopes and experiencing delivery delays. Another dimension of FAX was its ability to transmit not only text information but its image, for instance handwritten documents with signature, drawings and other black-and-white (B&W) photographs. We should note that even today, the FAX channel is still largely used for purchase certification (e.g., hotel booking) and document authentication (e.g., bill or estimate approvals).

For people living in France, the MINITEL initiative, a true precursor of the Internet, came into the home picture via the telephone line. This MINITEL offered a wide variety of on-line services, such as national phone directory consulting, "yellow-page" search/browsing by keywords and locations, train/airplane schedules and reservations, weather forecasts, text mailboxes and interactive chat. The difference between MINITEL and similar 1-900 pay services existing in the United States (for instance) was the $6'' \times 8''$ B&W monitor with its pop-up alphanumerical keyboard, a compact cube-like appliance leased for free to the requesting telephone subscriber. Although very slow (by current standards) and having a relatively poor image resolution (below photography quality), the pay-per-minute service was affordable and its use through the simple keyboard commands did not require any specific training. (Interestingly, this popular service has survived the Internet revolution; furthermore, it also has a Web site from which a "MINITEL emulator" can be downloaded for use with home-PC). In most developed countries, the development of cable-TV (CATV and coaxial) networks made it possible to bring local information services (weather, traffic, theater programs, community events...) into the home via the TV as opposed to the newspapers. Satellite dishes made it possible to receive practically any TV-channel worldwide, thus connecting foreign-speaking traditional communities and other expatriates with their home countries. This feature is of major family impact, from the children's development of their native language, to following cultural news and national sports events.

In most homes using a FAX, a second telephone line would be generally leased in order not to interrupt the "lifeline" telephone service. Software could be installed into home computers in order to be able to send/receive faxes directly from the data files without the mediation of paper and cumbersome printers (limitations being supplies shortages and paper jams). Home telephone appliances also progressed with the appearance of automatic "answering" machines, which could record and replay a limited number of voice messages (through tape, then memory chips). Finally, cordless handsets (analog, then digital with error-correction) made it possible to roam the house without being entangled by wires, to pick up a call without having to rush to a specific room, or simply to enjoy outdoor activities with the phone within immediate reach.

1.4.2 An Internet Intrusion

This previous description is indeed that of an age foregone. The intrusion of the Internet, initially via the home telephone line, has radically changed the concept of home-based telecommunications. Such an intrusion did not displace any of the traditional telephone uses, now augmented by the development of cheap portable/ cellular phones. People were not restricted to the fixed telephone line of their households to place and receive calls, while other family members, including young and older children, were browsing the Internet or checking their e-mails. See Chapter 3, Section 3.3 of Desurvire (2004) concerning a detailed description of "Internet jargon" to get a flavor of the cultural change and the variety of new services introduced by the Internet. Thus voice (telephone), video (TV) and data (Internet) have made their way into the individual's and family's intimate world to progressively shape their life patterns. More than just another home appliance, the home computer also became a powerful data-processing machine, allowing high-speed data transfer from and to the premises. This was made possible through a variety of wireline technologies: twisted-pair copper wires (ADSL/VDSL), coaxial cables (HFC) and optical fiber (FTTH/FTTC, PON). These technologies have literally changed the private house into an actual *network terminal*. While such a terminal is now essentially a computer with an Internet modem, tomorrow it will be a *control hub* with a fully developed home network, which we describe in this subsection.

Several studies have shown that some 40–50% of North-America homes are now equipped with PCs, one-fifth of which having at least two PC units. Predictably, the proportion of these PCs actually connected to the Internet (regardless of modem speed) is in the 90–100% range. The figures are similar for countries like Sweden, The Netherlands, United Kingdom or Korea and somewhat less in other European countries and Japan. The same national figures are also found when considering all broadband-access means (DSL, cable and FTTx). One should not confuse these optimistic "% household" figures with the realistic "% population" figures. For instance, a 40–50% household internet access may correspond to only 3–15% of a given national population. Furthermore, the number indicate averages, which may mask the fact that certain suburban or country areas may in fact have very low penetration figures, in favor of densely populated urban areas. Needless to say, these statistics only concern the most-developed countries, which means that PC/Internet penetration in the least-developed ones may just be near 0.1% (<1 for 1,000 habitants) or even less. However, it is a fact that many developing countries have already undertaken aggressive plans to modernize their telephone and data infrastructure, in order to service the largest number of business and residential areas, using satellite technologies to reach the most remote regions. With this ambition, then at least every school in the smallest country village could soon benefit from the Internet, as a first development phase before seeing percentage home coverage in the "upper tens."

In the inventory of today's home networks, the age-old fixed-telephone line access remains central. One of the key issues is the customer perception of an uninterrupted, five-nines (99.999% availability) service which does not depend upon home power. The view is that while storms or other catastrophes may discontinue the power/electricity service, the PSTN remains available for emergency in any situation. At least this is true if the fixed-telephone appliance does not rely upon home power, which is generally not the case with current digital/cordless units. In case of power failure, the tone is not available, in spite of handset batteries, which the readers can verify on their own appliance with a simple experiment (just unplug the telephone power cord). Such a "lifeline" reputation of the PSTN has somewhat evolved with the introduction/penetration of personal mobile/cellular phones. In case of an even minor catastrophe (e.g., house-fire, momentary power shutdown by storms), the mobile telephone actually represents the *only* dependable solution. If this view is correct, then house customers may become more trusting of digital voice services from FTTx/DSL offers, to replace the POTS.

The next layer in house networks concerns *data*. One may distinguish the wireline and wireless data network types. The wireline network is already a reality: the suite consists in a DSL or cable connection, a modem/splitter, and a PC with its own network of printers, scanner, fax, stereo and other peripherals. Such a PC-centric plant is soon to be replaced by distributed PC networking (several family users in a multi-PC network with shared peripherals, including multiple-port modems). Such a network may eventually turn into a complete house LAN, with the inclusion of *IP-based appliances*. These appliances may serve a broad variety of roles, from safety and security (fire and intrusion alarm systems), to well-being (temperature and moisture conditioning), to many other practical daily-life purposes. One potential application concerns the integration of the radio, stereo, TV and video recorder with the Internet as a controlling agent. According to preset individual and family profiles, the Internet could then select and customize various entertainment and educational programs, in addition to classic VoD services.

It is beyond the scope of this book to describe the home of 2010–2020, just suffice it to say that there is enough indication that many home appliances will some day soon exhibit Internet "intelligence" and remote-control/monitoring capabilities. Two key elements for IP-controlled house devices are *simplicity of installation and use*, and *interoperability of standards*. As the TV and video history has already illustrated, the multiplication of wireless remote-controls with complex commands, buttons and preprogramming functions are against the idea of a convivial home network. For this reason, most stereo and video appliances are now much more simplified and integrated. But the concept of an in-house data network goes far beyond entertainment applications. Such a network should first serve business purposes (home office) as many residential customers have become teleworkers (also uncertainly called telecommuters). This new category of residential customers need services far more advanced than mere ADSL Internet, depending upon their profession, data usage profile and the importance given to delay-sensitive communications. One possibility for broadband data networking is *Ethernet*, as carried through DSL or HFC. One of the organizations dedicated to standardize DSL-access networking is the *home phone network alliance*

(HomePNA). As we have seen earlier, FTTH/FTTC can also provide a means for broadband access from residential premises, but the cost could be prohibitively high for a small or isolated business (see previous section on FTTx). New service models may emerge where groups of independent residential customers could share the resource of this broadband access through a short-range LAN.

1.4.3 Wiring the House Network

It would be an understatement to say that houses are filled with all kinds of wires for the telephone, the TV, the stereo, the computer (serial and parallel bus cables) and electrical powering (fixed and extension cords), all representing a complex and intricate mesh. With the introduction of home networking, more wires may be needed, complemented with "wireless" radio links. Figure 1.22 provides a generic view of the home network and its various fixed/mobile appliances, which we shall describe in detail hereafter.

The wired approach to transport data inside the home premises can be based upon three supports: twisted-pair, coaxial, or cheap optical fiber such as *polymer*-based (POF). The major drawback of both coaxial and twisted-pair wires is their volume and appearance, unless they can be embedded in the wall or floor structure by some means. Another problem is the multiplication of parallel wires and branching boxes in the network development, which may rapidly become intractable or highly un-esthetical from a "home" point of view. In contrast, small-diameter (<1 mm) fiber-optic POF cables, especially packaged into bus ribbons, may be

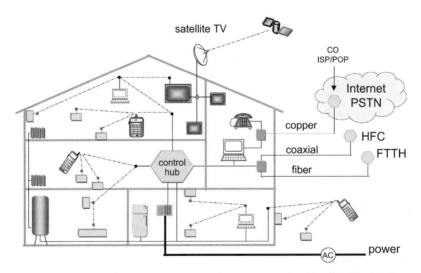

FIGURE 1.22 Generic view of home networking, showing (1) electrical powering, (2) connection to the network (Internet, PSTN, satellite) through copper-line, coaxial, fiber or dish antenna, (3) central control hub, (4) wired and wireless indoor/outdoor communication links, (5) fixed/mobile, remote-controlled appliances, and (6) fixed/mobile home-computer terminals.

invisibly deployed along the home walls and floors. Connectors for POF are also cheap and of small size. These advantages are to be balanced with the additional cost of required E/O and O/E interfaces, unless they are built into the appliances. Standards for twisted-pair telephone wiring have been developed by the HomePNA. These are similar to Ethernet with 1–10-Mbit/s capacities.

One of the cheapest solutions that can be envisaged for home networking is to use the embedded AC *electrical power lines*. Although electric wires have limited bandwidth, they can be used to convey RF signals over the few-meter distances separating rooms. The many electrical outlets available in the house may just be used as built-in network access points. Device-to-device communication can be done by high-frequency digital signaling (e.g., *frequency-shift keying* or FSK, see Chapter 1, Section 1.2 of Desurvire, 2004), using specific codewords for exclusive home usage. Network protection against noise and external parasitics (e.g., from neighboring home networks) can be achieved by specific electronic circuits which let the 50–60-Hz AC in but block high-frequency signals, acting like a low-pass filter. Operation at rates up to 1 Mbit/s is then possible. Other approaches are based upon *spread-spectrum*, which uses codes (or "chips") that are uniquely recognized between appliances. See the following on discussion on spread-spectrum coding when considering radio links. A final solution uses *amplitude-modulated* signals at 50–60 Hz, as sent through the AC current's zero-crossings. Such an approach is therefore limited to ON/OFF-signaling, which remotely activates or deactivates devices. In spite of all these advantageous possibilities, the lack of common standards and inter-operability between home appliances makes power-line communications difficult to implement.

Home appliances need not to be exclusively connected by wires. Indeed, *wireless links* can also be used to reach fixed and especially mobile appliances. Radio links are preferable because they can penetrate walls to some extent and are not sensitive to obstacles, unlike infrared links. Note that the capacity of wall penetration by radio signals is dependent upon the construction materials and the number of obstacles to traverse. In hard-stone buildings (such as found in Europe), it is unlikely that house networking could be exclusively based upon wireless links, unless some kind of radio relays/bridges were implemented. In any case, the extensive deployment of indoor radio links does not preclude a wired network infrastructure which can segment the different house-network functions, as illustrated in Figure 1.22.

As in most multiple-access wireless LAN (WLAN), protocols are either TDMA (*time-division multiple access*), FDMA (*frequency-division multiple access*) or CSMA/CA (*carrier-sense multiple access/collision avoidance*). Protocols specifically designed for house networking use *spread-spectrum* algorithms (see Chapter 4 of Desurvire, 2004). Two alternative and mutually-incompatible versions are *frequency-hopping spread-spectrum* (FHSS) and *direct-sequence spread-spectrum* (DSSS). In the first case, the changing frequency patterns (frequency hops) can be recognized between two machines, thus creating a logical communication link. The same principle applies in the second case, but based upon "pseudo-noise" code sequences (called *chipping codes*, or *chips* for short).

Short-range radio and infrared links can be managed by well-established protocols already used for PC networking. These are *Bluetooth* and *802.11b/WI-FI* (radio), and *IrDA* (infrared), as described in Chapter 4, Section 4.4 of Desurvire (2004). They were specifically developed to ensure wireless data communications between PCs and a wide range of other consumer electronic appliances, and for this reason do not include voice. The 802.11b Standard, which is sponsored by the *Wireless Ethernet Compatibility Alliance* (WECA), is based on DSSS. A competing protocol, which was devised by the *HomeRF* working group is the *standard wireless access protocol* (SWAP), is based upon a FHSS. Both 802.11b and HomeRF approaches provide up to 10-Mbit/s short-range connectivity, but they differ in power consumption and service features. For instance, digital voice and services are available only to the second (up to 6 voice channels). This illustrates the fact that at the current stage, wireless house networking is also a matter of choice between incompatible standards, which complicates the integration with terminal appliances.

1.4.4 House-Network Control

A key design issue for home networking is the type of *network control*. The control could be either centralized or distributed. In the first approach, a single "always-on" master PC is used. In the second, several intelligent machines can work in parallel with their own application domains (e.g., voice, entertainment, heating, security, kitchen ...). The drawback of a PC-centric house network (whether central or distributed) is its exposure to unavoidable "fatal" errors or system breakdown due to software imperfections, incompatibilities or viruses. The system must be able to reboot itself while holding in memory the network status and various tasks in operation. A radically different approach consists in using a central control hub (Figure 1.22). Such a hub would not only manage the house network but also act as an interface with the external world (telephone and Internet). It would be able to supervise the various house network functions, recognize and accept new appliances joining the network, while also serving as a voice/Internet gateway system. The advantage of this (futuristic) solution is the ability for certain routine functions to be externally managed by the Internet. Representative examples of such functions concern safety (fire, flood), security (intrusion), energy savings (control of light and temperature, regulation of kW consumption according to cost schedules), automatic meter-reading and remote maintenance of household machines by suppliers. Albeit sounding futuristic, other service applications of house networking that are now considered concern a variety of intelligent/automated services such as: managing food supplies (monitoring expiration dates, restocking), activation of cooking devices, caring of pets (during owner's absence) and many possible house-keeping functions. It is clear that the home of the future will look very different from the picture we now have, with fewer wires and appliance boxes and more intelligent functional integration. But it is not clear whether progress in this field will be so rapid and widely accepted by the home clients. This strongly depends upon their perception of advantages for making life more practical, safe and enjoyable, while at an affordable investment and operation cost. In addition, a decisive argument for the success of house networking is simplicity of installation and use, and multistandard appliance compatibility.

a MY VOCABULARY

Can you briefly explain each of these words or acronyms in the broadband wireline access context?

AAL (ATM)	CSMA (radio)	Foreign-NEXT	ISP
Access (network)	CWDM	Fourier transform	ISX
Adaptive equalization	DACS	Frame	ITU-T
	Dark fiber	FS-VDSL	IXC
ADSL	DAVIC	FSAN	LAN
AEX (ADSL)	DCS	FTTB	Last mile
Alien-Xtalk	DDM	FTTC	LCF (APON)
ANSI	Decibel	FTTCa	LEC
APON	Deep fiber	FTTH	LEPA-HDSL
Asymmetric subscriber line	Dense WDM	FTTN	LEX (ADSL)
	DFT	FTTO	Local (subscriber) loop
ASAM	DHSS (radio)	FTTP	
ASP	Dispersion	FTTx	Loss (power)
Attenuation (coefficient)	Distance learning	FTTU	LXC
	DLEC	g.lite	M-ary QAM
ATM	DMT (coding)	g.shdsl	MAN
ATM-DSLAM	DMUX	gauge	Metro(politan) network
ATU-C	Downlink	GFP (GPON)	
ATU-R	Downstream	GigE	MDU
AWG	DSLAM	GPON	MINITEL
Bearer (DSL)	Duplex	H/E (HFC)	MMS
BER	DWDM	HDSL	Modem
Bin (frequency)	DWMT	HDSL2	M-PAM
Bit/s/Hz	E/O (conversion)	HDSL4	M-PSK
BIP (APON)	e-mail	HDTV	MQAM
BLEC	Echo (cancellation)	HFC	MTU
Bluetooth (radio)	EFM (PON)	Hilbert pair (filters)	MUX
BMS	Encryption	HomePNA	MxU
BPON	EPON	HomeRF	NEXT
Broadcast	Equalization (frequency)	HTU-C	NTR
Byte		HTU-R	NVoD
CAP	Excess loss	IAD	Nyquist frequency
CATV	Fast byte	ICI	O/E (conversion)
CDMA	Fast-data buffer	IDSL	OAM
Chipping codes	FDD-DMT	IFFT	OAN
Chips	FDMA	ILEC	OFDM
Cipher	FEC	In-phase (component)	OLEC
CLEC	FEXT		ONU
Coarse WDM	FFT	Insertion loss	PC
Constellation	FHSS (radio)	Interleaved-data buffer	PCS (HFC)
CP	Fiber-rich/lean (topology)		PDH
CO		IP-DSLAM	Plain-text
Core (network)	FITB	IP-PON	PLOAM
CRC (code)	FITL	ISDN	POF
Crosstalk	Foreign-FEXT	ISI	PON

POP
POTS
Power splitter
PSK
PSTN
P2MP
P2P
QAM
QoS
QPSK
Quadrature
  (component)
RADSL
RFI
RS (code)
RXCF (APON)
S-CDMA (HFC)
SAN
SAR

SDH
SDMT
SDSL
Self-FEXT
Self-NEXT
Signal-to-noise
  ratio
Simplex
SNR
SOHO
Spectral efficiency
Splitting loss
SSP
STM
Superframe
SWAP (radio)
TDD-DMT
TDM
TDMA

Telecommuting
Teleconferencing
Teleworking
THP
Tomlinson–
  Harashima
ToS
Transmission loss
Transport (log-haul)
Tree-and-branch
  (topology)
Universal DSL
Uplink
Upstream
VDSL
Video conferencing
VoATM
VoD
VoDSL

VoFR
VoHFC
VoIP
VoMSDN
VoP
VPN
WAN
Wavelength
Wavelength routing
Wavelet
WDM
Web page
WECA (wireless)
WI-FI (radio)
WWDM
XDSL
Xtalk
Zipper (DSL/DMT)
2B1Q (format)

CHAPTER 2

Optical Fiber Communications, Components and Networks

In this chapter, we review the field of optical communications from basic physics, passive and active components to advanced WDM networking.

■ 2.1 OPTICAL COMMUNICATIONS: FROM WIRELESS TO WIRELINE

Before the emergence of the Internet, optical communications with their thin hair-sized light wires looked like a "technology of the future," mixing the concepts of colorful esthetics (probably because of relating the principle of fibers to certain types of lamp appliances) and of ultimate performance (from the telecommunications engineer standpoint). Such an intuitive to educated perception turned out to be fully correct, since optical, or *lightwave* communications, are both based upon a colorful and highly superior exploitation of bandwidth. Indeed, the use of "colors" to define individual optical-communication channels is what we now know as *wavelength-division multiplexing* (WDM), just as traditional radio systems used *frequency-division-multiplexing* (FDM), popularly known in FM and TV broadcasting as radio and TV channels. The light spectrum was sliced into individual WDM colors, although not perceptible as such by the human eye, unlike with the familiar sunlight rainbow. The use of optical bandwidth is "superior," because as this chapter will describe, optical frequencies can be coded at bit rates which are orders of magnitude above those of electric or radio systems.

Nowadays, these powerful features of *lightwave communications* have become mere *commodities* in the end-user Internet world, letting one forget too easily how extremely complex is the related technology, recently named "photonics." From the

---

Wiley Survival Guide in Global Telecommunications: Broadband Access, Optical Components and Networks, and Cryptography, by E. Desurvire
ISBN 0-471-67520-2   © 2004 John Wiley & Sons, Inc.

intuitive understanding of light rays being guided into a fiber, to the idea of particle-like *photons*, which carry light energy and which are at the root of laser sources, optical amplifiers and receivers, there is a huge conceptual gap to bridge. But as we attempt to convey in this chapter, there is no need for any advanced mathematics to grasp most of the elementary, yet fundamental principles that concern light and photonics technologies. Indeed, it is possible to explain with simple and intuitive arguments how light is generated (by laser-diode sources), guided (by glass optical fibers), regenerated (by optical amplifiers) and converted into electric signals (by semiconductor photo-diodes). This chapter provides many details on all these aspects of fiber-optic technologies, which can be understood without a scientific background.

2.1.1 The Early Times of Wireless Optical Communications

But before considering technologies, let us step back and look first back into human history. The discipline of *optical* communications, in which meaningful information symbols are transmitted by light (or via a "light channel"), is in fact as old as humankind. It was a "wireless" technology. Here, we shall overlook *image and pattern symbolism*, whereby fixed messages can be transmitted by mere "sight," without other support such as text or graphics, for immediate or traditional/delayed communications purposes. In the core of modern big cities, we are surrounded by this overall visual environment, which conveys everything from meaningful information (dynamic digital text/picture displays) to the most meaningless advertisement. But using optics as an effective communication channel for important messages is not new. For instance, the practice of waving a white cloth to indicate surrender in battle is one minimal, yet still relevant example in troubled areas. The unfortunate practice of flashing car lights to oncoming traffic in order to signal the presence of police patrols is another illustration. In all these "one-shot" communication events, the information is a single, but meaningful YES/NO bit for eyesight or binocular distances. More sophisticated, but equally rudimentary approaches can convey significantly more information over extended distances. For instance, the early Native Americans used smoke signals from wood fires to communicate coded information over desert/open landscapes, which could be several kilometers wide. Not only was it easier to make smoke visible to anyone in the area than sending a messenger, but it also represented both instantaneous and long-distance communications. In mountainous areas, the messages could be relayed from hilltop to hilltop by replicating the perceived/received patterns through the same coding technique. Interestingly, the codes could be made meaningful only to a limited number of people or initiated groups, thus protecting the actual information against "third-party" understanding. Here we stand on the border of *cryptography* where very meaningful messages can be actually communicated or encapsulated through minimal, *one-time keys* (see Chapter 3). If two parties have secretly agreed on a complex action plan (e.g., option A or option B), a single bit like a flashlight, is sufficient to communicate the information. It is left to reasonable speculation that such smoke-cloud communications could have also been bidirectional, meaning that some acknowledgment/reply mechanisms actually existed. Although the information rate of such smoke

2.1 Optical Communications: From Wireless to Wireline

signals was limited to a few symbols per minute (probably using repeating sequences for redundant "error correction"), the codes could be powerful enough to convey a variety of important messages. The same principle was used in 1800s/early-1900s maritime and military communications before the invention of radio (e.g., *semaphore*), and even after, with Morse-code signaling lamps. The two main advantage of these early optical-communication "channels" was to be instantaneous and to be able to reach all distances of interest, and in some case without third-party interception, just as with modern optical communications.

Another age-old use of optical signaling is the *maritime beacon*. As a source of light visible at night from very large distances (up to 50 km, as limited by the sea-surface curvature, weather and atmospheric conditions), the beacon identifies the location of city harbors (e.g., Alexandria, some 300 years B.C.), coastal features and dangerous reefs to avoid. For ages, beacon light was generated by various sources of fire, until the discovery of electricity and incandescent light in the late 1800s. Light bulbs with multikilowatt power and 1/2-foot-diameter were placed in the center of a rotating cylinder structure. This structure comprised gigantic glass lenses (known as Fresnel lenses) to form a parallel light beam. Several lenses were placed on the cylinder's perimeter according to certain predefined positions and aperture sizes, making it possible to generate a variety of code-word patterns (e.g., light on for short, long, and short times, followed by nothing for a few seconds and then again). Through the code-word, each beacon is given a recognizable identity, which becomes familiar to local sailors for navigation and positioning during night. Note that such coded-light information is completed by the angle formed by the beacon's line-of-sight and magnetic North. Using two beacons in sight with the compass reference, the position at sea can be pretty much estimated, which amounts to saying that two angles are sufficient to determine an absolute position on the local half-plane defined by the sea. Nowadays, radio/radar and GPS positioning techniques have taken over these primitive line of sight techniques. But the latter are still no replacement, especially in developing countries and in traditional fishing. It is also a fact that modern airplane and maritime communications still heavily rely upon beacon-light signaling, at least during night navigation and especially in case of accidental radio-link failure between an aircraft and its control tower. An example of such a light communication channel, which works anytime for emergency airplane operations in North America, is provided in Table 2.1. The table shows manual *"light-gun"* codes used by the control tower to provide safe instructions to either airborne or grounded airplanes when radio links are defective. The craft can acknowledge by blinking its own landing or navigation lights, or if on the ground, by moving the ailerons or rudder. The fact that such primitive visual codes remain valid in our current computer age, exemplify their importance as a "lifeline communications" means.

A last suite of wireless optical communications, which is highly relevant to our Internet times, is known as *free-space optics* (FSO), and is described in detail in Desurvire (2004), Chapter 4, Section 4.4. The principle of FSO is to convey high-bit-rate data (up to 2.5 Gbit and over) through the atmosphere between two points having a clear *line of sight* (LOS). The FSO systems, which use invisible infrared (IR) light as the carrier, are made of two self-aligning telescopes pointing

TABLE 2.1 Manual "light-gun" signal codes used in North American airports for unusual/emergency tower-to-aircraft instructions (indicative 1980s, for illustrative purposes only)

Color/Code and Meaning		Ground Aircraft Recipient	Airborne Aircraft Recipient
green	steady	clear for takeoff	clear to land
	flashing	clear for taxi	return to airfield wait for landing clearance
red	steady	stop	give way to traffic and keep circling
	flashing	taxi clear of landing/runway area in current use	airport not safe, do not land
white (flashing)		return to airport starting point	–
red and green (alternating)		general warning	exercise extreme caution

at each other. The telescopes' function is to both collimate and focus the IR light beam, in simultaneous transmit and receive modes. The first advantages of FSO over radio links and even DSL is *bandwidth* (multi-Gbit/s rates). A second advantage is the possibility of near-immediate installation/operation without need of any cable installation. The argument is particularly sound in environments that are already saturated by radio systems (no bandwidth slots available), or densely populated to such a degree that cable installation costs become prohibitive. FSO communications can come as a solution in environments subject to high electromagnetic interference (EMI). This is particularly true in certain military operations, where EMI is purposefully used for scrambling communications, but FSO is immune to EMI. Considering civilian applications, an extra bonus advantage is that standard FSO systems can be operated *without any license* or *exploitation fees*, unlike for radio, DSL or FTTx!

The other side of the FSO coin is that such systems are severely limited by two factors: light *diffraction*, the effect of beam expansion over distance, and light *attenuation*, the effect of beam absorption/diffusion by particles such as rain drops (rain and fog), snowflakes and dust storms (leaving birds and construction cranes as representing spurious events). Another impediment is caused by air *refraction* (beam bending) and turbulence (random refraction) caused by temperature gradients and atmospheric conditions. Yet it is a fact that FSO represents another powerful and cost-effective means for broadband access in the "last-mile."

2.1.2 The Conceptual Grounds of Wireline Optical Communications

The possibility of trapping light into waveguides was known from the eighteenth century's light fountains. The principle was to create a small glass window in the fountain's reservoir side and on the same axis as the water nozzle. Upon placing a light source (such as a candle) behind the window, the light going through the reservoir is trapped inside the output water jet. This is explained by an effect called *total internal reflection*, which is described in the next section. In spite of light being trapped inside, the water jet glows and projects light in every direction, because of its turbulent material flow and self scattering, in addition to escaping rays. Such a picture of light dissipation is wholly similar to that of optical fiber waveguides, except that the water is replaced by glass. While light is trapped indeed inside the fiber, material irregularities or impurities and minute geometrical imperfections in the core produce light scattering, whereby a fraction escapes to the outside. The comparison stops here, because guiding light in optical fibers is somewhat more complex. It is based upon two fundamental and nonintuitive principles. The conceptual transition from *wireless* optics to *wireline* optics requires one to pay closer attention to these underlying principles.

The first principle is that of *coherence*, a unique property of the light source which we know as *lasers*, and for which light is emitted in a specific set of *radiation modes*. The laser radiation mode is defined both by its frequency (a narrow, single tone in the frequency domain) and spatial geometry (an intense beam with a specific intensity-distribution pattern in the transverse plane). The purest form of a laser radiation mode is that of a beam having a circular intensity pattern, which we familiarly know as a laser pointer. Lasers, which were invented and developed in the 1960s can be realized in different materials, including gas, liquids and solids. The case of *semiconductor lasers* is of special interest, since efficient light emission can be achieved by simply applying a drive current on the material. Moreover, semiconductor lasers have miniature chip sizes (10–20 μm section, 200–250 μm length), which is advantageous for both massive integration and compact-size packaging.

The second principle to address is that of *waveguide modes*, meaning that light signals can propagate in waveguides only according to specific and well-defined "mode" patterns which can be excited by the laser source. In contrast, candlelight or sunlight are highly in-coherent, meaning that they exhibit an infinite number of radiation modes. It is possible to trap the energy of candle or sun light into an optical fiber, but only to the extent corresponding to the mode matching relation between the light source and the waveguide. The related concepts of source coherence and mode wave-guiding are explained by the dual *electromagnetic* and *quantum* nature of light. While wave-guiding only relies upon the electromagnetic properties, coherence relies upon quantum principles, which define how light is actually generated from atomic transitions, which is the subject of Section 2.2.

In Section 2.2 we review the field of *guided-wave optics*, which includes waveguides of various designs and geometries, namely *planar,* as in integrated-optics circuits, and *cylindrical* as in optical fibers. The case of *single-mode fibers* is of special interest, since they represent the unique, point-to-point propagation medium of optical communications. The case of *multimode fibers*, such as those based on polymers is also evoked briefly. The fiber design has evolved in order to enhance system capacity by changing the operating wavelength from 1.3 μm to 1.55 μm, then implementing massive WDM to more fully exploit fiber bandwidth. Special fiber components were developed in order to compensate for *fiber dispersion*, the velocity difference between tones, and *fiber attenuation*, the effect of signal absorption with transmission distance. When based upon material glass, fibers does not behave in the same way at low and high signal intensities. As the signal intensity increases past a certain threshold, new parasitic effects appear which are called *nonlinearities*. While some of these nonlinearities can be advantageously used for some applications (e.g., Raman amplification, soliton propagation), their combination with dispersion produces irreversible signal degradation. As shown in a later section of this chapter, both dispersion and nonlinearity can be *managed*, meaning that some compromises and modulation techniques have been developed to optimize the system capacity while obtaining reasonably low error-rates.

The above description leads to two sections that separately consider *passive* and *active* optical components. Passive and active components represent the technology arsenal performing a wide variety of essential functions that have made optical communications possible. Examples of passive components are *1 × N, N × N couplers and wavelength multi-/demultiplexers* used to split or combine signals from/into fiber trunks. Other examples are *dispersion compensators, optical isolators*, and *optical filters/equalizers*. Optical *fiber amplifiers*, used to regenerate light signals, also belong to this category. They are passive components in the sense that the temporal shape of the amplified signal is a faithful replica of the input, as in a linear transmission device. Examples of active components are *semiconductor laser sources* and *amplifiers, electro-optic modulators, N × N optical switches* and *photo-receivers*. The field of photonics thus represents a vast array of active and passive technologies to process lightwave signals. With such a background, we can address the *principles of lightwave transmission* (based upon some fundamentals described in Desurvire, 2004, Chapter 1), and *WDM networking*, which led to current multigigabit/s and multiterabit/s communications.

As a concluding remark, it is important to note that the field of optical communications is only 30 years old, representing a relatively recent event in our engineering history. Indeed, the first low-loss fibers and room-temperature semiconductor laser diodes were demonstrated only in the early 1970s. Yet, it took another 20 years or so to come to the point of massive commercial exploitation, in particular with the invention of practical optical amplifiers and related WDM technologies. What we know as the *global Internet* with its virtual capacity of instant communication between any two points on Earth, is actually based upon a global optical network infrastructure, as deployed across continents and oceans. Such a revolution is not even 10 years old, but the technologies have seemingly been reduced to mere commodities, as if *unlimited bandwidth access* always has been there and is a straightforward concept. As this chapter shows, such a view is both superficial and wholly incorrect! Optical bandwidth is huge, but it is finite, and so are its various access resources, especially in view of technology implementation, complexity and cost.

■ 2.2 BASIC PHYSICS OF LIGHT-WAVE SIGNALS

Here, we shall review various fundamental aspects of light waves, from their electromagnetic to quantum nature, their velocity in materials, their ray characteristics and their generation/amplification by laser sources and optical amplifiers. This is what we call the "basic physics" of lightwave signals, which is meant to provide a rapid understanding and familiarity with the fundamental concepts underlying photonics technologies. This task being fulfilled without, or with only very elementary mathematics! Note that this section makes use of various terms defining different *orders of magnitude* and *wave parameters*. The reader who is not familiar with these definitions might quickly refer first to Desurvire (2004), Chapter 1, Section 1.1, and in particular to Table 1.1, where these notions are explained in detail.

2.2.1 Light's Electromagnetic Nature

Light is made of a superposition of electromagnetic (EM) waves with different frequencies and amplitudes, just as the surface of the sea looks to a sailor or viewed from the sky. The distance between two crests (or two troughs) is known as the *wavelength* (λ), measured in *meters*. The number of crests traversing a fixed point of space in one second is known as *frequency* (f) as measured in cycles-per-second (s^{-1}) or *Hertz*. Light is simply defined as the domain of waves corresponding to a certain portion of the EM spectrum which spreads from *nanometer waves* ($\lambda = 10$ nm, $f = 3 \times 10^{16}$ Hz $= 30$ PHz $= 30,000$ THz), which is the domain of the *ultraviolet* (UV), to *millimeter waves* ($\lambda = 1$ mm, $f = 3 \times 10^{11}$ Hz $= 0.3$ THz $= 300$ GHz), which is the domain of the *far infrared* (far-IR).

The full EM spectrum is shown in Figure 4.5 of Desurvire (2004). As revealed from this figure, what we familiarly know as *visible light* (the sun rainbow) represents a relatively narrow slice of the EM spectrum, since it concerns waves ranging from 400-nm to 750-nm wavelengths. What we know as *light colors* represent discrete spectral regions which the eye's retina perceives differently. As the figure shows, 500 nm looks blue while 600 nm looks yellow-orange. It is in fact amazing to realize how much we don't see of the light spectrum, considering that it spreads over five orders of magnitudes (300 GHz–30 PHz), or over 100,000 times wider than the visible domain. But the same consideration applies to the radio spectrum which expands from far-IR ($\lambda = 1$ mm, $f = 300$ GHz) to extra-low frequencies (ELF) ($\lambda = 1,000$ km, $f = 300$ Hz).

EM waves (light or radio) are made of two time-varying fields, namely an *electric field* (E) and a *magnetic field* (H) which oscillate in phase while being perpendicular to each other (Figure 4.4 of Desurvire, 2004). The amplitude of this magnetic field, is proportional to the E-field, according to $H = \sqrt{\varepsilon}E$, where ε is a constant equal (or very nearly equal) to unity in vacuum (or air). The meaning of this constant is discussed in the following text. An intuitive description of E-fields and how they can be generated by radio antennas is also provided in said chapter. It remains an intriguing feature that E-waves are always accompanied by perpendicular H-wave companions, but we shall leave it here as a mystery that is confined to *Maxwell's* laws of electromagnetism. What is important to remember is that the E-field defines a unique orientation in space, which is referred to as its "*polarization*" direction, or polarization, for short. It is possible to combine two E-fields having different polarizations, for instance being orthogonal to each other. The resulting E-field polarization depends upon the phase difference between the two E-field components, as we shall explain further below. If the phase difference is random, the resulting light is said to be *unpolarized*, meaning that the orientation of the E-field at any given time is random. An example of unpolarized light is sunlight. Photographers use polarizing filters to block one of the polarizations and thus suppress unwanted reflections. The effect is not to block 50% of the light intensity (as may be inferred), but much more, based upon the fact that light reflection is not the same for the two polarizations. See more about polarization properties in Section 2.2.3.

2.2.2 The Speed of Light

A key characteristic of EM waves is their *velocity* (c), or speed of light. It is a universal constant common to all EM waves, regardless of frequency. According to Einstein's special relativity, no signal information can travel from one point to another at a speed exceeding that of light. The fact that light does not travel at infinite speed was first discovered in 1675 by the astronomer O. C. Römer, based upon the observation of anomalies in the eclipse cycles of Jupiter's satellites. He came up with $c \approx 3.5 \times 10^8$ m/s, a result which compares well with today's best estimate of $c = 2.99778 \pm 0.0001 \times 10^8$ m/s. For practical purposes, the convention is to use the approximation $c = 3 \times 10^8$ m/s. This extremely high velocity is hard to conceptualize, unless it is situated in the context of the universe. For instance, the Moon being located at 300,000-km distance from Earth is reached by a light signal in one second. The conversation between astronauts on the Moon and the Earth base station would take two seconds for each round-trip. The same exchange from Mars would take 522 seconds, or 8 mn 42 s, which is also close to the time for light to traverse the distance between the Sun and the Earth. Astronomy uses the unit of light-years (LY), which is the distance traversed by light over a full year. A LY is thus $(3 \times 10^8$ m/s$) \times 365.25$ days $\times 24$ h $\times 3,600$ s $= 9.46 \times 10^{15}$ m, or 9.46×10^{12} km, or 10 PM (Peta-meter), for short. The closest star, *Proxima Centauri*, is located at 4.3 LY, meaning that any round-trip communications would take 8.6 years to complete. This provides a sense of both the finiteness of the speed of light and the dimensions of our "immediately surrounding" universe. For addicts of *Arthur C. Clarke's* science-fiction novels, see Exercise 2.1 at the end of the chapter concerning the case of Jupiter.

Back to human dimensions, if the Earth were flat, or if light rays could follow its curvature, a signal could do the round-trip between two extreme or antipodal points (20,000 km separation) in 133 ms, close to 1/10 s. Curiously enough, this absolute delay is on the edge of human perception. The reader may have placed an international phone call which was relayed through a *geostationary satellite* (see Desurvire, 2004, Chapter 4, Section 4.3). He or she might have experienced an annoying delay in the conversation exchanges, as explained by the 36,000-km satellite altitude, which corresponds to a really perceptible time lag of 0.2 s or 200 ms or 1/5 s. As shown in the Exercises, this delay only applies to the Earth area that is "visible" to the satellite relay. Longer delays would be involved for antipodal stations, by use of three such satellites. Considering *submarine optical-cable networks*, which now cover the entire planet, and overlooking buffering/routing delays, one yet remains close to the 200 ms upper limit. This is because optical fibers introduce a 1.5-fold retardation factor, as we explain hereafter. See more on delay issues involved in satellite *constellation* systems in Desurvire (2004), Chapter 4.

Light propagates more slowly in physical matter. Materials that are "transparent" to light are called *dielectric*. This term refers to the property of matter that responds to EM light excitation with the effect of *polarization*. Medium polarization (P) is akin to a secondary or induced electrical field, which is proportional to the incident excitation field E, according to the direct relation $P = \varepsilon E$. One can represent the situation as the atomic/electric charges in the material being forced to move back and forth according to the incoming E-field excitation. This polarization effect literally slows down the light-wave propagation, yielding an effective or net speed of light of $c' \leq c$. One defines the dimensionless quantity $n = c/c'$, which is always greater than or nearly equal to unity, as the medium's *refractive index*. The refractive index is related to the constant ε according to $n = \sqrt{\varepsilon}\eta_0$, where $\eta_0 \approx 377\,\Omega$ is the *vacuum impedance*, as expressed in ohms (Ω).

Here, we don't need to worry too much about anything pertaining to the companion magnetic field. Suffice it to say that optics (and optical communications) can be satisfactorily reduced to a simple description of the E-wave characteristics and behavior (we shall however briefly come back to the magnetic component when considering waveguides). At this stage of the description, the most important result is that "light" propagates in materials at a reduced speed defined by $c' = c/n$. At ground level, the air refractive index is very close to pure vacuum, with $n = \sqrt{\varepsilon}\,\eta_0 = 1.0003$. This means that the speed of light in air is $1/1.0003 = 0.99970$ times slower, a minute correction that can be overlooked unless high accuracy is required for metrology calibration purposes. Most transparent/clear-glass materials, based on an amorphous or noncrystalline form of silicon oxide (SiO_2) have refractive indices of $n = 1.45$ at $\lambda = 1\,\mu m$ wavelength. The refractive index is weakly dependent upon wavelength: between $0.6\,\mu m$ to $1.6\,\mu m$, it decreases from $n = 1.46$ to 1.44. In semiconductor materials (see below), and at the wavelengths of interest, the indices fall into the 3.0–4.2 range (e.g., $n = 3.6$ for gallium arsenide or GaAs and $n = 3.5$ for indium phosphide or InP).

It is important to realize that in most materials, the refractive index is not defined by a unique number. The number may vary according to the incident polarization, or equivalently, to privileged orientation directions of the material, as is the case for crystals. The fact that light velocity depends upon direction is referred to as material *anisotropy*. Anisotropic materials are characterized by three different refractive indices, thus defining an index *ellipsoid* (a sphere with three different "radii" n_1, n_2, n_3). Considering light propagation in the direction perpendicular to a given material's plane, the description reduces to two index parameters, n_o, n_e, called *ordinary* (n_o) for the "rapid" direction of low index, and *extraordinary* (n_e) for the direction of high index, respectively. Such materials having two refractive indices are therefore said to be *birefringent*. This is opposed to *isotropic* materials, such as amorphous glass, which are characterized by a unique refractive index. Optical glass fibers are naturally isotropic, meaning that light propagates at the same speed regardless of the incident E-field signal polarization. If the fiber were perfect, this means that the signal polarization would remain the same from end to end, like traversing vacuum. As we shall see in Section 2.2.4, however, optical fibers do exhibit a small birefringence. This unwanted birefringence or anisotropy is due to residual stress in the material itself, or to the fiber being locally bent or twisted, and also packaged into various constraining cable layers. This material- or stress-induced birefringence is characterized by minute relative values of the order of $\delta = (n_e - n_o)/n_o \approx 10^{-5}\text{--}10^{-6}$. As a result, the E-field polarization is rapidly scrambled, since the speeds of its two components are different and vary randomly. For this reason, ordinary optical fibers *do not to maintain polarization*, meaning that the output signal polarization is generally undetermined and randomly fluctuates over time.

On the other hand, fibers can also be fabricated with built-in stress or geometrical anisotropy in order to be purposefully birefringent, with $\delta \approx 10^{-2}\text{--}10^{-3}$, as in the case of *polarization-maintaining fibers* (PMF). There is more on the polarization properties of light in Section 2.2, while polarization in optical fibers is reviewed in Section 2.3.

Optical fibers are made of different refractive-index layers, the index being higher near the center region (fiber core) and lower in the outer region (fiber cladding). As described in the next section, in such a structure light does not propagate like "rays" but like "modes." The mode also has a unique propagation speed, which we can call c_{eff}. One then refers to the *mode's effective refractive index* $n_{\text{eff}} = c/c_{\text{eff}}$. The effective refractive index always takes a value between the highest core index and the lowest cladding index. For light to stay confined in the fiber core and to be effectively guided, n_{eff} should have a value sufficiently higher than the cladding index. We see that the concept of speed of light (or refractive index) is central to fiber optics. We need now to understand why light can be effectively confined into a waveguide, "trapped" into a propagation mode.

2.2.3 Light Reflection, Refraction and Polarization

The basic properties of light rays called *reflection* and *refraction*, were known to the early Greeks (Euclid, Ptolemy). Refraction was formalized in the mid-1600s by Descartes and Snell. Consider a ray incident at an angle θ_1 on a surface separating two transparent media of refractive indices n_1, n_2, as shown in Figure 2.1. The law of

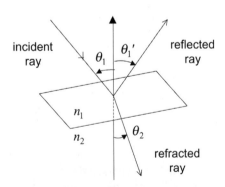

FIGURE 2.1 Reflection and refraction of light rays at an interface separating two different media of refractive index n_1 (top) and n_2 (bottom). The incident, reflected and refracted angles are $\theta_1, \theta_1' = \theta_1$ and θ_2, respectively. The value of θ_2 is given by Descartes' or Snell's law.

reflection states that the reflected ray is in the same plane as that of the incident ray, with a reflection angle θ_1' equal to the incident angle θ_1 (see Figure 2.1). Refraction concerns the ray traversing the separation surface and propagating through the second medium: the change of refractive index cause its direction angle θ_1 to change into θ_2. The law of refraction, called either *Descartes's* or *Snell's* Law (according to cultural preference) is then written:

$$n_1 \sin \theta_1 = n_2 \sin \theta_2 \tag{2.1}$$

Two important conclusions can be drawn from the above formula. First, if the ray has a normal incidence ($\theta_1 = 0$), then the refracted angle is also zero, meaning that the ray traversing the surface is continuous or unbroken. Second, if the two media have identical refractive index ($n_1 = n_2$), then the incident and refracted angles are equal, meaning that the ray is also unbroken. Third, when the two media have different refractive indices (i.e., $n_2 < n_1$), the refracted angle is $\theta_2 = \sin^{-1}[(n_1/n_2) \sin \theta_1]$, which is greater than the incident angle θ_1 since $n_1/n_2 > 1$. As illustrated in Figure 2.2 the maximum possible value for θ_2 is 90°, i.e., the refracted ray is parallel to the medium separation surface. According to equation (2.1), the *critical* incident angle θ_1^{crit} for which $\theta_2^{crit} = 90°$ is simply $\sin \theta_1^{crit} = (n_2/n_1) \sin 90° = n_2/n_1$, or

$$\theta_1^{crit} = \sin^{-1}(n_2/n_1) \tag{2.2}$$

It is seen from Figure 2.2 that all incident rays such that $\theta_1 \geq \theta_1^{crit}$ are completely reflected into the medium of higher index, meaning that no ray escapes into the second medium of lower index. Such a condition is called *total internal reflection*. Scuba-divers (or pool swimmers with goggles) know this effect well as the surface beneath the sea (or swimming pool) appears mirror-like when looked into at a certain angle away from vertical (the index of water being higher than that of air). See Exercises at the end of the chapter. As we shall describe in the Section 2.3, the effect of total internal reflection of light rays is at the root of the concepts of *waveguiding* and waveguide *modes*. As discussed, ray optics only provide an intuitive explanation of these concepts when considering waveguide structures of dimensions

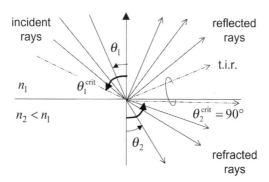

FIGURE 2.2 Effect of increasing the incident angle θ_1 when the second medium refractive index is lower ($n_2 < n_1$). A critical angle θ_1^{crit} exists for which the refracted angle θ_2 is 90°, corresponding to the condition of total internal reflection (t.i.r.).

comparable to light wavelength. In this case, the properties of light as an *EM wave* (as opposed to a ray) must be considered.

It is useful to mention here that the refractive index is generally wavelength-dependent, a property which is called *index dispersion*. Thus, incident light rays containing different wavelengths refract with different angles. Using a glass prism, dispersion causes a white-light ray to break out into a rainbow, as discovered by Isaac Newton in 1668 (a less-known story is that Newton also observed that the colors would not break again after passing into a second prism, leading to the first intuition of a fundamental "granularity" in light energy). Index dispersion naturally affects the properties of guided modes, which makes it a central concept in fiber optics and WDM transmission (see description in Section 2.3).

We consider next light *polarization*, a concept which was introduced in a previous subsection. Basically polarization is defined as the orientation of the E-field, or more accurately as the orientation of the plane within which it oscillates. If such an orientation is constant, one refers to the polarization as being *linear*. If a medium is isotropic, or having a unique refractive index, the polarization remains unchanged. At least this is true in the absence of external magnetic fields (see the following text on the *Faraday effect*). What happens to the polarization if the medium is anisotropic? In this case, the speed of light is different according to the material directions defined by the two indices which, as we have seen earlier, are called ordinary (n_o) and extraordinary (n_e) for the fast and slow directions, respectively. The two index directions are then referred to as the material's *birefringence axes*. When a linearly polarized E-field is aligned with (or parallel to) either axis, the polarization remains unchanged, as in the purely isotropic case.

The situation becomes more complex if the incident (linear) polarization forms some finite angle α with the birefringence axis, as illustrated in Figure 2.3. In this case, one can decompose the E-field into two components E_o, E_e which are parallel to the two birefringence axes. According to a previous statement, the polarizations of these two linearly polarized components remain unchanged. But the corresponding E-waves do not propagate at the same speed: for the E_o-field, the speed is c/n_o (fast birefringence axis), while for the E_e-field, the speed is $c/n_e < c/n_o$ (slow birefringence axis). Because the two speeds are different, the delay induced by

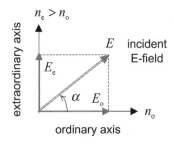

FIGURE 2.3 Decomposition of an incident E-field with polarization angle α into its components along the ordinary-index (E_o) and extraordinary-index (E_e) directions. The propagation direction of the E-field is normal to the page's plane.

the medium into each wave is also different. After a propagation distance x in a medium of index n, such a wave delay is expressed by the *phase* (φ), as defined by

$$\varphi = n \frac{2\pi x}{\lambda} \tag{2.3}$$

where λ is the E-field wavelength. Accordingly, the two components phases are

$$\begin{cases} \varphi_o = n_o \dfrac{2\pi x}{\lambda} \\ \varphi_e = n_e \dfrac{2\pi x}{\lambda} \end{cases} \tag{2.4}$$

The two phases are seen to increase linearly with distance, with a higher rate for the extraordinary component. The *relative phase* $\Delta\varphi$ is defined by the difference

$$\Delta\varphi = \varphi_e - \varphi_o = (n_e - n_o)\frac{2\pi x}{\lambda} \equiv \Delta n \frac{2\pi x}{\lambda} \tag{2.5}$$

where $\Delta n = n_e - n_o$ is, by definition, the medium birefringence. This expression makes it possible to analyze the polarization evolution with distance. The first observation is that the two E-field components are *in phase* ($\Delta\varphi = 2k\pi$, k = integer) at all distances $x = k\lambda/\Delta n$. These distances are integer multiples of a quantity $L_B = \lambda/\Delta n$, which is called the *beat length*. It is clear then that the E-field polarization comes back to its original input state at every beat-length multiple. Let's look now at what happens to the polarization within any beat-length distance. If the time-varying magnitude of the incident E-field shown in Figure 2.3 is defined by $E \cos(\omega t)$, the two E-field components are

$$\begin{cases} X = E_o \cos(\omega t + \varphi_o) \\ Y = E_e \cos(\omega t + \varphi_e) \end{cases} \tag{2.6}$$

where $E_o = E \cos\alpha$ and $E_e = E \sin\alpha$ are initial constants and X, Y are the time-/space-varying components of the E-field. Since the absolute phase is arbitrary and only the phase difference $\Delta\varphi = \varphi_e - \varphi_o$ matters, we can also write the components in the form

$$\begin{cases} X' = X/E_o = \cos\left(\omega t - \dfrac{\Delta\varphi}{2}\right) \\ Y' = Y/E_e = \cos\left(\omega t + \dfrac{\Delta\varphi}{2}\right) \end{cases} \tag{2.7}$$

where the amplitudes have been normalized to the initial E-field components, assuming $\alpha \neq 0$. It is possible to obtain a time-independent relation between the two components (see Exercises at the end of the chapter). The result is

$$(X')^2 + (Y')^2 - 2X'Y'\cos(\Delta\varphi) = \sin^2(\Delta\varphi) \tag{2.8}$$

The above equation is that of an *ellipse* which is inscribed into a square of side-length $l = 2$, as illustrated in Figure 2.4. In the nonnormalized system (X, Y), the ellipse is inscribed into a rectangle of sides $2E_o$ and $2E_e$, respectively. This curve, which is called the *polarization ellipse*, defines the maximum amplitude of the E-field at any point x of the medium corresponding to the relative phase $\Delta\varphi$.

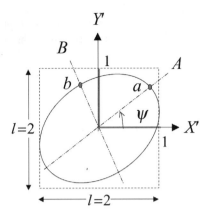

FIGURE 2.4 Polarization ellipse defined by the E-field amplitude upon propagation through a birefringent medium, with initial conditions of ordinary and extraordinary amplitude components $X'(0) = Y'(0) = 1$. The ellipse is defined by its major and minor axes lengths a and b and major-axis orientation angle ψ.

The definition of the polarization ellipse becomes much simpler in two elementary cases:

1. the phase delay is $\Delta\varphi = \pi/2$: in this case, we have $\cos\Delta\varphi = 0$ and $\sin\Delta\varphi = 1$, yielding

$$(X')^2 + (Y')^2 = 1$$

which is the familiar equation of a *circle* of unity radius; in the nonnormalized coordinate system, this equation becomes

$$(X/E_o)^2 + (Y/E_e)^2 = 1$$

which is the equation of an *ellipse* with axes of lengths $2E_o$ and $2E_e$, respectively; the same results are obtained for $\Delta\varphi = (2k+1)\pi/2$ where k is an integer; the occurrence $\Delta\varphi = (2k+1)\pi/2$ corresponds to the locations $x = (2k+1)L_B/4$, representing odd-integer multiples of one-quarter of the beat length L_B;

2. the phase delay is $\Delta\varphi = 0$ or π: in this case, we have $\cos\Delta\varphi = 1$ and $\sin\Delta\varphi = 0$, yielding

$$(X')^2 + (Y')^2 \pm 2X'Y' = 0 \leftrightarrow (X' \pm Y')^2 = 0 \leftrightarrow Y' = \pm X'$$

with (in the last equation) the "+" and "−" cases corresponding to $\Delta\varphi = 0$ and $\Delta\varphi = \pi$, respectively; the curve thus defined is a *straight line* with slope ± 1 in the normalized-coordinate system, or $\pm E_e/E_o = \pm\tan\alpha$ in the nonnormalized system; the case $\Delta\varphi = 0$ (or $2k\pi$) corresponds to the origin $x = 0$ or an integer number of beat lengths, $x = kL_B$; the case $\Delta\varphi = \pi$ (or $(2k+1)\pi$), corresponds to an odd-integer number of half beat lengths, $x = (2k+1)L_B/2$.

By definition, an E-field whose components describe a circle or an ellipse is said to be *circularly* or *elliptically polarized*. As we have seen, an E-field whose components describe a straight line is said to be linearly polarized. One also refers to *circular, elliptical* and *linear* polarizations as different possible *states of polarization* (SOP) for the E-field.

This demonstration has shown that the SOP evolves from *linear* ($Y' = X'$) at the origin to *circular* at one-quarter of the beat length, then back to *linear* with opposite slope ($Y' = -X'$) at half the beat length, then back to *circular* at three-quarters of the beat length, then finally back to the initial *linear* state at one full beat length. Such an SOP evolution corresponds to a transformation of the polarization ellipse according to coordinate x and associated phase delay $\Delta\varphi$, as illustrated in Figure 2.5 in the (X', Y') normalized-coordinate system. Figure 2.5 also shows intermediate points at multiples of $L_B/8$ where the SOP is elliptical. Note that in the (X, Y) system the SOP can be circular only with launching conditions where $E_o = E_e$ ($\alpha = 45°$) or where one of the components is identically zero ($\alpha = 0$ or $90°$). Any other conditions result in SOP that are either linear or elliptical, since the space in which the ellipse is inscribed is a rectangle. Thus a circular SOP is a particular case of a more general elliptical SOP. On the other hand, a linear SOP represents the limiting case of an elliptical SOP where one of the ellipse axes is of zero length.

Circular or elliptical polarizations are naturally associated with a *rotation* of the E-field's orientation, an effect which is illustrated by the arrows in Figure 2.5. The rotation direction can be either clockwise (*right-handed* rotation)

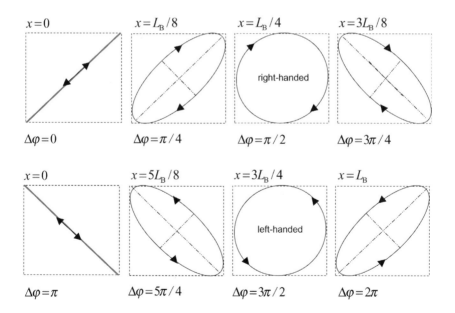

FIGURE 2.5 Evolution of polarization ellipse according to medium coordinate x and relative phase delay $\Delta\varphi$ over a full beat length L_B, showing transformation from linear to elliptic to circular and the reverse. The arrows indicate the rotation direction of the corresponding E-fields.

or counter-clockwise (*left-handed* rotation). Note that the appellations "right-handed" and "left-handed" are purely conventional (in fact, the right hand with the thumb being up rather defines counter-clockwise rotation, which is the convention used in physics). It is left to an exercise at the end of the chapter to show in which direction the E-field polarization rotates, given any phase delay $\Delta\varphi \neq k\pi$.

To complete the description of elliptical SOP, one needs to define the two ellipse parameters which are its orientation angle, ψ, and its axes ratio ($b/a \equiv \tan\chi$). To alleviate the burden of a lengthy demonstration, we shall just provide here the result, as in the two definitions:

$$\tan(2\psi) = \tan(2\alpha)\cos\Delta\varphi \equiv \tan(2\alpha)\cos\left(2\pi\frac{x}{L_B}\right) \quad (2.9)$$

$$\sin(2\chi) = \sin(2\alpha)\sin\Delta\varphi \equiv \sin(2\alpha)\cos\left(2\pi\frac{x}{L_B}\right) \quad (2.10)$$

These two definitions show that the ellipse's orientation and axes ratio evolve with coordinate x. At the origin ($x = 0$), we have $\Delta\varphi = 0$ thus $\psi = \alpha$. The ellipse is thus initially oriented with the same angle as the incident E-field. We also have $\chi = 0$ and $\tan\chi = b/a = 0$, or $b = 0$, meaning that the initial ellipse is reduced to a straight line of length $2a$ with orientation angle α, corresponding to a linear SOP. The two above expressions show that the SOP comes back to its initial position and geometrical shape at every integer beat-length multiple ($x = kL_B$), for which $\Delta\varphi = 0$. The case of a linear input SOP parallel to any of the axes ($\alpha = 0$ or $\alpha = \pi/2$) leads to $\psi = \chi = 0$, which means that the SOP remains linear and constant, consistent with the earlier statement. The case of a linear input SOP at 45° of the axes ($\alpha = \pi/4$) is problematic since $\tan(2\alpha) \to \infty$, corresponding to a singularity in definition in equation (2.10). However, the previous polarization-ellipse description, which led to Figure 2.5, is valid, with the (X, Y) ellipse being confined into a square as opposed to a rectangle. We see then that the ellipse orientation is either $+45°$ or $-45°$ while it remains undefined or is indifferent in the purely circular SOP ($\Delta\varphi = \pi/2$ or $3\pi/2$). The following description shows how the two parameters (ψ, χ) can in fact be used to fully characterize light SOP.

In the most general case, the incident E-field can have any SOP. It is then not easy to visualize the evolution of the polarization ellipse, unless some other representation means be introduced. The solution to this problem was devised in the mid-1800s by the physicist G.G. *Stokes* and in the 1890s by the mathematician H. *Poincaré*. These two names remain forever attached to the study of polarized light, as we shall describe next.

Stokes first showed that any SOP could be described though *four* parameters, called S_0, S_1, S_2 and S_3. These *Stokes parameters* are defined as follows:

$$\begin{cases} S_0 = E_o^2 + E_e^2 \\ S_1 = E_o^2 - E_e^2 \\ S_2 = 2E_oE_1 \cos\Delta\varphi \\ S_3 = 2E_oE_1 \sin\Delta\varphi \end{cases} \quad (2.11)$$

with the relation

$$S_0^2 = S_1^2 + S_2^2 + S_3^2 \qquad (2.12)$$

as left to the reader to verify. It is seen that S_0 represents the E-field *power* (the sum of the squared components being equal to the square of the E-field, see Vol. 1, Chapter 1). Since S_0 is also defined through the three other Stokes parameters, there are only three degrees of freedom. Thus one may conceive of a 3-dimensional space bounded by the condition of equation (2.12) as represented by a spherical surface with $R = S_0$ as the radius. Indeed, if one identifies the Stokes parameters S_1, S_2, S_3 as three independent coordinates $\hat{x}, \hat{y}, \hat{z}$, then equation (2.12) is that of a sphere ($\hat{x}^2 + \hat{y}^2 + \hat{z}^2 = R^2$), the *Poincaré sphere*. It is an easy exercise (see the end of the chapter) to show that the three Stokes coordinates are uniquely defined by the two ellipse angles (ψ, χ) and the power (S_0) according to:

$$\begin{cases} S_1 = S_0 \cos 2\chi \cos 2\psi \\ S_2 = S_0 \cos 2\chi \sin 2\psi \\ S_3 = S_0 \sin 2\chi \end{cases} \qquad (2.13)$$

We see from these definitions that the two ellipse angles (ψ, χ) correspond to a spherical coordinate system in which each SOP (S_1, S_2, S_3) has a unique representation point on the Poincaré sphere surface. It is now a game to locate the different SOP possibilities on the Poincaré sphere, which is illustrated in Figures 2.6 and 2.7. The first figure shows the correspondence between the spherical coordinates S_1, S_2, S_3 and the ellipse angles (ψ, χ). The second figure shows the SOP ellipses corresponding to different points of interest. Several observations can be made from these two complementary figures:

- The *Equator circle* is defined by the points for which $\chi = 0$ (or $\tan \chi = b/a = 0$), corresponding to purely linear SOP; in particular, the points

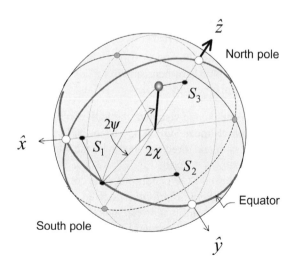

FIGURE 2.6 Poincaré-sphere representation of the state of polarization (SOP), showing its relation with the Stokes parameters.

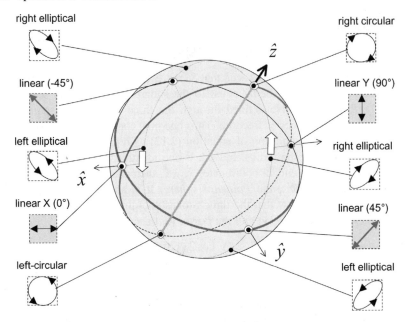

FIGURE 2.7 Correspondence between specific cases of linear, circular (right/left-handed) and elliptic SOP with locations onto the Poincaré sphere. Linear SOP are located exclusively on the Equator (shaded icons), right-handed SOP on the Northern hemisphere and left-handed SOP on the Southern hemisphere. The white arrows indicate the SOP path/direction corresponding to the polarization-ellipse evolution shown in Figure 2.5.

- $S_1 = S_0$, $S_2 = S_0$, $S_1 = -S_0$ and $S_2 = -S_0$ correspond to linear SOP at angles $0°$, $45°$, $90°$ and $-45°$ with respect to the birefringence axis, respectively;
- The *North* and *South* poles ($S_3 = \pm S_0$, $2\chi = \pm \pi/2$) correspond to the right-circular and left-circular SOP, respectively;
- Any SOP outside the Equator ($\chi \neq 0$) are elliptical, the right-handed ones being located on the Northern hemisphere ($\chi > 0$) and the left-handed ones on the Southern hemisphere ($\chi < 0$).

The Poincaré sphere thus helps one visualize the SOP evolution from a given initial state. For instance, an initial linear SOP at 45° from the birefringence axis evolves according to the circular path and direction shown by the white arrows in Figure 2.7; this is strictly the same path as described by Figure 2.5. We can now work out the SOP evolution given any initial SOP conditions. The corresponding path is defined by the *meridian* onto which this initial SOP is located.

This description of SOP evolution corresponds to an ideal birefringent medium with uniform characteristics. Such a condition is closely approached in highly birefringent optical fibers, also called *polarization-maintaining fibers* (PMF), as described in the next section. In the previous subsection, we mentioned that ordinary optical fibers do not maintain polarization, meaning that the SOP is rapidly scrambled with propagation. This can be explained by the fact that the birefringence axes have random orientations at any point in the fiber path. We can now visualize this effect as corresponding to a *random walk* of the SOP onto the Poincaré sphere. As discussed in the next section, this causes limitations in the rate at which individual light pulses can propagate in fibers, namely the fiber bandwidth. Intuitively, a short light pulse must eventually split up into its two fast/slow polarization components, an effect which is called *polarization-mode dispersion* (PMD). Thus PMD is associated with random signal distortion and bit-error-rate limitations.

To conclude this description of polarization property, it is interesting to mention the effect of *polarization rotation* due to two physical phenomena called *optical activity* and *Faraday effect*. Polarization rotation due to *optical activity*, also called *rotatory power,* naturally occurs in most optical materials. Formally, it is explained by a velocity difference between the right- and left-handed circular polarizations (the explanation of which being beyond the scope of this description). We have seen earlier that in birefringent media, a circularly polarized E-field is the superposition of two orthogonal, linearly polarized E-fields which propagate at different velocities. Conversely, one may view a linearly polarized E-field as the superposition of two opposite, circularly polarized E-fields. If the two circular-polarization components have the same velocities, the E-field direction remains constant, while it must rotate in the opposite case, since one component is faster than the other. If one calls n_+ and n_- the refractive indices associated with the right and left circular polarizations (respectively), it can be shown that after a propagation distance x, the polarization plane rotates by an angle ρ, as defined by the amount

$$\rho = \pi \frac{x}{\lambda}(n_- - n_+) \tag{2.14}$$

where λ is the wavelength. For instance, if the right-circular polarization component is faster than the left-circular ($n_+ < n_-$), the rotation angle is positive, meaning that the E-field rotates in the same direction as this component (clockwise). The material thus acts as a natural polarization rotator. The same rotation effect occurs if light travels in the opposite direction, which means that the net rotation angle is zero after a round-trip.

Polarization rotation due to the *Faraday effect* is caused by applying a *static magnetic field* (B) whose direction is aligned to that of the light ray. In this case, the rotation angle is defined as $\rho = VBx$, where V is a material characteristic called the *Verdet constant*. The intriguing feature of the Faraday effect is that the rotation angle is *independent* of the direction of light propagation, contrary to optical activity where the angle reverses with direction. Thus, a round-trip inside a Faraday rotator of length L corresponds to a double rotation angle of $\rho = 2VBL$. If the Faraday rotator is designed to produce a one-way rotation of exactly $45°$, the round-trip will cause a $90°$ rotation, or a linear polarization that is orthogonal to the incident one. As described in Section 2.4 concerning "passive" components, such an effect can be exploited to realize *optical/Faraday isolators* and *optical circulators*. The function of *isolators* is to pass incoming signals while blocking those coming from the opposite direction, such as signal reflections or light noise. They are essential components in the realization of laser-source and optical-amplifier modules. *Circulators* are three-port devices which have separate paths between each of their endpoints. They are essential components in a variety of telecom applications, for instance to add or to drop signal wavelengths from a WDM aggregate (OADM). See further on isolators and circulators in Section 2.4.

2.2.4 Classical and Quantum Natures of Light

The description of light as an EM wave, which is fully governed by *Maxwell's laws* of electromagnetism is sufficient to explain all possible properties and phenomena observed in ray optics and waveguide structures. But as soon as we consider light *interaction with matter*, another perspective opens where the EM wave picture becomes incomplete if not wholly inadequate. We can immediately derive a sense of this inadequacy by considering the effect of EM-wave radiation by antennas. In full accordance with Maxwell laws, EM waves are radiated by the motion of electrons. As we know, matter is made of atoms having a certain number of electrons (of negative charge) clustered around a nucleus (of positive charge). Since charges of opposite sign attract each other, the electrons must be in some kind of revolving motion in order not to collapse into the nucleus, like planets with the Sun. Thus electrons should constantly radiate EM waves. But this radiation corresponds to a dissipation of energy per unit time proportional to the square of the E-field, E^2. Then it is clear that after some time, there should be no energy left for the electron to resist the nucleus attraction, which would result in the eventual collapse of the atom. Since atomic matter is observed to be stable (overlooking the unrelated effect of atomic decay by radioactivity), something must be wrong with such a description!

The answer to the above dilemma was provided by Albert Einstein in the early 1900s, following the discovery of the *photo-electric effect* in the late 1880s. In this effect, a current can be generated through a vacuum tube when shone with a light beam having a frequency above a certain threshold value. Einstein's explanation of the effect relied upon the earlier Planck's *theory of quanta*. According to Planck's theory, light absorption or emission by matter can occur only by discrete steps, called *quanta*, to which correspond a characteristic energy $h\nu$ (ν is the light frequency and $h = 6.6262 \times 10^{-34}$ J.s is a constant). Einstein postulated that in addition to its classical EM-wave nature, light should have an intrinsic, localized, particle-like structure or granularity, which would correspond to the energy quantum, and which he called "*photons*." Thus, each event of an electron crossing the vacuum tube was explained by the capture of a single light photon from the negatively charged electrode.

To better visualize such energy granularity consider a light beam having an optical power P. By definition, this power corresponds to a particle flux of $n = P/h\nu$ photons per second. It is a simple exercise (see the end of the chapter) to determine the photon flux for different frequencies. The result shows that even a billionth of a Watt in a beam of visible light corresponds to a flux of billions of photons per second. It can also be shown (see Saleh and Teich, 1991) that the Earth is under a constant "rain" of millions to trillions of photons: from starlight, moonlight, indoor light to sunlight, the photon flux density (flux per unit area) varies from $\Phi = 10^6, 10^8, 10^{12}$ to $\Phi = 10^{14}$ photons/s/cm^2. These examples illustrate that the photon granularity is in fact extremely small. When the photon number

2.2 Basic Physics of Light-Wave Signals 111

is this large, light can therefore be conceived of (or be represented as) a pure EM wave. It is then what is called *classical* light. At relatively small photon numbers, however, for instance just considering single light/atom interaction events, the energy granularity takes effect. In this case, light should be seen as being of a *quantum* nature, which obeys very different and complex laws of physics under a new field called *quantum electrodynamics* (QED). At this scale, QED thus takes over Maxwell's *classical electromagnetism*. One also refers to *quantum optics* as describing all phenomena associated with the quantum, or "photonic" nature of light.

It is beyond the scope of this book to convey even a superficial view of what light QED is about, but we shall just provide a simple illustration. For instance, QED predicts that *it is impossible to simultaneously determine the energy (number of photons) and the phase of an EM wave with perfect or absolute accuracy*. This is true even when the number of photons becomes identically close to zero! This is explained by an intriguing phenomenon called *vacuum noise*, or *vacuum fluctuations*, which represent a random E-field always present in vacuum. With this concept in mind let's look at Figure 2.8. The figure shows plots of the E-field amplitude associated with quantum light (as calculated from QED) according to increasing number of photons, i.e., from $n = 0$ to $n = 1,000$. This E-field amplitude is only a statistical variable which takes random values, as visible in the figure from the scattering of data. The amplitude's average value is given by $\langle E \rangle = \sqrt{n} \sin(\omega t + \varphi)$, which points to a deterministic waveform. In the absence of any photons ($n = 0$), this mean amplitude is identically zero. But the E-field amplitude is finite, as seen in the figure. Such a case corresponds to the aforementioned vacuum-noise or vacuum fluctuations, for which there is no distinguishable E-wave nor measurable E-field signal. But as the number of photons increases, it is seen that the E-wave progressively takes shape and becomes less and less noisy. At large photon numbers (e.g., $n = 1,000$) the quantum E-field resembles the classical E-field, with deterministic amplitude $E = \sqrt{n} \sin(\omega t + \varphi)$. This example, which gives a small flavor of QED, is also meant to illustrate that classical EM light and its laws are real and applicable at normal or macroscopic scale. A more complete and finer view of light is to see it as a stream of particle-like photons *associated* with an EM wave. Whether the observed behavior of light is that of a particle stream or an EM wave depends upon the measurement apparatus in use. Such an ambiguous and two-fold representation is known as the *wave-particle duality*. As another QED hint, *all* elementary particles (such as electrons) can be represented by associated waves. Again, their particle-like or wave-like behavior depends upon the type of measurement or effect. In this broader and unified perspective, it is seen that the photon constitutes just one example of elementary particle, and for which the associated wave is an EM wave.

When considering effects happening at the atomic scale, and trying to understand how light and matter interact, one absolutely needs the photon-representation concept. Light/matter interaction lies at the core of effects called *optical amplification* and *laser-light* generation, which are described in the next two subsections. This is why we need to further describe some of Einstein's other findings (1916). In these early times of quantum physics, it had been already well established (at least by experimental observation) that electrons could occupy different possible atomic

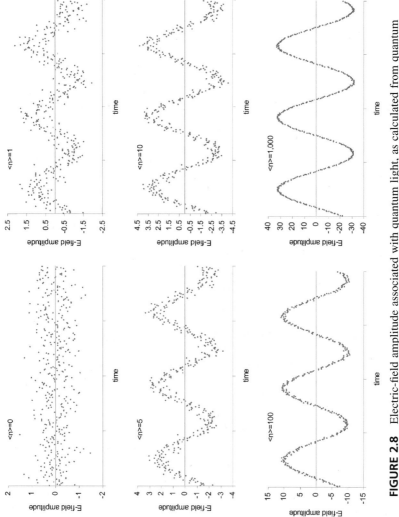

FIGURE 2.8 Electric-field amplitude associated with quantum light, as calculated from quantum electrodynamics (QED) for different photon numbers $n = 0, 5, 10, 100$ and $1,000$, with E-field phase of $\varphi = -\pi/4$ (after E. Desurvire, *Erbium-Doped Fiber Amplifiers, Device and System Developments*, © Wiley, 2003).

2.2 Basic Physics of Light-Wave Signals

energy levels, representing many *excitation states* for the atom. According to the theory of quanta, if an electron jumps from one atomic level of energy E_1 to another of energy $E_2 > E_1$, such an event corresponds to the *capture* of an incident photon by the atom. For this photon capture (or absorption) to happen, the incident-photon energy must be exactly $\Delta E = E_2 - E_1$, corresponding to the light frequency $\nu = \Delta E/h$. The atom must also be in the lower-state level. Conversely, when an electron jumps from an upper level (E_2) to a lower level (E_1), such an event is accompanied by the *release* of one photon having the frequency $\nu = \Delta E/h$. For this photon release (or emission) to happen, the atom must be in the upper-level or excited state.

The capture or release of photons by atoms represented a first and basic model to explain light absorption or emission by matter. But Einstein developed the analysis further by defining three possible light/atom interaction processes, as illustrated in Figure 2.9. These three processes are called *absorption*, *spontaneous emission* and *stimulated emission*, respectively:

- *Absorption*: Same as previously described for photon capture;
- *Spontaneous emission*: Same as previously described for photon release, the event being "spontaneous" or unpredictable; the direction/polarization/phase characteristics of the EM wave associated with this photon are *purely random*;
- *Stimulated emission*: A photon is released by atomic de-excitation, but the event is "stimulated" by the passing of an incident photon; the direction/polarization/phase characteristics of the EM wave associated with this photon are *strictly identical* to that of the EM wave associated with the incident photon.

The last two definitions point to the existence of two independent light-emission processes, i.e., spontaneous and stimulated. It is important to note that the EM wave

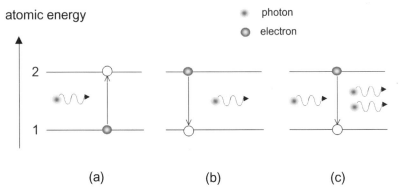

FIGURE 2.9 The three fundamental light/atom interaction processes: (a) absorption of an incident photon, moving an atom electron from initial state "1" to excited state "2", (b) spontaneous emission, with atom de-excitation from state "2" to state "1" accompanied by photon emission with random E-field characteristics, and (c) stimulated emission, with forced atom de-excitation by incident photon and emission of second photon with replicated E-field characteristics.

resulting form the first process has random direction/polarization/phase characteristics, while in the second process the EM wave is a "clone" of the incident one. One also says in this last case that the incident and resulting waves are *coherent* with each other, and this effect is at the root of the generation of "coherent light" through the *laser* effect (see Section 2.2.7).

As an additional postulate in Einstein's analysis, the rates of absorption (W_{abs}) and stimulated emission (W_{stim}) should be equal and proportional to the incident photon flux density (as defined as the number of photons per second per unit area, Φ). The rates $W_{abs} = W_{stim} = const \times \Phi$ define the number of absorption or stimulated-emission events per second, as expressed in (s^{-1}) units. The constant must be a function of the atomic characteristics and the energy-level pair involved. It must also be a function of the frequency ν. Since the unit of Φ is $(s^{-1}cm^{-2})$, the unit of the proportionality constant must be a surface [cm^2], which we call the *atomic cross-section* $\sigma(\nu)$. The fundamental result of this analysis is therefore

$$W_{abs} = W_{stim} = \Phi \sigma(\nu) \qquad (2.15)$$

The atomic cross-section should exhibit a maximum (σ_0) at the frequency $\nu_0 = (E_2 - E_1)/h$ and should rapidly vanish at other frequencies. Such a condition is consistent with the previous description of light absorption and emission, which assumed that the photon frequency and the energy-level separation should match to some extent. But such a matching condition is not absolute, since the absorption/stimulated-emission rates remain finite within the vicinity of the peak frequency ν_0. See more on this issue below when considering the actual definition of $\sigma(\nu)$.

The rate of spontaneous emission events should be a characteristic constant of the atom or more specifically, of the upper energy level involved in the de-excitation process. We can thus define $W_{spon} = 1/\tau$, where τ is a time constant called the *spontaneous emission lifetime*, or alternatively, the *fluorescence lifetime*. One can better visualize spontaneous emission by considering indeed the effect of *fluorescence* (also called *luminescence*). Everyone is familiar with those materials that glow in the darkness after having been exposed to light for a short duration. This is the effect of fluorescence. One has also observed that the glow intensity decreases over time, with a more rapid decay at the beginning. Thus, one can model the light-glow intensity as a time-decaying exponential. The probability of observing a "glow" photon is then proportional to $\exp(-t/\tau)$, where τ is the fluorescence "lifetime." This term means that at $t = \tau$, the intensity has decreased by the factor $\exp(-1) = 1/e = 0.36$, representing 36% of the initial intensity. At longer times $t = 2\tau, 3\tau\ldots$, etc., the intensity is $\exp(-2) = 1/e^2 = 0.13$, $\exp(-3) = 1/e^3 = 0.04$, etc., showing that the intensity vanishes. The effect of spontaneous emission is strictly the same as fluorescence: the probability of observing an atom in the excited state at a time t after the excitation event is indeed proportional to $\exp(-t/\tau)$.

There is a relation between the atomic cross-section $\sigma(\nu)$ and the fluorescence lifetime τ. Without going through any formal demonstration, suffice it to consider the spectrum of the E-field associated with spontaneously emitted light. This E-field has the form $E(t) = E_{max} \exp(-t/\tau) \sin(2\pi\nu_0 t)$. It can be shown that the associated spectral line-shape is of the form

$$L(\nu) = \frac{1}{1 + 4\left(\dfrac{\nu - \nu_0}{\Delta \nu}\right)^2} \tag{2.16}$$

where $\Delta \nu = 1/(2\pi\tau)$. This type of function is called *Lorentzian*. It has a bell-like, symmetrical shape with a peak ($L = 1$) at the resonance (or matching) frequency $\nu = \nu_0$, while it rapidly decays ($L \to 0$) as the difference $|\nu - \nu_0|$, or frequency mismatch, increases. The finite bell's width $\Delta\nu$ corresponds to an effect of *line broadening*. Very long lifetimes ($\tau \to \infty$) correspond to narrow line-shapes ($\Delta\nu \to 0$) and the reverse. This finite line width thus defines the frequency range over which spontaneous-emission events occur and their corresponding probability (being maximum at the resonant frequency ν_0). It makes sense that the atomic cross-section, which defines the stimulated-emission rate probability, be also proportional to this line-shape function. Thus we should have

$$\sigma(\nu) = \sigma_0 \times L(\nu) \tag{2.17}$$

where σ_0 is the peak or resonant cross-section. As it conveniently turns out, the calculation result is simply $\sigma_0 = \lambda^2/(4\pi^2\tau\Delta\nu) \equiv \lambda^2/(2\pi)$, where $\lambda = c/\nu_0$ is the peak wavelength. Thus, the absorption/stimulated-emission rates are also most simply defined by $W_{\text{abs}} = W_{\text{stim}} = \Phi\lambda^2/(2\pi)$, meaning that the associated probability rates (within a 2π factor) are equal to the product of the *incident photon flux* times the *square of wavelength*, a result that is surprisingly elementary, and should we say, "physically elegant."

The above description corresponds to a most fundamental case and is meant for a rapid familiarization with the effect of light/matter interaction. But according to any expectations, the physical reality happens to be far more complex, even if this description remains wholly accurate for simple atomic systems. Additional effects to take into account in the realistic calculation of the line-shape characteristics are the following:

- Competing de-excitation events to other lower-energy levels;
- De-excitation processes which do not generate light (called *nonradiative*);
- Energy-level splitting into sublevels due to static E-field (called the *Stark effect*);
- Random variations, from atom to atom, of the energy-level separation ΔE (called *inhomogeneous broadening*);

While the above effects belong to another level of complexity (which fully explains observed atomic line-shapes and their characteristics), Einstein's elementary description of absorption and spontaneous/stimulated emission remains central to the understanding of *optical amplification* and the associated process of generating *coherent light*, known as the "laser" effect, to be discovered some 45 years later.

2.2.5 Optical Amplification

The Section 2.2.4 provided the conceptual tools needed to analyze the effects of *light amplification* and *coherent-light (laser) generation*, without which there would be no such thing as our current lightwave telecommunications!

Consider first *light amplification*. We have seen that an atom in the excited state can be stimulated to de-excite itself by the passing of an incident photon, when it emits a "clone" photon with the same EM-characteristics as that of the incident one. The rate at which such events happen is $W_{\text{stim}} = \Phi\sigma$, where Φ is the number of photons per second and unit area (photon flux density). As we have seen, the atom can also spontaneously de-excite, with a rate $W_{\text{spon}} = 1/\tau$. After either stimulated or spontaneous de-excitation, the atom now stands in its lower-energy state, and is therefore able to absorb any incident photon with the rate $W_{\text{abs}} = \Phi\sigma$, incidentally equal to W_{stim}. Since the rates for absorption and stimulated emission are equal, nothing more would happen than this atom jumping up and down between its two energy states, and the resulting output photon flux would be left unchanged in the average. But the situation is wholly different when considering a large number of atoms present in the material. Call N_1 or N_2 the *populations* of atoms found in the lower state or upper state, respectively, as expressed in number of atoms per unit volume (or *density* in m^{-3}), such that $N = N_1 + N_2$ represents the total atomic density. Consistently, the total absorption, spontaneous-emission and stimulated-emission rates (now in units of s^{-1}m^{-3}) are given by

$$\begin{cases} W_{\text{abs}} = \Phi\sigma N_1 \\ W_{\text{spon}} = N_2/\tau \\ W_{\text{stim}} = \Phi\sigma N_2 \end{cases} \quad (2.18)$$

In a slice of the medium of thickness Δz, the total rates become

$$\begin{cases} W_{\text{abs}} = \Phi\sigma N_1 \Delta z \\ W_{\text{spon}} = \eta(N_2/\tau)\Delta z \\ W_{\text{stim}} = \Phi\sigma N_2 \Delta z \end{cases} \quad (2.19)$$

The factor η, which was introduced in the definition of W_{spon} represents the fraction of spontaneous-emission events which produce EM-waves traveling in the same direction as the signal, but within some unspecified cone angle (there is no need to consider spontaneous EM waves emitted in larger angles or other directions). Since the units in equation (2.19) are now that of (s^{-1}m^{-2}), the above rates represent actual changes in the photon-flux density Φ. However, we must count as negative the contribution of absorption (W_{abs}) and as positive the two contributions of emission (W_{spon}, W_{stim}). The resulting weighted sum provides the net rate of change of the photon flux:

$$\Delta\Phi \equiv W_{\text{stim}} - W_{\text{abs}} + W_{\text{spon}} = \left[\Phi\sigma(N_2 - N_1) + \frac{\eta N_2}{\tau}\right]\Delta z \quad (2.20)$$

It is immediately seen from this result that the change is positive if the condition $\Delta N = N_2 - N_1 > 0$ is satisfied. In this case, there are more atoms in the upper-state than in the lower-state ($N_2 > N_1$). Such a specific condition is referred to as *population inversion*. The term "inversion" is justified by the fact that in ordinary conditions, or in the absence of any external process for atomic excitation, the majority of atoms stand in their low-energy state, or $N_1 > N_2$.

Equation (2.21) is of the form $\Delta \Phi / \Delta z = f(\Phi)$, which, if we chose Δz sufficiently small, represents the rate of change of Φ per unit length at any coordinate z in the medium. It then becomes a "differential" equation which can be exactly solved for any distance z. If one overlooks the spontaneous-emission contribution in N_2/τ, the equation is of the form $\Delta \Phi / \Delta z = g\Phi$, where $g = \sigma(N_2 - N_1)$, and its solution is

$$\Phi(z) = \Phi(0) \exp(gz) = \Phi(0) e^{\sigma(N_2 - N_1)z} \tag{2.21}$$

where $\Phi(0)$ is the photon flux input in the medium. It is seen from this result that the photon flux exponentially increases or decreases with distance, according to whether the condition $g > 0$ or $g < 0$ is realized. In the case where the atomic populations are equal ($N_2 = N_1$), we have $g = 0$ and $\Phi(z) = \Phi(0)$, meaning that the photon flux remains constant. The quantity $G(z) = \exp(gz)$ is called the *amplifier gain*, where g is referred to as *the gain coefficient* (units of m^{-1}, sometimes cm^{-1}). When $g > 0$ or $G > 1$, the input light is amplified, while in the contrary case ($g < 0$ or $G < 1$) it is attenuated. We thus conclude that *population inversion is a fundamental prerequisite for signal-light amplification*. The techniques to achieve such an inversion are described further below.

In the general case, the amplifying medium exhibits a finite background loss, as characterized by an *attenuation coefficient* (α). Since the medium transmission in absence of any other effect is $T = \exp(-\alpha z)$, we can define the *net gain* as $G_{net} = GT = \exp[(g - \alpha)z]$. The condition for light amplification ($G_{net} > 1$) is therefore $g > \alpha$ or $N_2 - N_1 > \alpha/\sigma$.

It is useful to introduce here another unit for the amplification gain, which is the *decibel* (dB) and is used in telecommunications engineering. Since decibels concern powers, let's first convert the photon-flux density Φ into an optical power P. The relation is simply $P = h\nu \Phi A_{eff}$, where A_{eff} represents an effective or equivalent area over which the light beam propagates. The amplifier equation for optical power at $z = L$ is thus:

$$P(L) \equiv G_{net}P(0) = P(0)e^{(g-\alpha)L} \equiv P(0)e^{\bar{g}L} \tag{2.22}$$

where $\bar{g} = g - \alpha$ is the net gain coefficient (in the following, we shall just use the notation "g"). By definition, the decibel gain is

$$G_{dB} = 10 \log_{10}\left(\frac{P(L)}{P(0)}\right) \tag{2.23}$$

One can also express the powers in relative units of decibel-milliwatt (dBm), which gives

$$G_{dB} = 10\log_{10}\left(\frac{P(L)}{1\,\text{mW}}\frac{1\,\text{mW}}{P(0)}\right)$$

$$= 10\log_{10}\left(\frac{P(L)}{1\,\text{mW}}\right) - 10\log_{10}\left(\frac{P(0)}{1\,\text{mW}}\right) \equiv P_{dBm}(L) - P_{dBm}(0) \quad (2.24)$$

and shows that the net amplifier gain is just defined by the dBm-power difference between output and input signals. It is also possible to express the gain coefficient in units of decibel-per-meter (dB/m). The amplifier gain is then provided by the relation $G_{dB} = g_{dB/m} \times L_m$. The relation between dB/m and m^{-1} gain coefficients is simply $g_{m^{-1}} \approx 0.2302 \times g_{dB/m}$. See exercises at the end of the chapter for proof and for other illustrative application examples.

We consider next the second term due to spontaneous emission in equation (2.21), which we have neglected so far. Without demonstration, we define the fraction of spontaneous emission falling into the signal cone angle as $\eta = 2\Delta\nu\sigma\tau/A_{\text{eff}}$, where $\Delta\nu$ is the line-shape width, A_{eff} is the effective area of the signal beam, and where the factor "2" accounts for the two possible polarizations. Converting next the flux equation into an optical-power equation ($P = h\nu\Phi A_{\text{eff}}$), and introducing $P_{\text{noise}} = 2h\nu\Delta\nu$ which has the dimension of a power (J.ss^{-1}m^2 m^{-2} s^{-1} = Js^{-1} = W), we obtain

$$\frac{\Delta P}{\Delta z} \equiv g(P + n_{sp}P_{\text{noise}}) \quad (2.25)$$

where

$$n_{sp} = \frac{N_2}{N_2 - N_1} \quad (2.26)$$

represents a "relative degree" of inversion assuming $N_2 \neq N_1$. Equation (2.25) is of the form $\Delta P/\Delta z = g(P + const)$, and has for solution

$$P(L) \equiv GP(0) + n_{sp}(G-1)P_{\text{noise}} \equiv GP(0) + P_{\text{ASE}} \quad (2.27)$$

The intriguing result is that if there is no input signal ($P(0) = 0$), our amplifier still produces an output power equal to $P_{\text{ASE}} = n_{sp}(G-1)P_{\text{noise}}$. As we have seen, such a power comes from the effect of spontaneous emission, or more accurately, its amplification through the medium. Therefore one refers to this output as the *amplified spontaneous emission*, or ASE. The ASE is *noise*, since it represents the superposition and amplification of all signals generated by spontaneous-emission events, each initially having a random E-field. We observe that this ASE power is nearly proportional to the amplifier gain G. At high gains, we have $G - 1 \approx G$, and the ASE is $P_{\text{ASE}} \approx Gn_{sp}P_{\text{noise}}$. This result shows that it is as if the amplifier is input with an *equivalent noise signal* of power $n_{sp}P_{\text{noise}} = 2n_{sp}h\nu\Delta\nu$. Equivalently, this input noise signal corresponds to the power of $2n_{sp}$ photons in bandwidth $\Delta\nu$, or n_{sp} photons in each polarization. Note that since n_{sp} is not generally an integer, its value corresponds to average photon-numbers. Consistently, we may now say that $n_{sp} = N_2/(N_2 - N_1)$ is *the spontaneous-emission factor*. In the usual population-inversion conditions ($N_2 - N_1 > 0$) this factor is always greater than unity. In conditions of highest-possible inversion, where the number of atoms in the lower state becomes negligible ($N_1 \to 0$), this factor reaches a minimum value of unity, $n_{sp} \to 1$, which also corresponds to a minimal ASE-power generation. The minimum ASE is thus equivalent to one input noise photon per polarization. Note that when population inversion is not achieved ($N_2 - N_1 < 0$) the factor n_{sp} is negative, but since the gain becomes less than unity ($G < 1$) the ASE power $P_{ASE} = n_{sp}(G - 1)P_{noise}$ is positive. In this case, the ASE power can be reduced to a zero limit ($P_{ASE} \to 0$ as $N_2 \to 0$, $n_{sp} \to 0$ and $G \to 0$), but this corresponds to a useless amplifier with no gain and high loss. Instead, the idea is to *maximize the gain* and *minimize the ASE noise*, which is achieved in conditions that are as close as possible to *complete population inversion* ($N_1 \approx 0$). The important result of this demonstration is that *the minimum amplifier noise is equivalent to one input noise photon per polarization*.

We address next the issue of how population inversion can be achieved and optimized. The principle of population inversion consists not only in bringing the atoms into their excited-state but in keeping the population of such excited atoms as high as possible. As we have seen, atoms de-excite through the spontaneous and stimulated emission processes, which corresponds to a decay in the excited-state population. We thus need a means to constantly bring back the atoms to the excited-state at a rate which overcomes this emission decay rate. Figuratively, this action is equivalent to pumping water into a leaking tank. The means to "pump" atoms into the upper state consists in illuminating the medium with an auxiliary optical signal, called the *pump*, whose frequency ν_p is higher than that of the signal ν_s. The pump is then absorbed to bring the atom into some high energy level. Figure 2.10 summarizes different pumping schemes, according to the number of relevant energy levels involved.

Consider first the *3-level pumping scheme*. We see from the figure (diagram at left) that the absorption of a pump photon at ν_p brings the atom from level 1 to level 3. If level 3 has a very short lifetime (τ'), the atom rapidly decays to level 2, which should have a long lifetime (τ). Then emission events at signal frequency ν_s bring the atom to the lower state at level 1. Since the pumping rate (R) is assumed much higher than the emission rates (W_{spon}, W_{stim}), the population of level 1 is always lower than that of level 2, which is the condition for population inversion ($N_2 > N_1$) and signal gain. Alternatively, the top level, now called level 2, can be characterized by a long lifetime, and the intermediate level, now called level 1, a short lifetime (Figure 2.10, center). This feature also makes possible population inversion between the two levels, or $N_2 > N_1$. If the decay rate of level 1 is sufficiently high, then the pumping rate does not need to be as strong since the inversion condition $N_2 > N_1$ will be always satisfied with minimum pumping.

The *4-level pumping scheme* is similar to the previous case (see Figure 2.10, right), except that an additional level 4 with short lifetime (τ'') introduces an extra decay step in the process.

The rapid decays from levels having very short lifetimes are not associated with photon emission; the energy is released in the form of thermal-excitation

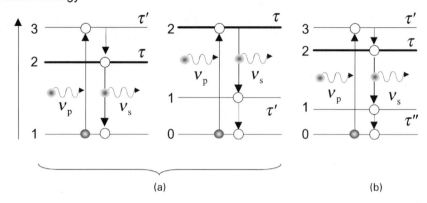

FIGURE 2.10 Different pumping schemes for achieving population inversion, according to the atomic energy-level systems (a) three-level and (b) four-level. The full-line arrows pointing up or down correspond to pump-photon absorption at frequency ν_p, or signal-photon emission absorption at frequency ν_s, respectively. The dashed arrows correspond to fast decays between atomic levels with short lifetimes τ', τ''. The upper-level (2) of the de-excitation responsible for signal emission, shown in bold, is characterized by a relatively slow decay of long lifetime $\tau \gg \tau'$, τ''.

quanta called *phonons*. In this case the de-excitation process is therefore called *nonradiative*. The slow decays from levels having comparatively long lifetimes are usually purely *radiative*, i.e., the de-excitation is exclusively made though photon emission without phonons. It is clear that the condition of purely radiative decay is a prerequisite for achieving efficient pumping (maximum population inversion at minimum pump power) and obtaining high amplification gains.

The energy-level *positions* and associated *lifetimes* required to achieve population inversion though 3- or 4-level pumping, with *purely radiative* de-excitation at a *desired signal frequency* ν_s, are pretty demanding conditions to be realized at once! These conditions are indeed far too rare to be obviously (or luckily) found "as is" in materials. This is in spite of the phenomenal abundance of material compositions, in gas, liquid or solid forms, the latter in either crystalline or amorphous varieties.

Some natural gas molecules (e.g., carbon-dioxide or CO_2) or compositions (e.g., helium and neon, or He-Ne) happen to meet these requirements for $\lambda_s = 10.4$ μm (infrared) and $\lambda_s = 0.632$ μm (red), respectively. Such materials must be excited or pumped through electric discharge in the gas tube. We are most familiar with the *He-Ne laser*, which is used to read barcodes in the store and as "laser pointers." Overlooking the case of liquids (organic dye molecules), most materials eligible for population inversion are made as *doped crystals* or *doped amorphous glasses*. The doping consists in introducing elemental "defects" into the otherwise uniform crystal or glass matrices. The matrix alters the electronic distribution and the energy-level positions of the dopant, which creates new varieties of atomic systems with the desired properties. The first 3-level case shown in Figure 2.10 corresponds to that of *erbium-doped glass*, which has a fluorescence lifetime near $\tau = 10$ ms and an emitting signal wavelength near $\lambda_s = 1.55$ μm. The Er:glass system can be pumped at $\lambda_p = 098$ μm or $\lambda_p = 1.48$ μm, for instance, which correspond to different energies for the upper-level 3. Incidentally, the pumping configuration at $\lambda_p = 1.48$ μm is also likenable to that of a 4-level system, with levels 1–2 and 3–4 being very close to each other, which explains why the pump and signal wavelengths are within only 0.07 μm (70 nm) of each other. Doping the core of glass fibers with erbium has led to the realization of *erbium-doped fiber amplifiers*, or EDFA, which are now extensively used in lightwave WDM telecommunications. See more on EDFA in Section 2.4 on "passive" optical components. Another type of glass-fiber amplifier is based on the effect of *stimulated Raman scattering* (SRS). Although SRS is not based upon atomic energy levels, one can model Raman amplification according to the second 3-level process shown in the middle of Figure 2.10. The effect has also been exploited for 1.3 μm/1.55 μm telecommunications in *Raman fiber amplifiers*, or RFA, as described in the section on passive components.

2.2.6 Light and Photo-Current Generation in Semiconductors

Another category of synthetic materials of high interest for their optical properties is that of *semiconductors*. Semiconductors are crystal compounds made of two to four elements selected from specific compatible groups. While semiconductors are largely known for applications as transistors and integrated circuits (e.g., silicon, silicon/germanium, etc.) they also represent a key technology in optical telecommunications as *laser-diode* transmitters, *photo-diode* receivers and *semiconductor optical-amplifier* gates (see Section 2.5 on active components).

In semiconductors, the effect of light/matter interaction is a bit more complex to analyze, but the principles of photon absorption, spontaneous and stimulated emission remain strictly the same. In such materials, the discrete electronic energy levels are very closely spaced and the set can therefore be seen as forming a continuum. Such a continuum is however interrupted by an intermediate region or "forbidden zone" called the *bandgap*, as defined by an energy separation E_g. The energy-continuum region above the bandgap ($E \geq E_g$) is called the *conduction band*, while that under the bandgap ($E \leq E_g$) is called the *valence band*, as illustrated in Figure 2.11. In the absence of thermal excitation (absolute zero temperature), all electrons are located inside the valence band and the conduction band is completely empty. At finite temperature, however, a certain number of electrons have escaped into the conduction band, leaving as many *holes* in the valence band. Such an escape can be caused by *photon absorption*, as shown in the figure. The condition for photon absorption to occur is $E_{\text{photon}} = h\nu \geq E_g$. The *valence electron* thus promoted to the conduction band becomes a *conduction electron*. The generation of conduction electron leaves as many holes in the valence band. If an electrical field is applied, the conduction electrons and the valence holes drift in opposite directions, generating an electrical current. For this reason, electrons and holes are called *charge carriers* or *carriers*. Conduction electrons are also called *free carriers*. As the figure illustrates, a free carrier may spontaneously jump back

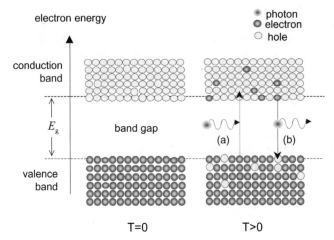

FIGURE 2.11 Energy diagram of semiconductor materials showing valence and conduction bands separated by energy bandgap. T = 0: valence band is completely occupied by electrons while conduction band is empty or completely occupied by holes; T > 0: electrons and holes are present in the conduction and valence bands, respectively; (a) valence electrons jump up to the conduction band by absorption of photons; (b) conduction electrons jump down to the valence band through electron-hole or carrier recombination associated with photon emission.

into the valence band and fill up a hole. In the process, the corresponding electron-hole energy difference is released through the emission of a photon. This process is referred to as *electron-hole recombination*, or *carrier recombination*. Other recombination processes can be essentially non-radiative, in which case the semiconductor is not eligible as a light amplifier. The natural "lifetime" associated with the conduction band electrons, which is also the *carrier-recombination lifetime*, is of the order of $\tau = 1$ ns (10^{-9}s).

A straightforward way to pump a semiconductor amplifier is to drive an electrical current through the material. The current injects electrons in the conduction band (or holes in the valence band). These extra carriers then recombine to produce photons at frequencies $\nu \geq E_g/h$. The cross-section line-shape profile, $\sigma(\nu)/\sigma_0$ [equation (2.17)], is different from Lorentzian. But all the properties which have been previously described for absorption and emission rates, gain and noise strictly apply, with $\Delta N = N_2 - N_1$ now representing the population of free carriers. The photon-emission rate Φ resulting from this "electrical pumping" is proportional to the current I. This is because the current is equal to the number of electron charges injected per second (recalling that 1 ampère = 1 coulomb/s = 6.25×10^{18} electrons/s). More accurately, the photon rate is proportional to the current density, I/A_{eff}, where A_{eff} is the effective area where the carriers are confined. The free-carrier density (in m^{-3}), which generates optical gain ($G = \exp(\sigma \Delta N L)$), is provided by the simple formula:

$$\Delta N = \frac{\tau}{el} \frac{I}{A_{\text{eff}}} \quad (2.28)$$

where τ is the free-carrier lifetime of the semiconductor, $e = 1.6 \times 10^{-19}$ C is the electron's charge and l the thickness of the device's recombination region. The carrier confinement is achieved by realizing complex semiconductor structures with mutually isolating boundaries. Thus the production rate of free carriers, and hence the output photon rate can be significantly increased by current confinement. Note that optical pumping of semiconductors is also possible, but this requires an external light source and optimized conditions of illumination of the recombination region, which makes the approach less practical than electrical pumping.

Consider now a different experiment, where the semiconductor material is traversed not by an electrical current but by a *static electric field*, such as generated by the potential difference from a simple battery. What happens if one illuminates the semiconductor with a light beam of frequency $\nu \geq E_g/h$? The answer is exactly the reverse process! Incident photons are absorbed, generating free carriers in the conduction band and holes in the valence band (Figure 2.11). The carriers then move under the force generated by the static E-field, thus generating a current. Such an optically induced current is called *photo-current*. This should not be confused with the aforementioned "photo-electric effect," which concerned photo-induced current through vacuum tubes. But a similarity exists to the extent that photon quanta are absorbed to free electrons otherwise bounded into a low-energy state, which generates current. Consistently, the photo-current I_{phot} is proportional to the incident photon flux Φ.

$$I_{\text{phot}} = \eta e \Phi A_{\text{eff}} = \frac{\eta e P}{h\nu} \qquad (2.29)$$

where P is the incident optical power, and η is the fraction of this power captured by the recombination region. This effect of photo-current generation constitutes the basic principle of a *photo-detector*, where light signals are converted into current signals. See more on photo-detectors in the active components in Section 2.5.2.

The possibility of efficiently generating signal light (or signal current) by direct injection of current (or light) in semiconductor devices is at the root of practical optical telecommunications and their phenomenal technology evolution over the last 60 years. But the central element in this picture is the *LASER*, an acronym meaning the effect of *light amplification by stimulated emission of radiation*, which we describe next.

2.2.7 Lasers and Coherent Light Generation

The invention and realization of the *laser* in 1958–1960 (by A. L. Schawlow and C. H. Townes, T. H. Maiman) represents a conceptual extension at light frequencies of a previous device called the MASER, for *microwave amplification by stimulated emission of radiation*. In the early days, lasers were called "*optical masers*" and in the Russian literature "*optical quantum generators (OKG)*" or "*quantum generators of light*." The extension of the maser to optical frequencies was initially regarded as an intriguing curiosity of physics. Reportedly some skeptical technology critics would have even called the laser "a solution in search of a problem"! It would take only a few more years (1966–1970) to conceive and demonstrate the first low-loss optical fibers, to inaugurate the field of *lightwave* telecommunications as one of the most successful applications of lasers. The invention of the laser also had a tremendous impact in fields such as medicine, defense, materials processing, data storage and audio/video entertainment, to quote just a few.

The principle of the *laser effect* is to exploit stimulated emission to generate a standing EM wave inside a certain type of *optical cavity*. An optical cavity is made of two parallel mirrors placed at a certain distance L. If the mirror's reflection coefficients are a bit less than 100%, it is possible for a light beam at certain wavelength λ to enter this cavity. One can then chose the cavity length to be equal to an integer number of half wavelengths, i.e., $L = k\lambda/2$, as illustrated in Figure 2.11 (top). The mirrors must also be weakly transmitting for light to be able to enter the cavity. Such an arrangement is called a *Fabry–Pérot* (FP) cavity, or equivalently, a *FP interferometer*. Signal light entering from mirror M_1 travels through the cavity to be strongly reflected by mirror M_2 back to its entering point. There, it is strongly reflected again, and due to the condition $L = k\lambda/2$, the reflected signal is exactly in phase with the incoming signal. Since the reflected wave is in phase with the incoming wave, a standing-wave oscillation occurs, just like a musical string oscillates between it's two instrument bridges (referred to as "first harmonic" oscillation). Considering that the light wavelength is very short (of the order of 10^{-6} m or 10^{-3} mm) this phase-matching condition requires that the mirrors' parallelism be extremely accurate. The accuracy of the angle alignment should be such that all possible phase discrepancies within the light beam-area remain negligible, say of the order of $\lambda/10$ or better. In such conditions, the beam inside the FP cavity forms a standing EM-field wave with a very accurately defined wavelength. For a given FP cavity length, the possible oscillation wavelengths or frequencies are $\lambda_k = 2L/k$ or $\nu_k = kc'/(2L)$, where $c' = c/n$ is the speed of light in the FP cavity, which defines many discrete possibilities for *oscillation modes*. By definition, one refers to these as *longitudinal modes* or *FP modes*. The frequency separation between two adjacent FP modes is constant and equal to $\delta \nu = \nu_{k+1} - \nu_k = c'/(2L)$. The FP mode-spacing characteristic is called the *free spectral range* (FSR). We thus conclude that the FP cavity/interferometer acts as a *frequency-selective device* which only allows a certain set of discrete frequencies ν_k to oscillate. We note that no restriction applies to the polarization state of the signal; the FP modes are thus said to be *polarization-degenerate*.

The next trick is to *amplify* this standing wave using the effect of stimulated emission, using the fact that one (or a close subgroup) of these discrete frequencies could match the atomic energy $\Delta E/h$ associated with some materials. This can be achieved by implementing a *laser cavity*.

A *laser cavity* is a FP cavity which contains an active medium with a pumping source, as illustrated in Figure 2.12. Unlike in the "passive" FP case, there is no need to introduce light to see an effect. Indeed, as the pumping mechanism is turned on, some atoms immediately jump into the excited state. As these excited atoms spontaneously decay within their fluorescence lifetime (τ), many spontaneous photons are emitted. As mentioned above, the E-fields associated with spontaneous emission events have random directions, polarizations and phases. But if accidentally a group of spontaneously emitted photons happens to have similar (associated) characteristics of direction, polarization and phase while meeting on one of the cavity mirrors, a new process is initiated. Because of the large number of spontaneous

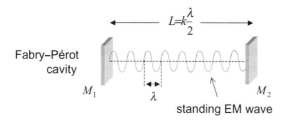
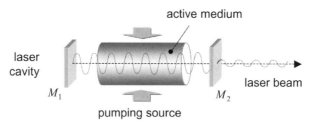

FIGURE 2.12 Top: principle of a Fabry–Pérot (FP) cavity, as formed from two reflecting mirrors (M_1, M_2) separated by a an integer-multiple of half-wavelengths, showing the formation of a standing electromagnetic (EM) wave. Bottom: principle of a laser cavity, as made of a FP cavity which contains an active medium with a pumping source; the reflectivity of one of the mirrors (M_2) is set to less than 100% so that a fraction of the EM-wave escapes as a beam aligned with the FP-cavity axis.

emission events, this accident actually occurs only within a few τ. Upon reflection, this "peer" photon group traverses again the active medium. Stimulated emission then generates as many identical photon "peers." After a few round trips, the peer group has multiplied itself into a huge number of photons. Looking at Figure 2.8, a large number of photons ($n > 1,000$) having identical E-field characteristics corresponds to the "classical" E-field, as defined by a unique sine wave. A standing EM wave has thus been generated in the cavity. Its power is maintained constant through the balance between the pumping process (which drives the atomic excitation) and the stimulated-emission process (which converts the atomic excitation into light emission). The reason why the EM-wave power stabilizes itself is analyzed below.

As Figure 2.12 indicates, one of the cavity mirrors can be set to be partially transmitting so that a fraction of the standing EM-wave can escape the cavity (the other mirror is usually 100% reflective). Understandably, the output beam of laser light is *highly directional*, as aligned with the FP-cavity axis. This is the reason why laser light appears in the form of a narrow and intense light beam. Note that for pumping-optimization purposes, the active medium of the laser cavity can also include a light waveguide, which increases the signal light intensity (or photon flux density) and the stimulated-emission rate. This is the principle of *semiconductor laser sources* and *fiber lasers*, as discussed in Section 2.4.4 and more specifically, in Section 2.5.1 on active components.

Our previous description showed that the laser wave builds up after a few round-trips in the cavity, the wave being amplified at each passing through the gain medium. Since the number of round-trips is unlimited, and since the wave cannot build up to an infinite power, there must exist some power-limitation mechanism which leads to equilibrium or "steady-state" conditions and clamps the laser power to some maximum level. We also need to determine how much pumping is required to make this laser-emission process happen. The answer to both issues is in fact quite simple to analyze, as we shall see.

Let us define Q as the *pumping rate*, which represents the number of atomic-excitation events per second due to the pumping process. Given the total laser flux Φ_{in} inside the cavity (as propagating in both directions), it is possible to analyze the steady-state values of the atomic-level populations N_1, N_2 corresponding to any of the atomic systems shown in Figure 2.10 and determine the population-density difference $\Delta N = N_2 - N_1$ at equilibrium. We shall skip here the details of this simple analysis to provide the result, which takes the form

$$\Delta N = \begin{cases} N \dfrac{Q\tau}{1 + \Phi_{in}\sigma\tau} \\ N \dfrac{Q\tau - 1}{1 + \Phi_{in}\sigma\tau} \end{cases} \quad (2.30)$$

where (to recall) N is the total atomic density in m^{-3}. The upper case corresponds to the two atomic systems at center and right in Figure 2.10, whose terminal level "1" is not the lowest-possible level. The lower case corresponds to the atomic system shown at left in the figure, for which the terminal level "1" is the lowest possible. To simplify the description, we will only consider the upper case. We observe from the definition that for a constant pumping rate Q the population difference decreases as the signal flux Φ_{in} increases. Consider next the fact that the laser cavity is lossy, since some of the power is allowed to escape through the output mirror M_2. If the mirror reflectivity is $R < 1$, the associated power loss per round-trip is $\exp(\log R) < 1$. Neglecting any other causes of cavity loss, the net gain per round-trip is thus

$$G_{net} = G^2 R = e^{2\sigma\Delta NL + \log R} \quad (2.31)$$

The condition for reaching steady-state or equilibrium is that the net gain reduces to unity, or $G_{net} \to 1$, meaning that the laser cavity has somehow become "transparent" to the laser signal. If we replace the result of equation (2.30) into that of equation (2.31) with the condition $G_{net} = 1$, we get:

$$\Phi_{in} = \frac{1}{\sigma\tau}\left[\frac{Q}{Q_{th}} - 1\right] \quad (2.32)$$

with

$$Q_{th} = \frac{1}{\tau}\frac{\log(1/R)}{2N\sigma L} \quad (2.33)$$

This result shows that the laser effect or the generation of an oscillating wave inside the cavity ($\Phi_{in} > 0$) can only be obtained under the condition that the pumping rate (Q) be above a certain *threshold* value (Q_{th}). Under this threshold, there is no oscillation or $\Phi_{in} = 0$. Above threshold, the laser power linearly increases as a function of the ratio Q/Q_{th}. We can also express the laser power as $P_{in} = h\nu\Phi_{in}A_{eff}$, or

$$P_{in} = \frac{h\nu A_{eff}}{\sigma\tau}\left[\frac{Q}{Q_{th}} - 1\right] \equiv P_{sat}\left[\frac{Q}{Q_{th}} - 1\right] \qquad (2.34)$$

where P_{sat} represents an intrinsic saturation power. Such a power defines the slope of the linear power increase.

One observes from the threshold definition in equation (2.33) that for a given cavity length (L), the laser threshold can be minimized by choosing materials with long fluorescence lifetimes (τ), high cross-sections (σ) and high active-atom concentrations (N), while setting the mirror reflectivity as close as possible to $R = 1$ or 100%. Independently of the material's choice, if one could just use a perfectly reflecting/loss-less mirror ($R \to 1$), then the threshold would vanish ($Q_{th} \to 0$) and the laser flux generated inside the cavity would theoretically be unlimited or "infinite" ($\Phi_{in} \to \infty$). In reality, such a situation is not possible, since physical laser cavities always have some finite loss. Even in the perfect loss-less case, pumping indefinite amounts of energy into such a closed cavity would simply result into a catastrophic burn-out or melt-down of the entire apparatus!

This defines the total laser flux *inside* the cavity, Φ_{in}. The laser output flux, Φ_{out}, is given by the half-flux propagating towards the output mirror M_2. Since the mirror has a finite transmittance $T = 1 - R$, the output laser flux is therefore $\Phi_{out} = (1 - R)\Phi_{in}/2$. Using the last two definitions while assuming $Q/Q_{th} \gg 1$, we can approximate the laser output flux as $\Phi_{out} \approx NLQ(1 - R)/\log(1/R)$. It is easily verified that the optimum reflectivity is obtained for $R \to 1$, but this limit is nonphysical ($\Phi_{in} \to \infty$). We can escape this dilemma by considering that realistic laser cavities have a finite amount of loss. This cavity loss can be modeled through the additional factor $T = \exp(-2\alpha L)$ in equation (2.31), where α is an effective absorption coefficient. The same basic calculation as previously leads to the corrected threshold value:

$$Q_{th} = \frac{1}{\tau}\frac{\log(1/R) + 2\alpha L}{2N\sigma L} \qquad (2.35)$$

and we get $\Phi_{out} \approx NLQ(1 - R)/[2\alpha L + \log(1/R)]$. We leave it to an exercise at the end of the chapter to show that, in practical conditions, the optimum reflectivity value falls near $R = 85-90\%$.

At this point of the description, we have established that above a certain pumping threshold, a laser cavity can generate a highly directional laser beam with an output flux Φ_{out}, corresponding to the output power $P_{out} = h\nu\Phi_{out}$. We have also determined that the standing EM-wave frequency must correspond to a single or to several *Fabry–Pérot* (FP) modes, as defined by the discrete values $\nu_k = kc'/(2L)$, where k is an integer and c' the speed of light in the FP cavity. At the onset of the laser oscillation, all FP modes whose frequencies fall into the gain cross-section $\sigma(\nu)$ will experience some amplification gain and thus will "compete" to build up their own laser oscillation and monopolize the highest fraction possible of the available pumping power. If we call (σ_k, α_k) the cross-section and cavity-loss coefficient at frequency v_k, the condition of laser oscillation for this mode is:

$$G_{net}(v_k) = e^{2\sigma_k \Delta NL - 2\alpha_k + \log R} = 1 \tag{2.36}$$

which corresponds to the individual "lasing" pump threshold

$$Q_{th}(v_k) = \frac{1}{\tau} \frac{\log(1/R) + 2\alpha_k L}{2N\sigma_k L} \tag{2.37}$$

It is clear that several FP modes located near the peak-frequency of the gain line-shape can fulfill the first oscillation condition, and find themselves in the above-threshold situation as well ($Q > Q_{th}(v_k)$). The result is that all these modes oscillate simultaneously, making the laser output a superposition of discrete "lasing" frequencies. In this case, the laser is said to be *multimode*. This denomination, which actually means "*multilongitudinal mode*," should not be confused with the other concept of "*spatial*" or "*transverse*" *modes* to be discussed next. We must also recall that the FP modes are polarization-degenerate, so in principle the laser output could have any state of polarization. In reality, there always exists a difference in cross-section and loss coefficient between the two polarization states; as a result, the polarized FP mode having the highest cross-section and/or the lower loss will experience a higher net gain and therefore take over the laser oscillation. The gain and loss experienced by each FP mode also fluctuates in time. Thus the FP mode excitation randomly vary, an effect referred to as model *partition noise* (see Section 2.5 on laser sources).

The frequency range for which the FP modes are above the lasing threshold corresponds to the *laser spectral width*, Δv_{laser}. It can be shown (*Shawlow–Townes* formula) that this width is proportional to the factor $n_{sp}(\alpha + \log(1/R))^2/P_{in}$. This result first indicates that the laser width reaches an ideal zero limit for a loss-less cavity ($\alpha = 0$, $R = 1$), which is also a "useless" laser from which no power can be extracted, and is also unrealistic. Another observation is that the laser width narrows as the inversion increases, with a lower limit given by $n_{sp} = N_2/(N_2 - N_1) \to 1$. Finally, the laser width narrows in inverse proportion to the laser power $P_{in} \propto Q/Q_{th} - 1$, which linearly increases with the relative pumping rate Q/Q_{th}. The conclusion to retain is that given the cavity loss parameters, the minimum achievable laser width is given by the ratio $Q_{th}/Q \approx 1/P_{in}$.

Given the cavity's free spectral range (FSR = $\delta v = c/(2nL)$), the number of FP modes included in the laser spectral width is $p = \Delta v_{laser}/\delta v$. The laser spectrum and its finer comb-like FP structure, is shown in Figure 2.13. It is seen that within the boundary of the laser width (Δv_{laser}), only a few FP modes are "allowed" to oscillate, while there are an infinity of other FP modes remaining in the spectrum tails. If the FP modes are sufficiently far apart (corresponding to a large cavity FSR), only one FP mode will fall into the width Δv_{laser} and be able to oscillate. Such a "single-moded" laser is characterized by a unique frequency line and is therefore called "*single-frequency*." Basically, a single-frequency laser is a source of "*monochromatic*" light, which corresponds to near-perfect sinusoidal EM-wave.

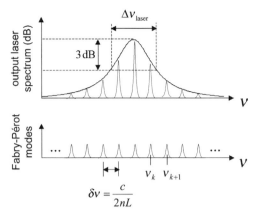

FIGURE 2.13 Bottom: spectrum of longitudinal modes corresponding to a Favry–Pérot (FP) cavity of length L and index n, with free-spectral range $\delta\nu$. Top: output laser spectrum (dB) corresponding to laser width $\Delta\nu_{\text{laser}}$, showing selective oscillation of lasing FP modes and inner spectral structure.

One should note that *long-cavity lasers* have closely spaced FP modes (small FSR), which allows a relatively large number of modes to oscillate within the laser width $\Delta\nu_{\text{laser}}$. In contrast, short-cavity lasers have largely spaced FP modes (large FSR), which allows the oscillation of fewer modes, down to a single mode. Shortening a laser cavity from 1 m to 1 mm corresponds to a thousand-fold increase of the FSR. For instance, semiconductor lasers have miniature cavity lengths of $1/4$ mm (250 μm) typically, representing a FSR of 200 GHz, or about 1.6 nm at $\lambda = 1.55$ μm ($n = 3$). Under typical operating conditions with output power near $P_{\text{in}} \approx 1$ mW, these lasers have typical spectral widths of $\Delta\lambda_{\text{laser}} \approx 10\text{–}20$ nm, which allows several FP modes to oscillate. To reduce the oscillation to a single mode, one must then resort to more complex mode-selection techniques. One possibility is to concatenate two FP cavities having different FSR, the second cavity acting as a narrow frequency filter. Another approach is to use a *Bragg grating* (see next section) in place of the cavity back mirror (M_1), for which the reflection coefficient is highly frequency-dependent. One of the most elegant solutions consists in creating the Bragg grating directly into the active medium, which is the case of *distributed-feedback* (DFB) semiconductor lasers. See more on these DFB lasers in the section on active components.

The above description has shown that laser light has a narrow spectrum made of single or multiple FP tones, forming an intense, unidirectional beam. We have also shown that the beam size is characterized by some effective cross-sectional area A_{eff}. The remaining question is now: what are the actual E-field and intensity distributions in the beam's transverse plane (x, y)? This is a complex question which requires Maxwell's electromagnetic formalism as applied to the three-dimensional laser-cavity geometry, possibly including additional lens effects introduced by mirror curvature. If we assume a cavity of length $L \gg \lambda$, having two rectangular mirrors with side-lengths (d_1, d_2) placed at points $\pm L/2$ on the z axis, it can be shown that, inside the cavity volume thus defined, the E-field amplitude can be approximated by:

$$E(x, y, z) = E_0 \sin\left[\left(x + \frac{d_1}{2}\right)\frac{p\pi}{d_1}\right] \times \sin\left[\left(y + \frac{d_2}{2}\right)\frac{q\pi}{d_2}\right]$$
$$\times \sin\left[\left(z + \frac{L}{2}\right)\frac{r\pi}{L}\right] \quad (2.38)$$

where p, q, r are integer numbers, while $E = 0$ outside the cavity. It is seen from this result that the E-field vanishes at any point $x = \pm d_1/2$, $y = \pm d_2/2$ and $z = \pm D/2$, showing that it is a three-dimensional standing wave. The beam exiting the cavity has the same amplitude definition. At any distance where the z-component of the E-field is maximum, and assuming square mirrors for simplicity ($d_1 = d_2 = d$) the distribution reduces to

$$E(x, y) = E_0 \sin\left[\left(x + \frac{d}{2}\right)\frac{p\pi}{d}\right] \times \sin\left[\left(y + \frac{d}{2}\right)\frac{q\pi}{d}\right] \quad (2.39)$$

which defines the beam's transverse amplitude. Note that the E-field polarization is parallel either to the x or the y axis. We see from this expression that the E-field amplitude oscillates in the transverse plane, with discrete spatial frequencies determined by the two integers (p, q). These different integer values define as many

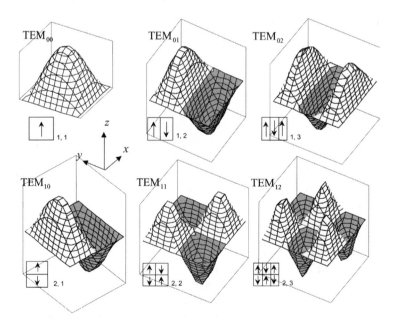

FIGURE 2.14 Transverse distribution of E-field amplitude corresponding to discrete oscillation modes TEM_{XY} (X, Y = 0 ... 2) of a laser cavity having square mirrors. Insets show one of the two possible polarization directions and the relative phase of the different lobes. The values of (p, q) involved in the E-field amplitude definition are also shown.

possible E-field oscillation modes, which we now call *transverse modes*. Figure 2.14 shows the first few transverse modes obtained from the above definition ($d = 1$) with $p = 1, 2$ and $q = 1, 3$. The figure also indicates the TEM$_{lm}$ (*transverse electromagnetic*) names which were attributed to these different modes. In this nomenclature, the index l, m indicates the number of E-field nodes in the x, y axes of the transverse plane, respectively. It is seen that TEM$_{00}$ exhibits a single lobe pattern, which we may call the *fundamental mode*. Other modes exhibit several lobes, which we then call the *higher-order modes*. Note that given the same polarization orientation, the modes TEM$_{01}$ and TEM$_{10}$ are not equivalent: the same modes also exist in the perpendicular polarization. This is why transverse modes are called *polarization-degenerate*. Figure 2.15 shows the nine TEM$_{lm}$ modes defined by $l, m = 0, 2$, as viewed from the longitudinal or propagation axis. The same mode family is also shown in the case of circular cavity mirrors. Due to the cylindrical symmetry of the cavity, the corresponding E-field definition is different from that for square mirrors, but the principle remains similar. The difference introduced by this symmetry is that the amplitude lobes are defined by the number of zero crossings in radial and angular coordinates instead of x, y coordinates. Note the similarity of the TEM$_{00}$ fundamental modes in the square and circular cases.

It is clear that given cavity-loss parameters and pumping conditions, any of the TEM$_{lm}$ modes may have a chance to oscillate on its own. The laser output beam then

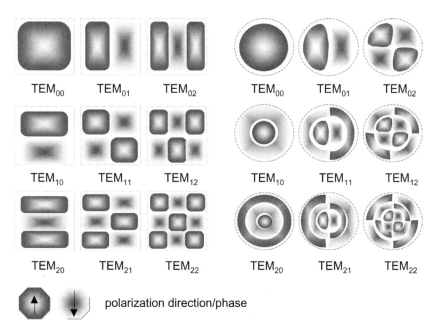

FIGURE 2.15 Transverse modes TEM$_{XY}$ (X, Y = 0 . . . 2) of laser cavity having square (left) or circular (right) mirrors, as corresponding to a single polarization direction and viewed from the longitudinal propagation axis. In both cases, the two polarization phases are shown through positive and negative lobe shadows.

represents a linear superposition of all these transverse modes. Since the lobes in each individual mode have different phases and envelopes, the modes interfere, which produces even more complex amplitude, polarization and intensity patterns. For a very large number of oscillating modes, the interference pattern becomes extremely complex, resulting into a characteristic, granular-like aspect, which is called the laser *speckle*. If the number of modes becomes "infinite," the overall interference pattern is nearly continuous, which results in a Gaussian or bell-shape intensity distribution of unpolarized light. In order to operate the laser in its fundamental TEM_{00} mode, a simple solution is to introduce a small square or circular *aperture* on the cavity axis. Since for the higher-order modes most of the radiation energy is located away from the cavity axis (Figure 2.14), such an aperture introduces substantial modal loss, except for the fundamental TEM_{00} case, which is then left alone to oscillate. The fundamental mode then chooses the polarization direction which has the lowest loss or the highest net gain. This eventual polarization selection is a combined effect of polarization dependences in the gain medium itself (as defined by its crystalline orientation), in the cavity mirrors, and in other possible elements introduced into the cavity for this purpose.

What can be concluded from the overall description of laser-lightwave generation is that lasers can oscillate according to both discrete *longitudinal* and *transverse* modes. The longitudinal or Fabry–Pérot modes define *spectral* purity (laser linewidth), while transverse modes define *spatial* purity (beam intensity envelope). If the laser's operating conditions only allow a single FP mode to oscillate in the fundamental TEM_{00} mode, then the resulting lightwave is of the highest possible purity order. One also refers to frequency and spatial purity as *temporal coherence* and *spatial coherence*, respectively. As further described in this chapter, the use of narrow-linewidth, fundamental-mode laser sources is central to optical telecommunications applications. The familiarity that we have now acquired with transverse *cavity modes* will be very helpful in addressing the higher concept of (transverse) *guided modes*, which is covered in the following section.

■ 2.3 OPTICAL WAVEGUIDES

The description in the previous section showed that coherent laser light can be generated in the form of TEM_{lm} *cavity modes*, which define the E-field amplitude distribution in the plane transverse to the light beam. The *fundamental mode*, TEM_{00}, was seen to correspond to the most coherent form of laser beam pattern. It can be viewed as an orderly group of parallel E-field waves propagating in the same direction and sharing the same frequency and phase. The E-field amplitude however decreases with the wave's distance from the center axis. The approximated formula used in the previous section implied that the E-field is identically zero outside an region defined by the cavity-mirror's cross-sectional area. In fact, the real solution provided by Maxwell's equations shows that the E-field exponentially vanishes with radial distance, meaning that it is never exactly zero at any point of space, but this more correct view does not alter the essential picture of transverse-modes.

The next conceptual step is that of an E-field wave propagating inside a *waveguide structure*. As the name indicates, such a structure is able to guide the wave along a certain longitudinal direction and according to certain discrete transverse patterns, which are also called *modes*. These guided modes are similar to cavity modes, except that the wave's boundary conditions are determined by a transverse structure (the waveguide) as opposed to a longitudinal structure (the laser cavity). As shown in the section on active components, it is also possible to realize active *laser-cavity waveguides*, which provide both effects of transverse and longitudinal confinements. Passive waveguides can also include active elements, such as in the case of *fiber amplifiers* (see Section 2.4.4 on passive components). But their mode properties remain unaffected by the amplification process. Here, we will be only concerned with the case of passive waveguides made of purely transparent materials and defined by their transverse index-structure and geometry.

This section provides a simple but detailed description of optical waveguiding. The description is based on elementary examples of *planar waveguide structures*. We first analyze how light rays bounce inside and refract outside planar index-layer structures. Then we introduce the EM-wave properties of light, observing that the ray analysis has some intrinsic limitations. We first analyze the simpler case where reflecting mirrors are used for the waveguide boundaries. Then we look at waveguides made from a dielectric or multilayer index medium. Once this whole conceptual effort has been duly performed, we can finally approach with due serenity and ease the case of those cylindrical waveguides called *optical fibers*.

2.3.1 Ray Propagation in Index Layers and Waveguiding

To immediately visualize how any waveguiding effect actually works, consider the structure shown in Figure 2.16. The structure is made of three horizontal medium regions. The inner one is a layer of thickness d with refractive-index n_1. It is surrounded on top and bottom by two outer regions of lower index $n_2 < n_1$. Recalling the principle of reflection and refraction described in Section 2.2 (Figure 2.1), we observe that any incident light ray forming a finite angle θ with the horizontal axis

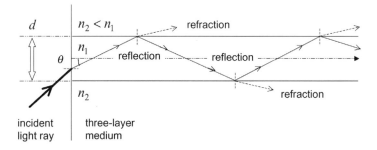

FIGURE 2.16 Propagation of a light ray incident at angle θ inside a three-region medium (thickness d) of higher index n_1 (inner layer) and lower index n_2 (outer regions), showing multiple reflection and refraction effects.

is forced to reflect inside the inner slab. At each reflection point, some fraction of the light-ray's power is also refracted in the two outer regions. Therefore, the effect of multiple refractions is to cause the power to eventually vanish.

We should now recall that below a certain incident angle θ_{crit}, the light ray can be totally reflected at the index-layer interfaces (remember, the condition for total internal reflection is $\theta_{\text{crit}} = \sin^{-1}(n_2/n_1)$, where $n_1 > n_2$). Since multiple reflections occur inside the structure, the condition of total internal reflection means that the ray is definitely "trapped" in the inner layer. In the absence of any propagation/reflection loss effect, this ray is forced to reflect and propagate *ad infinitum* inside the inner structure.

But what applies to a ray is not necessarily true for a wave. As we saw earlier, a uni-directional light wave can be viewed as a group of parallel rays having the same phase and frequency. When launched into the previous structure below a certain angle θ, the wave consistently follows the multiple-reflection effect. Figure 2.17 shows the corresponding wavefronts (or locus of identical E-field phase) after two reflections inside the structure. The condition for constructive interference is that the ray-path difference $\Delta = BC - BB'$ be an integral multiple of the wavelength. An elementary analysis shows that this path difference is equal to $\Delta = 2d \sin\theta$. Therefore, the condition $\Delta = m\lambda$ (m = positive/nonzero integer) leads to

$$\begin{cases} \sin\theta = m\dfrac{\lambda}{2d} \\ \leftrightarrow \theta_m = \sin^{-1}\left(m\dfrac{\lambda}{2d}\right) \end{cases} \quad (2.40)$$

This result shows that E-field waves can constructively interfere by self-replication in the index-layer structure only under the condition that the ray direction be defined by any of the discrete angles $\theta_m = \sin^{-1}[m\lambda/(2d)]$. The additional condition $\theta_m \leq \pi/2 - \alpha_{\text{crit}}$ (α_{crit} = critical angle at index interface), which corresponds to *total internal reflection*, makes it possible for the ray to indefinitely propagate inside the inner structure layer without refraction loss (see below).

For index-layer structures whose thickness d is large compared to the wavelength ($d \gg \lambda$), the definition shows that there is a large number of possible ray

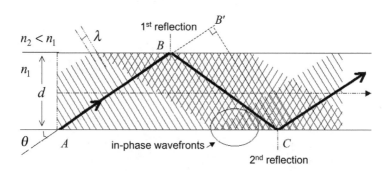

FIGURE 2.17 Ray propagation in slab index-layer structure (thickness d), showing one possible ray angle θ for which multiple reflections at wavelength λ provide in-phase incident and reflected wavefronts.

angles. The maximum number of allowed rays is given by the integer M which is closest to $2d/\lambda$ (i.e., satisfying the requirement $M\lambda/(2d) \leq 1$). In the opposite case, the number of allowed rays rapidly decreases as the thickness approaches the wavelength (see exercise at the end of the chapter). Given a layer thickness, the E-field wavelength must verify $\lambda < \lambda_{\max} = 2d$ in order to ensure that $M \geq 1$. The maximum wavelength $\lambda_{\max}$ is referred to as a "*cut-off*" wavelength above which no ray is allowed to propagate in the structure. This conclusion only concerns rays characterized with finite angles $\theta > 0$. The case of a purely horizontal ray ($\theta = 0$) is irrelevant since it is not subject to any reflection or refraction effects.

From the above analysis, we conclude that a lightwave can be made to indefinitely propagate between two index layers provided its initial ray angle falls into a set of discrete values. This is evidence of discrete *propagation modes*, which in turn suggests the existence of *guided modes* (as will be described later). In thick innerlayer structures, it is clear that the set of angles or propagation modes is a near-continuum. If all these modes were simultaneously present, they would form an interference pattern very similar to the "speckle" of cavity modes (see Section 2.2). At the opposite limit, what happens if the inner-layer thickness d becomes of the order of a few wavelengths or even less? As we have seen, the condition $d = \lambda/2$ represents a limiting case for which no mode can be supported. The only way for light to propagate into the structure is to follow a strictly horizontal path ($\theta = 0$), but since propagation is indifferent to the layer's thickness, we cannot consider this limiting case to correspond to a "guided" mode *per se*. On the other hand, if we substitute this ray with a real E-field wave, then the structure boundaries have the effect of a vertical FP cavity. The thickness condition $d = \lambda/2$ corresponds to a standing wave in the transverse direction. If the upper and lower layers were 100% reflective, an EM wave launched at zero incidence would propagate indefinitely in the structure. In the general case, a fraction of the wave energy would then be expected to leak in the transverse direction, but we can't tell how and to what extent. This simple view however suggests the possibility of a *fundamental waveguide mode*, which could be supported by such a three-layer structure, under certain index difference and inner-layer thickness conditions.

We must abandon at this point the ray picture and make use of the wave properties of the EM field. Although the formal analysis of EM wave propagation in index structures, as based upon Maxwell's equations, is rather complex, we can still use the same simplified approach as for the analysis of laser cavity modes (Section 2.2). The difference between the index-structure and the laser cavity is that we must find the transverse E-field amplitudes corresponding to light rays having finite incidence angles with respect to the horizontal axis. In order to be able to analyze how the EM waves can constructively interfere inside the structure, it is necessary to introduce the notion of *wavevector*. The wavevector indicates the propagation direction of the EM-wave. Its magnitude has the dimension of an inverse length: at wavelength λ and in a medium of refractive index n, it is defined through

$$k = n\frac{2\pi}{\lambda} \equiv nk_0 \quad (2.41)$$

where $k_0 = 2\pi/\lambda$ is the wavevector magnitude corresponding to vacuum ($n = 1$).

The wavevector is a most fundamental characteristic for the EM wave. Not only does it point to the propagation direction of the wave, but it also allows one to determine its absolute phase. Indeed, the E-field and M-fields are defined according to

$$\begin{cases} E(t, L) = E_0 \sin(\omega t + \varphi) \equiv E_0 \sin(\omega t - k \cdot L) \\ H(t, L) = H_0 \sin(\omega t + \varphi) \equiv H_0 \sin(\omega t - k \cdot L) \end{cases} \quad (2.42)$$

with $\omega = 2\pi \nu = 2\pi c/\lambda$ being the wave's (circular) frequency. In the above expression, the term $k \cdot L$ is the "scalar" product $k \cdot L = k_x L_x + k_y L_y + k_z L_z$, where (k_x, k_y, k_z) and (L_x, L_y, L_z) are the three-dimensional coordinates of the wavevector and the physical distance traversed, respectively.

We can then analyze the principle of waveguiding by evaluating the phase delays experienced by reflected rays through this wavevector definition. For this purpose, we shall only consider the E-field, knowing that the conclusions are similar for the M-field (see discussion below). To simplify notations, we shall set $E_0 = 1$, and overlook the time dependence ($t = 0$) in the description.

Consider now the index-structure shown in Figure 2.18. The E-field is assumed polarized in the plane perpendicular to the page. By definition, this configuration is called "*transverse electric*" or TE. Since the E-field wavevector k forms an angle θ with respect to the horizontal axis of this structure, the wave can be seen as the superposition of two components:

1. A longitudinal component propagating in the horizontal direction (z), whose associated wavevector is $k_z = k \cos \theta \equiv \beta$
2. A transverse component propagating in the vertical direction (y), whose associated wavevector is $k_y = k \sin \theta \equiv \gamma$

with the relation $k^2 = k_z^2 + k_y^2 = \beta^2 + \gamma^2$. By definition, the wavevector projection over the horizontal direction (which is called β) is the mode's *propagation constant*. We now introduce the fact that the allowed direction angles must verify the condition $\sin \theta_m = m\lambda/(2d)$. We thus obtain for the two

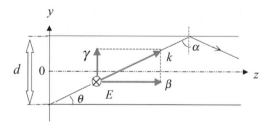

FIGURE 2.18 Decomposition of E-field wavevector into longitudinal (β) and transverse (γ) components. The E-field (E) is assumed to be polarized in the plane orthogonal to the page (TE mode).

wavevector components:

$$\beta^2 = k^2 \cos^2 \theta_m = k^2(1 - \sin^2 \theta_m) = k^2\left[1 - \left(m\frac{\lambda}{2d}\right)^2\right] = k^2 - \left(\frac{2\pi}{\lambda} m \frac{\lambda}{2d}\right)^2$$

$$\longleftrightarrow \beta_m \equiv \sqrt{k^2 - \frac{m^2 \pi^2}{d^2}} \tag{2.43}$$

and

$$\gamma = k \sin \theta_m = \frac{2\pi}{\lambda} m \frac{\lambda}{2d} \leftrightarrow \gamma_m \equiv m\frac{\pi}{d} \tag{2.44}$$

In the following two subsections, we shall assume two configurations for each outer-layer interface: (a) it is made of mirrors having 100% reflectivity; or (b) it is made of a reflecting dielectric medium having an index $n_2 < n_1$. We will refer to these two planar configurations as the *mirror waveguide* and *dielectric waveguide*, respectively.

2.3.2 Mirror Waveguide

The two constants β_m, γ_m defined previously characterize the phase delays experienced by the x-polarized E-field wave as it propagates in the longitudinal and transverse directions, respectively. Since there are two possible angles ($\pm \theta_m$), the total E-field wave can be viewed as the superposition of two independent x-polarized E-field waves which are moving upward and downward, respectively. These upward ($\uparrow$) and downward ($\downarrow$) wave components can then be expressed as follows:

$$E_x(\uparrow) = \frac{1}{2} \sin\left[-\gamma_m\left(y + \frac{d}{2}\right) - \beta_m z\right] \tag{2.45}$$

$$E_x(\downarrow) = \frac{1}{2} \sin\left[+\gamma_m\left(y + \frac{d}{2}\right) - \beta_m z\right] \tag{2.46}$$

In this result, the constant phase terms $\phi_m^{\pm} = \pm \gamma_m d/2$ were introduced, so that the ray path corresponding to the transverse direction at locations $y = \pm d/2$ is an integer number of half wavelengths. These phase terms account for the effect of wave reflection on the mirror interfaces. The relative phase shift between the upward and downward waves at $y = 0$, $\delta \phi = \phi_m^+ - \phi_m^- = \gamma_m d \equiv m\pi$ also reflects the fact the upward and downward waves recombine on the z axis after a half round-trip. The recombination on the z axis is thus in-phase ($\delta \phi \equiv 2\pi$) for even values of m and out-of-phase ($\delta \phi \equiv \pi$) for odd values.

The sum of the two waves is then easily calculated using the formula $\sin a + \sin b = 2 \sin[(a+b)/2] \cos[(a-b)/2]$. Overlooking a minus sign in the result, we obtain:

$$E_x(y, z) = E_x(\uparrow) + E_x(\downarrow) = \cos\left[\gamma_m\left(y + \frac{d}{2}\right)\right]\sin(\beta_m z)$$
$$\equiv \cos\left[\frac{m\pi}{d}\left(y + \frac{d}{2}\right)\right]\sin(\beta_m z) \tag{2.47}$$

We recognize in the above result the same TEM mode pattern as obtained in the case of laser cavities [equation (2.38)], but reduced to a single transverse dimension (y). The wave also propagates in the $+z$ direction, but is not bounded by a longitudinal cavity. It is straightforward to generalize this result to the case of a two-dimensional index structure with a rectangular cross-section. Indeed, assuming the same thickness in the x and y directions, we can write the E-field as

$$E(x, y, z) \equiv \cos\left[\frac{p\pi}{d}\left(x + \frac{d}{2}\right)\right]\cos\left[\frac{q\pi}{d}\left(y + \frac{d}{2}\right)\right]\sin(\beta_{pq} z) \tag{2.48}$$

with (p, q) being integers corresponding to the wavevector projections $\gamma_{px} = p\pi/d$, $\gamma_{qy} = q\pi/d$, and with the propagation constant β_{pq} being given by the conservation relation $\gamma_{px}^2 + \gamma_{qy}^2 + \beta_{pq}^2 = k^2$. The corresponding TEM mode patterns, which are the same as in the laser cavity, are shown in Figure 2.14. Note that the cases $p = 0$ or $q = 0$ (or both) are non-physical or irrelevant, as discussed in the previous ray-propagation model. We must then conceive of the *fundamental mode* as the one for which the two integer are minimum, i.e., $p = q = 1$, corresponding to TEM$_{00}$. It is important to note that this fundamental mode is subject to the wavelength-cutoff condition $\lambda < \lambda_{max} = \lambda_c = 2d$, under which the minimum wave angle ($\theta_0 = \sin^{-1}[\lambda/(2d)]$) exists and is different from zero.

This description assumed the E-field to be x-polarized, corresponding to *transverse-electric* or TE modes. According to Maxwell's laws, the associated M-field is polarized in the (y, z) plane, with both transverse (y) and longitudinal (z) components. An alternate mode configuration, called *transverse magnetic* (TM), corresponds to the case where the M-field is polarized in the x direction, putting the E-field in the (y, z) plane. In this case, it is the E-field which has transverse and longitudinal components. Concerning these E-field amplitudes, the analysis leads to the same result as in equation (2.47) except in the transverse case where the result is multiplied by a factor $1/\tan\theta_p$ (with $\sin^{-1}(\lambda/2d) \leq \theta_p < \pi/2$). The conclusions for TM modes are the same as for TE modes.

2.3.3 Dielectric Waveguide

We assume next a dielectric planar structure with ($n_1, n_2 < n_1$) for the inner and outer refractive indices, respectively. At the index interfaces, the EM wave is both reflected back to the inner layer and refracted into the outer layers. Under certain incidence conditions, however, the wave can be totally reflected. As we have seen the critical incidence is defined as $\alpha_{crit} = \sin^{-1}(n_2/n_1)$, corresponding for the ray angle reference to $\theta_{crit} = \cos^{-1}(n_2/n_1)$. The key difference with the previous case of planar-mirror waveguides is that each reflection at the dielectric interface is characterized by a certain *phase jump*, which we call ϕ_r. It can be shown that in the mirror case, this phase jump is constant and equal to $\phi_r = \pi$, regardless of the incident angle α (this condition ensuring that the sum of the incident and reflected E-fields is identically zero at the mirror interface). Note that in previous analysis, we purposefully overlooked this effect for simplicity's sake, which did not affect the result.

In the dielectric case, the phase jump ϕ_r due to reflection on an index-layer interface depends upon the E-field polarization and other conditions, according to the following rules (see for instance Born and Wolf 1980, or Saleh and Teich 1991):

TE wave (E-field polarized parallel to layers): φ_r is identically zero for $\alpha < \alpha_{\text{crit}}$ and is otherwise given by the formula:

$$\tan^2(\phi_r/2) = \frac{\sin^2 \alpha - \sin^2 \alpha_{\text{crit}}}{\cos^2 \alpha}$$

TM wave (E-field polarized perpendicular to layers): φ_r is identically zero for $\alpha < \alpha_{\text{crit}}$, and is otherwise given by the formula:

$$\tan^2(\phi_r/2) = \frac{\sin^2 \alpha - \sin^2 \alpha_{\text{crit}}}{\cos^2 \alpha \sin^2 \alpha_{\text{crit}}}$$

We observe that the phase jump difference between the TE and TM cases differ in the $\tan^2$ definition only by a multiplying factor. Therefore, it suffices to analyze the TE case considering that the conclusions will be similar.

Looking back to Figure 2.23, we now recall that the physical path difference between incident and twice-reflected TE waves is $\Delta = BC - BB' = 2d \sin \theta$. Since there a phase jump φ_r at each interface reflection, we must rather consider the net phase-delay difference Φ between these two paths. The condition for constructive (in-phase) interference now becomes

$$\Phi = k\Delta - 2\phi_r = 2m\pi$$

$$\leftrightarrow \frac{2\pi}{\lambda} 2d \sin \theta - 2\phi_r = 2m\pi \qquad (2.49)$$

$$\leftrightarrow \frac{\phi_r}{2} = \pi \frac{d \sin \theta}{\lambda} - m\frac{\pi}{2}$$

Note that in the above expression we used for notation simplicity the wavevector definition $k = 2\pi/\lambda$ instead of $k = 2\pi n_1/\lambda_0$ (λ_0 = wavelength in vacuum), which is the same as defining the wavelength in the medium of index n_1 as $\lambda = \lambda_0/n_1$.

Using the correspondence $\theta = \pi/2 - \alpha$ between the two angles (α, θ), the TE phase jump can also be expressed as a function of the angles θ, θ_{crit}:

$$\tan^2(\phi_r/2) = \frac{\sin^2(\pi/2 - \theta) - \sin^2(\pi/2 - \theta_{crit})}{\cos^2(\pi/2 - \theta)}$$

$$= \frac{\cos^2(\theta) - \cos^2(\theta_{crit})}{\sin^2(\theta)} \equiv \frac{\sin^2(\theta_{crit})}{\sin^2(\theta)} - 1 \quad (2.50)$$

Combining the two expressions of equations (2.49) and (2.50), we finally obtain the constructive-interference condition for the angle θ:

$$\left(\frac{X_{crit}^2}{X^2} - 1\right)^{\frac{1}{2}} = \tan\left(\pi\frac{d}{\lambda}X - m\frac{\pi}{2}\right) \quad (2.51)$$

where $X = \sin\theta$ and $X_{crit} = \sin\theta_{crit}$. Given the waveguide thicknesss (d), critical angle (θ_{crit}) and wavelength (λ), this equation in X can be solved numerically for each value of m. This equation can also be easily solved graphically, using the parameter X for the horizontal axis and plotting both LHS and RHS (as determined for different m). The LHS/RHS intersection points provide the solutions X and then the corresponding angle θ. An illustrative example of such graphical solution is provided in Figure 2.19, where the value $X_{crit} = 5\lambda/(2d)$ was assumed. With the normalized variable $x = X/X_{crit} = \sin\theta/\sin\theta_{crit}$, the functions plotted in the figure

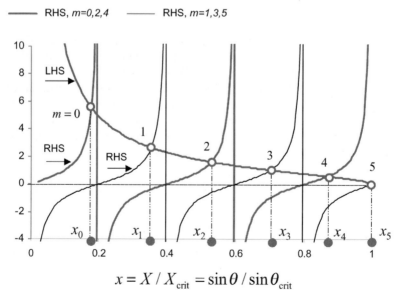

$$x = X/X_{crit} = \sin\theta/\sin\theta_{crit}$$

FIGURE 2.19 Graphical solution of equation (2.51) or equation (2.52), for different values of the integer parameter m, showing LHS and RHS intercepts (open circles) and corresponding discrete solutions $x_m = \sin\theta_m/\sin\theta_{crit}$ on the x axis (dark circles). The values of x_m are listed in Table 2.2.

are simply defined by

$$\left(\frac{1}{x^2} - 1\right)^{\frac{1}{2}} = \tan\left[\frac{\pi}{2}(5x - m)\right] \quad (2.52)$$

with [0, 1] being the interval of interest for x, since the argument in the square root must be strictly positive. It is seen that the value $x = 1$ (or $\theta = \theta_{\text{crit}}$) corresponds to $m = 5$, which sets an upper limit to the parameter m. We observe from the figure that the LHS is a monotonic decaying function while the RHS is periodic with period $T_x = (1/2)2\pi/(5\pi/2) \equiv 0.4$. For each value of m ($m = 0\ldots 5$), the two types of curves have a unique intercept in the upper plane, yielding many discrete solutions $x_m = \sin\theta_m / \sin\theta_{\text{crit}}$, where θ_m is the corresponding solution angle. The exact numerical values of x_m are listed in Table 2.2.

We shall now use Table 2.2 in order to derive a concrete interpretation of the above results. To establish physical correspondence between the x_m solutions and the mode angles θ_m, we must attribute some value for the parameter $\sin\theta_{\text{crit}}$ (so far assumed equal to $\sin\theta_{\text{crit}} = 5\lambda/(2d)$ without any other specification). Recall that the critical angle is given by the definition $\cos\theta_{\text{crit}} = n_2/n_1$, where n_2 and $n_1 > n_2$ are the refractive indices of the outer and inner layers, respectively. Thus we have $\sin\theta_{\text{crit}} = \sqrt{1 - (n_2/n_1)^2}$. We can then assume n_2/n_1 varies from 0.9 to 0.9999, corresponding to a relative index difference (also called *relative index*):

$$\Delta = \frac{n_1 - n_2}{n_1} = 1 - \frac{n_2}{n_1} \equiv 10^{-1} \text{ to } 10^{-4} \quad (2.53)$$

For each relative index Δ (or n_2/n_1 ratio), we can therefore determine the family of angle solutions θ_m corresponding to the x_m values listed in Table 2.2. We first observe from results listed in the table that the angle θ_m increases with the mode order m, as expected. Second, it is seen that for relatively large index differences ($\Delta = 10^{-1}$), the angle values spread between 4° and 26°, corresponding to a

TABLE 2.2 Values of $x_m = \sin\theta_m/\sin\theta_{\text{crit}}$, corresponding to the graphical solutions shown in Figure 2.19

		θ_m ($\cos\theta_m$)			
m	x_m	$\Delta = 10^{-1}$	$\Delta = 10^{-2}$	$\Delta = 10^{-3}$	$\Delta = 10^{-4}$
0	0.175	4.36 (**0.997**)	1.41 (**0.999**)	0.44 (**0.9999**)	0.14 (**0.99999**)
1	0.355	8.88 (**0.998**)	2.86 (**0.998**)	0.89 (**0.9998**)	0.28 (**0.99998**)
2	0.530	13.3 (**0.973**)	4.28 (**0.997**)	1.33 (**0.9997**)	0.42 (**0.99997**)
3	0.705	17.8 (**0.952**)	5.70 (**0.995**)	1.77 (**0.9995**)	0.56 (**0.99995**)
4	0.870	22.2 (**0.925**)	7.04 (**0.992**)	2.19 (**0.9992**)	0.69 (**0.99992**)
5	1.00	25.8 (**0.900**)	8.10 (**0.990**)	2.52 (**0.9990**)	0.80 (**0.99990**)

The waveguide solution angles θ_m (degrees) and their cosine (in bold) are also listed while assuming different relative-index values: $\Delta = 10^{-1}$, 10^{-2}, 10^{-3} and 10^{-4} (corresponding to $\sin\theta_{\text{crit}} = 0.435$, 0.141, 0.044 and 0.014, respectively).

broad range of possible ray incidences. But for very small index differences ($\Delta = 10^{-4}$), the angle spread is reduced to a $0.1°–0.8°$ range, which is associated with very shallow ray incidences. We can then conclude that the ray incidence of the mode is controlled by the relative index parameter: the smaller the relative index, the shallower the mode-ray incidence. In the limit $\Delta \to 0$ (but with $\Delta \neq 0$), all modes propagate like near-parallel rays.

Table 2.2 also lists the mode-angle cosines, i.e., $\cos \theta_m$, which are associated with the modal *propagation constants* β_m. Recall indeed that β_m is given by the projection of the mode wavevector on the longitudinal direction, i.e.,

$$\beta_m = k \cos \theta_m \equiv n_1 k_0 \cos \theta_m \tag{2.54}$$

It is then possible to define an *effective mode index* according to $n_{\text{eff}}(m) = n_1 \cos \theta_m$, so that the mode propagation constant is equal to

$$\beta_m \equiv n_{\text{eff}}(m) k_0 \tag{2.55}$$

Since $\cos \theta_m$ is bounded according to $n_2/n_1 = \cos \theta_{\text{crit}} \leq \cos \theta_m < 1$, we also have $n_2 \leq n_{\text{eff}}(m) = n_1 \cos \theta_m < n_1$, meaning that the effective index falls somewhere between the inner-layer index (n_1) and outer-layer index (n_2). For shallow rays ($\cos \theta_m \approx 1$) the effective index is close to n_1. As the ray incidence increases, the effective index becomes closer to n_2, until one reaches the value $\theta_m = \theta_{\text{crit}}$ where its is exactly equal to n_2. This feature is illustrated by the results shown in Table 2.2 for $\cos \theta_m$. Considering for instance $\Delta = 10^{-1}$, we see that the cosine decreases from 0.997 ($m = 0$) to 0.990 ($m = 5$), corresponding to an effective-index range of $n_{\text{eff}}(0) = 0.997 \times n_1 < n_1$ to $n_{\text{eff}}(5) = 0.990 \times n_1 = n_2$. The last case, where $n_{\text{eff}}(5) = n_2$ or $\beta_5 \equiv n_2 k_0$, corresponds to a regime where the mode's effective index is identical to that of the surrounding layers of the waveguide, and the propagation constant is the same as in a medium of uniform index n_2. For this reason it does not correspond to a "guided" mode, even if it is a solution of the self-replicating wave problem. The only guided modes are the solutions for which $n_2 < n_{\text{eff}}(m) < n_1$, corresponding to $m = 0 \ldots 4$. The key conclusion of the analysis is that for guided modes, the propagation constant β_m always verifies the double inequality:

$$n_2 k_0 < \beta_m < n_1 k_0 \tag{2.56}$$

The previous example used as a working assumption (and for practical illustration purposes) the ad hoc relation $\sin \theta_{\text{crit}} = 5\lambda/(2d)$, which lead to the identification of 5 guided-mode solutions corresponding to $m = 0 \ldots 4$. In the general case, given the angle θ_{crit} and the wavelength λ, what is the number of modes supported by the waveguide? To answer this question, we first note that the RHS in equation (2.51) has asymptotes ($\tan(k\pi/2) = \infty$, $k = 1, 2, 3 \ldots$) every time $\sin \theta$ is a multiple of $\lambda/(2d)$. We also know that there is a mode solution $\sin \theta_m$ between two such asymptotes. Since the maximum allowed value for $\sin \theta$ is $\sin \theta_{\text{crit}}$ (included), the

2.3 Optical Waveguides

total number of mode solutions is given by

$$M = \frac{\sin\theta_{\text{crit}}}{\frac{\lambda}{2d}} = \frac{2d}{\lambda}\sqrt{1-\cos^2\theta_{\text{crit}}} = \frac{2dn_1}{\lambda_0}\sqrt{1-\left(\frac{n_2}{n_1}\right)^2} \equiv \frac{2d}{\lambda_0}\sqrt{n_1^2 - n_2^2} \quad (2.57)$$

where M stands for the closest integer immediately above the real value taken by the RHS of the expression (e.g., if RHS = 0.8, 1, or 1.5, then $M = 1, 2,$ or 2). This rule can be verified graphically by assuming $\sin\theta_{\text{crit}} = u\lambda/(2d)$ with u integer or real ($u > 0$). We then introduce a new waveguide parameter called *numerical aperture* defined by:

$$NA = \sqrt{n_1^2 - n_2^2} \quad (2.58)$$

which gives for the number of modes

$$M = \frac{2d}{\lambda_0} NA \quad (2.59)$$

The important result of this analysis is that the dielectric structure always supports at least one mode ($m = 0$, $n_{\text{eff}}(0) > n_1$) for any given wavelength λ and numerical aperture NA. This is true even at the limit $NA \to 0$ or $n_2 \approx n_1$ and $n_{\text{eff}}(0) \approx n_1$. At such a limit, however, the mode is not really "guided" by the structure, even if the ray angle θ_0 is identically zero. The requirement for the structure to support only one mode is $M < 1$, or

$$\lambda > \lambda_{\text{cutoff}} = 2dNA \quad (2.60)$$

where λ_{cutoff} is by definition the *cut-off wavelength*. For wavelengths above the cut-off value, the waveguide is said to be *single-mode*. One also refers to the first mode solution ($m = 0$) as the *fundamental mode*. See Exercises at the end of the chapter for examples. It is also left as an exercise to show that the numerical aperture is related to the waveguide's *acceptance angle*, namely the angle under which an outside ray must be launched into the waveguide in order to be captured or totally internally reflected.

The previous analysis has made it possible to determine the set of TE modes that can be supported by a planar dielectric waveguide, given their thickness and relative-index characteristics (d, Δ). We have also determined the mode propagation constants $\beta_m \equiv n_1 k_0 \cos\theta_m$ corresponding to the ray-angle solutions θ_m. To complete the picture, we must finally determine the mode E-field envelope, $E(y, z)$, corresponding to the waveguide's inner and outer layers.

We first recall from the previous subsection that the TE mode is made of the sum of two TE-field waves, $E_x(\uparrow)$, $E_x(\downarrow)$, travelling upwards and downwards with longitudinal and transverse wavevectors defined by $k_z(m) = k\cos\theta_m \equiv \beta_m$ (propagation constant) and $k_y(m) = k\sin\theta_m$, respectively. A second property (already discussed in the previous subsection) is that when recombining on the z-axis, the upward and downward waves exhibit an intrinsic phase difference of $\delta\phi = m\pi$. This is due to the fact that each wave recombines after having traveled over only a half round-trip (the full round-trip corresponding to $2\delta\phi = 2m\pi$). Therefore, the wave recombination on the z-axis is either in-phase ($\delta\phi \equiv 2\pi$) for even m values, yielding an E-field maximum, or out-of-phase ($\delta\phi \equiv \pi$) for odd m values, yielding an E-field minimum or a wave node. Consistently with these properties, the TE-fields in the inner waveguide layer can be expressed as follows:

$$E_x(\uparrow) = \frac{1}{2}\sin[-k_y(m)y + m\pi/2 - \beta_m z] \quad (2.61)$$

$$E_x(\downarrow) = \frac{1}{2}\sin[+k_y(m)y - m\pi/2 - \beta_m z] \quad (2.62)$$

with $|y| \leq d/2$. Using the formula $\sin a + \sin b = 2\sin[(a+b)/2]\cos[(a-b)/2]$ the net sum (in absolute value) of the two fields is identically equal to:

$$E_x(y, z) = E_x(\uparrow) + E_x(\downarrow) = \cos\left[\frac{2\pi\sin\theta_m}{\lambda}y + m\frac{\pi}{2}\right]\sin(\beta_m z) \quad (2.63)$$

noting that $\lambda = \lambda_0/n_1$. Detailing the result for even and odd values of m, we get:

$$E_x^{\text{in}}(y, z) = \begin{cases} \cos\left(\dfrac{2\pi\sin\theta_m}{\lambda}y\right)\sin(\beta_m z), & m = 0, 2, 4 \ldots \\ \sin\left(\dfrac{2\pi\sin\theta_m}{\lambda}y\right)\sin(\beta_m z), & m = 1, 3, 5 \ldots \end{cases} \quad (2.64)$$

where the superscript "in" recalls that the definition applies to the E-field located inside the waveguide's inner layer. From this last result, we can observe the following features:

- The E-field amplitude oscillates in the transverse plane, as in the mirror-waveguide case, but the corresponding oscillation periods are not harmonic ($T_m = \lambda/\sin\theta_m$);
- On the waveguide axis ($y = 0$), the amplitude is either maximum or zero, according to the mode order m being either even or odd;
- The amplitude does not vanish at the waveguide boundaries $y = \pm d/2$, unlike in the the mirror-waveguide case;
- The even-mode amplitude distributions (cosine) are symmetric with respect to the axis, while the odd-mode amplitude distributions (sine) are antisymmetric.

By definition, the even- or odd-numbered modes are called *symmetric* or *antisymmetric*, respectively.

Since the E-field does not vanish at the waveguide's interfaces, the amplitude distribution must extend into the outer layer. We expect this E-field extension to be of finite range, since the modes are effectively guided in the structure. We may also expect it to vanish more rapidly for the modes of lower order, corresponding to rays of shallower incidence, and the reverse for modes of higher order. The analysis of this problem requires one to use Maxwell's laws of electromagnetism. Here, we shall only provide the result, which is remarkably simple. The E-field amplitude in the outer layer $|y| \geq d/2$ is given by

$$E_x^{\text{out}}(y, z) = A_m^{\pm} \exp(-\mu_m y) \sin(\beta_m z) \quad (2.65)$$

with μ_m being a decay constant defined by

$$\mu_m^2 = \beta_m^2 - n_2^2 k_0^2 \equiv n_1^2 k_0^2 \cos^2 \theta_m - n_2^2 k_0^2$$

$$\leftrightarrow \mu_m = k_0 \sqrt{n_{\text{eff}}^2(m) - n_2^2} \quad (2.66)$$

Since the effective index, $n_{\text{eff}}(m)$, is strictly greater than the outer index, n_2, the decay constant is always non-zero. Moreover, it increases with $n_{\text{eff}}(m)$, which means that the lower-order modes ($n_{\text{eff}}(m) \to n_1$) experience a more rapid decay than the higher-order ones ($n_{\text{eff}}(m) \to n_2$). In other words, the lower-order modes are mode confined into the waveguide, while the higher-order ones extend further in the waveguide's outer region. For all guided modes, the E-field thus extends more or less in the outer region with the amplitude decaying as an exponential. This portion of the mode is called an *evanescent wave*.

In the above E-field definition, the constant $A_m^{\pm}$ is chosen so that the TE-field amplitude is continuous at each the two waveguide interfaces, i.e., $E_x^{\text{in}}(y = \pm d/2, z) = E_x^{\text{out}}(y = \pm d/2, z)$. For symmetric modes, the value is the same for the lower and the upper layer, while for anti-symmetric modes, the two values are of opposite sign.

We consider the previous example whose ray angles θ_m have been listed in Table 2.2, in the specific case $\Delta = 10^{-1}$ ($n_2 = 0.9n_1$, $\theta_{\text{crit}} = 25.8°$). In order to plot the corresponding E-field amplitude distributions defined in Equations (2.64) and (2.65), we need to specify certain parameters:

- Recall that these angle solutions were obtained under the condition $\sin \theta_{\text{crit}} = 5\lambda/(2d)$, which fixes the ratio λ/d; if we assume $d = 10\,\mu\text{m}$, the wavelength must be $\lambda = 2d \sin \theta_{\text{crit}}/5 = 1.741\,\mu\text{m}$;
- Another free parameter is the index n_1; let assume $\lambda_{\text{cutoff}} = 2dNA = 10\,\mu\text{m}$, which gives $NA = 0.5 = n_1\sqrt{1 - (n_2/n_1)^2}$ and consequently $n_1 = 1.1470$ and $n_2 = 1.0323$; for reference, the vacuum wavelength $\lambda_0 = n_1 \lambda = 1.9968\,\mu\text{m} \approx 2\,\mu\text{m}$.

Having specified all required parameters, we can proceed to calculate the E-field amplitude $E_x^{\text{in,out}}(y, z = 0)$ for each mode m as a function of the transverse coordinate y. For doing this, we need to evaluate the propagation constant β_m, damping constant μ_m, the cosine/sine argument $u_m = (2\pi/\lambda) \sin \theta_m$ and the amplitudes $A_m^{\pm}$ corresponding to each mode m. These different parameters are listed in Table 2.3 for the reader's reference. Figure 2.20 shows the plots of the mode's amplitude distributions, as separated into groups of symmetric and anti-symmetric modes. Several features can now be observed from this result. First, the E-field remains confined inside the waveguide, while having rapidly decaying evanescent tails. As the mode-order m increases, the evanescent tails are seen to spread further in the outer index region, although with more rapid decay (see Table 2.3) for corresponding values of the decay constant μ_m). As expected, the highest confinement is achieved

TABLE 2.3 Values obtained for the propagation constant β_m, the damping constant μ_m, the cosine/sine argument $u_m = (2\pi/\lambda)\sin\theta_m$ and the amplitudes $A_m^\pm$ for each mode m, based upon the angle-solutions θ_m listed in Table 2.2 for $\Delta = 10^{-1}$

m	θ_m (deg)	$\sin\theta_m$	β_m (μm^{-1})	μ_m (μm^{-1})	u_m (rad)	$E_x^{in}(d/2)$	A_m^+
0	4.36	0.076	3.597	1.545	0.2742	0.198	448.32
1	8.88	0.154	3.564	1.467	0.5557	−0.355	−544.22
2	13.3	0.230	3.511	1.333	0.8300	−0.533	−418.11
3	17.8	0.305	3.485	1.249	1.1007	0.703	362.34
4	22.2	0.377	3.340	0.778	1.3605	0.868	42.45

The other parameters are $\lambda_0 = 2\,\mu m$, $d = 10\,\mu m$, $\lambda_{cutoff} = 10\,\mu m$, $NA = 0.5$, $n_1 = 1.1470$ and $\theta_{crit} = 25.8°$.

with the fundamental mode ($m = 0$), which is characterized by a single lobe. The mode power or intensity distributions can be obtained by taking the square of these E-field envelopes. The result is that all modes but the fundamental one exhibit multiple lobe patterns, the number of lobes being $m + 1$. Note that the

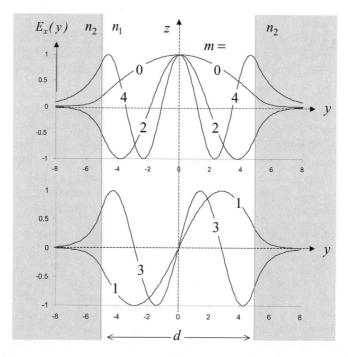

FIGURE 2.20 Transverse-electric field-amplitude distributions in planar dielectric waveguide of thickness $d = 5\,\mu m$ at wavelength $\lambda_0 = 2\,\mu m$, corresponding to the angle solutions obtained in Figure 2.19 and parameters listed in Table 2.3. The top and bottom curves represent symmetric ($m = 0, 2, 4$) and antisymmetric ($m = 1, 3$) modes, respectively.

same family of modes exists for the TM case, for which the M-field is polarized parallel to the transverse x axis and the E-field parallel to the (y, z) plane.

This analysis has been conceptually intensive. It made it possible to understand how EM waves can propagate as self-replicating modes inside dielectric waveguides. We have seen that the E-field extends in the outer region as an evanescent wave, while most of the energy is confined in the center of the index structure. Each mode is characterized by an equivalent ray angle and a propagation constant which represent the projection of the wavevector on the longitudinal axis. Equivalently, one can associate to each mode an effective refractive index whose value falls between the inside and outside index values. The effective index of the lowest-order modes is closer to the inside (higher-value) index, while that of the higher-order modes is closest to the outside (lower-value) index. The effect of waveguiding stops for the last mode solution for which the effective index is exactly equal to the outside index, which corresponds to the angle boundary for total internal reflection. Finally, we have seen that the waveguide is characterized by a numerical aperture (NA) and a cut-off wavelength (λ_c). At wavelengths above this cutoff value, the waveguide only supports the fundamental mode, which corresponds to the lowest ray incidence. Remarkably, all these important results were obtained by using a relatively elementary model of rays and wavevector composition. The only extra assumption concerned the evaluation of the evanescent mode tails, which required the use of a result from Maxwell's equation analysis. In any case, we now have all the main conceptual tools to understand the properties of other types of dielectric waveguides.

It is easy to generalize the results previously obtained with *planar* waveguides to the case of *strip waveguides*. A strip waveguide is defined as a two-dimensional index structure, i.e., with a central region of higher index (called the *core*) surrounded in the transverse plane (x, y) by an outer region of lower index (called the *cladding*). A light ray traversing this structure's core can be decomposed into two projections: one on the horizontal plane (x, z) with wavevector k_x and one on the vertical plane (y, z) with wavevector k_y. Consistently, the three wavevector cooordinates $(k_x, k_y, k_z = \beta)$ must verify $k^2 = k_x^2 + k_y^2 + \beta^2$ (with $k = n_1 k_0 = 2\pi n_1/\lambda_0$). In order for the wave to self-replicate after two internal reflections, the transverse wavevector components must verify the same discretizing condition as in equation (2.44), but now in two dimensions, namely

$$\begin{cases} k_x = m\dfrac{\pi}{d} \\ k_y = n\dfrac{\pi}{d} \end{cases} \quad (2.67)$$

where (m, n) are two positive integers and the waveguide cross-section is assumed square with sides of length d. The range of permitted values for (m, n) is defined by the condition $k_x^2 + k_y^2 < k^2$, for which there exists a nonzero propagation constant $\beta = \beta_{mn}$. Since the transverse wavevectors increase by integer multiples of π/d, the number of mode possibilities M (for TE or TM) is approximately given by the number of unit cells with surface $s = (\pi/d)^2$ that fit inside a quarter-circle of radius k (surface $S = \pi k^2/4$). The result is thus

$$M \approx \frac{S}{s} = \frac{\pi}{4}\left(\frac{2\pi}{\lambda}\right)^2 \bigg/ \left(\frac{\pi}{d}\right)^2 = \frac{\pi}{4}\left(\frac{2d}{\lambda}\right)^2 \qquad (2.68)$$

If we recall that in the one-dimensional case the number of modes is about $M \approx 2d/\lambda$, we see from the above result that the second degree of freedom allows a number of modes approximately equal to the square of this value. If we now assume that the waveguide interfaces are not 100% reflective, the mode-ray incidences must remain under a certain critical ray angle θ_{crit}. The existence condition becomes $k_x^2 + k_y^2 < k^2 \sin^2 \theta_{\text{crit}}$, which changes the circle radius into $k \sin \theta_{\text{crit}} = k\sqrt{1 - \cos^2 \theta_{\text{crit}}} = (2\pi n_1/\lambda_0)\sqrt{1 - (n_2/n_1)^2} = (2\pi/\lambda_0)NA$. The corresponding number of TE or TM modes is therefore

$$M \approx \frac{\pi}{4}\left(\frac{2d}{\lambda_0}\right)^2 NA^2 \qquad (2.69)$$

As we have seen earlier, the condition for single-mode operation is $M < 1$. This condition corresponds to $\lambda_0 > \lambda_{\text{cutoff}}$, where the cut-off wavelength is now defined as $\lambda_{\text{cutoff}} = dNA\sqrt{\pi}$. We can also introduce a dimensionless parameter called *V-number*, which is defined as follows:

$$V = \frac{2\pi a}{\lambda_0} NA \qquad (2.70)$$

with $a = d/2$ being the half-width of the waveguide, the condition $M < 1$ is equivalent to

$$V < \sqrt{\pi} = 1.77 \qquad (2.71)$$

This V-number, also called *normalized frequency*, which defines the condition for single-mode operation, is another important parameter for characterizing waveguides. As described in the next section, the V-number for single-mode fibers is defined the same way as in equation (2.70), but with a representing the core radius. While the mode analysis is far more complex than in the above model, we will see that the condition for single-mode operation in fibers is $V < 2.405$. The V-number definition and the 2.405 limiting value are both simple to memorize.

Different basic possibilities for strip-waveguide geometries are illustrated in Figure 2.21. The realization of a strip waveguide requires a planar substrate into which a core/cladding index structure can be built. The one-dimensional index structure, previously called the *planar waveguide*, is also called a *slab waveguide*. Strip waveguides can be buried in the substrate at different depths (*buried waveguide*). The index difference and confinement can be obtained by various techniques of material-layer deposition and regrowth, ion/impurity doping or beam implantation. Alternatively, the strip can be built on top of the substrate to form a *ridge waveguide*. This can be realized for instance after etching away the cladding materials immediately surrounding the core, in which case the air acts as a natural cladding. Buried and ridge waveguides are based on *planar technologies*.

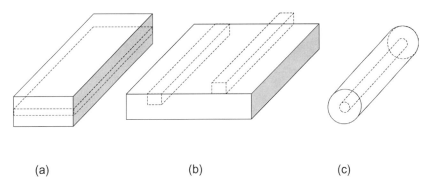

FIGURE 2.21 Different types of dielectric-waveguide geometry: (a) slab, (b) strip: buried (left), ridge (right), and (c) fiber. In each case, the region of higher index (waveguide core) is delimited by the dashed line.

Several examples of planar-technology applications are provided in the section concerning both passive and active optical components. The substrate's material can be chosen to exhibit additional properties, such as index-modulation by an external E-field (*electro-optic effect*) and gain generation by current injection (*semiconductor laser diode*). The special case of waveguides having a *cylindrical geometry*, corresponds to *optical fibers*, which we describe next.

2.3.4 Glass Fiber Waveguides

Optical fibers are cylindrical waveguides made of low-loss materials such as *glass*, which is mostly based on *fused-silica* (SiO_2) compounds, and in specific applications of heavy-metal *fluoride glasses* (X_nF_m, X = Zr, Ba, Al, Na, ...) and even *polymers*. The fiber is characterized by a *core* of high index (n_1), and a *cladding* of lower index (n_2). It is always surrounded by a polymer *jacket*, which provides mechanical strength against bending (the fiber would be otherwise brittle). Another purpose for the jacket is to keep the glass away from *water*, which otherwise causes surface defects and diffuses through the material causing significant absorption-loss increase. Finally, the jacket of higher refractive index forms a second cladding waveguide which strips away any unwanted *cladding modes*. Excluding the jacket, the standard outside diameter (OD) of fibers is 125 μm or 1/8 mm. With the jacket, the OD is closer to 1 mm. Such very small dimensions make it possible to bundle tens to hundreds of fibers within a region of one-centimeter cross-section, allowing the realization of lightweight and very dense communication cables.

The dimensions of the fiber core are not uniquely defined: *multimode* (MM) *fibers* are characterized by relatively large core diameters (50 or 62.5 μm) while *single-mode* (SM) *fibers* have core diameters of the order of 10 μm or less. The basic approach to fabricating an optical fiber is to realize a meter-long glass rod with OD = 2–5 cm, with core and cladding regions of different refractive indices. To modify the index, certain dopants are added to the glass composition of the core. Germanium (Ge), phosphorous (P) and aluminum (Al) are the most current elements used to raise the glass index. Such *index-raising codopants* readily integrate into silica glass as oxides (GO_2, P_2O_5 and Al_2O_3). An alternative approach is to *decrease* the silica index using elements such as boron (B) or fluorine (F) in the fiber cladding. The relative index difference $\Delta = 1 - n_2/n_1$ can be augmented by combining both approaches, i.e., using index-raising and index-lowering codopants in the core and in the cladding, respectively. The concentrations of such codopants must be kept relatively low (a few %) in order not to significantly alter the glass purity and cause excess material absorption (see following text). Another issue associated with doping is the change of expansion coefficient (especially with phosphorus); excessive differences between the core and the cladding's coefficients may result in unwanted stress and fiber cracks.

The transverse variation of the refractive-index is called the *index profile*. The index-profile characteristics are not only important for waveguiding and defining the number of modes, but also for controlling the fiber *dispersion*. The dispersion is an important parameter which defines the signal velocity as a function of wavelength. The index profile also corresponds to a specific numerical aperture (NA), which defines the acceptance cone angle for launching rays into the fiber (see Exercises at the end of the chapter). The fiber NA must therefore be carefully chosen according to the type of laser source and its characteristic beam cone angles. Finally, the index profile controls the mode's transverse dimensions or the power confinement into the fiber core. This confinement is important for minimizing the effect of loss due to fiber bending. In certain applications such as fiber lasers and amplifiers, a high power confinement corresponding to high intensities (or power/surface ratio) is a prerequisite for an efficient pumping operation. Different types of index profiles are shown in Figure 2.22. The profiles exist in two main categories: *step index* and *graded index*. In the first case, the profile is just defined as an index discontinuity forming a core of radius a. The index difference $n_1 - n_2$ can be increased by

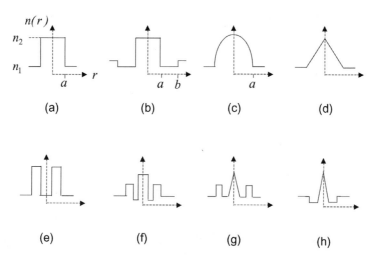

FIGURE 2.22 Different types of fiber refractive-index profiles: (a) step, (b) step with depressed cladding, (c) graded, (d) triangular, (e) annular-ring, (f) W or double-clad, (g) triangular with annular ring and (h) triangular with depressed cladding.

depressing the cladding index in a region defined by the radii a, b with the external radius b chosen sufficiently large to avoid waveguiding effects from the second cladding ($r > b$) of intermediate index. The graded-index profile consists in a continuous index variation from a lower value n_2 at some radius $r = a$ to a peak value n_1 at $r = 0$. The profile can be defined according to the power law

$$n(r) = n_1 \sqrt{1 - 2\left(\frac{r}{a}\right)^p \Delta} \qquad (2.72)$$

for $r \leq a$ and $n(r) = n_2$ otherwise. In this definition, the exponent parameter p determines the profile slope. *Parabolic* and *triangular* profiles are obtained for $p = 2$ and $p = 1$, respectively (see Figure 2.22, cases (c) and (d)). The step profile corresponds to the limit $p \to \infty$. The exponent is chosen to target specific dispersion properties, for instance to obtain the dispersion at the 1.55 μm telecommunications wavelength. Other possible index-profiles are of the *annular* and "W" type (cases (e) and (f) in figure). Many derivatives of these profiles exist, including for instance a combined triangular/annular shape (case (g) in figure). Such profiles provide extra degrees of freedom when designing fibers for special dispersion properties. Note that the W profile should not be confused with the depressed-cladding one. In the W case indeed, the multiple transverse variations of the index have dimensions comparable to the mode size, which strongly affects the propagation constant and its wavelength dependence (dispersion).

The analysis of EM-wave propagation in fiber waveguides follows an approach quite similar to that previously described in the case of planar waveguides. The specific cylindrical geometry of fibers makes the analysis a bit more complex. We can conceive indeed that in step-index fibers, the rays bounce on the core-cladding interface like a tennis ball launched with any arbitrary angle would do inside a large pipe. But in the case of graded-index the ray is continuously bent according to Descartes/Snell's law: $n \sin \theta = (n + \delta n) \sin(\theta + \delta \theta)$, where δn is the small change of index on any point in the ray path, and $\delta \theta$ the corresponding angle deviation. Because of the continuous index gradient, these rays do not bounce inside the core but rather follow a helical or cork-screw path. They are called *skewed rays*. We may leave here the ray-picture to focus on the EM properties of the guided waves.

The analysis of EM-wave propagation in graded-index (or in any arbitrary index structure) is done through Maxwell's equations. While the TE/TM mode analysis is quite tedious, even in the most basic case of step-index profiles, suffice it to say that it can be simply handled by routine computer software, just as in the calculation of antenna radiation patterns and other problems linked to three-dimensional EM-wave propagation. Since the background concepts have already been introduced with planar waveguides, we shall only provide the main results of interest (for an advanced and thorough description, see G. Keiser (1991) for instance).

As we have seen earlier, the modes in two-dimensional waveguides are indexed by two integer numbers (m, n). Considering step-index fibers, the (m, n) mode solutions with corresponding propagation constants $n_2 k_0 < \beta_{mn} < n_1 k_0$ form a prolific set of possibilities which break out into four distinct families. These families are called TE_{mn}, TM_{mn}, EH_{mn} and HE_{mn}, respectively. The TE and TM modes correspond to the pure transverse-electric and magnetic field polarizations, similar to that obtained in two-dimensional planar waveguides. In the TE case, the E-field is perfectly orthogonal to the z axis, while the M-field (itself perpendicular to the E-field) has a small component along the z axis. The reverse applies for the TM case. In the case of EH and HE modes, both E- and M-fields have components along the z axis, with corresponding rays following a helical path. They are referred to as *electric-helical* (EH) or *magnetic helical* (HE) according to the dominance of their z-component (electric or magnetic).

Any mode in these four categories can be supported by the fiber waveguide provided its associated V-number (see Section 2.3.3) is greater than some cut-off value. Although there is no logic to be found in the mode-name sequence, the order according to which these modes "show up" as the V-number is increased is the following:

HE_{11} ($V = 0$)

TM_{01} and TE_{01} ($V \approx 2.405$)

HE_{21} ($V \approx 2.6$)

EH_{11} ($V \approx 3.8$)

HE_{12} ($V \approx 3.9$)

HE_{31} ($V \approx 4.1$), etc...

From this result, we conclude that HE_{11} must be the fundamental mode (no cut-off), and furthermore, that the condition $V < 2.405$ ensures that only this mode can exist, which corresponds to the operation of a *single-mode fiber*. The EM-field amplitude and polarization patterns associated with each of these fiber modes are quite complex and difficult to visualize in 3-dimensional space, even for the lowest-order types. The E-field lines (to which the E-field is locally tangent) corresponding to the lowest-order modes are shown in Figure 2.23. When computing the propagation constants in fibers having very small index differences ($\Delta \leq 10^{-2}$), it was observed that some of the solutions exhibit very similar propagation constants, $\beta/k_0 = f(V)$. In such *weakly guiding fibers*, it thus made sense to group together the mode solutions having these similarities, defining modes not as pure solutions

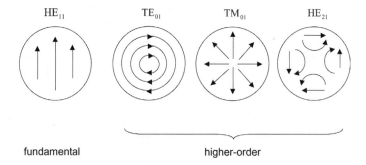

FIGURE 2.23 E-field lines corresponding to lowest-order step-index fiber modes: (a) fundamental or HE_{11}, (b) higher-order modes TE_{01}, TM_{01} and HE_{21}.

but as the hybrid superposition of different mode components. This analysis led to a new denomination called *linearly polarized* (LP$_{lm}$) modes. The principle of such a mode hybridization into LP modes is illustrated in Figure 2.24, considering for instance a horizontal polarization. The LP$_{01}$ (fundamental) mode is just the same as HE$_{11}$ counted twice. The LP$_{11}$ (higher-order) mode is made from two possible hybridizations, which result in two possible polarization patterns. Accounting for the two possible polarization directions, there are two LP$_{01}$ modes and four LP$_{11}$ modes. Overlooking the evanescent tail, the lobe patterns of the LP$_{lm}$ modes are similar to that of the TEM$_{pq}$ cavity modes previously shown in Figure 2.15 for circular mirrors. This similarity helps one to visualize the higher-order LP$_{lm}$ cases. As a rule, the index l represents the number of zero crossings as the azimuthal angle describes a full circle (like q in TEM$_{pq}$ modes), while the index m describes the number of zero crossings in the radial direction (like p in TEM$_{pq}$ modes).

The lesson to retain is that optical fibers have a fundamental mode (LP$_{01}$) in each polarization direction. Higher-order LP$_{lm}$ modes appear as the V-number increases from the first cut-off value $V = 2.405$. The condition for achieving single-mode operation is to have $V = (2\pi a/\lambda)NA < 2.405$, which can be realized for a desired signal wavelength (λ) by a proper combination of fiber radius (a) and numerical aperture (NA). The cut-off wavelength is thus simply defined as $\lambda_c = 2\pi aNA/2.405$, and the fiber is single-mode for any signal wavelength $\lambda > \lambda_c$.

Why should we want to get rid of the higher-order modes? The reason is that each LP$_{lm}$ mode has a different effective index, $n_{\text{eff}}(l, m) = \beta_{lm}/k_0$, meaning that the mode propagates at its own individual velocities of $c' = c/n_{\text{eff}}$. A signal pulse whose power would be randomly divided into different LP$_{lm}$ modes could not keep its integrity. The pulse would randomly spread in time and space according to the relative mode velocities. This effect is referred to as *intermodal dispersion*.

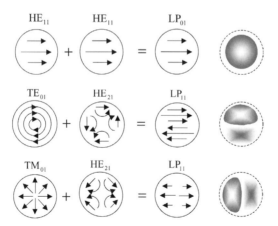

FIGURE 2.24 Construction of LP$_{01}$ and LP$_{11}$ modes (horizontal polarization) from HE$_{11}$, TE$_{01}$, TM$_{01}$ and HE$_{21}$ modes. The shadowed patterns on the right correspond to the E-field amplitude distributions, showing a π phase shift between the LP$_{11}$ lobes. The same construction exists for the vertical polarization.

Focusing exclusively on single-mode fibers, we just need to have a useful definition of the LP$_{01}$ E-field amplitude in the radial direction. Considering the core ($r \leq a$) and the cladding ($r > a$) regions, the E-field amplitude is defined as

$$E(r) = \begin{cases} AJ_0\left(U\dfrac{r}{a}\right), & r \leq a \\ BK_0\left(W\dfrac{r}{a}\right), & r > a \end{cases} \quad (2.73)$$

where $J_0(x)$ and $K_0(x)$ are the Bessel and Hankel functions of order zero and U, W are mode constants (also incorrectly called "transverse" propagation constants) defined by

$$\begin{cases} U = a\sqrt{n_1^2 k_0^2 - \beta^2} \\ W = a\sqrt{\beta^2 - n_2^2 k_0^2} \end{cases} \quad (2.74)$$

which also implies the relation $U^2 + W^2 = V^2$. In equation (2.73), the multiplying constants A, B ensure continuity of the E-field at the core/cladding interface ($r = a$). If we chose $A = 1$, we can immediately establish that $B = J_0(U)/K_0(W)$. The above definitions indicate that given the V-number, we only need to know the corresponding parameter U to determine the E-field envelope. As it turns out, there is an approximate analytical relation between U and V which is useful:

$$U = \frac{(1 + \sqrt{2})V}{1 + (4 + V^4)^{0.25}} \quad (2.75)$$

We leave it to an exercise at the end of the chapter to determine the LP$_{01}$ E-field and power mode envelope, as calculated for different V-numbers in a practical example. The results are shown in Figure 2.25. It is seen that the LP$_{01}$ mode has a bell-shaped envelope which is very similar to a *Gaussian distribution*. This feature will be used to simplify the mode-size evaluation (see following text). From Figure 2.25 we can also conclude the following:

- The E-field and power extend relatively far into the cladding: at the core-cladding interface and with respect to their peak values, the amplitude is about half to one-third, while the power is about 20%;
- The mode size (i.e., its cross-section, diameter, or radius at half-maximum) is sensitive to wavelength; as the wavelength decreases, the mode is more confined to the fiber core.

Most fibers are not based upon the principle of step-index profiles. In the case of graded-index fibers, however, it is generally possible to determine an *equivalent step-index* (ESI) fiber whose NA, V-number, cutoff wavelength and mode-size characteristics are very closely similar. Other index profiles (e.g., W) do not lend themselves to this relatively easy characterization, requiring the support of advanced computing software. This is particularly true if the profile is defined by several parameters which represent many degrees of freedom or dimensions from which to find an optimum configuration.

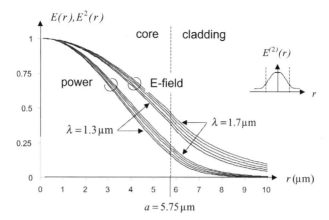

FIGURE 2.25 Radial E-field amplitude and power distributions of fundamental LP$_{01}$ in step-index fiber (radius $a = 5.75\,\mu$m, numerical aperture $NA = 0.1$), as plotted for different wavelengths on both sides of the cut-off wavelength $\lambda_{\text{cutoff}} = 1.5\,\mu$m. The inset on the right recalls that the distributions are symmetrical with respect to the core axis at $r = 0$.

The *mode size*, ω_0, also called *mode spot size*, and not to be confused with the core radius, a, is an important parameter in fiber-optics. One also refers to the mode-field diameter (MFD), as MFD $= 2\omega_0$. The spot size is associated with several features:

- How efficiently light signals can be launched into a fiber (laser-source coupling efficiency);
- Compatibility between fibers of different types (fiber-to-fiber coupling loss);
- Sensitivity to nonlinearities as related to signal intensity (high for low mode-size, low for large mode-size), see following text;
- Pumping/power-conversion efficiency (for optical amplifiers), see next section.

It is possible to simplify the mode-envelope/size definition by using the so-called *Gaussian (mode) approximation*. In this case, the E-field is defined according to

$$E(r) = E_0 \exp\left(-\frac{r^2}{\omega_0^2}\right) \quad (2.76)$$

where the spot size ω_0 is the mode radius at $E = E_0/e \equiv 0.367 E_0$, also called $1/e$ size. Knowing the V-number and core radius of the (real or ESI) fiber, the spot size is given by the well-known *Marcuse's approximation formula*:

$$\begin{aligned}\omega_0 &= a\left(0.65 + \frac{1.619}{V^{3/2}} + \frac{2.879}{V^6}\right)\\ &= a\left[0.65 + 0.434\left(\frac{\lambda}{\lambda_c}\right)^{3/2} + 0.0149\left(\frac{\lambda}{\lambda_c}\right)^6\right]\end{aligned} \quad (2.77)$$

This formula very accurately predicts the E-field *coupling efficiency* from a LP_{01} mode into an equivalent Gaussian-envelope mode. When considering either signal *power* or *intensity*, as defined by $P(r) \propto I(r) \propto E^2(r) = E_0^2 \exp(-2r^2/\omega_0^2) \equiv E_0^2 \exp(-r^2/\omega_I^2)$, the mode radius reduces to $\omega_I = \omega_0/\sqrt{2}$, which is narrower by the factor $1/\sqrt{2} = 0.707$. This parameter is called power mode size, in contrast to the E-field mode size. The above formula can thus be applied for either E-field or power mode sizes. As a lesser-known feature, however, is that the Gaussian envelope does not accurately match the LP_{01} envelope (or its square) at all radius points, especially inside the fiber core, unless the signal wavelength is close to cut-off. As a result, the calculated signal intensity inside the core can be erroneous by a significant factor (e.g., 10–20%). A more accurate approximation, which is valid at all wavelengths, also uses a Gaussian envelope but with a modified power-mode size given by the following formula (Desurvire's Gaussian-mode approximation):

$$\omega_I = a \frac{VK_1(W)}{UK_0(W)} J_0(U) \qquad (2.78)$$

with the mode-power envelope defined as $P(r) \propto E_0^2 \exp(-r^2/\omega_I^2)$. The mode power area A, defining the *mean signal intensity* $(I = P/A)$ is thus $A = \pi \omega_I^2$. Because the associated mode profile is accurate throughout the core region where the local intensity $I(r) = P(r)/A$ is the highest, this definition (and not Marcuses'), should be used when considering *doped fiber amplifiers* (see Section 2.3.5).

The fiber's *effective area* is the surface defined as $A_{\text{eff}} = \pi \omega_{\text{eff}}^2$ where ω_{eff} is an effective E-field-mode-size close to ω_0. The term "effective" comes from the fact that this concept is associated with *nonlinear interactions*. Such interactions involve different signals generally having different mode sizes (see below). In particular, the effective intensity $I = P/A_{\text{eff}}$ determines the gain factor of *Raman fiber amplifiers* (see Section 2.3.5). Because the definition is based on E-field mode sizes, it should not be confused with the mode power area A previously discussed. The need to reduce the deleterious effects of nonlinear interactions has led to the development of special *large-effective area* (LEA) fibers. Unlike ordinary single-mode fibers, whose effective-area is of the order of $A_{\text{eff}} = 50 \ \mu m^2$ ($\omega_0 \approx 4 \ \mu m$), LEA fibers have effective areas between 70 μm^2 and 200 μm^2 ($\omega_0 \approx 5-8 \ \mu m$).

Because single-mode fibers are characterized by relatively small core diameters and mode sizes, the corresponding signal intensities can be extremely high. Considering a typical fiber with power-mode area $A = \pi(4 \ \mu m/\sqrt{2})^2 = 25 \ \mu m^2$ for instance, a launched power of $P = 1$ mW or 0 dBm (typical laser-diode output) corresponds to an intensity of $I = P/A = 1 \times 10^{-3}/(25 \times 10^{-8} \ cm^2) = 4 \ kW/cm^2$. Since special laser diodes can also deliver output powers as high as $P \approx 2$ W or +33 dBm, the corresponding core intensities can be of the order of $I \approx 10 \ MW/cm^2$. These two examples clearly illustrate the effect of power confinement in single-mode fibers.

2.3.5 Fiber Loss and Dispersion

Section 2.3.4 addressed the analysis of optical fibers as *single-mode waveguides*. Here, we shall expand the perspective and conceive of fibers as a means of transmitting light signals over long distances, in view of realizing high-speed communication systems. By "long-distance" we mean any length from a kilometer to 100 kilometers. Although we now take it for granted, it was not obvious at all for the early fiber-optic engineers that single-mode fiber waveguides with lengths equivalent to 10^9 wavelengths (1 km = 10^9 μm) and minute index-profile features could ever be fabricated, let alone manufactured with near-perfect accuracy and reproducibility. To understand how this has been possible, a key concept is that of the *fiber preform*. Basically, a fiber preform is a glass rod about 1 m long and 1.5 cm wide. This rod, which is characterized by an inner index profile, in fact represents the macroscopic version of the fiber. Upon melting inside a vertical furnace, the glass material can be pulled down into a miniature fiber thread of 125-μm diameter, which is immediately coated with a polymer jacket and coiled on a reel. The index-profile in the fiber is the exact reproduction of the preform's, with 125 μm / 1.5 cm ≈ = 100× as a reduction factor. In this pulling process, an important parameter to control is the fiber *outside diameter* (OD). If this OD is controlled within a micron accuracy, for instance using laser-interferometry and feedback in the pulling speed, then the fiber's index profile is also accurately reproduced over indefinitely long lengths of fiber material. Once the whole fabrication process has been mastered, *tens of kilometers* of the same fiber can be realized from a single preform rod (see exercise at the end of the chapter). Many patented processes exist to realize fiber preforms of more or less complex index-profile structure and glass volume, which is beyond the scope of this book to describe. Suffice it to mention that the purity of the material constituents (silica-glass, co-dopants) is absolutely essential. In particular, the elimination of *water* and its ionic derivative (OH) is also critical. In the realization of optical fibers, the first obstacle that had to be overcome is *loss* or *signal attenuation* due to material absorption. This effect is characterized by the *attenuation or loss coefficient* (α), now expressed in dB/km or *decibels per kilometer*.

The first attempts to realize optical fibers took place in the late 1960s. The fibers exhibited loss coefficients of the order of $\alpha = 1$ dB/m. If such attenuation corresponds to a transmission of 80% over one meter length, it is also equivalent to 1,000 dB/km, representing a transmission as incredibly low as 10^{-100}! Significant progress in controlling material purity and eliminating OH traces led Japanese laboratories and then Corning Glass to achieve loss coefficients of 100 dB/km (10^{-10}), then 20 dB/km (10^{-2}). This last result, for which 1% of light signals could survive traversing a kilometer of fiber, is considered to have marked the birth of optical telecommunications. Nowadays, the lowest attenuation coefficients are close to 0.15 dB/km, which corresponds to a transmissions of 70% over 10 km and 3% over 100 km. This lowest attenuation performance is achieved at the wavelength of $\lambda = 1.55$ μm, as illustrated in Figure 2.26. The figure shows that the loss coefficient in silica-glass fibers is determined by a combination of several factors. If one overlooks the effect of UV absorption, the two main contributing factors are Rayleigh scattering and IR absorption. Rayleigh scattering, which is due to microscopic and random variations in the refractive index of the glass, rapidly decreases with wavelength according to a $1/\lambda^4$ law. The IR absorption tail is due to fundamental Si-O, Ge-O and P-O molecular vibrations, which have strong peaks in the 8–11-μm region (10^{10} dB/km). It is seen that the intercept of Rayleigh and IR losses defines an absolute attenuation minimum ($\alpha \approx 0.15$ dB/km) at the $\lambda = 1.55$ μm wavelength. The other features of the attenuation spectrum concern the OH vibration peak near 1.39 μm and other overtones at shorter wavelengths. To provide an idea of the effect of OH impurities, only a part per million (ppm) is sufficient to generate a peak loss of 40 dB/km at this wavelength. In order to obtain an attenuation spectrum similar to that shown in the figure, the OH concentration should be kept to levels well below 0.1 ppm. There are many other secondary effects contributing to fiber loss, such as *microbending* (deformation of index profile due to curvature-induced stress) and *waveguide attenuation* (especially in the case of depressed-cladding index profiles) both resulting in some fraction of power leakage in the cladding. Two other causes of loss are the need to periodically connect/splice fiber trunks into long chains, sometimes with different nonmatching types.

Overall, it is important to realize that until the advent of optical amplifiers in the late 1980s, optical communications were primarily limited by the effect of fiber loss. Indeed, a minimum amount of signal power is required at the end of the fiber link in order to achieve a target bit-error-rate (BER) performance. This minimum power is usually expressed in terms of *photons/bit*, i.e., the number of required bit photons on average to achieve said BER (see Section 2.5, this chapter). Since the best transmission corresponds to an attenuation of 20 dB or more per 100 km, corresponding to a 99% loss of signal photons, it is clear that the achievable distances had to be kept well below this limit. It is left as an exercise at the end of the chapter to show that increasing the signal power at the transmitter level is definitely not a practical solution for increasing the system transmission distance or haul. This is because an increase of transmitter power by orders of magnitude only *linearly* increases the haul. Before the development of practical optical amplifiers, the only solution

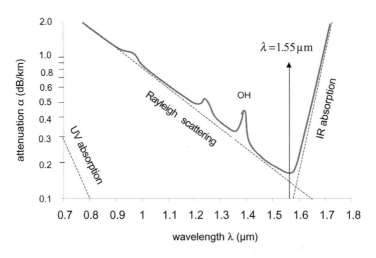

FIGURE 2.26 Typical wavelength-dependence of attenuation coefficient in silica-glass fibers, showing contributions from UV and IR absorption, Rayleigh scattering and OH impurity, yielding a minimum at $\lambda = 1.55$ μm.

available for achieving long-haul (e.g., 1,000-km) systems was to periodically regenerate the signal by means of in-line *electronic "repeaters"* (see Sections 2.3 and 2.5 concerning optical amplifiers).

Regardless of any other considerations, the loss spectrum of optical fibers defines a *transmission window*, which is centered near $\lambda = 1.55$ μm. It is sometimes referred to as the "third" transmission window because previous optical system generations were severely limited by the OH peak and overtones, which set the minimum attenuation at 1.3-μm (second window) and 0.8-μm (first window) wavelengths, respectively. If we arbitrarily define the maximum acceptable loss between two amplifiers/repeaters to be 0.3 dB and assume complete OH-peak suppression, Figure 2.26 shows that the allowed wavelengths should fall between $\lambda_{\min} = 1.25$ μm and $\lambda_{\max} = 1.65$ μm. This limits the actual *fiber bandwidth* to $\Delta\lambda = \lambda_{\max} - \lambda_{\min} = 400$ nm, or to the frequency range $\Delta\nu = c/\lambda_{\min} - c/\lambda_{\max} \approx 60$ THz. With signals modulated with spectral efficiencies of 1–10 bit/s/Hz, this fiber bandwidth is thus equivalent to a transmission capacity of 60–600 Tbit/s or 60,000–600,000 Gbit/s. Such a bandwidth is huge indeed, but it is not "infinite," as often wrongly stated in the press and even technical articles. Furthermore, other limiting effects contribute to significantly shrinking the potential of this "huge" bandwidth reservoir, as discussed next and also in Section 2.5. On a concluding note, it is important to keep in mind that there are other transparent materials (e.g., BaF_2, fluoride-based glasses such as the so-called ZBLAN, and $ZnCl_2$) which exhibit attenuation coefficients in the 0.001–0.01 dB/km range at 2–4-μm wavelengths. The extreme if not excruciating difficulties in realizing good-quality fibers from such complex and unstable glass materials should not obscure the potential, considering from where silica fibers really started! Although there is no current market perspective for such "ultra-low-loss" fibers, it is the author's conviction that this potential will be exploited in full in some other technology phase and age, in particular to alleviate the need of optical amplifiers in continental or transoceanic systems (10^{-3} dB/km ↔ 3 dB/1,000 km or 30 dB/10,000 km). Because fused silica, which is at the basis of current optical fibers, is such a cheap and easily tractable material, the prospect for such an evolution however remains indefinitely remote.

Loss is not the only limiting issue in fiber-optic systems: the second obstacle to be overcome is *fiber dispersion*. As previously described, dispersion is the effect of wavelength dependence of the refractive index. This property means that EM-waves of different wavelengths or frequencies travel at different velocities. This velocity difference causes a light pulse made of an initial "group" of frequencies to spread over time and space, like runners in a marathon. Since guided modes are characterized by an effective refractive index, n_{eff}, they also propagate at different and well-defined mode velocities ($c' = c/n_{\text{eff}} \equiv v_g$), an effect referred to as *intermodal (or modal) dispersion*. This unwanted intermodal dispersion can be alleviated by ensuring by design that the fiber is single-mode (LP_{01}) at the signal's center or mean wavelength. But the guided signal also has a spectrum of finite spectral width: a light pulse of temporal width Δt must by definition have a spectral width of $\Delta\nu = const/\Delta t$, where the constant depends upon several factors, including the degree of coherence (see earlier subsection) and the pulse temporal shape. For Gaussian-shaped pulses, the constant is equal to 0.44, meaning that the spectral width and time width is about half of the temporal width reciprocal. The different frequency components of the LP$_{01}$ mode must then experience some amount of velocity dispersion, to the extent that the pulse is narrower or the spectrum is broader. One refers to this concept as the mode's *group-velocity dispersion*, or GVD. Considering a single frequency component of the mode, the associated wave's velocity is simply:

$$v_p = c' = \frac{c}{n_{\text{eff}}} = \frac{c}{\beta/k_0} = \frac{2\pi c}{\lambda_0} \frac{1}{\beta} = \frac{2\pi\nu}{\beta} \equiv \frac{\omega}{\beta} \tag{2.79}$$

Such a definition is called *phase velocity*. When considering an EM-wave group made of a distribution of carrier frequencies (as in a light pulse guided by a fiber mode), the overall wave speed is given instead by the ratio:

$$v_g = \frac{\Delta\omega}{\Delta\beta}$$

$$\longleftrightarrow \frac{1}{v_g} = \frac{\Delta\beta}{\Delta\omega} \equiv \frac{\Delta\beta}{\Delta V}\frac{\Delta V}{\Delta\omega} = \frac{\Delta\beta}{\Delta V}\left(\frac{aNA}{c} + \frac{a\,\Delta NA}{c\,\Delta\omega}\right) \tag{2.80}$$

which by definition is the mode's *group velocity*. The second definition ($1/v_g$) illustrates the fact that there are two components in the group velocity. The first is singly associated with the waveguide parameters (a, NA, $\Delta\beta/\Delta V$), and the second is also a function of the bulk's refractive-index dispersion ($\Delta NA/\Delta V$). While this group-velocity definition is wavelength-dependent, the associated mode is characterized by a unique propagation speed. This speed is simply defined by the value of v_g at the mode's center wavelength. It is like defining the overall speed of a group of marathon runners from the average center of the group, regardless of the fact that due to the different runners' abilities, the group spreads in time and space over the race track. It is also possible to define the group velocity according to a *group-index*, i.e., $v_g = c/n_g$. The *dispersion of the group velocity* (or GVD) is given by the rate of change $\Delta(1/v_g)/\Delta\omega \propto \Delta n_g/\Delta\omega$, which can be calculated from the previous definition. Without specifying the details, suffice it to state that the result is the sum of two contributions: the *waveguide dispersion* (function of waveguide parameters) and the *material dispersion* (function of the bulk's refractive index). Material dispersion is fixed by the choice of bulk-glass material, and the possibilities for controling or changing it are limited. In contrast, waveguide dispersion is very sensitive to the index-profile shape (e.g., gradient, W, annular) and the waveguide parameters. This is an important result, because the net GVD, also called "*chromatic*" dispersion (to evoke its "color" dependence) can be controlled and optimized. If the idea is to *cancel* the dispersion at a given signal wavelength, then the waveguide must be optimized in order for the waveguide dispersion to be exactly of the opposite value of the material dispersion. This is precisely the principle of dispersion-shifted fibers (DSF), as described in the following text.

We can more simply define the GVD or chromatic dispersion using the concept of *group delay*. In a light pulse, a given frequency component at ω travels at the group velocity $v_g(\omega)$. Its arrival time (group delay) after a propagation distance L is therefore $T = L/v_g$. Consider now two frequency components: with frequency separation $\omega - \omega' = \Delta\omega$: the difference in their arrival times, $\Delta T = T - T'$ is

$$\Delta T = \frac{\Delta T}{\Delta\omega}\Delta\omega = L\frac{\Delta(1/v_g)}{\Delta\omega}\Delta\omega \equiv L\beta_2\Delta\omega \qquad (2.81)$$

In this definition, we have called β_2 the rate of change $\Delta(1/v_g)/\Delta\omega$. This single parameter (also called β'') defines the whole effect of fiber GVD. The key result is that the arrival time difference between the two frequency components is proportional to the frequency separation $\Delta\omega$, the distance L and the GVD, β_2. A nonzero arrival time ($\beta_2 \neq 0$, $\Delta T \neq 0$) corresponds to an effect of *pulse broadening*. But if we chose a wavelength for which the GVD is identically zero ($\beta_2 = 0$), there is no pulse broadening ($\Delta T = 0$). Alternatively, we can cancel the effect of pulse broadening by using a short length L_{comp} of a fiber having an opposite GVD with the appropriate value $\beta_{2comp} = -(L/L_{comp})\beta_2$. This approach is referred to as *dispersion compensation*. The associate fiber type is called *dispersion-compensating fiber* or DCF. We note that if the DCF length must be relatively short ($L_{comp} \ll L$) its absolute GVD must be relatively high ($|\beta_{2comp}| \gg \beta_2$).

Since optical systems are usually described in terms of wavelengths rather than frequencies, it is useful to define GVD as the rate of change $D = \Delta(1/v_g)/\Delta\lambda$. The difference between D and β_2 is just the proportionality factor $\Delta\omega/\Delta\lambda = -2\pi c/\lambda^2$, or $D = -(2\pi c/\lambda^2)\beta_2$. Using this new GVD definition, a pulse having a spectral width $\Delta\lambda$ broadens after distance L by the amount

$$\Delta T = \frac{\Delta T}{\Delta\lambda}\Delta\lambda = L\frac{|\Delta(1/v_g)|}{\Delta\lambda}\Delta\lambda \equiv L|D|\Delta\lambda \qquad (2.82)$$

The dispersion D is conveniently defined in units of *picosecond-per-nanometer-per-kilometer* (ps/nm/km or ps/nm-km), where 1 ps = 10^{-12} s.

This description leads us to a very important conclusion about bandwidth in fiber-optic systems. Indeed, if the fiber link has some finite amount of GVD ($D \neq 0$), signal pulses must then broaden by the factor $\Delta T = L|D|\Delta\lambda$. This broadening should be kept sufficiently small for the signal pulses not to overlap but to remain distinguishable from each other. At a data rate of B (number of bits per second), a sufficient condition to be satisfied is $B\Delta T < 1$, or using this definition:

$$BL|D|\Delta\lambda < 1 \qquad (2.83)$$

In this result, the bandwidth B is in Hertz (s^{-1}), the spectral width $\Delta\lambda$ is in nm, the length L in km, and the dispersion is defined according to $|D|_{\text{s nm}^{-1}\text{km}^{-1}} = 10^{-12}|D|_{\text{ps nm}^{-1}\text{km}^{-1}}$. For light pulses with finite spectral width $\Delta\lambda$, the bandwidth $\times$ length (BL) performance limit of a fiber-optic system can then be expressed as:

$$BL(\text{THz} \times \text{km}) < \frac{1}{|D|_{\text{ps nm}^{-1}\text{km}^{-1}}\Delta\lambda_{\text{nm}}} \qquad (2.84)$$

We can also introduce the fact that the signal spectral width $\Delta\nu$, now expressed in frequency, is also related to the bandwidth, according to $\Delta\nu \geq B$. Since $\Delta\lambda = (\lambda^2/c)\Delta\nu$, we have $\Delta\lambda \geq (\lambda^2/c)B$. By substituting this property into the previous expression, we can also express the bandwidth-distance limit in terms of the product B^2L as follows:

$$B^2L(\text{THz}^2 \times \text{km}) < \frac{0.3}{|D|_{\text{ps nm}^{-1}\text{km}^{-1}} \lambda^2_{\mu\text{m}}} \tag{2.85}$$

(see Exercises at the end of the chapter for tricky unit conversion).

The above result indicates that the system bandwidth-length product could be made arbitrarily high (or "infinite") by reducing the dispersion to arbitrarily small value ($|D| \to 0$) or by choosing a signal wavelength for which $|D| \approx 0$. With realistic fiber-optic systems, however, dispersion cannot be made exactly zero. This is because the link is made of a concatenation of fibers having minute dispersion variations. The resulting net dispersion is usually of the order of 1 ps/nm/km. If we assume a signal wavelength of $\lambda = 1.55$ μm, we get from the above formula: $B^2L < 0.12\,\text{THz}^2 \times \text{km}$. For a system with 100-km length, the upper bandwidth limit would under these conditions be $B^2 = (0.12/100\,\text{km})\text{THz}^2$ or $B \approx 35\,\text{GHz}$, corresponding to a maximum bit rate of $B = 35$ Gbit/s for a given wavelength channel.

As a matter of fact, the actual calculation of bandwidth limits due to fiber dispersion/GVD is far more complex than in the above estimations. The only merit of that simplified description was to illustrate the dispersion concept and also to show what the dispersion limitations actually were from the viewpoint of early times. The modern reality is more complex and dispersion-tolerant, because pulse broadening can in fact be allowed to a large extent. This is true if one can implement dispersion compensation (such as through DCF) along the line and/or at the transmitter- and receiver-terminal levels. Another consideration is that real systems share the bandwidth between multiple signal wavelengths, a technique referred to as *wavelength-division multiplexing* (WDM); see Section 2.6. Such WDM systems rely heavily upon dispersion compensation. As described in that later section, the approach is based upon quite sophisticated or elegant combinations of in-line and terminal dispersion compensation. But even when channels are compensated one by one, with the greatest care possible, bandwidth limits still exist because it is generally not possible to strictly achieve a regime where $|D| \approx 0$ for all channels. And as closely as one may be able to approach this zero-dispersion regime, other limits come into the picture, namely *loss* (or equivalently, amplification noise; see Section 2.5) and *fiber nonlinearities* (see subsequent text in Section 2.3). The next subsection provides more details on the different dispersion properties of single-mode fibers.

2.3.6 Single-Mode Transmission-Fiber Types and Dispersion Compensation

The early *single-mode fibers* (SMF) used for 1.3 μm telecommunications were based on *step-index* (also called matched-cladding) and *depressed-cladding* index profiles. In the first case, the core index is raised by including a dopant, which creates a uniform index difference between the core and the cladding. In the second case, the rationale is to maximize the index-difference by lowering the cladding index given the maximum achievable core-index value. Another reason not to further increase this core index is to minimize the excess loss due to index-raising codopants such as germanium or phosphorus. In either cases, the contribution from waveguide dispersion is relatively small ($-3\,\text{ps/nm/km}$ at 1.3 μm), meaning that the net GVD is similar to that of the bulk silica-glass material.

The typical wavelength dependence $D(\lambda)$ of the SMF dispersion is shown in Figure 2.27. It is seen that the dispersion is negative at wavelengths below $\lambda_0 = 1.3$ μm and positive above this value. At 1.55-μm wavelength the SMF dispersion is $+17\,\text{ps/nm/km}$, a value to memorize for future reference. The fact that the SMF dispersion exhibits a "natural" zero-point at 1.3 μm is also at the origin of the concept of a "second transmission-window" and its associated system generation. This observation is true, notwithstanding the fact that $\lambda_0 = 1.3$ μm also corresponds to a local absorption minimum, as defined on the left side of the OH peak (see Figure 2.26). It is just coincidental that SMF could have zero dispersion and a local absorption minimum at this wavelength. But since the absolute minimum of absorption is located at the longer 1.55-μm wavelength, one would have to develop a new type of fiber with zero-dispersion wavelength at $\lambda_0 = 1.55$ μm, called a *dispersion-shifted fiber* or DSF. The DSF can also have a step-index-profile, but with a substantially higher index ($\Delta \approx 1.2$), which is explained by the need to keep the LP_{01} mode narrowly confined at this longer wavelength. It is also possible to minimize this high index-difference requirement through a triangular/ring or triangular/depressed-cladding profile design with $\Delta \approx 1.0$ in the center (see Figure 2.22, case (g)–(h)). Finally, DSF can also be realized by a graded-index having a Gaussian shape.

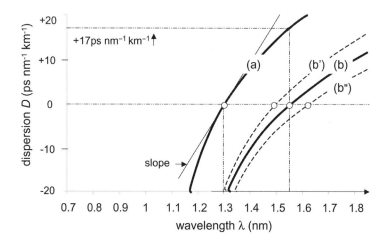

FIGURE 2.27 Typical wavelength dependence of chromatic fiber dispersion (GVD) in two basic types: (a) single-mode fiber or SMF, and (b) dispersion-shifted fiber of DSF, with definition of the dispersion slope. The dashed curves correspond to special cases of DSF called non-zero-dispersion-shifted fibers (NZDSF$^\pm$), which have either positive (b') or negative (b'') dispersion at 1.55 μm, respectively. In all cases, the open circles define the locations of the zero-dispersion wavelengths.

Obviously, DSF are by all means single-mode fibers, but oddly enough the term SMF remained to exclusively define the old $\lambda_0 = 1.3$-μm fibers. One also refers to SMF as being "*standard*" single-mode fibers, in contrast with "nonstandard" DSF, due to historic reminiscence rather than logic. Now, the SMF and DSF have been re-named, by the ITU-T standard groups, as C.652 and G.653, respectively (no logic involved either in these numbers), which represent strictly equivalent definitions. The wavelength dependence of the DSF dispersion with its zero-crossing at $\lambda_0 = 1.55$ μm is also shown in Figure 2.27. Since the 1.55-μm absorption minimum is relatively broad, the zero-dispersion wavelength λ_0 of DSF can be shifted to somewhat shorter values (positive dispersion at 1.55 μm) or longer values (negative dispersion at 1.55 μm). Such fibers are called *nonzero-DSF* or NZDSF$^\pm$ (or G.655 for ITU-T), according to the sign of their dispersion at the reference wavelength 1.55 μm; see Figure 2.27. These NZDSF have been specifically designed for broadband WDM transmission, where under certain conditions the overall performance can be improved by having a small amount of either positive or negative dispersion at some channel wavelengths. Another type of fiber (not shown in the figure) is the *dispersion-flattened fiber* (DFF). The DFF are designed through either double-clad/W or quadruple-clad index profile, which produces a nearly constant ("flat") dispersion between 1.3–1.4 μm and 1.6 μm. In order to obtain such a dispersion flattening effect, the waveguide dispersion in this region of interest must be very accurately controlled so that it closely takes the opposite value of the material dispersion. Several drawbacks of DFF (and of any other special fibers having such complex index-profiles) are the difficulty in accurately reproducing the dispersion characteristics in large-scale production, their higher transmission loss and, in some cases, their sensitivity to microbending. So far, and possibly for historical reasons as well, the DFF approach has not been as successful as the DSF/NZDSF in the investigation and commercial development of broadband WDM systems.

The design of broadband WDM systems thus relies upon the control of fiber dispersion, as determined by the curves shown in Figure 2.27. A parameter of import is the *dispersion slope*. As shown in the figure, the slope is the rate of dispersion increase ($D' = \Delta D/\Delta\lambda$) as measured at the zero-dispersion wavelength λ_0, and as expressed in units of $ps/nm^2/km$. This parameter is also called *third-order dispersion*. Consistently, the slope of the slope, ($D'' = \Delta D'/\Delta\lambda$, which defines the dispersion curvature is called *fourth-order dispersion*. Near λ_0 wavelength, the dispersion can be modeled according to the "expansion" formula:

$$D(\lambda) = (\lambda - \lambda_0)D' + \frac{(\lambda - \lambda_0)^2}{2}D \qquad (2.86)$$

Given the zero-dispersion wavelength (λ_0) and the third- and fourth-order dispersion values (D', D''), the above formula provides the dispersion at any wavelength. For wavelengths close to λ_0, the second contribution (D'') can be neglected and the dispersion can be approximated by the line $D(\lambda) \approx (\lambda - \lambda_0)D'$. We see that the dispersion increases (approximately) with the wavelength separation $\Delta\lambda = \lambda - \lambda_0$. If one wants to achieve a near-identical performance for all WDM channels, it is important that the slope (D') be reduced to a minimum. Typical slope values at 1.55-μm wavelength are: $D' = 0.057$ ps/nm^2/km for SMF and $D' = 0.070$ ps/nm^2/km for DSF and NZDSF. Some NZDSF$^+$ have been specially designed to have a reduced slope of $D' = 0.045$ ps/nm^2/km. See Exercises at the end of the chapter for two illustrative applications.

Because the generation of 1.3-μm systems has preceded that of 1.5-μm systems, most of the fiber which has been deployed over the last two decades is of the SMF type. At least this is true for *terrestrial systems*, in contrast with *submarine systems*. In submarine systems indeed, the fiber type and operating wavelength can be changed at each new cable deployment, making it possible to use 1.55-μm/DSF technologies. Even if the DSF was actually designed for a transition to a new 1.5-μm system generation in terrestrial systems, the enormous capital of already-installed SMF (*millions of kilometers* worldwide) could not be abandoned or wasted. Since the minimum loss of SMF is also at 1.55 μm, designers had then to consider operating the older SMF plant with this new signal wavelength. In addition, practical optical amplifiers had just been developed for 1.55-μm applications, which made it possible to extend the haul of the older, electronically repeated SMF trunks. One refers to such a system evolution as *SMF plant upgrade*, or 1.55-μm upgrade. This approach contrasts with the deployment of brand-new terrestrial systems based upon (NZ)DSF, inherently more expensive from both considerations of fiber/cable cost and its installation in the ground. Thus, SMF upgrade has been and is still a key issue in the evolution and commercial aspects of lightwave networks.

But one major problem of using 1.55 μm for signal wavelength in SMF trunks is the relatively high dispersion ($D = +17$ ps/nm/km). The solution was to develop dispersion-compensating elements, to place regularly along the signal path. Such elements should exhibit a high negative dispersion in order to compensate the group delay accumulated over the preceding fiber segment. One of the solutions is the *dispersion-compensating fiber* (DCF), as previously described. The typical dispersion value for DCF is -80 ps/nm/km to -100 ps/nm/km. Thus, DCF lengths of 0.21 to 0.17 times the length of the SMF segments can exactly cancel the cumulated dispersion. Thus, the dispersion of 100 km of SMF ($+1,700$ ps/nm/km) can be compensated with 17–21 km of DCF. But the price to pay is the extra loss introduced by the DCF segments, which also have significantly higher attenuation coefficients ($\alpha_{DCF} = 0.35$–0.6 dB/km). Periodic dispersion compensation can also be made through low-loss *fiber gratings* (IFBG), as discussed in the next section on passive optical components. It can also be done by alternating fiber segments (A and B) having strictly opposite dispersions and slopes, i.e., as verifying $D_{fiberA}(\lambda) = -D_{fiberB}(\lambda)$. If SMF is the fiber A, then fiber B is called the *reverse-dispersion fiber* (RDF). Note that the SMF/RDF dispersion-compensation approach is not related to SMF upgrade (since a new fiber must be installed over 50% of the trunk length) but rather to submarine networks (see Section 2.6). The issue of dispersion compensation is described in Section 2.4.

We conclude this subsection by briefly discussing other special types of optical fibers, which are of interest for a more limited (yet important) range of telecom applications. The first is the *polarization-maintaining fiber* (PMF) and the second is the *polymer optical fiber* (POF).

Consider first the case of PMF, also called *polarization-preserving* fibers. As discussed earlier through Section 2.1, light polarization (the absolute orientation of the E-field) is invariant or "conserved" when the dielectric medium is characterized by a uniform refractive index. The case of a birefringent medium, which has two different refractive indices (n_x, n_y) according to the transverse plane axes, is more tricky. In birefringent media, linear polarization is conserved only if the signal is launched with E-field parallel to either axis. If not, the state of polarization (SOP) evolves according to certain patterns, which are well described by the *Poincaré sphere* representation (see Section 2.1). Ordinary single-mode fibers such as SMF or DSF, which are made from high-purity glass with near-perfect cylindrical geometry still exhibit some residual and randomly-oriented birefringence. As discussed in the next subsection, this weak/random birefringence is responsible for an unwanted effect *called polarization-mode dispersion* (PMD). With PMF, the approach is radically different: the fiber is fabricated with the intent of yielding a relatively strong and well-controlled birefringence. If a linear-SOP signal is launched along one of the two PMF axes, then its SOP is conserved upon propagation over extended lengths (meters to kilometers) with negligible SOP ellipticity or random degradation. One defines the PMF *polarization crosstalk* (CT) as the power ratio between the two polarization modes found at some distance L, when a single polarization mode is initially excited (say parallel to the x axis), or

$$\text{CT} = 10\log_{10}\left[\frac{P(\text{LP}_{01}, y)}{P(\text{LP}_{01}, x)}\right] \tag{2.87}$$

Typical PMF have polarization crosstalks of -30 to -40 dB/km, meaning that the fraction of power which leaks (or "couples") into the unwanted polarization is as small as $1/1{,}000$ to $1/10{,}000$. Single-polarization fibers which support only one LP_{01} mode have crosstalks near -30 dB, a value which is length-independent above a few hundred meters.

As described in Section 2.1, birefringence is characterized by a *beat length*, which is defined through $L_B = \lambda/\delta n$, where $\delta n = n_x - n_y$. Weakly birefrigent fibers such as SMF or DSF have beat lengths of a few centimeters to a few meters ($L_B = 10$ cm–10 m), corresponding to very small index differences of $\delta n = 10^{-5} - 10^{-7}$ ($\lambda = 1$ μm). In contrast, PMF are characterized by beat lengths of 1 mm or less corresponding to very high index differences of $\delta n \geq 10^{-3}$. The effect of polarization maintenance starts with a minimum of $\delta n \geq 10^{-4}$ for which a fiber can be called specifically PMF (or "*HI-BI*" fiber) as opposed to a merely "birefringent" fiber. Birefringence is realized by using one of two effects: waveguide geometry or stress. The first consists in making the core highly elliptical so as to lift the propagation constant degeneracy between the two orthogonal LP_{01} modes. Since the effect of such a *geometrical birefringence* is weak, large core/cladding index differences are required, which causes unwanted excess loss. The other approach is based upon *stress-induced birefringence*. Stress is introduced along one direction of the transverse plane. This can be done by realizing an elliptical cladding having a high thermal-expansion coefficient, for instance based on boron or phosphorus doping. Due to unmatched expansion coefficients, a residual stress is formed upon fiber solidification, this stress being different along the short and the long axes of the ellipse. An alternative approach consists in surrounding the core sides in a given plane with materials of different glass composition and expansion coefficients. The fiber cross-section may look either like a bow-tie or (figuratively) a panda face, hence the names of *bow-tie* or PANDA PMF. These fibers can be made to cause very high excess loss on one of the polarization modes. In that case, one refers to *single-polarization* PMF. Finally, stress birefringence can be obtained by twisting the fiber around its axis, at a rate of 5 turns per meter. The result is *elliptical birefringence*, for which the left- or right-handed circular polarizations are being maintained (see Section 2.1). Because of their complexity and realization cost, and also because of their higher attenuation ($\alpha \geq 0.23$ dB/km), PMF are not used in optical transmissions. Another reason is the high modal birefringence which corresponds to a large GVD difference between the two polarized LP_{01} modes (the effect can be however cancelled by periodically rotating the PMF axes by $90°$ at each splicing). A potential remains for applications to *coherent systems*, where maintaining the signal polarization from end to end could reduce the receiver-design complexity. One of the most significant uses of PMF in optical systems is the physical connection between laser sources and external E/O modulators (such as based upon *lithium niobate* or $LiNbO_3$ waveguides, see Section 2.5). Since the characteristics of such modulators are polarization-dependent, the SOP of the incident laser signal should be well defined (namely linearly polarized along some given direction), hence the need for PMF to connect lasers and external modulators. Another application of PMF concerns the mitigation of polarization-mode dispersion (PMD), as discussed in the next section on passive components. Other useful applications of PMF concern the domain of fiber sensors.

We briefly consider next the case of *polymer fibers* (POF), also called all-plastic fibers. The POF are highly-multimode fibers based upon step-index (SI-POF) or graded-index (GI-POF) profiles. The plastic materials are *polystyrene*, *polymethyl-metacrylate* (PMMA), *polycarbonate* or *perfluoro-polymers*. The distinct advantages of POF are their very low production cost, large core diameters (120 μm to 1.5 mm) and large numerical apertures (NA = 0.4–0.6), which makes them simple and inexpensive to connectorize. This is one of the main selling arguments of POF against standard glass optical fibers (referred to from this perspective as "GOF"). An argument in support of optical links versus electrical links, whether GOF or POF, is the absence of RF radiation or the difficulty of eavesdropping, which represents a key asset in security issues. The security argument is relevant when considering small private networks such as in residential/business LAN and home networking applications, which are now primarily focussed on wireless solutions (see Chapter 1, Section 1.4).

A first drawback of POF is their high chromatic dispersion which is dominated by intermodal GVD difference (highly multimode regime) and the material dispersion itself at the wavelengths used. But the major drawback of POF is their intrinsic absorption, whose minimum value between several absorption peaks near $\lambda = 0.65-0.68$ μm is in the range $\alpha = 20-50$ dB/km! While such a transmission loss is prohibitive for normal telecom applications, the potential is still reasonably good for short range systems (e.g., 50–500 m, corresponding to 1–25 dB loss budgets). Such applications could concern intraoffice computer LAN, and home networking, with the key advantage of using inexpensive red *light-emitting diodes* (LED) for signaling. Recent progress (1999) with perfluoro-POF has also opened the potential for 1.25-Gbit/s, 2.5-Gbit/s and 10-Gbit/s infrared transmission systems operating at 1.3-μm wavelength and covering 1,000-m, 500-m and 100-m distances, respectively. Similar performance can be achieved at 840-nm wavelength with 1.25-Gbit/s WDM operation with 1, 2, and 4 channels, respectively. Although such capacity × distance performance remains modest in comparison to the rival GOF champions, the favorable economic argument in the aforementioned LAN/house applications could some day prove successful, especially when used to supplement wireless (radio) and wireline (DSL, FTTx) technologies. Another interesting potential application of POF for telecoms is the realization of *solar-pumped* doped-fiber lasers, as signal sources for space and intersatellite link applications.

Finally, it is worth mentioning the special case of *microstructured POF* (MPOF), based upon the principle of guiding light though periodic air-gap structures, also referred to as *photonic crystals*. Such photonic-crystal fibers remain to be qualified for low-loss optical transmission, and are therefore described in the section on passive optical components. It makes sense that the MPOF losses should be much lower than those of POF, since in such waveguides light is guided by the air structure rather than by the polymer-material index difference. The corresponding absorption coefficient could potentially be as low as $\alpha = 0.01\,\text{dB/km}$. But this loss-reduction effect should be first proven by the more rapid progress in glass-based, photonic-crystal fibers, which still remain as a future promise. Another key advantage of the MPOF approach is the possibility of realizing single-mode waveguides, which is not possible in the current POF technology.

The case of *planar polymer waveguides*, which are of high interest in a variety of WDM applications because of their sensitivity to UV as a source for generating index-difference structures, is discussed in the section on passive components.

2.3.7 Polarization-Mode Dispersion (PMD)

As recalled in the previous subsection and discussed in Section 2.1, single-mode fibers such as SMF and DSF are weakly birefringent. This birefringence is of both intrinsic and extrinsic origins. At microscopic scale, the bulk refractive-index distribution is not absolutely uniform, as it is associated with randomly oriented residual stress generated by the glass solidification process. Geometrical/structural variations due to manufacturing imperfections such as core ellipticity (typically 1% diameter fluctuations) also cause significant modal birefringence or change in the propagation constants. At macroscopic scale, stress-induced birefringence is caused by several effects, from fiber twist to fiber bending, spooling, packaging or cabling. Consequently, to each fiber-length coordinate correspond a set of birefringence axes, defining the local slow and fast index components, which are randomly oriented. Since the axes' orientation changes at both microscopic and macroscopic scales, the initial state of polarization (SOP) is not immediately scrambled. Rather, it evolves according to a circular path on the Poincaré sphere with a periodicity corresponding to the birefringence *beat length* (Section 2.1).

Over relatively short segments, the fiber birefringence (beat length) and birefringence-axis orientation can be considered as approximately uniform. The two LP_{01} modes polarized along x or y have a difference in propagation constant which is equal to $\delta\beta = (2\pi/\lambda)\delta n = (\omega/c)\delta n$, where $\delta n = n_x - n_y$ is the effective-index difference between the two modes. We note from earlier definition of $\Delta\beta/\Delta\omega$ in equation (2.80) that the rate of change of $\delta\beta$ with frequency ω, namely $\Delta\delta\beta/\Delta\omega$, corresponds to the difference of inverse group-velocities $1/v_o - 1/v_e$, where v_o, v_e are the mode group-velocities. Consider now a light pulse whose linear polarization is oriented at 45° to the fiber axes. As illustrated in Figure 2.28, this pulse can be decomposed into two individual pulse components, each having a linear polarization parallel to one axis and being associated with one of the LP_{01} modes. Upon propagation, the difference in group velocities between the two modes causes their pulse components to move apart from each other, causing the initial pulse to split up. The time separation between the two components is called *differential group delay* (DGD). After a propagation distance L, the DGD is given by

$$\delta\tau = L\left(\frac{1}{v_o} - \frac{1}{v_e}\right) = L\frac{\Delta\delta\beta}{\Delta\omega} = L\frac{\Delta(\omega\delta n/c)}{\Delta\omega} \equiv L\frac{\delta n}{c} + L\frac{\omega}{c}\frac{\Delta(\delta n)}{\Delta\omega} \qquad (2.88)$$

In this expression, the first term ($\delta n/c$) is due to the mode velocity difference and the second ($\Delta\delta n/\Delta\omega$) is related to the dispersion of the index difference which can be neglected in this analysis. If we then make the approximation $\delta\tau \approx L\delta n/c$, we find that for a distance corresponding to the beat length L_B, the DGD is simply $\delta\tau_B \approx L_B \delta n/c = (\lambda/\delta n) \times \delta n/c = \lambda/c = 1/v = T$, meaning that for each beat-length distance, the pulse splits by one E-field oscillation period T. If the fiber were of the PMF type, it is clear that the above relations would apply for any length, giving $\delta\tau = (L/L_B)T$, which corresponds to a purely deterministic DGD.

In the case of weakly birefringent fibers, the birefringence axes randomly change over microscopic distances, and the above result is not applicable. Yet we

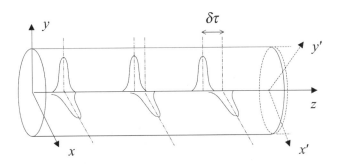

FIGURE 2.28 Definition of differential group delay (DGD) in a short length of birefringent fiber (x = fast axis, y = slow axis). The dashed lines at left correspond to a succeeding fiber segment with different axis orientation x', y'.

can continue the above reasoning up to the next fiber segment. As Figure 2.28 illustrates, this fiber segment has different axes orientations (x', y'). Thus, each pulse component coming from the first segment decomposes itself into two polarization components along these new axes. If the new axis (x', y') happen to be at exactly $0°$ with respect to the previous ones (x, y), it is easy to see that the net DGD corresponding to the two segments will be doubled, while if the new axes are rotated by $90°$, the net DGD will be zero (the second segment canceling the pulse-splitting effect of the first). In the general case where the new axes orientation is random, the net DGD is therefore random. We can better visualize this randomization effect between the two polarization modes by considering their power evolution. Assume that fiber loss is independent of the polarization. Call P_x and P_y the mean powers associated with each of the polarization modes. It is clear that after some distance (corresponding to many random rotations of the birefringence axes), each polarization mode should be nearly equally excited and have nearly identical powers, or $P_x - P_y \to 0$. If the initial launch condition is $P_x = P$ and $P_y = 0$ (only the x-polarization mode being excited), then one can define the (mode) *coupling length* L_c as the length for which the power difference has reduced to $P_x - P_y = 1/e^2$. This coupling length can vary from $L_c = 0.5-1$ m to $L_c = 0.5-1$ km, depending upon the fiber packaging conditions. The first case corresponds to tight fiber spooling (as used in the laboratory), while the second applies to fiber cabling. The case $L_c \to \infty$ would correspond to a perfect PMF, where no cross-coupling between the two modes could occur because of the high birefringence.

In the short-length regime, $L \ll L_c$, the deterministic result obtained for the DGD, $\delta\tau = (L/L_B)T$, is fully valid, showing that the DGD increases linearly with distance. In the long-length regime, $L \gg L_c$, the DGD is not deterministic. It can be shown that in this regime, the random mean-square value (rms) of the DGD is defined by

$$\sqrt{\overline{\delta\tau^2}} = T\sqrt{\left(\frac{2L_c}{L_B}\right)\frac{L}{L_B}} \equiv const \times \sqrt{L} \qquad (2.89)$$

This important result is that the DGD increases as the *square root* of the fiber length L or distance, a property to be used later.

The analysis of DGD can be further refined to take into account the fact that the random axis orientation also randomly fluctuates over time! This means that DGD is a random time-varying parameter which should be characterized not by its *instant value* $\delta\tau = x$, but by its *time-averaged value* $\langle \delta t \rangle = \langle x \rangle$. By definition, the time-average DGD, as measured at a given signal wavelength is called *polarization-mode dispersion* or PMD. For this reason, the DGD is also called *instantaneous PMD*. Since the birefringence (beat length) is wavelength-dependent, it is also possible to define the PMD through the wavelength-averaged DGD value at a given time, PMD'. Remarkably enough, the two definitions turn out to be the same, giving PMD' = PMD. A refined statistical analysis predicts that the probability

$P(\mathrm{DGD} = x)$ of mean $\langle x \rangle = \mathrm{PMD}$, corresponds to the *Maxwellian* distribution:

$$P(\mathrm{DGD} = x) = \frac{32}{\pi^2}\left(\frac{x}{\mathrm{PMD}}\right)^2 \exp\left\{-\frac{4}{\pi}\left(\frac{x}{\mathrm{PMD}}\right)^2\right\} \qquad (2.90)$$

Figure 2.29 shows the above Maxwellian distribution plotted as a function of the DGD, x, for different values of the PMD parameter, both being expressed in ps. Physically, this distribution would correspond to a histogram of a large number of DGD measurements over time (or over wavelength at a given time), which are asymmetrically distributed around a certain mean value called the PMD. If the system fiber was a HI-BI PMF, the DGD and PMD would exactly match, as a fully predictable and single parameter. It is important to note from the shape of the distribution that the instant DGD is unbounded. This means that there is always a finite possibility that it exceeds, by several times, the mean (PMD) value.

The natural unit for the mean DGD or PMD, when measured in actual fiber-optic systems, is the picosecond (ps). Since the DGD increases with the square root of the system length (see earlier discussion) it is more convenient for comparison purposes to define a length-independent *PMD coefficient* in $\mathrm{ps}/\sqrt{\mathrm{km}}$ units. Thus, the net system PMD should be expressed in ps consistently with a given $\mathrm{ps}/\sqrt{\mathrm{km}}$ fiber-PMD coefficient. To illustrate the use of such units, a 100-km system having a net PMD of 1 ps has a PMD coefficient of $1\,\mathrm{ps}/\sqrt{100\,\mathrm{km}} = 0.1\,\mathrm{ps}/\sqrt{\mathrm{km}}$. This is the value achieved by some of the current, best-performance SMF/DSF fibers. In contrast, the SMF fibers of the earlier generation, which belong to our "legacy," have PMD as high as $0.8 - 2.0\,\mathrm{ps}/\sqrt{\mathrm{km}}$. At the time, PMD was not an issue since both bit rates and "unrepeatered" transmission distances were relatively limited.

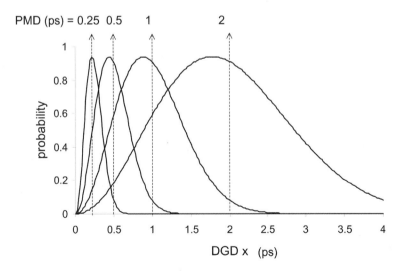

FIGURE 2.29 Maxwellian probability distribution of the differential group delay (DGD) corresponding to different values of polarization-mode dispersion (PMD), as expressed in picoseconds. Each dashed line shows the position of the mean DGD, which is equal to the PMD.

One should note that PMD is not an additive system parameter. If the transmission line is made of several trunks or optical components of different lengths ($L_1, L_2, \ldots$) and PMD ($\text{PMD}_1, \text{PMD}_2, \ldots$), the resulting total PMD is given by the formula:

$$\text{PMD}_{\text{total}} = \sqrt{L_1 \times (\text{PMD}_1)^2 + L_2 \times (\text{PMD}_2)^2 + \cdots} \qquad (2.91)$$

if the trunk PMD are expressed in ps/$\sqrt{\text{km}}$, or

$$\text{PMD}_{\text{total}} = \sqrt{(\text{PMD}_1)^2 + (\text{PMD}_2)^2 + \cdots} \qquad (2.92)$$

if the trunk PMD are expressed in ps. Note that these formula are not accurate if the trunk fibers or line optical components exhibit *polarization-dependent loss* (PDL), i.e., a difference in attenuation coefficient between the two polarization modes. In this case, the total PMD must be calculated though elaborate numerical simulations.

Consider now the effect of PMD in system bandwidth. Like chromatic dispersion or GVD, PMD is a limiting bandwidth factor, since it causes a single data bit to broaden over time. More accurately, PMD causes the bit pulses to broaden and eventually split into two individual/polarized components by an amount equal to the DGD. A small broadening effect causes intra-pulse interference and envelope distortion. Large broadening effects cause the pulse to split and/or overlap with the neighboring bits, which is a source of bit errors. The key difference between GVD and PMD broadening is that in the second case, the DGD randomly fluctuates over time and is *wholly unpredictable*, even if the mean DGD value, the PMD may be known. Performance calculations have shown that for a given system, the power penalty (see definition in Section 2.6) is kept within 1dB if the total PMD (ps) does not exceed 14% to 30% of the bit period. See two exercises at the end of the chapter for illustrative examples. One example shows that the maximum allowable distance at 40-Gbit/s bit rate dramatically drops from 1,200 km to 20 km, when the PMD is reduced from 0.1 ps/$\sqrt{\text{km}}$ (current fibers) to 0.8 ps/$\sqrt{\text{km}}$ (legacy fiber). At 10-Gbit/s rates, the same PMD range corresponds to distances of about 20,000 km to 300 km. These numbers illustrate the high limitation impact of PMD as the system bit rate and haul increase. When being so severe as to prevent 10–40-Gbit/s transmission over 100–500-km hauls, certain techniques of *PMD-compensation*, also called *PMD mitigation*, must be implemented (see Section 2.4.3 on passive components). In the case of WDM systems, compensation or mitigation requires a per-channel processing, since the instant DGD is different for each wavelength and shows little or no correlation between distant channels. A second and important observation is that in practical systems, the (mean) PMD also fluctuates over time. This can be due to temperature or vibrational effects, such as caused by day/month cycles and proximity to highways and railways with variable traffic patterns. Another consideration is that PMD compensation can only function within a predefined range of min/max DGD. As we have seen, DGD is truly unbounded. Thus, extreme (albeit rare) occasional fluctuations of the instant DGD outside this compensation range may cause a temporary system failure, also called *system outage*. This explains that PMD "compensation" in fact consists in PMD "mitigation." Such a mitigation only represents an attempt to bring the DGD at all times as close as possible to a zero value, and also to reduce to a maximum the probability of temporary system outages.

For advanced or future reference, or just mere scientific curiosity, it is worth briefly evoking three important concepts which are central to a finer analysis of PMD. These are:

- The *principal states of polarization* (PSP);
- The *PMD vector*;
- The *higher-order (or second-order) PMD*.

It can be shown that in a system having PMD, one can find for each wavelength, a special pair of orthogonal states of polarization (SOP), called *PSP*. If either of these two SOP is used for launching conditions, the SOP is reproduced undistorted at the system's receiving end. By "undistorted," one means here that the associated Stokes vector (see Section 2.1) just undergoes a smooth rotation on the Poincaré sphere. With a DGD being identically equal to zero, the signal pulses launched into any of these PSP come out undistorted, one arriving slower than the other (defining slow and fast PSP). In the case of a PMF link, the two PSP coincide with the orthogonal, linearly polarized SOP. In other non-PMF link types, the PSP represent an instant set of SOP launching conditions which remain unaffected by the system PMD. Since the system PMD characteristics fluctuate over time, the PSP are not fixed but rather correspond to instantaneous parameters. The second concept is the *PMD vector*, called τ. Essentially, τ is the Stokes vector (p) defining the SOP of the slow PSP at any fiber coordinate, but with the DGD ($\delta\tau$) as magnitude, or $\tau \equiv p\delta\tau$. As previously stated, the evolution of the PMD vector is defined by a rotation on the Poincaré sphere. It is from the evolution properties of this PMD vector that one can fully analyze the system PMD, and in particular its properties in the wavelength or frequency domain. Just as GVD was defined as a wavelength-dependent expansion formula with higher-order coefficients (see equation (2.86), the PMD vector can be expanded according to frequency as follows:

$$\tau(\omega + \Delta\omega) = \tau(\omega) + \frac{\Delta\tau}{\Delta\omega}\Delta\omega + \cdots \quad (2.93)$$

In the previous RHS expression, the second term ($\Delta\tau/\Delta\omega$) which defines the frequency correction is called *second-order PMD*. This term along with any other contributions in the expansion is called *higher-order PMD*. Since $\tau = p\delta\tau$, we have

$$\frac{\Delta\tau}{\Delta\omega} = \frac{\Delta(p\delta\tau)}{\Delta\omega} \equiv \frac{\Delta p}{\Delta\omega}\delta\tau + p\frac{\Delta(\delta\tau)}{\Delta\omega} \quad (2.94)$$

which shows that the second-order PMD breaks into two contributions: The first is the rate of change (or dispersion) of the PSP Stokes vector, and the second is the rate of change (or dispersion) of the DGD. By definition, the first contribution is called *PSP depolarization* (due to the change of the Stokes vector with frequency), while the second is called *polarization-dependent chromatic dispersion* or PCD (due to the change of DGD with frequency).

These concepts of PSP, PMD vector and second-order PMD just provide a flavor of the intriguing (and should one say frightening) complexity and elegance of PMD analysis. Suffice it to say that PMD has become a field in itself, with enough maturity to fill entire books. This observation is also true for *fiber nonlinearities*, which we shall describe, in the same spirit of physical accuracy but with mathematical/formalism restraint, in Section 2.3.8.

2.3.8 Fiber Nonlinearities

Optical fibers are not passive linear materials, unless relatively low signal powers or intensities are involved. When considering high signal intensities, indeed, an increase of the input signal is generally not followed by a proportional increase of the output signal. Rather, some fraction of the incoming signal may be back-scattered, frequency up/down-converted, or transferred to other incoming signals. These intensity-dependent power changes are also accompanied by self-induced or mutually induced phase distortion, resulting in spectral broadening or frequency shifts. One refers to such behavior as the fiber's *nonlinear response*, and to any of these effects as *fiber nonlinearities*.

Previous subsections have described the bandwidth limitations associated with fiber dispersion (GVD, PMD). We can now include as a second source of bandwidth limitations: the effect of fiber nonlinearities. Unlike dispersion, nonlinearities cannot be compensated or reversed. The combination of dispersion *and* nonlinearity is usually detrimental, because the second alters the quality and exactness of dispersion compensation, which is a linear technique. But in some cases, associating dispersion and nonlinearity can prove beneficial. This is the case for *soliton-pulse* propagation (see following text), where the self-induced nonlinear phase changes compensate dispersion, resulting in a stable transmission regime. Dispersion can also be used as a means to alleviate nonlinearity, breaking the coherence in signal cross-modulation effects. As seen earlier, one milliwatt of signal power in the core of a single-mode fiber corresponds to an intensity of the order of 1 kW/cm^2 (the light of about 10 halogen-lamp appliances concentrated over a single square inch!) which explains that nonlinearity thresholds can be rapidly approached as the signal power is increased. Furthermore, optical fibers offer extend interaction distances, which range from a few hundred meters to hundreds of kilometers, and with optical amplifier/repeaters, up to 1–10,000 km. Over such distances, it is clear that even extremely weak nonlinearities may manifest themselves and cause new types of transmission limitations. Finally, certain nonlinearities can also be advantageously exploited, for instance to prevent pulse broadening (see self-phase modulation, below) or to amplify signals as they propagate in the fiber (see Raman effect, below). Nonlinear effects are not observed exclusively in optical fibers. Rather, they form a special category in the vast field of *nonlinear optics*, for which exists an unbounded or unfinished list of nonlinear materials and complex intensity-dependent phenomena. Thanks to its natural transparency, the silica glass making up fiber waveguides is one of the weakest possible nonlinear materials. But as we have seen, the combined effects of intensity confinement in the core and ultra-long interaction distances make it possible to observe nonlinearity in this material at relatively low signal powers. In this subsection, we shall first provide some background on the physical origin of nonlinearity and, we believe, a very easy mathematical description. Then we shall briefly review the different types of fiber nonlinearity, from the viewpoint of both their specific physical origins and communications-system implications.

The origin of *nonlinearity* lies in the perturbation by the incident electromagnetic (EM) field of the *medium polarization*. Developing further the previous description in Section 2.1, the EM field causes the atoms to "polarize" themselves, meaning that gravity centers of positive and negative atomic charges, initially coincident (since matter is neutral), are forced to move apart. This atomic charge separation is called a *dipole*. The distance over which these charges move apart is limited by the attractive force they exert upon each other, acting as a pull-back force. Since the EM field reverses itself over one light cycle, the dipole is also forced to oscillate in synchronicity. At each oscillation cycle, the energy captured by the atomic dipole (corresponding to the work of the force) is radiated in the form of an identical EM field, as in the antenna effect. At low incident E-field amplitudes (E), the medium polarization (P) is simply defined as the linear relation $P = \varepsilon E$, where the constant ε is proportional to the square of the refractive index (n^2). As the E-field amplitude is increased above a certain threshold, the charges are further stretched away from their equilibrium position and the resulting motion is no longer proportional to the driving force. The resulting dipole oscillation (and radiated EM field) exhibits the same periodicity as in the incident EM-field, but is not purely sinusoidal. Rather, it can be seen as the superposition of elementary sinusoids oscillating at different harmonic frequencies. To formalize this effect, we write the atomic polarization in the form of a development in E-field powers:

$$P = \chi_1 E + \chi_2 EE + \chi_3 EEE + \cdots \quad (2.95)$$

By definition, the coefficients $\chi_1, \chi_2, \chi_3, \ldots$ associated to each of the E-field powers ($E, EE, EEE, \ldots$) are called the material *susceptibilities*. Clearly, the first-order coefficient (χ_1) is the material's *linear susceptibility* (polarization proportional to the E-field) while the other coefficients ($\chi_2, \chi_3, \ldots$) represent *nonlinear susceptibilities* of the second-order, third-order, and so on. The polarization can also be expressed as the sum $P = P_L + P_{NL}$, where a nonlinear polarization component (P_{NL}) comes as a perturbation of the linear polarization ($P_L = \chi_1 E$). It is assumed that the higher-order susceptibilities ($\chi_2, \chi_3, \ldots$) are very small, so that at low E-field amplitudes, the corresponding nonlinear polarization (P_{NL}) only represents a negligible perturbation. But since this perturbation increases as the square or the cube of the E-field amplitude, it is expected that its contribution cannot be neglected or may even become dominant in some incident-power regimes.

This expression of the total medium polarization is key to the understanding of the origin of nonlinear effects. As we shall see, the simple description immediately following, like an exploratory game, makes a real wealth of possible nonlinear effects appear!

Consider first the *second-order nonlinearity*. Define the E-field at frequency ω according to $E = E_0 \cos \omega t$. Then the nonlinear polarization is

$$P_{\text{NL}} \equiv \chi_2 EE \equiv \chi_2 (E_0 \cos \omega t)^2 = \chi_2 E_0^2 \cos^2 \omega t \equiv \chi_2 \frac{E_0^2}{2}(1 + \cos 2\omega t) \quad (2.96)$$

In the previous text, we applied the elementary formula $\cos^2 x = (1 + \cos 2x)/2$. The result is that the net medium polarization ($P = P_{\text{L}} + P_{\text{NL}}$) now includes two new components: one oscillates at zero frequency ($\cos(0t) = 1$) and the other at double the second-harmonic frequency 2ω. The first effect is called *optical rectification*. It is amazing that nonlinearity is thus able to change a light wave into a mere *static* E-field, just as a DC transformer does for 50–60-Hz alternating current! The second effect is called *second-harmonic generation* (SHG). It is a technique to double the frequency of a light source, for instance to generate *visible* light (e.g., $\lambda = 0.5$ μm green wavelength) out of an IR laser (e.g., $\lambda = 1$ μm wavelength).

The game continues if we now mix two E-fields having different frequencies ω_1 and ω_2 and amplitudes E_{01} and E_{02}. For simplicity, we assume that the two fields remain in phase at all times, corresponding to negligible differences in fiber dispersion or group index at the two frequencies. The total field is the sum $E = E_{01} \cos \omega_1 t + E_{02} \cos \omega_2 t$. Consistently, the second-order nonlinear polarization becomes:

$$\begin{aligned} P_{\text{NL}} &\equiv \chi_2 EE \equiv \chi_2 (E_{01} \cos \omega_1 t + E_{02} \cos \omega_2 t)(E_{01} \cos \omega_1 t + E_{02} \cos \omega_2 t) \\ &= \chi_2 \{E_{01} E_{01} \cos^2 \omega_1 t + E_{02} E_{02} \cos^2 \omega_2 t + 2 E_{01} E_{02} \cos \omega_1 t \cos \omega_2 t + \} \\ &= \frac{\chi_2}{2} \left\{ \begin{array}{l} E_{01} E_{01}[1 + \cos 2\omega_1 t] + E_{02} E_{02}[1 + \cos 2\omega_1 t] \\ 2 E_{01} E_{02}[\cos[(\omega_1 + \omega_2)t] + \cos[(\omega_1 - \omega_2)t]] \end{array} \right\} \end{aligned} \quad (2.97)$$

where we applied the elementary formula $2 \cos x \cos y = \cos(x + y) + \cos(x - y)$. We recognize in the above expression the two optically rectified E-fields (terms in "1" at zero frequency), the two SHG tones (terms oscillating at $2\omega_1$ and $2\omega_2$), and two new tones oscillating at $\omega_1 \pm \omega_2$ (difference to be taken as absolute value). Such tones correspond to an effect called *frequency sum/difference mixing* or *parametric mixing*. For instance mixing two IR waves can produce a visible-light signal, while two visible-light waves can produce an IR signal. The sum/difference tones are also called *mixing products*. Another possibility is to use a static field for one of the inputs (e.g., $\omega_2 = 0$). This can be done by placing a waveguide between two electrodes and applying a voltage. The two tones in $\omega_1 \pm \omega_2$ then oscillate at ω_1, with a total amplitude equal to $2\chi_2 E_{01} E_{02}$, which is proportional to the static field E_{02}. This special application of second-order nonlinearity, called the *Pockels effect*, is used for phase/amplitude modulation applications in LiNbO$_3$ waveguides/interferometers (see Section 2.5 on active components).

It can be shown from symmetry considerations that amorphous materials, such as glass optical fibers, do not ordinarily exhibit second-order or even-order nonlinearities, or $\chi_2, \chi_4, \chi_6 \ldots = 0$. The first nonlinearity occurring in fibers is therefore of the third-order type. Since the fifth-order nonlinearity in glass is extremely weak, *third-order nonlinearity is the only cause of nonlinear effects in fibers.*

Although the above restriction applies to all types of telecommunications fibers, it is however possible to "artificially" generate second-order nonlinearity in Ge-doped fibers, specifically. This is done by creating molecular bonding defects in the glass material, by means of intense laser exposure with visible or UV light. As a result SHG can be routinely obtained in such specially prepared fibers, but there has been no application to telecommunications so far. The technique of generating microscopic defects in Ge-doped fibers is however central to the realization of *fiber Bragg gratings* (see Section 2.4 on passive components), which have essential applications in WDM transmission systems.

Consider next *third-order nonlinearity*. Assume three E-fields with frequencies ω_1, ω_2 and ω_3 with amplitudes E_{01}, E_{02} and E_{03}. The total E-field is now $E = E_{01} \cos \omega_1 t + E_{02} \cos \omega_2 t + E_{03} \cos \omega_3 t$. The third-order nonlinear polarization becomes

$$P_{\rm NL} \equiv \chi_3 EEE \equiv \chi_3 \left\{ \begin{array}{l} (E_{01} \cos \omega_1 t + E_{02} \cos \omega_2 t + E_{03} \cos \omega_3 t) \\ \times (E_{01} \cos \omega_1 t + E_{02} \cos \omega_2 t + E_{03} \cos \omega_3 t) \\ \times (E_{01} \cos \omega_1 t + E_{02} \cos \omega_2 t + E_{03} \cos \omega_3 t) \end{array} \right\} \quad (2.98)$$

It is a very simple (but a bit tedious) algebraic exercise to develop the 27 product terms involved in above expression and then reduce them into single-frequency tones. We leave it as an Exercise at end of chapter to show that the result is

$$P_{\rm NL} = \frac{\chi_3}{4} \left\{ \begin{array}{l} E_{01}E_{01}E_{01}[3\cos\omega_1 t + \cos 3\omega_1 t] + E_{02}E_{02}E_{02}[3\cos\omega_2 t + \cos 3\omega_2 t] \\ + E_{03}E_{03}E_{03}[3\cos\omega_3 t + \cos 3\omega_3 t] \\ + 3E_{01}E_{01}E_{02}[2\cos\omega_2 t + \cos(2\omega_1+\omega_2)t + \cos(2\omega_1-\omega_2)t] \\ + 3E_{01}E_{01}E_{03}[2\cos\omega_3 t + \cos(2\omega_1+\omega_3)t + \cos(2\omega_1-\omega_3)t] \\ + 3E_{02}E_{02}E_{01}[2\cos\omega_1 t + \cos(2\omega_2+\omega_1)t + \cos(2\omega_2-\omega_1)t] \\ + 3E_{02}E_{02}E_{03}[2\cos\omega_3 t + \cos(2\omega_2+\omega_3)t + \cos(2\omega_2-\omega_3)t] \\ + 3E_{03}E_{03}E_{01}[2\cos\omega_1 t + \cos(2\omega_3+\omega_1)t + \cos(2\omega_3-\omega_1)t] \\ + 3E_{03}E_{03}E_{02}[2\cos\omega_2 t + \cos(2\omega_3+\omega_2)t + \cos(2\omega_3-\omega_2)t] \\ 6E_{01}E_{02}E_{03}[\cos(\omega_1+\omega_2+\omega_3)t + \cos(\omega_1+\omega_2-\omega_3)t \\ + \cos(\omega_1-\omega_2+\omega_3)t + \cos(\omega_1-\omega_2-\omega_3)t] \end{array} \right\}$$

(2.99)

This impressive collection of oscillating terms just represents the wealth of nonlinear effects that can be generated by third-order parametric mixing, also called *four-wave mixing* (FWM). We note that certain terms appear several times ($1\times, 3\times, 6\times$) which is referred to as a multiplying *degeneracy factor* (see following) Let's make an inventory of the different tones involved, using the indices i, j, k to group them by similarities:

- Tones at frequency ω_i, with amplitude $E_{0i}E_{0i}E_{0i} \equiv P_i E_{0i}$ (P_i = power at ω_i): a nonlinear polarization is induced at the same frequency as the original wave "i," with strength proportional to the wave's power P_i: this effect corresponds to a *self-induced nonlinearity* and is called the *Kerr effect* or *self-phase modulation* or SPM (see below);

- Tones at frequency ω_i, with amplitude $E_{0k}E_{0k}E_{0i} \equiv P_k E_{0i}$: a nonlinear polarization is induced at the same frequency as the original wave "i," with strength proportional to another wave "k" with power P_k: this effect corresponds to a *cross-induced nonlinearity* and is called either *cross-phase modulation* (XPM), or *parametric gain* (see following);
- Tones at frequencies $2\omega_i + \omega_j$ or $2\omega_i - \omega_j$, with amplitude $E_{0i}E_{0i}E_{0j} \equiv P_i E_{0j}$: yet other type of mixing products;
- Tones at frequencies $\omega_i + \omega_j + \omega_k$ or $\omega_i + \omega_j - \omega_k$, with amplitude $E_{0i}E_{0j}E_{0k}$: yet other type of mixing products;
- Tones at frequency $3\omega_i$, with amplitude $E_{0i}E_{0i}E_{0i} \equiv P_i E_{0i}$: this corresponds to an effect of frequency tripling, also referred to as *third-harmonic generation* (THG).

All third-order mixing products described above have been observed in optical glass fibers, including THG. But the mixing products of interest for 1.5-μm optical communications are those which fall into the WDM bandwidth. We must now concentrate on the actual effect that this third-order polarization induces on the WDM signals. So far in the description, we have just assumed the existence of a characteristic susceptibility χ_3, which is at the very origin of the fiber nonlinearity. At this point, we must introduce as a postulate that this nonlinear susceptibility can be of two types:

1. "Real," which leads to signal *refractive-index changes*;
2. "Imaginary," which leads to signal *amplitude changes*.

The odd term of "imaginary" comes from the mathematics of so-called complex numbers, which are made from the sum of real and imaginary parts. It is beyond this book's scope to expand further into the topic of complex numbers. Suffice it to state that third-order nonlinearity corresponds to two distinct types of effect. The first one produces a modulation of the fiber refractive index (more accurately, of the propagation constants), also called *Kerr effect*, and is associated with *self-phase modulation* (SPM) and *cross-phase modulation* (XPM). The second one produces a modulation of the E-field amplitudes, which causes power transfers between WDM channels, as associated with *parametric amplification*. The single and relatively simple formalism developed here for third-order nonlinearity actually applies to a broad range of effects each having quite different physical origins. Traditionally, the expressions of "parametric mixing" and "four-wave mixing" are most often used to refer to the second type (parametric amplification), although the first type (phase modulation) is no less concerned with the concepts. In the forgoing, we shall describe the different nonlinear effects which have been inventoried in optical fibers, and stress their relevance to WDM system impairments or applications.

Four-Wave Mixing (FWM)

One can better visualize the different parametric-mixing possibilities between WDM channels from the graphs shown in Figure 2.30. We first observe that

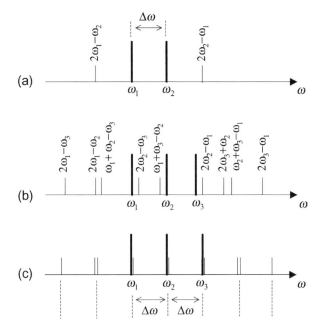

FIGURE 2.30 Effect of third-order nonlinearity, showing the generation of multiple side tones by frequency mixing: (a) two WDM channels at frequencies ω_1 and ω_2 separated by $\Delta\omega$, (b) three WDM channels at frequencies ω_1, ω_2 and ω_3 with unequal frequency spacing and (c) three WDM channels with equal frequency spacing. (In E. Desurvire, *Erbium-Doped Fiber Amplifiers: Principles and Applications*, Wiley, New York, 1994).

two channels at frequencies ω_1 and ω_2 generate two side-tones (or side-bands) at frequencies $2\omega_1 - \omega_2$ and $2\omega_2 - \omega_1$, respectively. This means that in a comb of equally-spaced WDM channels, each channel pair generates parasitic side-tones in the two nearest channels. The figure also shows the mixing products generated by either equally spaced or unequally spaced WDM channels. The same observation can be made in the equal-spacing case, i.e., the tones have the same frequency locations as the channels, generating many mutually-induced perturbations. In the uneven-spacing case, the tones fall outside the channel frequencies, which makes it possible to alleviate the perturbation. But unequally spaced WDM combs are non-standard, and therefore the first effect cannot be avoided. This illustrates that parametric mixing, also called *four-wave-mixing* (FWM) is one potential source of limitations in WDM systems. As we shall see next, the effect can be alleviated by exploiting fiber dispersion.

The previous derivation of the third-order nonlinear polarization [equation (2.99)] assumed that all E-field waves propagate at the same velocity in the fiber medium, which explains the absence of phase terms in the development. The effect of phase is, however, easily introduced. Indeed, assume that any E-field at frequency ω_i is of the form $E_i = E_{0i} \cos(\omega_i t - \beta_i z)$, where $\beta_i(\omega_i)$ is the corresponding propagation constant, which is dependent upon ω_i. To each type of nonlinear polarization at frequency $\Omega = \omega_i \pm \omega_j \pm \omega_k$, as defined by the mixing product $E_{0i}E_{0j}E_{0k}\cos(\Omega t - \varphi)$, one can associate the following phase delay φ:

- Tone ω_i, amplitude $E_{0i}E_{0i}E_{0i}$: $\varphi = (\beta_i - \beta_i + \beta_i)z = \beta_i z$
- Tone ω_i, amplitude $E_{0k}E_{0k}E_{0i}$: $\varphi = (\beta_k - \beta_k + \beta_i)z = \beta_i z$
- Tone $2\omega_i \pm \omega_j$, amplitude $E_{0i}E_{0i}E_{0j}$: $\varphi = (\beta_i + \beta_i \pm \beta_j)z = (2\beta_i \pm \beta_j)z$
- Tone $\omega_i + \omega_j \pm \omega_k$, amplitude $E_{0i}E_{0j}E_{0k}$: $\varphi = (\beta_i + \beta_j \pm \beta_k)z$
- Tone $3\omega_i$, amplitude $E_{0i}E_{0i}E_{0i} \equiv P_i E_{0i}$: $\varphi = (\beta_i + \beta_i + \beta_i)z = 3\beta_i z$

The importance of this phase delay makes immediate sense if one considers that the E-field at Ω is characterized by a propagation constant $\beta(\Omega) \equiv \beta_\Omega$. Thus the E-field oscillates according to $\cos(\Omega t - \varphi')$, with $\varphi' = \beta_\Omega z$, while the nonlinear polarization oscillates at $\cos(\Omega t - \varphi)$. Clearly, the nonlinear polarization should oscillate coherently with the E-field in order for the interaction to produce any effect and build up, yielding the *phase-matching condition* $\phi' = \varphi$. The case of the third harmonic ($\Omega = 3\omega$) provides an eloquent illustration. Indeed, and after the above list, the phase-matching condition is $\beta(3\omega) = 3\beta(\omega)$, or $\Delta\beta = \beta(3\omega) - 3\beta\omega) = 0$, where $\Delta\beta$ is called the *phase mis-match*. There is strictly no *a priori* reason that the condition $\Delta\beta = 0$ be verified in any given waveguide (this is why THG is difficult to generate in fibers, in addition to the effect of high absorption loss). A second example, which is now relevant to WDM systems, is the mixing product at $\Omega = 2\omega_1 - \omega_2$. According to the above list, the phase-matching condition is

$$\beta(2\omega_1 - \omega_2) = 2\beta(\omega_1) - \beta(\omega_2)$$
$$\leftrightarrow \Delta\beta = \beta(2\omega_1 - \omega_2) - 2\beta(\omega_1) + \beta(\omega_2) = 0 \qquad (2.100)$$

In WDM systems where the wavelength channels are closely spaced (e.g., 0.2–0.8 nm or 25–100 GHz), such a phase-matching condition has every chance of being very nearly satisfied. Assume indeed that $\omega_2 = \omega_1 + \delta$, where δ is a small frequency interval. We have $2\omega_1 - \omega_2 = \omega_1 - \delta$, and the above phase-matching condition is written:

$$\Delta\beta = \frac{\beta(\omega_1 - \delta) + \beta(\omega_1 + \delta)}{2} - \beta(\omega_1) = 0 \qquad (2.101)$$

Since in the close vicinity of ω_1 the propagation constant $\beta(\omega_1)$ varies approximately linearly with frequency, it is clear that the phase-matching condition is fulfilled. For advanced reference, we shall provide here a general approximation formula for the phase mismatch of any mixing product at frequency $\Omega = \omega_i + \omega_j - \omega_k$:

$$\Delta\beta = A\Delta\omega_{ik}\Delta\omega_{jk}\left(D + \frac{\Delta\omega_{ik} + \Delta\omega_{jk}}{2\pi}AD'\right) \qquad (2.102)$$

where $\Delta\omega_{ik} = |\omega_i - \omega_k|$, $\Delta\omega_{jk} = |\omega_j - \omega_k|$, $A = 2\pi c/\Omega^2$ and D, D' are the second- (GVD) and the third-order dispersions at Ω, respectively (see, for instance, Desurvire, 1994, Desurvire, 2002 and references therein).

In near phase-matching conditions ($\Delta\beta = \varepsilon \approx 0$), the nonlinear polarization and the E-field are in phase, which builds up the detrimental effect of four-wave mixing. The effect is characterized by the production of parasitic side tones, as previously illustrated in Figure 2.30, which for the WDM channels both represents detrimental power loss and noise. The FWM noise power at frequency $\Omega = \omega_i + \omega_j - \omega_k$ which is generated by three WDM channels at $\omega_i, \omega_j, \omega_k$ with input powers P_i, P_j, P_k is given by the simple formula:

$$P_{\text{FWM}}(\Omega) = \eta A P_i P_j P_k \tag{2.103}$$

where A is a proportionality constant defined by

$$A = \frac{1024\pi^6}{n^4 \alpha^2 \lambda^2 c^2 A_{\text{eff}}^2} (m\chi_3)^2 \tag{2.104}$$

(n = bulk refractive index, α = fiber absorption coefficient, A_{eff} = mode effective area, m = degeneracy factor equal to 3 or 6 if $i = j$ or $i \neq j$, respectively), and where $\eta' = \eta T(1-T)^2$ with $T = e^{-\alpha L}$ being the system transmission after distance L. The parameter η is the FWM *conversion efficiency*. According to this result, the FWM noise power thus increases as the *cube* of the input WDM-channel power ($P_i = P_j = P_k$) with a certain conversion efficiency η and a damping factor $T(1-T)^2 \approx e^{-\alpha L}$ due to fiber loss. The fact that FWM is cubic in channel power explains that the effect can be substantial above some power level, no matter how weak the third-order nonlinear susceptibility. But the FWM build up also requires some finite conversion efficiency, which we expect to be 100% in phase-matched conditions ($\Delta\beta \to 0$) and to rapidly decay otherwise. Its precise definition is the following:

$$\eta = \frac{\alpha^2}{\alpha^2 + \Delta\beta^2} \left\{ 1 + \frac{4T}{(1-T)^2} \sin^2 \frac{\Delta\beta L}{2} \right\} \tag{2.105}$$

We observe indeed that the efficiency is maximum ($\eta = 1$) in the limit $\Delta\beta \to 0$ and that it rapidly decreases as $1/\Delta\beta^2$. Furthermore, the efficiency oscillates with increasing phase mis-match (period $L_p = 2\pi/\Delta\beta$), which reflects an effect of periodic coherence between the FWM E-field $E(\Omega)$ and its third-order polarization source $P_{\text{NL}}(\Omega)$. Figure 2.31 shows plots of FWM efficiency as a function of system length up to $L = 50$ km ($\alpha = 0.2$ dB/km or 0.046 km^{-1}), as obtained for different values of $\Delta\beta$. We observe that for a phase mis-match corresponding to the effective "coherence length" $1/\Delta\beta = 10$ km, the FWM efficiency is 50% or higher over the 50-km distance. When this coherence length is reduced to a kilometer scale ($1/\Delta\beta \leq 1$ km), then the efficiency falls into the 1%–0.1% range after about 10 km. Note that the FWM power is proportional to $\eta(1-T)^2$, therefore it is identically zero at the origin ($T = 1$) despite the fact that the efficiency is maximum.

This example clearly illustrates the importance of keeping the phase mis-match between adjacent WDM channels sufficiently high in order to prevent FWM noise. This condition appears to be in contradiction to the previous requirement according to which signals should be transmitted in the zero-dispersion regime in order to maximize fiber bandwidth. Such a contradiction is lifted if one implements *periodic dispersion compensation*. Thus, at any point in the link the fiber dispersion can be made sufficiently high to ensure strong phase mis-match and prevent FWM, while

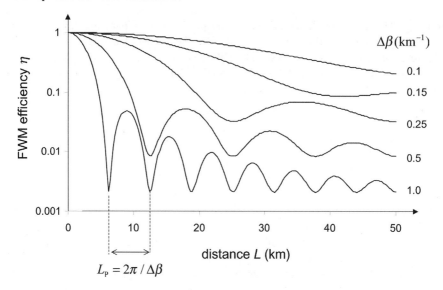

FIGURE 2.31 Four-wave mixing efficiency as function of fiber transmission distance for different values of the phase mis-match $\Delta\beta$. The definition of the oscillation period, $L_P = 2\pi/\Delta\beta$ is also indicated in the lowest curve.

the net or cumulated dispersion is periodically brought back to zero. Such a technique, which is described in more detail in Section 2.6.2 is referred to as *dispersion management* (DM).

Stimulated Raman and Brillouin Scattering (SRS, SBS)

Stimulated scattering is another third-order parametric effect, which consists in the coherent modulation of light and energy transfer between signals. As with FWM, stimulated scattering causes power-dependent signal loss. With *stimulated Raman scattering* (SRS), a fraction of the incident light at frequency v_p (referred to as the "*pump*" wave) is converted into a signal at lower frequency v_s (referred to as the *Stokes* wave). The Stokes wave is emitted in all directions of space, except in optical fibers where it can propagate in only two directions: *forwards* (same direction as the pump) or *backwards* (opposite direction to the pump). The frequency difference between pump and Stokes waves, $\Delta v_R = v_p - v_s$, is called the *Raman shift*. This shift corresponds to specific molecular vibration modes and is therefore characteristic of the medium. The SRS process can be viewed as the spontaneous or stimulated generation of a vibrational energy quanta $h\Delta v_R$, called a *phonon*, which is exactly equal to the energy difference, $hv_p - hv_s$, between the pump and the Stokes wave. In an amorphous medium such as silica glass, the vibration modes concern the Si-O-Si and Ge-O-Ge molecular bonds. In such bonds, the heavier Si/Ge atoms act as fixed nodes while the lighter oxygen atoms (the glass network formers) act as the oscillating pendulum mass, with three types of possible motion: bending, stretching and rotation. Since the angle formed by any Si-O-Si or

Ge-O-Ge bond is random, the characteristic vibrational frequency is also randomly distributed, which generates a continuum of possible Raman shifts, as opposed to a single and well-defined spectral peak. In silica-glass fibers, the Raman-shift continuum is maximum near $\Delta \nu_R = 10\text{--}15$ THz, which represents a wavelength difference of $80\text{--}120$ nm at 1.55-μm wavelength. The continuum width is also quite broad, representing about 8 THz (or 65 nm) at half width.

Since the Raman frequencies fall into the optical-domain range, one refers to the phonons as *"optical" phonons*. This is in contrast with *stimulated Brillouin scattering* (SBS), which is an effect quite similar to SRS, but involves *"acoustic" phonons*, as associated with much smaller *Brillouin frequency shift*. In silica glass, the Brillouin shift has a peak value $\Delta \nu_B = 10\text{--}11$ GHz. Another difference is that, unlike SRS which is due to static molecular vibrations, SBS is caused by a dynamic sound wave. The Brillouin shift ($\Delta \nu_B$) is directly related to the sound velocity in the medium (V_s) according to $\Delta \nu_B = 2\pi V_s/\lambda$. Since the material's sound velocity is well-defined, the Brillouin-shifts continuum is much narrower than in SRS. In optical fibers, the material composition (core codopants) varies with radial coordinate, which causes a velocity spread for this sound wave, corresponding to a Brillouin-shift continuum width of about $\delta \approx 50$ MHz at half-maximum. As with SRS, light scattering by SBS is omnidirectional, but the shift varies with the relative pump/Stokes direction angle: it is zero in the forward direction and maximum in the backward direction. This is why SBS in optical fibers translates to an effect of *signal back-scattering* with 10–11-GHz frequency downshift.

With both SRS and SBS, it is spectacular to observe that a substantial fraction of the power incident in the fiber is eventually converted into this frequency-shifted Stokes wave. One can define a *pump power threshold* P_{th} for which the SRS or SBS effects become significant, for instance half of the initial pump power being converted into Stokes power. For a fiber of length L, the Raman or Brillouin thresholds are approximately defined by the relations

$$\begin{cases} g_R \dfrac{P_{th}}{A_{eff}} L_{eff} = 16 \\ g_B \dfrac{P_{th}}{A_{eff}} L_{eff} = 21 \end{cases} \quad (2.106)$$

where g_R, g_B are the *Raman or Brillouin gain coefficients*, A_{eff} the (mode-field) effective area and $L_{eff} = (1 - T)/\alpha_p$ is the fiber's *effective (interaction) length* corresponding to a transmission $T = e^{-\alpha_p L}$ at the pump frequency ν_p. This effective length, which can be approximated by the "absorption length," $L_p = 1/\alpha_p$ (≈ 22 km at $\lambda = 1.55$ μm), reflects the fact that the pump power exponentially decays with transmission distance, which sets a limit to the nonlinearity strength and range. In silica fibers, and at $\lambda = 1.55$-μm wavelength, the peak Raman gain coefficient ($\Delta \nu_R = 10\text{--}15$ THz) is equal to $g_R^{peak} \approx 6.5 \times 10^{-14}$ m/W. This value is wavelength-dependent and decreases as $1/\lambda$ (e.g., $g_R^{peak} \approx 6.1 \times 10^{-14}$ m/W at $\lambda = 1.65$ μm). Since there is no SRS in the polarization orthogonal to the pump wave, the Raman gain coefficient represents a polarization-averaged value as obtained in non-polarization-maintaining fibers (SMF or DSF). This value should therefore be multiplied by two if a PMF is used instead. For Raman shifts $\Delta\nu \leq \Delta\nu_R$, the Raman gain coefficient can be approximated by a linear law, i.e., $g_R = g_R^{peak} \times \Delta\nu/\Delta\nu_R$.

The peak gain-coefficient value for SBS ($\Delta\nu_B = 10\text{--}11\,\text{GHz}$) is $g_B^{peak} \approx 5 \times 10^{-11}\,\text{m/W}$, or three orders of magnitude greater than that of SRS. It is left as a small exercise at the end of the chapter to calculate the SRS/SBS power thresholds in optical fibers such as SMF and DSF. The results indicate that for SRS, the threshold power is in the 0.5–0.8-W range, while for SBS, this range is 1–1.5-mW. We can accurately conclude from this analysis that both SRS and SBS define an absolute upper limit for signal powers to be used in fiber-optic transmission systems. The bad news is that SBS would reduce such a power to a mere milliwatt! But fortunately, the SBS conversion process is characterized by a relatively narrow spectrum ($\delta \approx 50\,\text{MHz}$ at half-maximum), while the spectrum of modulated signals is significantly broader (2.5–40 GHz). Thus the small SBS threshold must be interpreted as a power relative to the narrow spectral width δ, which raises the actual threshold value to over one order of magnitude. In high-power system applications where SBS is a limiting effect, the countermeasure is to further broaden the signal spectrum right from transmitter level.

The effect of SRS can be advantageously exploited to amplify signals at the Stokes wavelength (peak of the Raman-shift continuum) or in that spectral vicinity (i.e., within half Raman shift from the peak). Given the signal wavelength (e.g., $\lambda_s = 1,550\,\text{nm}$), one can thus realize a *Raman fiber amplifier* (RFA) from any type of silica-glass fiber by sending a strong pump wave at the appropriate wavelength (i.e., $\lambda_p = 1,430\text{--}1,470\,\text{nm}$), and in either propagation direction relative to the signal. When the pump is sent codirectionally or contradirectionally with the signal, the process is referred to as *forward* or *backward pumping*, respectively. Another possible configuration is to pump *bidirectionally*, which allows one to use lower pump powers at both fiber ends. Upon propagation over a distance L of fiber, the net Raman gain experienced by the signal is given by the expression:

$$G = \exp\left[g_R \frac{P_p}{A_{\text{eff}}} L_{\text{eff}} - \alpha_s L\right] \quad (2.107)$$

where P_p is the initial pump power launched at one fiber end. This result assumes that there is no gain saturation effect due to the pump-to-signal power conversion. At the fiber output, the signal power is defined by

$$P_s(L) = GP_s(0) + 2n_{sp}(G-1)P_{N0} \quad (2.108)$$

which is the sum of the amplified signal $GP_s(0)$ and the noise due to *amplified spontaneous scattering*. As described in Section 2.2, this spontaneous noise is equivalent to the amplification of n_{sp} fictitious input photons in each polarization mode (hence the factor of 2 in the above), corresponding to the equivalent input signal power $2n_{sp}P_{N0} \equiv 2n_{sp}h\nu_s\Delta\nu_s$, and where $\Delta\nu_s$ is the signal bandwidth. In laser amplifiers (such as doped fibers), the spontaneous emission factor, n_{sp}, refers to an effective degree of medium inversion. Although SRS amplification is not a laser but a scattering process, the spontaneous noise generation is wholly similar. In the SRS case, n_{sp} is equal to

$$n_{sp} = \frac{n_{th} + 1}{1 - \frac{\alpha_s A_{eff}}{g_R P_p}} \qquad (2.109)$$

where n_{th} is the mean population of optical phonons at the Stokes frequency $\Delta\nu_R$, which is both frequency- and temperature-dependent. In the study of thermal noise in radio signals we have provided a definition for this quantity [See Equation (4.28), Section 4.1.6 in Desurvire, 2004]. Substituting $f = \Delta\nu_s = 10-15$ THz and $T = 290$ K (17°) in the definition yields $n_{th} = 0.09-0.23$. The above definition shows that at high pump power the RFA spontaneous emission factor reduces to $n_{sp} \approx n_{th} + 1$, or $n_{sp} \approx 1.09-1.23$. This result illustrates that RFA are *essentially low-noise amplifiers*, equivalent to fully inverted laser amplifiers. The amount of pump power required to achieve a target gain G can be calculated by using equation (2.107). Clearly, the required pump is lower for DSF which has an effective area smaller than SMF. See Exercises at the end of the chapter for illustrative examples. These examples show for instance that using a 30-km long RFA made from DSF, Raman gains of $0/+3/+10/+20$ dB can be achieved with approximately $80/100/200/300$ mw of 1.55-μm pump power. Although this power requirement is relatively high by device or system standards, it can be met with specially designed high-power diodes. It is possible to reduce the power requirement per diode source if two pumps are combined in one fiber end with orthogonal polarizations (referred to as polarization-multiplexed pumping) and if two such pump modules are used in the bi-directional pumping configuration. See further on this issue in *repeaterless systems*, in Section 2.6.

Since RFA can be made to operate at any arbitrary signal wavelength (unlike doped fiber amplifiers) they make it possible to open new transmission windows for WDM systems, or at least extend the range previously accessible to doped fiber amplifiers (see Section 2.4 on passive optical devices. Another important application of RFA is their use as *distributed amplifiers*, as opposed to *lumped amplifiers*. A lumped amplifier is a device which boosts the signal power prior to or after launching into a dispersive trunk segment. In contrast, a distributed amplifier generates gain on the signal's overall transmission path. Since SRS can be implemented in standard communication fibers (SMF, DSF...), the effect of *distributed Raman amplification* can be used to cancel the transmission loss experienced by signals as they propagate between two trunk/repeater stations (corresponding to $G = 0$ dB in equation (2.107). The principles of both lumped and distributed (Raman) amplification are illustrated in Figure 2.32. The figure shows the power evolutions when the two path-averaged powers are chosen equal. We observe that lumped amplification corresponds to a large power excursion, starting with a high level at the beginning of the trunk and finishing with a low level at the end of the trunk. In contrast, *distributed amplification is characterized by a reduced power excursion*, the best configuration being provided by bidirectional pumping. For a given path-averaged power requirement, the reduced power excursion is therefore able to suppress (or at least

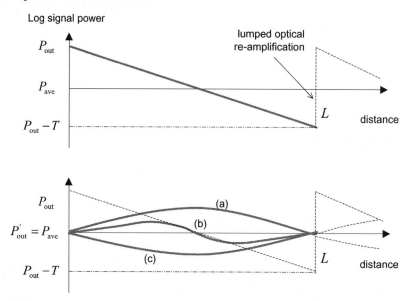

FIGURE 2.32 Top: principles of lumped amplification with trunk segment of length L and transmission $T_{dB} = \alpha_{dB/km} L_{km}$, showing signal power evolution with distance from initial launched value P_{out} and path-averaged power P_{ave}. Bottom: principle of distributed Raman amplification, showing signal power evolution with distance from initial launched value $P'_{out} = P_{ave}$; the three curves (a),(b) and (c) correspond to configurations of forward pumping, bidirectional pumping and backward pumping, respectively.

alleviate) unwanted nonlinearities, unlike in the lumped-amplifier case where the power levels at the beginning of the trunk are always significantly higher. The important conclusion is that SRS, as a nonlinearity, can thus be advantageously exploited in distributed amplification in order to combat other unwanted fiber nonlinearities. Note that in long-haul WDM systems ($L > 500$ km), both lumped and distributed amplification must be implemented together, because of the fact that exact loss compensation cannot be ensured for all wavelength channels. Yet the transmission improvement introduced by some amount of distributed amplification can be significant, because power excursion and unwanted nonlinearities can then be reduced to some *ad hoc* minimum.

Dense and high-capacity WDM systems can also suffer from the unwanted effect of *self-induced SRS* (SI-SRS) There are three primary reasons:

1. Since the WDM signals are optically repeated every 50–100 km, the actual interaction length for SRS is multiplied by a factor corresponding to the number "k" of repeated trunks, or kL_{eff};
2. In such systems, the number of WDM channels can be quite large (e.g., 32–64–128), which corresponds to higher transmitted powers, meaning that each channel subgroup can act as a Raman pump for another channel subgroup at longer center wavelengths;

3. The bandwidth occupied by WDM system (e.g., 32 nm or 4 THz) falls well within the Raman gain bandwidth (peak at 10–15 THz), meaning with previous arguments that Raman gain and pump/Stokes power transfers from short-wavelength channels towards long-wavelength channels could be significant.

The principle of SI-SRS is illustrated in Figure 2.33 for a set of M channels spaced by $\Delta\lambda$ with individual powers P_0. Consider the SRS effect of channel "0" on channel "k," which is spaced by $k\Delta\lambda$. Consistent with an earlier definition, the associated Raman coefficient is given by the proportionality relation $g_R = g_R^{peak} k\Delta\nu/\Delta\nu_R$ or $g_R = g_R^{peak} k\Delta\lambda/\Delta\lambda_R$ (with $\Delta\lambda_R \approx (\lambda^2/c)\Delta\nu_R$). Neglecting signal loss, the corresponding Raman gain is $G = \exp(kU)$ with $U \equiv (\Delta\lambda/\Delta\lambda_R)g_R^{peak} P_0 L_{eff}/A_{eff}$. If we assume the SRS effect to be minute or the gain very close to unity, we have $G \approx 1 + kU$. Thus channel "k" gains the extra power $P_0 kU$, corresponding to the equivalent power loss for channel "0." Summing the effect over all channels 1 to $M-1$, the fraction of power lost by channel "0" is

$$\frac{\Delta P}{P_0} = U(1 + 2 \cdots + M - 1) \equiv \frac{M(M-1)}{2} \frac{\Delta\lambda}{\Delta\lambda_R} g_R^{peak} P_0 \frac{L_{eff}}{A_{eff}} \quad (2.110)$$

The same result applies to all other channels by decreasing the integer M all the way down to unity (consistently, the last channel does not experience SRS since $\Delta P = 0$). It is thus the first channel "0" which experiences the highest SI-SRS loss. At the receiving system end, such a power loss is translated into a *power penalty* (see Section 2.6), as defined in decibels by $\eta_{dB} = 10 \log_{10}(1 - \Delta P/P_0)$. If such a penalty must be kept

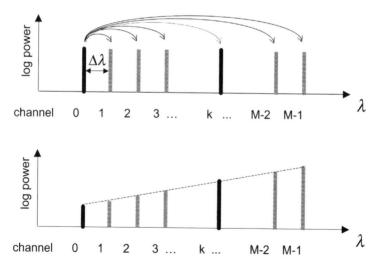

FIGURE 2.33 Top: principle of self-induced stimulated Raman scattering (SI-SRS) in WDM transmission systems: each channel (0 to M-1) acts as a Raman pump for the channels at longer-wavelengths, resulting in net power transfers. Bottom: result of SI-SRS after some transmission distance, showing power loss at short wavelengths and amplification at long wavelengths.

below 1 dB (for instance), we should have $10\log_{10}(1 - \Delta P/P_0) \leq 1 \leftrightarrow \Delta P/P_0 \leq 0.205$. It is a small exercise at the end of the chapter to prove that for transmission systems based upon DSF, this condition is equivalent to the following:

$$M(M-1)\Delta\lambda_{sm}P_0^{mW}L_{eff}^{km} \leq 42{,}000_{mW\,nm^{-1}km^{-1}} \quad (2.111)$$

For 1.55-μm WDM systems, this condition translates into a bandwidth × power limiting performance, considering indeed that: (1) $L_{eff} \approx q \times 22$ km (q = number of amplified trunks); (2) $\Delta\lambda_{nm} \leftrightarrow 125\Delta\nu_{GHz}$; (3) the full WDM (nm) bandwidth is approximately equal to either $(M-1)\Delta\lambda_{nm}$ or $M\Delta\lambda_{nm}$; and (4) the total WDM power is $P_{tot} = MP_0$. Taking into account these relations, the limiting condition is written

$$B_{tot}^{GHz} \times P_{tot}^{W} \leq \frac{250}{q_{GHz.W}} \quad (2.112)$$

To illustrate this result, consider a system with 1,000-km length and 100-km repeater spacing ($q = 10$). For an operating bandwidth of $B = 10$ THz (or 10^4 GHz), and using $B \times P = 25$ W $= 2.5 \times 10^4$ mW, the maximum total/WDM power should therefore not exceed 2.5 mW or +4 dBm. This example shows yet another bandwidth limitation due to fiber nonlinearity. One solution to increase the bandwidth × power performance, by a two-fold to four-fold factor, is to use *large effective-area fibers* (LEA), as previously described. The approach may be completed by the implementation of forward error correction (FEC), which alleviates to a finite (however salutary) extent the above power penalty.

Self- and Cross-Phase Modulation (SPM, XPM)

Another source of fiber nonlinearity is the effect of intensity-dependent refractive index change, also called *nonlinear refraction*, or *optical Kerr effect*. Recall that in the development of the third-order nonlinear polarization P_{NL}, we obtained two types of tone of the form $E_{0i}E_{0i}E_{0i} = P_i E_{0i}$ and $E_{0k}E_{0k}E_{0i} = P_k E_{0i}$. These tones not only oscillate at the same frequency ω_i as one of the E-field signals, but they are also phase-matched since the associated phases verify $\varphi = (\beta_i - \beta_i + \beta_i)z = (\beta_k - \beta_k + \beta_i)z \equiv \beta_i z$. Therefore, these nonlinear polarizations correspond to effects of power-dependent self-modulation and cross-modulation, respectively. We have stated that if the corresponding susceptibility χ_3 is purely real (as a complex number), this modulation impacts the phase of the E-field. It is possible to show that under this assumption, and given these two nonlinear polarizations, the E-field at distance z takes either one of the following forms

$$\begin{cases} E_i = E_{i0}\cos(\omega_i t - \beta_i z - \gamma P_i L_{eff}) \\ E_i = E_{i0}\cos(\omega_i t - \beta_i z - \gamma P_k L_{eff}) \end{cases} \quad (2.113)$$

with the constant γ being defined by $\gamma = k_i(2\chi_3)/A_{eff}$ ($k_i = 2\pi/\lambda_i$ = E-field wave-vector in vacuum) and $L_{eff} = (1 - e^{-\alpha z})/\alpha$ (α = attenuation coefficient at either signal wavelengths λ_i or λ_k). The factor 2 in γ only appears in the cross-modulation case in order to reflect the degeneracy factor ratio between the two effects [see development of nonlinear polarization in equation (2.99)]. Over a small propagation distance, we can make the approximation $L_{\text{eff}} \approx z$ and the E-fields become

$$\begin{cases} E_i = E_{i0}\cos[\omega_i t - (\beta_i + \gamma P_i)z] \\ E_i = E_{i0}\cos[\omega_i t - (\beta_i + \gamma P_k)z] \end{cases} \quad (2.114)$$

Introducing the notation $n_2 \equiv \chi_3$, it is seen that the nonlinearity is equivalent to an intensity-dependent change of the propagation constant according to

$$\begin{cases} \beta_i' = \beta_i + n_2 k_i \dfrac{P_i}{A_{\text{eff}}} \\ \beta_i' = \beta_i + 2n_2 k_i \dfrac{P_k}{A_{\text{eff}}} \end{cases} \quad (2.115)$$

which we define as representing the distinct effects of *self-phase modulation* (SPM) and *cross-phase modulation* (XPM, also called CPM), respectively. Recalling that the propagation constant is associated with the mode's effective index according to $\beta_i = n_i^{\text{eff}} k_i$, the effects of SPM and CPM are strictly equivalent to the intensity-dependent changes of refractive index:

$$\begin{cases} \text{SPM:}\ n_i' = n_i^{\text{eff}} + n_2 \dfrac{P_i}{A_{\text{eff}}} \\ \text{XPM:}\ n_i' = n_i^{\text{eff}} + 2n_2 \dfrac{P_k}{A_{\text{eff}}} \end{cases} \quad (2.116)$$

which also corresponds to the definition of the self-induced or cross-induced Kerr effects, respectively. The constant n_2, which has the dimension of m^2/W is called the *nonlinear (refractive) index*. In silica fibers, the nonlinear index varies between $n_2 = 2.1 \times 10^{-20}$ m^2/W and $n_2 = 2.5 \times 10^{-20}$ m^2/W. Such experimental values take into account the effect of polarization scrambling occurring in the fiber (the nonlinear index of an ideal, nonbirefringent silica fiber is obtained by multiplying the index by the corrective factor 9/8). Note finally that in the general E-field expression shown in equation (2.113), the effective length L_{eff} is substituted to the distance z to reflect the effect of attenuation in the modulating signal power (P_i or P_k), which asymptotically limits SPM or XPM to the absorption length $L_a = 1/\alpha$. The *nonlinear phase* associated with SPM and XPM is defined according to $\phi_{\text{NL}}^{\text{SPM}} = n_2 P_i L_{\text{eff}}/A_{\text{eff}}$ and $\phi_{\text{NL}}^{\text{XPM}} = 2n_2 P_k L_{\text{eff}}/A_{\text{eff}}$, respectively. It is also customary to use the expression $\phi_{\text{NL}}^{\text{SPM}} = \gamma P_i L_{\text{eff}}$ or $\phi_{\text{NL}}^{\text{XPM}} = 2\gamma P_k L_{\text{eff}}$, where γ is a nonlinear constant including the effect of the mode area A_{eff}, as expressed in units of km^{-1}W^{-1}. We leave it to an exercise at the end of the chapter to show that for SMF and DSF, this constant is of the order of $\gamma \approx 1$ km^{-1}W^{-1}.

Consider next a WDM system with M channels of equal power P_0. Each individual channel is subject both to its own SPM and to the XPM exerted by all the $(M-1)$ other channels. If we assume the worst case where all channels have a "1" bit (as opposed to a "0" bit associated with zero power), the total nonlinear phase is simply:

$$\phi_{\text{NL}} = \gamma P_0 L_{\text{eff}}[1 + 2(M-1)] = \gamma P_0 L_{\text{eff}}(2M-1) \quad (2.117)$$

For this nonlinear phase to remain comparatively negligible, the condition $\phi_{\mathrm{NL}} \ll 2\pi$ should be satisfied, or equivalently $(2M-1)\gamma P_0 L_{\mathrm{eff}} \ll 2\pi$. At 1.55-μm wavelength and using $L_{\mathrm{eff}} \approx 1/\alpha = 22\,\mathrm{km}$, $\gamma \approx 1\,\mathrm{km}^{-1}\mathrm{W}^{-1}$, we obtain the power-limiting condition

$$MP_0^W \ll 100 \tag{2.118}$$

Thus, a 100-channel WDM system ($M = 100$) should not have a per-channel power exceeding 1–10 mW (such that $MP_0^W \approx 10^{-3} - 10^{-2}$).

Since most commercial fiber-optic systems are based upon the principles of intensity modulation and direct detection (IM-DD), which are both phase-insensitive, it may superficially appear that neither SPM nor XPM can affect signal transmission performance. This conclusion is (unfortunately) erroneous, because these two types of phase modulation are also time-dependent, which causes instant frequency shifts. The time-dependence comes from the fact that the signal data is transmitted in the form of *pulse patterns*. Such patterns are either formed by individual pulses (return-to-zero or RZ modulation format), or rectangular-pulse sequences or variable lengths (nonreturn-to-zero or NRZ modulation format), see Chapter 1 in Desurvire (2004). To understand how the pulse time-dependence causes instant frequency shifts, consider the rate of change of the SPM nonlinear phase with time:

$$\Delta\phi_{\mathrm{NL}} = \frac{\Delta\phi_{\mathrm{NL}}}{\Delta t}\Delta t \equiv \gamma L_{\mathrm{eff}} \frac{\Delta P_i}{\Delta t}\Delta t \tag{2.119}$$

It is seen from this result that for any time change Δt and associated signal-power change ΔP_i, the nonlinear phase changes by the factor $\gamma L_{\mathrm{eff}}\Delta P_i/\Delta t$. For infinitesimal time changes δt, we can write $\phi_{\mathrm{NL}}(t) = \delta\omega_i t$, with $\delta\omega_i \equiv \gamma L_{\mathrm{eff}}\delta P_i/\delta t$ being an "instant" frequency shift. Since, according to equation (2.113), the E-field oscillates as $\cos(\omega_i t - \beta_i z - \phi_{\mathrm{NL}})$, we see that the net signal frequency is equal to $\omega_i' = \omega_i - \delta\omega_i$, which corresponds to an effect of instant frequency shift. The frequency is either increased (or blue-shifted) or decreased (or red-shifted) according to the opposite sign of $\delta\omega_i \propto \delta P_i/\delta t$. On the pulse's leading edge, the power increases with time ($\delta P_i/\delta t$ positive), which causes a red shift, while on the pulse's trailing edge, the power decreases with time ($\delta P_i/\delta t$ negative), which causes a blue shift. If the center of the pulse has a constant-power envelope (case of NRZ), there is no SPM effect in this region ($\delta P_i/\delta t = 0$). Figure 2.34 shows the frequency distribution inside a NRZ pulse (111... data sequence) when subject to SPM, with corresponding red/blue shifts in the leading/trailing edges. Taking now into account the effect of fiber dispersion, we see that the group velocity is different for the pulse center and the two edges. In the case of positive GVD, the (leading) red component travels slower than the pulse center and the (trailing) blue component travels faster. Such a GVD difference results in pulse narrowing and edge steepening due to interference, as also shown in the figure. This interference pattern is associated with greater time-dependent power changes and thus with locally enhanced SPM. When the pulse traverses another fiber segment with the opposite GVD, this nonlinear distortion is not reversed but rather continues to

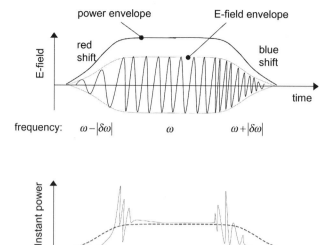

FIGURE 2.34 Top: effect of self-phase modulation (SPM), showing instant frequency change of the E-field with red and blue shifts, due to time-dependent power change according to NRZ pulse envelope. Bottom: combined effects of SPM and positive group-velocity dispersion, showing pulse steepening and distortion in the pulse's leading and trailing edges.

accumulate. A minor distortion effect would correspond to a relatively small increase of bit errors in the leading and trailing bits of the 1111... data sequence. But a greater effect would correspond to higher bit error rates and eventually the impossibility of any receiver correctly retrieving the initial bit sequence, regardless of signal power level. We can conclude from this overall analysis that the combination of SPM and dispersion is detrimental to the signal-transmission performance. To alleviate this impairment, a solution is to reduce the signal power, under the condition that there is sufficient power margin in the system (see Section 2.6). Another solution would be to lower the dispersion. But we have seen that in WDM systems, low dispersion is associated with efficient FWM, another nonlinear impairment. One may thus determine the right compromise between SPM and FWM limitations, which essentially consists in appropriately managing the line dispersion and finding an optimum operating point for the signal power.

The subject of fiber nonlinearities cannot be concluded without describing the most intriguing effect known as *soliton propagation*. Fundamentally, solitons are short light pulses which simultaneously exploit dispersion and nonlinearity in order to propagate without any broadening, keeping a constant temporal width. The name "soliton" is a short for "solitary wave" (which traces back to earlier observations in hydrodynamics) and also evokes a self-contained, particle-like behavior. To understand how solitons are generated in optical fibers, it is necessary to first briefly analyze how the dispersion acts on a single light pulses in absence of any nonlinearity.

A light pulse is principally characterized by three inter-related parameters: its *peak power*, (P_{peak}), its *temporal width*, also called *pulse width*, as defined for instance by the half-maximum duration ($\Delta\tau$), and its frequency width ($\Delta\nu$). Another important characteristic of the pulse is its envelope shape in either time or frequency domains. Such an envelope can take any shape from rectangular, triangular, Gaussian, or can follow a more complex definition. The pulse's *energy* (E) is given by the time-domain envelope surface. For rectangular or triangular pulses, this energy is simply $E = P_{\text{peak}} \times \Delta\tau$. The *Fourier transform* is a mathematical operation which makes it possible to convert any time-domain envelope into a frequency-domain one, and the reverse, which establishes a correspondence between $\Delta\tau$ and $\Delta\nu$. The quantity $\Delta\tau \times \Delta\nu$ is referred to as the *time-bandwidth product* or TBP. The spectral or temporal widths are thus given by the reciprocity relations $\Delta\nu = \text{TBP}/\Delta\tau$ at $\Delta\tau = \text{TBP}/\Delta\nu$, respectively. This shows that short pulses have a broad spectrum and long pulses have a narrow spectrum. The limit $\Delta\nu \rightarrow 0$ of purely monochromatic light corresponds to an infinitely-long pulse ($\Delta\tau \rightarrow \infty$), which corresponds to an unbounded E-field wave. If the frequency distribution inside the pulse is uniform, the TBP reaches a minimal value, and the pulse is said to be *Fourier-transform-limited* (FTL). In this case, the duration $\Delta\tau = \text{TBP}/\Delta\nu$ is also minimal, corresponding to the shortest pulse achievable. For square envelopes, the TBP is 0.7 (meaning $\Delta\tau \times \Delta\nu = 0.7$), while for Gaussian pulses TBP = 0.44. Taking this last example, a Gaussian FTL pulse of 1-GHz bandwidth has the minimal time width of $\Delta\tau = 0.44/10^9$ Hz = 0.44 ns = 440 ps, for instance. Finally, we must define the pulse width in the wavelength domain ($\Delta\lambda$), also called linewidth, as given by the relation $\Delta\nu = c\Delta\lambda/\lambda^2$ where λ is the wavelength. The relation between linewidth and time width is thus:

$$\Delta\lambda = \frac{\lambda^2}{c}\frac{\text{TBP}}{\Delta\tau} \qquad (2.120)$$

This formula is of fundamental value, since it determines the minimum linewidth of any signal transmitted under the form of (RZ) pulses, and corresponding to the data rate $B < 1/\Delta\tau$ (see Exercises at the end of the chapter for an illustrative example).

Since light pulses have a finite linewidth, they must broaden in the presence of fiber dispersion. According to equation (2.82), the time broadening corresponding to a finite dispersion $D(\lambda)$, at wavelength λ, and fiber propagation distance L, is $\Delta T = L|D|\Delta\lambda$. The special case where the pulse's center wavelength is at the zero-dispersion point ($D(\lambda_0) = 0$), also yields pulse broadening according to the approximation $\Delta T \approx 2L|D(\lambda_0 + \Delta\lambda/2)|\Delta\lambda$, which takes into account the GVD difference between the short- and long-wavelength regions of the pulse's spectrum. We conclude from this analysis that any light pulse, including in FTL conditions, must broaden to some extent because of its finite linewith.

If we introduce the effect of SPM, the above conclusion is not applicable anymore. This is because (as explained earlier) SPM generates self-induced frequency shifts, which grow in proportion to the distance travelled. If the dispersion is positive, SPM causes the pulse leading edge to travel slower than the pulse trailing edge, corresponding to an effect of pulse narrowing. Such a narrowing due to SPM nonlinearity competes with the broadening due to dispersion. If we manage to make the two effects exactly compensate each other, the pulse duration remains constant, which is how a *soliton pulse* is actually generated. One also refers to this specific condition (GVD compensated by SPM) as the *soliton-propagation regime*. The detailed analysis of soliton propagation is in fact quite complex, nonintuitive and mathematically intensive. Here, we shall just provide results obtained in the most simple propagation regime, where the fiber dispersion is constant and loss can be neglected over short propagation distances. Such conditions are referred to as the *fundamental-soliton* (or $N = 1$) propagation regime. It can be shown that the fundamental soliton pulse has a specific power envelope of the form $1/\cosh^2(1.76t/\Delta t)$, also called "hyperbolic secant," associated with TBP = 0.315. A laser pulse does not need to have such refined characteristics to transform itself into a soliton: as both theory and experiment show, a soliton can build up from any initial pulse shape (e.g., Gaussian, square) as long as the TBP is near the minimum (FTL) value.

Since SPM is power-dependent, there must exist a unique peak power for which a soliton pulse can be sustained. The detailed analysis shows that regardless of the initial FTL pulse shape, a soliton pulse of width $\Delta\tau$ progressively forms itself with propagation when the peak power verifies

$$P_{\text{peak}} \approx \frac{3\lambda^3 A_{\text{eff}}}{4\pi^2 c n_2} \frac{D}{\Delta\tau^2} \qquad (2.121)$$

or, in practical units:

$$P_{\text{mW}}^{\text{peak}} \approx 8.1 \times \lambda_{\mu m}^3 A_{\mu m^2}^{\text{eff}} \frac{D_{\text{ps nm}^{-1}\text{km}^{-1}}}{\Delta\tau_{\text{ps}}^2} \qquad (2.122)$$

We see from these definitions that the required soliton power scales linearly with dispersion and effective area, but is strongly dependent upon wavelength ($\propto \lambda^3$) and pulse width ($\propto 1/\Delta\tau^2$). See Exercises at the end of the chapter for an illustrative example concerning SMF- and DSF-based systems at 1.55-μm wavelength. From the results it is established that DSF systems with low dispersions such as 0.2–0.4 ps nm^{-1}km^{-1} can transmit solitons with peak powers in the 0.5–1-mW range, while the power is about one order of magnitude (50–100 mW) in SMF systems. This brings us to the interesting conclusion that in any propagation regime where dispersion is positive, SPM nonlinearity may naturally produces soliton pulses from RZ signals.

The world of solitons is very rich in most unique and intriguing effects. For instance, if the soliton's peak amplitude is increased according to the rule $A = N\sqrt{P_{\text{peak}}}$ ($N =$ integer), higher-order solitons form. These N-solitons exhibit self-replicating envelope-oscillation patterns which are of interest for laser pulse-compression applications. Solitons interact with their neighbors, causing an effect of periodic merging and separation. In a single-channel communication system, such a soliton/soliton interaction is a cause of bit-rate limitations. To maximize the spacing between adjacent solitons, a possibility is to use very narrow pulses for which $\Delta\tau \ll T_{\text{bit}}$ where $T_{\text{bit}} = 1/B$ is the bit period and B the bit rate. But a very narrow pulse corresponds to a very broad spectrum. The spectrum can then be sufficiently broad to cover the Raman continuum (see Section 2.3.7). As a result, the soliton is subject to *self-induced Raman scattering* (SI-SRS), also known as *soliton self-frequency shift* (SSFS). Such a frequency shift, undesirable in any transmission system, sets a lower limit to the RZ pulse width. In multi-channel/WDM systems, frequency shifts are mutually induced betwen solitons belonging to different channels. This is due to the FWM effect of soliton pulses passing through each other with different velocities, which is referred to as *soliton collision*. Such collisions induce permanent frequency shifts which the fiber dispersion translates into random time delays or arrival times (timing jitter), which causes bit errors in the receiver. Solitons can also exist under conditions of power loss with periodic reamplification, as so-called *path-averaged* (or *"guiding-center"*) solitons. The interaction between optical amplifier noise (ASE) and the soliton E-field also causes random frequency shifts and timing-jitter limitations (Gordon–Haus effect). Different techniques have been implemented to reduce the above limitations, for instance the use of periodic, narrow-band optical filters whose effect is to "guide" the soliton center frequency according to fixed or evolving specifications. Another solution is *all-optical regeneration*, which is described in Section 2.5. The relatively recent introduction of dispersion-management techniques, where fiber dispersion and sign are periodically changed, has produced even more developments in the already-rich soliton field. Against all expectations, soliton pulses can form when the path-averaged dispersion is zero or even negative. This new type of soliton-like pulse is called *dispersion-managed (DM) soliton*. Two peculiarities of DM solitons are that their envelope shape is Gaussian (unlike the "fundamental" soliton) and that their pulse width oscillates with the periodic dispersion changes along the line. The deleterious effects of soliton/soliton interactions, soliton collisions and other impairments turn out to be considerably alleviated.

The technique of soliton propagation, which exploits nonlinearity as an inherent advantage (but subject to other nonlinearity limitations!) corresponds to a radically-different approach for high-speed transmission systems. The performance of "soliton systems" has often been compared to that of "conventional systems" where nonlinearity is essentially avoided, not without some amount of controversy. Such a controversy was eventually solved when investigators found the evidence that in comparable WDM system conditions, DM solitons and conventional RZ pulses exhibit very similar temporal or spectral behaviors. Although intellectually satisfying, such a reconciliation should not hide the fact that the soliton field represents a higher form of mathematical and physical understanding of the combined phenomena of fiber dispersion and SPM/FWM nonlinearities.

■ 2.4 PASSIVE OPTICAL COMPONENTS

This section describes the variety of *passive optical components* used in lightwave systems. The word "passive" can be understood as meaning a function which neither requires high-frequency currents or electronics, nor involves any type of time-dependent signal processing, in contrast with *active* optical components. Another possible definition for passive components is that they are insensitive to signal modulation (bit contents and bit rate) and respond only to steady-state signal characteristics (polarization, frequency, wavelength, phase, power). The

2.4 Passive Optical Components

range of basic functions covered by passive components is extremely broad. Such functions can be grouped into four families, which concern signal *connectivity*, *filtering*, *compensation/equalization* and *amplification*, which will be covered here in this order. Passive components can also be classified according to their number of *device ports*, which can be either two (e.g., optical filter, isolator), three (e.g., pump/signal multiplexer, Y branch, optical circulator), four (e.g., 2 × 2 coupler) or multiple (e.g., star coupler, MUX/DMUX, OADM). It should be noted that in WDM-systems, the device ports are not only defined by the physical signal path, but also by the signal wavelength.

The technologies involved in the making of passive components are of three types: *bulk* (free-space optics), *waveguide* (integrated/planar or fiber) or *hybrid*. It is not unusual that several possible technologies exist to perform the same function. This is particularly the case for optical filters, wavelength multiplexers or dispersion compensators. As for any technology, the choice is guided by the four-fold consideration of *system compatibility*, *performance*, *unit cost* and *reliability*. Another important device characteristic is *loss*, which impacts upon both the system power budget and the signal-to-noise ratio. The loss characteristics are three-fold: *excess loss* (from inherent attenuation/absorption), *insertion loss* (from coupling ends) and *return loss* (from reflected input signal). Since passive optical devices receive signals from optical fibers, their characteristics must be *polarization-insensitive* (i.e., independent of the signal state of polarization or SOP), except in certain applications where PMF is required. The device loss characteristics may still exhibit a slight polarization dependence (e.g., the loss being higher for a given linear SOP). This is referred to as *polarization-dependent loss* of PDL.

From a manufacturing standpoint, a key consideration is the *yield*, or number of device units or modules that can be successfully produced from a single process or during a single phase of processing. The complexity of the fabrication, integration and assembly process widely varies: for some devices it is almost entirely automated and for others it remains purely manual. Other manufacturing considerations include *compactness* (volume reduction and in some cases, miniaturization), *robustness* (resistance to accidental mechanical shocks) and *temperature insensitivity* (typically $-20°$ to $+90°$). This is because in terrestrial systems, the devices are ultimately packaged into racks, shelves and bays, where space and ground "footprint" must be minimized, and where temperature conditions can largely vary. In sub-marine systems, the line components to be placed at periodic locations on the cable path are densely packaged into cylindrical repeater casings, with the same overall requirements but also a fault-free operation guaranteed for 20–25 years. Understandably, the cost of repeater repairs, involving complex marine operations, is extremely high, in contrast with terrestrial systems where faulty racks can be immediately accessed. For the above reasons, passive optical devices must be developed and manufactured with an in-depth knowledge of all possible causes of *aging* and *failure*.

In the following, we shall provide a brief description of some of the most important passive devices utilized in WDM systems. Because of the diversity of manufacturers and the rapid progress in this field, some device characteristics are provided only for illustrative purposes and should not be taken as absolute reference.

2.4.1 Connectors, Couplers, Splitters/Combiners, Multiplexers/Demultiplexers, and Polarization-Based Devices

In this subsection, we shall review a variety of components involved in system connectivity, i.e., which define the overall signal path. These concern *connectors, couplers, splitters/combiners* and *multiplexers/demultiplexers* on one hand, and polarization-based devices such as *polarizing beam-splitters, optical isolators* and *optical circulators*, on the other hand.

One of the past questions about single-mode versus multimode fibers was whether it would ever be possible to achieve practical low-loss connections between them, considering the relatively small core sizes (10 μm or 1/100 mm) and requirements in spatial/angular alignment accuracy. Such an argument was in favor of multimode fibers, which are more tolerant to such alignment issues. But with a bit of (smart) engineering, it turned out that single-mode fibers can be efficiently jointed by a variety of practical and cost-effective techniques. The first one, *Fusion splicing,* consists in permanently jointing the fiber ends through an electrical arc. This technique provides stable connections with very low loss (<0.2 dB to <0.1 dB), including in the extreme case of dissimilar single-mode fibers (e.g., doped fiber with SMF/DSF leads). Prior to fusing, the two fiber ends must be properly *cleaved*. This is done by locally removing the polymer jacket, then scribing/scoring the fiber side with a cutting tool (e.g., diamond or tungsten-carbide), then breaking it from this initial crack. To obtain low-loss splices, fiber alignment in both lateral and angular dimensions is extremely critical. The two fiber cores must also have very low eccentricity (e.g., <1%) relative to their axes. The fact that dissimilar fiber types (with different index profiles and core dimensions) can in some cases provide good-quality fusion splices is explained by the effect of longitudinal codopant diffusion in the melting glass. When properly controlled, such a diffusion creates a smooth index-profile transition from one fiber core to the other. Fusion-splicing is an essential technique to realize fiber-optic cables with lengths between 15 km and 100–120 km. The splicing is generally made directly in the factory, before fibers are integrated into multiple-pair cables. It can also be performed on the ground and outdoors, for *ad hoc* cable connections, reconfigurations, upgrades or repairs.

The second jointing technique consists in using *mechanical connectors*, also called *mechanical splices*. Such connectors provide the advantage of rapid, interchangeable or non-permanent fiber joints. The connection techniques are manifold, as illustrated in Figure 2.35. A first and most economical apparatus consists of a self-aligning plastic sleeve in which the two cleaved fibers are *butt-coupled*. Butt-coupling must include an index-matching compound in order to avoid reflections at the two fibers' air/glass interfaces. A second approach (see Figure 2.35) is based on a more complex fiber-end assembly, which involves an alignment sleeve and ferrules inside which the fibers are permanently glued. The ferrules' ends are then polished at a precise 90° angle. The alignment sleeve, which can be cylindrical or conical, makes it possible to bring the two polished ferrules in contact with the required parallelism and positioning accuracy, and maintain them in this position through various

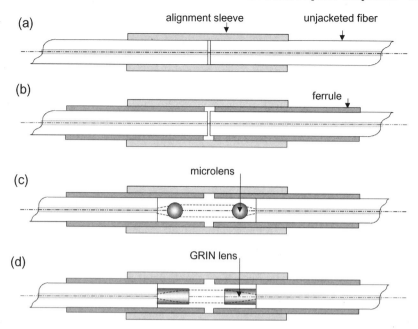

FIGURE 2.35 Different principles of fiber-to-fiber mechanical connectors: (a) butt joint with self-alignment sleeve, (b) same with precision ferrules surrounding the fiber, (c) beam-expanded with spherical microlenses, and (d) beam-expanded with graded-index (GRIN) rod lenses. The complex ferrule and casing structures for steady mechanical alignment and connection are not shown.

locking mechanisms. The insertion loss of mechanical connectors can be as low as 0.2 dB, as an average reference value. Repeated connections/disconnections increase the connector loss, primarily because of the effect of dust and scratching. A third approach (see Figure 2.35) is based on the principle of an *expanded-beam* connector. Spherical *microlenses* (2.5-mm to 250 μm diameter) are placed near the fiber ends at an exact focal distance. The two lensed ends are made to face each other at a few millimeters distance through an alignment sleeve. Upon passing through the first microlens, the signal beam from the input fiber is collimated (meaning that the light rays are parallel). The collimated beam is then focused by the second microlens into the core of the output fiber. A fourth approach (see Figure 2.35) is to use miniature *graded-index* (GRIN) rod lenses (0.5–2-mm diameter) directly attached onto the fiber ends for the beam collimation. In either microlens/GRIN-lens case, the lenses and fiber ends must be *anti-reflection coated* in order to minimize deleterious effects of back reflections and associated return loss. Typical beam-expanded connector losses range from 0.4 to 0.7 dB. Finally, it is worth mentioning the case of multi-fiber connectors, which have been developed to join *fiber ribbons*. The main approach consists in using silicon *V-groove arrays* in which fibers are embedded at 250-μm (two diameters) intervals. A single connector can thus join up to twelve single-mode fibers at once with loss of 0.8 to 0.4 dB, depending upon the technology.

198 *Optical Fiber Communications*

The first major category of passive optical components concerns multiple-port and multiple-path devices called *couplers, splitters* and *combiners*, which operate in a certain wavelength range (wavelength-selective devices to be considered thereafter). The two basic coupling structures are the Y-branch or 1 × 2 coupler (3-port) and the X-junction or 2 × 2 coupler (4-port), as illustrated in Figure 2.36. Note that the terms "1 × 2 coupler" and "X-junction" are never used, although they make sense. Both coupling structures can be realized with single-mode fibers and planar circuits, although a real Y-junction is difficult to make in the first case. The figure shows that both couplers are realized by making single-mode waveguides merge into each other (Y-junction) or approach parallel to each other in close vicinity (2 × 2 coupler). In the contact or vicinity regions, *the waveguide structure is no longer single mode*, which is conceptually important to understand the coupling effect. Indeed, it can be shown that when two single-mode waveguides are sufficiently close to each other, to the point that their evanescent mode-fields overlap, the structure supports two modes, which are called *normal modes*. The mode for which the E-field envelope forms two lobes having identical phase is said to be *symmetric*. The other mode for which the E-field envelope forms two lobes having opposite phases is said to be *antisymmetric*. This notion of symmetric/antisymmetric mode has previously been described in Section 2.1 concerning dielectric waveguides. Here, the difference is that the waveguide structure

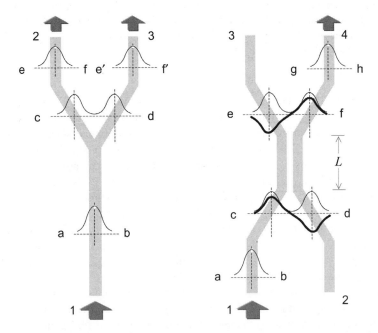

FIGURE 2.36 The two basic types of coupler structure: 3-port "Y-junction" (left) and 4-port "X-junction" or 2×2 coupler (right). The normal modes excited at different locations of these structures, assuming a signal input at port "1", are also shown (the E-field envelopes of the symmetric and antisymmetric modes are represented by thin and bold lines, respectively).

(upper region and center region of the Y- and X-junctions, respectively) has two individual cores. The E-field envelopes of the two normal modes are seen in the center region of the X-junction (Figure 2.36). Note that as for any waveguide, the normal modes exist in two orthogonal polarizations. Thus the Y- and X-junctions actually support four modes or two sets of linearly-polarized normal modes having orthogonal polarizations.

Consider first the Y-branch of Figure 2.36, which is input from the bottom port 1. This port is single-mode, as shown by the single E-field lobe (a–b). At the point where the waveguide core splits up, the structure has two normal modes. But as shown in Figure 2.36 (c–d), only the symmetric mode is excited. The reason for this is that the initial (symmetric) single mode cannot "couple" into the antisymmetric mode. Formally, one says that the two E-field envelopes (f_{sym}, $f_{antisym}$) have zero overlap, or mathematically that the net surface/integral of the function $f_{sym} \times f_{antisym}$, itself antisymmetric, is identically zero. As a result, the incident power is equally split into the two upper branches and ports 2 and 3. As the branches are moving apart with distance, the symmetric/normal mode has a wider intermediate region with zero E-field, which corresponds to two individual single-mode waveguides, as illustrated in Figure 2.36 (e–f and e'–f'). The Y-branch thus behaves similarly to an electrical-wire split which divides the initial current into two components of half intensity. We should however note that with Y-branches, *the power relation between input and output cannot be exactly 100%*. This is because the input and output mode envelopes are inherently different (just considering that the symmetric mode has a null in the center axis, unlike the fundamental mode). This means that the input/output mode overlap is always less than unity. However, the conversion efficiency can be brought reasonably close to unity by choosing a relatively narrow splitting angle (e.g., $<5°$), which narrows the E-field center null. In contrast, an angle as wide as $120°$, which would create a truly symmetrical Y-branch, would result in a dramatically poor mode overlap and very high power loss. This explains the fact that in integrated-optics, Y-branches are always characterized by narrow split angles. Another reason for this design approach is that the angles involved in the path breaks (bringing the output waveguides into parallel positions) should also be kept small. This is because a path break or change of direction is equivalent to a local change in core diameter and angle misalignment, to which corresponds a nonunity mode-overlap and thus a certain amount of power loss.

The analysis of Y-branches becomes more tricky when the device is input from the top port 2 (or equivalently, port 3), as shown in Figure 2.37. The initial single mode is seen to equally excite the two normal modes of the structure, with an in-phase relation for port 2 and out-of phase relation for port 3. At the point where the two input waveguides merge into a single output waveguide, the symmetric component is converted into the output single mode, while the antisymmetric component is lost by radiation into the bulk material. Such a loss, corresponding to 50% of the initial power, is due to the fact that the single-mode output waveguide cannot support an antisymmetric mode, or equivalently that the "overlap integral" between the input and output mode solutions is identically zero. If two different signals were input from ports 2 and 3 simultaneously, their

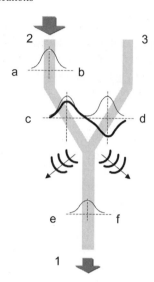

FIGURE 2.37 Sending a signal into port 2 of the Y-junction, showing initial single-mode propagation (a-b), followed by the equal excitation of the normal modes with symmetric/antisymmetric components (c-d), and back coupling into a single mode (e-f). This coupling is associated with a loss of half the signal power, corresponding to the antisymmetric component.

superposition would excite a set of normal modes, resulting in a net 50% power loss. The only way to avoid this 50% loss would be that the two signals are exactly in phase. As a result, the antisymmetric mode would not be excited leaving all the power to the symmetric mode towards an efficient conversion into the single output mode. Such a configuration (of academic interest) is equivalent to a time-reversed version of that in Figure 2.36, where the input/output arrows would be flipped. The main lesson to retain is that *Y-branches equally split signal input from port 1 with minimal excess loss, and equally combine signals input from ports 2 and 3 with at least 50% excess loss.*

After going through the preceding description, the case of X-junctions (Figure 2.36) is easier to understand. The geometry of the two waveguides defines a center region of approximate *interaction length L*, where the structure supports two normal modes, as shown in the figure. A key difference is that the normal modes do not have the same propagation constants. As a result, their mutual phase relationship evolves with propagation distance. Thus the net E-field sum of the sidelobes in the left and right waveguides evolves with distance: two in-phase lobes correspond to a power maximum in the waveguide, while two out-of-phase lobes correspond to a null. In the example shown, the lobe interference at distance L is such that the output power is steered into the upper-right waveguide output, while the initial signal is being launched in the lower left waveguide input. It is easy to conceive that if the device length were doubled ($L' = 2L$), the output power would be found in the upper-left waveguide. We refer to the length $L' = 2L$ as the device's *coupling length*. The same effect is produced for any device having an integer number of coupling lengths. If the length is halved ($L'' = L/2$) or equal to an odd-number of half-coupling lengths, the output ports 3 and 4 have identical output powers. In this case, the device is a *50–50 power splitter*, also called a *3-dB coupler*. Any interaction lengths other than in the two above examples will provide unequal powers at the output ports This effect can be used to make a *power tap*, where for instance the powers in the output ports approximately represent 95% and 5% of the input, respectively. The 5% output port can thus be used to monitor the power or spectral characteristics of the throughput signal (running through the 95% port). Both principles of 3-dB splitters and power taps are illustrated in Figure 2.38A. The definition of the total device loss must include the effect of power splitting, referred to as *splitting loss*. In these two examples, the splitting loss is -3 dB and -0.2 dB, respectively.

The 3-dB coupler performs a function analogous to that of the Y-branch coupler. However, two major differences between the two are that (1) the splitting effect is independent of the choice of input port (1 or 2 $\rightarrow$ 3 and 4 works the same as 3 or 4 $\rightarrow$ 1 and 2), hence the name of *2 × 2 coupler*, and (2) the input/output power conversion efficiency is close to 100% as opposed to 50%. Such a fundamental difference explains why it is not possible to construct a 4-port (2 × 2) coupler using two appended Y-branches ($>-<$), even if the principle would work in

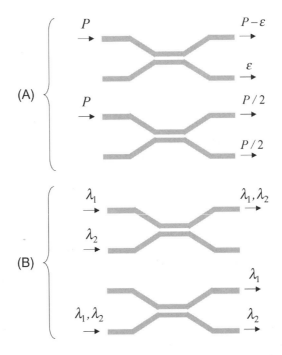

FIGURE 2.38 Four basic applications of 2×2 couplers (fiber or planar) concerning: (A) 50–50 power splitter or 3-dB coupler (top) or power tap with sampling ratio ε/P (bottom); (B) wavelength-selective couplers for wavelength combining/multiplexing (top) or wavelength splitting/demultiplexing (bottom).

electricity or hydraulics. The only workable solution (>=<) for a 2 × 2 coupler is to have a smooth conversion between a single-mode input excitation and the two normal modes of the structure, which approaches the 100% conversion efficiency regardless of the output splitting ratio. The finite coupler excess loss is caused by irregularities in the waveguide structure (such as angle breaks and geometrical imperfections), by fundamental mode-overlap considerations (as previously explained), by the material's excess loss and finally by the device's coupling loss and return loss. For planar devices, the loss may also be polarization-dependent to some extent, due to the difference in normal-mode tails and mode-coupling efficiencies according to the signal SOP. The lowest loss is achieved by 2 × 2 silica-fiber couplers, which can be 0.2–0.5 dB, depending upon the realization technique. The technique consists in fusing together two single-mode fibers in parallel over a few millimeters length, after having removed the jackets. During the fusion, the fibers are smoothly pulled to stretch the fiber material. This stretching brings the cores to a vicinity of a few wavelengths and increases the length of the coupling region (interaction length). Launching a signal from any port (e.g., 1), the output power from the two output ports (e.g., 3 and 4) can be monitored during the process. The fusion and pulling operations can thus be stopped when the desired result is obtained (e.g., 50–50 power splitter, 5% fiber tap, etc.).

For both fiber and planar 2 × 2 couplers, the interaction length which defines the amount of power obtained at the two output ports is a function of the signal wavelength. This is because the normal-mode propagation constants are strongly wavelength-dependent. It is then possible to realize a wavelength-selective coupler (WSC) if the device length corresponds to an integer multiple of coupling lengths at one wavelength and an odd-multiple of half-coupling lengths for the other wavelength. As illustrated in Figure 2.38, the effect can be used to either *combine* signals at different wavelengths into a single output waveguide, or *split* them into different output waveguides. If both signals carry data, the action of wavelength combining or wavelength splitting is called *multiplexing* or *demultiplexing*, respectively. Consistently, the corresponding devices, which work either way, are called MUX or DMUX. In optical (fiber) amplifiers, WSC are used to combine the pump (wavelength λ_p) with the signal or signals to be amplified (wavelength in a band centered at λ_s). By abuse of language, this application is widely referred to as a *pump multiplexer*, which is an inaccurate definition. Indeed, the pump is not a communications channel carrying data; it is just selectively "combined" with the data channels. One should note that the functions of power splitting and wavelength selection can also be performed with other types of planar and fiber waveguides, for instance with *Mach–Zenhder interferometers (MZI)* and *arrayed-waveguide gratings or AWG* (in the first case) and with *in-fiber Bragg gratings or IFBG* (in the second case). See more on MZI, AWG and IFBG in Section 2.4.2. To be complete, we must also mention that similar functions can be performed by bulk-optics devices, according to a wide variety of approaches and solutions. The principle is to expand the signal beam through microlenses or GRIN lenses. Three-port devices can be realized by placing 50%-reflecting or wavelength-selective mirrors in the signal path. Planar solutions are however preferable because of their miniature sizes and volume production. The main drawback is that fibers must be connected to the planar chip, which requires precision alignment and involves coupling loss. For some planar technologies, another drawback is the possibility of polarization-dependent loss (PDL) and higher material/waveguide excess loss. All-fiber devices occupy more volume, but are easier to manufacture, have negligible excess loss and PDL, and overall, are straightforward to fusion-splice/connect with other transmission-system elements.

The above description has made us understand some of the elegant and non-intuitive concepts of waveguide couplers, as based upon the concept of normal modes. We can now put this concept aside and focus on different possibilities of network-system applications. Consider first the case of *local-area networks* (LAN) with optical fibers as the transmission medium and a single or multiple signal wavelength for the communications channels. Several types of network topology are described in Desurvire (2004), Chapter 2, Section 2.5. Figures 2.21 and 2.22 show examples of *bus*, *ring*, *star* or *mesh* LAN topologies, and their evolving integration into a larger-size network. It is clear that the bus and ring LAN, whether uni-directional or bi-directional, only need power taps (single wavelength) or WSC (multiple wavelengths) for each station to pick up the signal(s) along the trunk. But the star LAN, which has a central connection node or hub, requires a more complex means to split or combine the signals going to or coming from all the N network stations, which corresponds to an $N \times N$ (also called $N:N$) coupler. Consider also the case of *passive optical networks* (PON), which are described in Chapter 1, Section 1.3, and whose topology is that of a *tree*. The most basic PON configuration is based upon a cascade of $1 \times N$ (or $1:N$) power splitters, as illustrated in Figure 1.16. A more sophisticated PON version is based upon wavelength splitting or DEMUX, as illustrated in Figure 1.18. In the following, we will analyze first the simpler case of star LAN and PON based upon a single-wavelength channel. We thus need to use our 2×2 couplers as building blocks to efficiently construct either $1 \times N$ (tree) and $N \times N$ (star) couplers while using a minimal number of elements. Figure 2.39 shows the optimized arrangements in the case $N = 4$. The 1×4 tree-coupler arrangements involving either three Y-branches or three 2×2 couplers are basic, and generalization to 1×8, $1 \times 16, \ldots 1 \times 2^Q$ is straightforward. In contrast, the case of the 4×4 star-coupler and its generalization to $2^Q \times 2^Q$ is far from intuitive. In the simplest case, the requirement for equal power splitting from any of the input ports into any of the output ports is met if we make a straight-through crossing between the two pairs of top and bottom 2×2 elements. Such a coupling algorithm is called the *perfect shuffle*. The perfect shuffle states that the required number of 2×2 elements to realize an $N \times N$ star (with $N = 2^Q$) is $n = (N/2) \log_2 N$. The number of couplers traversed for any signal path, which defines the overall transmission loss of the star coupler, is equal to $m = \log_2 N = Q$. We already know at least that these two statements are true for $Q = 1$ (2×2 couplers) and $Q = 2$ (4×4 couplers). The fact that the perfect-shuffle algorithm is optimal for any higher order is illustrated by the two 8×8 possible implementations shown in Figure 2.40. The first is based upon using four 4×4 modules, which represent sixteen 2×2 couplers and four 2×2 coupler crossings. It is easy to conclude that according to this implementation, an $N \times N$ star represents the crossing of $N/2$ individual 2×2 coupler units. The second implementation,

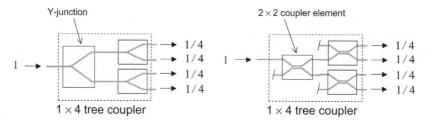

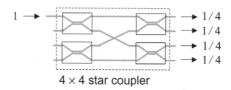

FIGURE 2.39 Optimized arrangements for 1×4 tree couplers (top), as based upon Y-branches (left) and 2×2 couplers (right) elements, and for 4×4 star couplers (bottom) as based upon 2×2 coupler elements, showing the effect of equal power splitting between input and output leads.

based upon the perfect shuffle, only requires twelve 2×2 couplers and three ($= \log_2 N = \log_2 2^3$) 2×2 coupler crossings. Thus the overall loss only grows as the base-2 logarithm of the number of ports N. An exercise at the end of this chapter provides an illustrative example on how to calculate the net signal loss experienced by the signal and the benefits of the perfect-shuffle approach as N becomes large.

When the number of ports exceeds $N = 8$, however, star couplers based upon the perfect-shuffle approach (or any other nonoptimized algorithms) become difficult to realize in practice. This is for several reasons. First, each 2×2 element occupies a length between one and two centimeters (because of the shallow angles required for the coupler structure) which, because of the size limits in the planar substrate (<10 cm) restricts the number of possible 2×2 crossings. The second reason is that the perfect shuffle requires complex connectivity patterns (see Figure 2.40) to which correspond steep angle changes in waveguide directions and multiple crossings. Both angle changes and waveguide crossings cause excess loss. In materials used for planar integration (e.g., silica), where the absorption loss is as low as 0.02 dB/cm, the 2×2 coupler excess loss can in fact be relatively high (e.g., 0.5–1 dB). In contrast, the all-fiber implementation based upon fusion-spliced, 2×2 fiber couplers has been for a long time the only practical solution for star couplers with virtually unlimited number of ports. The situation was changed around 1990 with the invention of the *free-space planar star coupler*, whose layout is shown in Figure 2.41. This device is made from two converging and diverging arrays of tapered waveguides which are disposed along a circle arc.

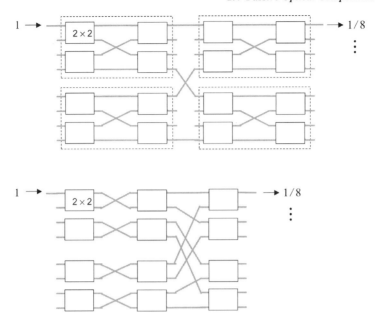

FIGURE 2.40 Implementation of an 8×8 star coupler from 2×2 coupler elements, as based upon 4×4 coupling modules (top) or the perfect-shuffle algorithm (bottom).

The intermediate zone defined by the two arcs is a free-space slab made of the same high-index material as the waveguides cores. The remarkable feature of such a structure is that with proper design, the illumination of any input port results in near-perfect equal power splitting into all output waveguides, as illustrated in the figure. By means of a short explanation of this phenomenon, the arrayed waveguide structure supports the so-called *Bloch modes*, which represent an infinite superposition of E-field waves traveling at the different angles defined by the

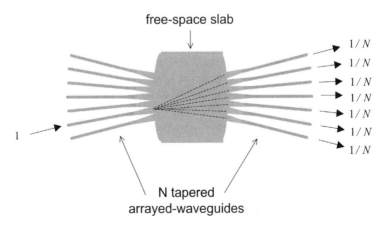

FIGURE 2.41 Layout of free-space planar star coupler.

array. A single-mode input from one of the waveguides progressively excites a set of in-phase Bloch modes having the same propagation constant, while the energy remains confined in the original waveguide. As the free-space slab interface is approached, the propagation constants diverge and the Bloch modes are out of phase, making the energy spread to all waveguides. The end result is the formation of an E-field wave with a unique k-vector (plane wave), which equally illuminates all the output waveguides located at the other extremity of the free-space slab. As shown below in this subsection, a key application of the free-space planar star coupler is integrated MUX/DMUX devices called AWG.

We consider next passive optical devices that perform the two functions of *multiplexing* (MUX) and *demultiplexing* (DMUX). The MUX concept is to combine into a single waveguide N channels at different wavelengths from N different waveguides, and the DMUX concept is the reverse operation which simply corresponds to changing the signal's propagation direction. Two key requirements for MUX/DMUX devices is that all WDM channels experience the same throughput loss and that such a loss be minimal. Solutions from bulk-optics are manifold. The two principal approaches are based upon the use of *reflection gratings* or *dielectric thin-film filters (TFF)*, as illustrated in Figure 2.42. The grating, which can be realized in glass or silicon materials, is an optical device which acts as a prism,

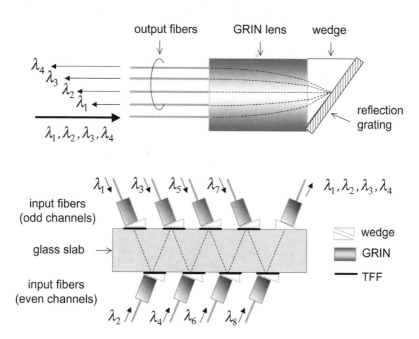

FIGURE 2.42 Examples of bulk-optics implementation of 1×4 or 1×8 wavelength multiplexer/demultiplexer (MUX/DMUX) based upon arrangements using: a reflection grating (top) or thin-film bandpass filters (bottom) on a glass slab. Both devices can act as MUX or DMUX by switching the signal directions or arrows (GRIN = graded-index lens, TFF = thin-film filter).

decomposing incident light into its spectral components. The input wavelength components incident upon the grating are thus reflected with different angles. A GRIN lens (or any other type of bulk-optics lens) directs the channels towards an array of output fiber waveguides. Alternatively, *concave* gratings or mirrors can be used to produce the lensing effect. The second apparatus shown in Figure 2.42 is based on the principle of multiple channel reflections inside a glass slab whose surface has been coated by different TFF. As explained in Section 2.4.2, a TTF is characterized by a narrow-band transmission spectrum centered about a specified wavelength λ_i. This means that only the channel at λ_i can pass through this TFF, while all other channels are 100% reflected. Starting from the left, the channel at λ_1 is first transmitted by its matched TTF, then reflected seven times until it reaches the output fiber. The second channel at λ_2 follows strictly the same path with six reflections, and so on until the last channel (here at λ_8). A key advantage of the approach is that it is independent of polarization. A drawback is the buildup of excess loss as the number of WDM ports is increased, due to material absorption, beam diffraction, filter imperfections and fiber-alignment inaccuracy (small angle errors resulting in substantial beam displacements at the output point). The net loss of bulk MUX/DMUX devices (generally called "insertion" loss) can widely vary according to several factors: the number of WDM channels and their separation/spacing, the material absorption and maximum path lengths, and the alignment accuracy.

Next, we turn to the realization of MUX/DMUX devices through planar technologies. The key component to achieve a desired wavelength selectivity is the *Mach–Zehnder interferometer* (MZI, or MZ for short). As shown in Figure 2.43, an MZI is a 4-port device made by connecting together two 3-dB couplers. A signal input in port 1 is then split by the first coupler into two signals

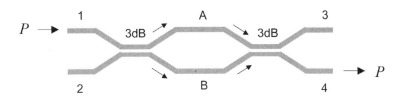

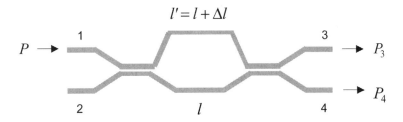

FIGURE 2.43 Principle of Mach–Zehnder interferometers (MZI), with equal-length/balanced arms (top) or unequal-length/unbalanced arms with path difference Δl (bottom).

following the paths (A, B) shown, which are then recombined by the second coupler. As shown in the figure, the path lengths can be made exactly equal (balanced MZI) or dissimilar with a difference Δl (unbalanced MZI). For now, let us concentrate on the balanced case. Looking at the figure, it is seen that one part of the signal goes through path A after a straight-through transmission through the input coupler, while the other part goes through path B after crossing this coupler. Upon traversing the second coupler, each of these signals can do a straight-through transmission or a crossing, which makes up four possibilities. A fundamental property of 3-dB couplers (which we must take here as a postulate), is that the two output ports exhibit a phase difference of $\pi/2$, the upper value corresponding to the cross-coupled path. With the balanced MZI structure, the four possible path histories are associated with the following phases (overlooking identical phase retardations due to propagation):

Path 1 to A to 3: two straight-through, with phase $\phi = 0 + 0 = 0$

Path 1 to B to 3: two crossings, with phase $\phi = \pi/2 + \pi/2 = \pi$

Path 1 to A to 4: one straight-through and one crossing, with phase $\phi = 0 + \pi/2 = \pi/2$

Path 1 to B to 4: one crossing and one straight-through, with phase $\phi = \pi/2 + 0 = \pi/2$

This elementary analysis shows that the two signals recombining in port 3 are out of phase ($\Delta\phi_3 = \pi$), while those recombining at port 4 are in phase ($\Delta\phi_4 = 0$). Thus there is no power at port 3 and all the signal exits from port 4, as illustrated in Figure 2.43 (top). Note that if the 3-dB couplers are wavelength-independent, the signal always exits through port 4 regardless of the wavelength.

Consider next the case of the *unbalanced* MZI, whose upper path A has an excess length Δl (Figure 2.43, bottom). Given the signal propagation constant β, the same analysis as before yields the four signal-path possibilities:

Path 1 to A to 3: two straight-through, with phase $\phi = 0 + \beta(l + \Delta l) = \beta l + \beta\Delta l$

Path 1 to B to 3: two crossings, with phase $\phi = \pi/2 + \beta l + \pi/2 = \beta l + \pi$

Path 1 to A to 4: one straight-through and one crossing, with phase $\phi = 0 + \beta(l + \Delta l) + \pi/2 = \beta l + \beta\Delta l + \pi/2$

Path 1 to B to 4: one crossing and one straight-through, with phase $\phi = \pi/2 + \beta l + 0 = \beta l + \pi/2$

The new result is that the phase differences are $\Delta\phi = \beta\Delta l + \pi$ for port 3 and $\Delta\phi = \beta\Delta l$ for port 4, respectively. If the excess length is chosen so that $\Delta l = \pi/\beta$ (or an odd integer multiple thereof), it is seen that $\Delta\phi_3 = 2\pi$ and $\Delta\phi_4 = \pi$, meaning that there is no power in port 4 and all the signal power exits port 3, representing the reverse situation of the balanced MZI. Alternatively, if we chose $\Delta l = m\pi/\beta$ where m is an even integer, the result is the same as in the balanced case. Since the signal propagation constant β is wavelength-dependent, only the signals at wavelengths λ_k, for which the condition $\beta(\lambda_k)\Delta l \equiv m\pi$ is met, will have a maximum output at either port 3 or port 4 (depending upon m being either an odd or even integer, respectively). This shows *that the unbalanced MZI is another type of wavelength-selective 2 × 2 coupler.* It is easily established that the MZI response is wavelength-dependent. Indeed, combining the two E-field signals at ports 3 and 4 while taking into account their relative phase:

$$\begin{cases} E_3 = \cos(\omega t + \beta \Delta l + \pi) + \cos(\omega t) \\ \quad = 2\cos\left(\omega t + \frac{\beta \Delta l}{2} + \frac{\pi}{2}\right)\cos\left(\frac{\beta \Delta l}{2} + \frac{\pi}{2}\right) \equiv 2f(t)\sin\left(\frac{\beta \Delta l}{2}\right) \\ E_4 = \cos(\omega t + \beta \Delta l) + \cos(\omega t) \\ \quad = 2\cos\left(\omega t + \frac{\beta \Delta l}{2}\right)\cos\left(\frac{\beta \Delta l}{2}\right) \equiv 2g(t)\cos\left(\frac{\beta \Delta l}{2}\right) \end{cases} \quad (2.123)$$

where $f(t)$, $g(t)$ are the time-oscillating components. The corresponding power is obtained by taking the time-averaged value of the square of the E-field, $P_i = \langle E_i(t) \rangle$, using $\langle f^2(t) \rangle = \langle g^2(t) \rangle = 1/2$, which gives

$$\begin{cases} P_3 = \sin^2\left(\frac{\beta \Delta l}{2}\right) = \sin^2\left(\frac{n_{\text{eff}} \Delta l}{2c}\omega\right) = \sin^2\left(\frac{m\pi}{2\omega_0}\omega\right) \\ P_4 = \cos^2\left(\frac{\beta \Delta l}{2}\right) = \cos^2\left(\frac{n_{\text{eff}} \Delta l}{2c}\omega\right) = \cos^2\left(\frac{m\pi}{2\omega_0}\omega\right) \end{cases} \quad (2.124)$$

This last result establishes that the MZI response is wavelength-dependent, since the propagation β is a function of wavelength. For clarity, we have also made the approximation $\beta(\lambda) = (2\pi/\lambda)n_{\text{eff}} = (\omega/c)n_{\text{eff}}$, where the effective index is assumed to be the same for all channels. Thus, we observe that the argument in the sine or cosine is proportional to the frequency ω. If we assume that the condition $\beta(\omega_0)\Delta l \equiv m\pi$ is satisfied for some frequency ω_0, then the MZI response is seen to oscillate with a cyclic-frequency period $\Omega = 2\omega_0/m$ (or a frequency period $F = \Omega/2\pi$). Such a property can be used either for periodic frequency-filtering (see next subsection) or for MUX/DMUX applications, as we shall discuss next.

To achieve a MUX/DMUX function for two given wavelengths λ_1, λ_2, the path imbalance must simultaneously verify $\Delta l \equiv p\pi/\beta(\lambda_1)$ and $\Delta l \equiv q\pi/\beta(\lambda_2)$ where p is even and q is odd or the reverse. Replacing these two conditions in equation (2.124) shows that $P_3(\lambda_1) = P_4(\lambda_2) = 0$ while $P_3(\lambda_2) = P_4(\lambda_1) = 1$, or the reverse. The two wavelengths being input at port 1 of the MZI then exit the device through the separate ports 3 and 4, which is the DMUX function (the inverse path corresponding to the MUX function). In the case of WDM systems, the channels are equally spaced in frequency by some interval $\Delta\omega$. It is thus possible to design the MZI in such a way that its frequency period Ω be equal to *two* channel spacings $\Delta\omega$. In this case, the MZI splits the odd-numbered channels ($\omega_1, \omega_3, \omega_5 \ldots$) and the even-numbered channels ($\omega_2, \omega_4, \omega_6 \ldots$) into port 3 and 4 (or the reverse). Since odd- or even-numbered channels are now separated by $2\Delta\omega$, they can be split again with the MZI having a period of $\Omega' = 4\Delta\omega$, and so on. See the Exercises provided at the end of the chapter for an illustrative numerical application. In a WDM system with $N = 2^Q$ channels, the number of MZI sequences required to achieve full

MUX/DMUX operation is Q, e.g. $Q = 4$ for $N = 16$. As easily established, the total number of MZI elements in the MUX/DMUX sequence is $M = 2^Q - 1$, e.g., $M = 15$ for $N = 16$. Although MZI devices can be integrated in large numbers on silicon substrates, the number of MZI that can be cascaded in sequence remains limited due to their finite (cm) sizes. As we have seen, the path imbalance can be made by an asymmetric MZI geometry (Figure 2.43, bottom). Alternatively, the imbalance can be generated by introducing a *phase shifter* in one of the two MZI paths, which permits more accurate control. In passive materials such as silica, a phase shift can be generated by locally applying *heat*, which is realized through thin-film electrodes. Silica-based devices with 3-stage MZI with thin-film electrode control have been realized for 1 : 8 MUX/DMUX at 10-GHz (0.08-nm) channel spacing.

Since the number of channels in the WDM systems is typically $N = 16, 32, 64$ (or possibly more), the problem has long been to find a practical way to realize massively-integrated MUX/DMUX devices. A clever solution is provided by the combination of a *multipath unbalanced MZI* using *two free-space star couplers* as multiple-port $N \times N$ couplers. Such a configuration is indifferently referred to as either an *arrayed-waveguide grating* (AWG), or a *waveguide grating router* (WGR). The AWG arrangement is shown in Figure 2.44. Basically, the AWG is a generalization of the unbalanced MZI having N different signal paths, with each path differing from its neighbors by an amount Δl. At the input edge of the first free-space star coupler, each input WDM channel is thus split into N paths. Let n_S be the refractive index of the free-space couplers, and $\beta = 2\pi n_{\text{eff}}/\lambda$ the arrayed-waveguide propagation constant (assuming negligible wavelength dependence in the WDM bandwidth). Consider one possible path "*ijk*," as shown in Figure 2.44. In this definition, i/k represents the input/output port numbers, and j the number of the MZI branch out of N possibilities. In the input/output free-space couplers, the signal traverses some distance $\delta l_i^{\text{in}}, \delta l_k^{\text{out}}$ as an unguided/plane

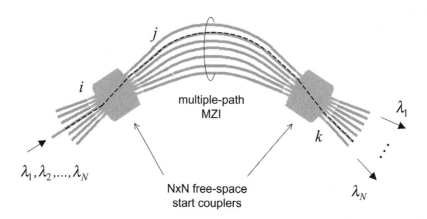

FIGURE 2.44 Layout of integrated arrayed waveguide grating (AWG) coupler for MUX/DMUX applications. The dashed line, indexed by the letters "ijk," shows one out of the possible signal paths followed by a given WDM channel.

wave. In the MZI region, the net path-length difference is simply defined by $l_j = j\Delta l$. Upon recombining at the output edge of the second free-space star coupler, the phase channel path is therefore

$$\phi_{ijk} = n_S \frac{2\pi}{\lambda} \delta l_i + j\beta \Delta l + n_S \frac{2\pi}{\lambda} \delta l_k$$

$$\equiv \frac{2\pi}{\lambda}[n_S(\delta l_i + \delta l_k) + j n_{\text{eff}} \Delta l] \qquad (2.125)$$

It is possible to design the free-space couplers with an appropriate arc curvature which makes the distance δl_m ($m = i, k$) defined by $\delta l_m = \delta l + m \delta l'$. We thus obtain:

$$\phi_{ijk} = \pi[U + (i+k)V + jW] \qquad (2.126)$$

where $U = 4n_s \delta l/\lambda$, $V = 2\delta l'/\lambda$ and $W = n_{\text{eff}} \Delta l/\lambda$ are real constants. We then observe that given the geometrical parameters $\delta l, \delta l', \Delta L$ (normalized to λ) and the indices n_S, n_{eff}, as represented by the three constants U, V, W, the phase is determined by three integers i, j, k. Given an input and output port pair (i, k), it can be shown that with the proper choice of the parameters $\delta l, \delta l', \Delta L$, one can show that all the N possible ϕ_{ijk} phases ($j = 1 \ldots N$) differ by even or odd multiples of π, corresponding to effects of constructive or destructive interference, respectively. The nice result is that a configuration exists for which, given the input port i, the interference is constructive for only one output port k, and is destructive for all others. Thus the WDM channels entering from different input ports can be made to exit from a single output port (MUX), or the reverse (DMUX), as illustrated in Figure 2.44. Compared to other MUX/DMUX technologies, the AWG exhibits higher insertion loss (<6 to <10 dB) and PDL (<1 dB). Part of the loss is due to the fiber-substrate coupling loss, which is higher in silicon materials compared to InP and Silica-based devices. The insertion loss can be compensated by placing the MUX/DMUX at appropriate locations in the terminals, namely near optical preamplifiers or booster amplifiers. The PDL, which is caused by the waveguide's rectangular shape, can be alleviated by implementing a *polarization-diversity* scheme, as further described below with the topic of optical circulators. Alternatively, PDL can be reduced by improved waveguide geometry and other in-device compensation techniques. Current technologies make it possible to produce AWGs with 0.8-nm spacing (100 GHz at 1.55 μm) and up to 64 channels.

Specific wavelength splitting/combining patterns can be obtained by customary AWG design. As an illustration, Figure 2.45 shows the case of a *static wavelength router*, in which the four output waveguides contain all possible wavelength permutations of the four input waveguides. Such static routers have applications in star-based WDM LAN. Another degree of freedom is provided by using input and output star couplers having a different number of ports, which makes it possible to realize a wide variety of novel $N \times M$ wavelength-selective devices.

The preceding text has illustrated some of the possible MUX/DMUX technologies and their implementation issues. Other approaches, based upon *circulators* and *fiber gratings* are also possible, which will be further described below. As we have

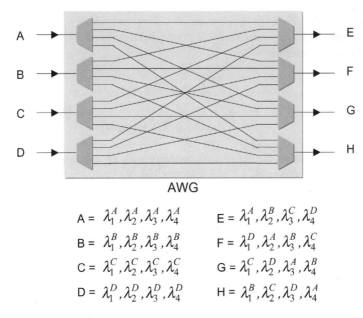

FIGURE 2.45 Principle of static wavelength router with equivalent routing connections, as based upon customary-designed AWG. In this example, the routing table corresponds to complete wavelength-channel permutations.

mentioned, two key parameters are the device's insertion loss and PDL. For actual WDM system implementation, the following other criteria must also be taken into account (Figure 2.46):

Passband: the width of the window defining WDM channel selection, usually characterized as -3-dB width and -1-dB;

Center frequency and frequency spacing: must accurately match the WDM wavelength comb;

Curvature: the shape at the center of the passband, which should ideally be as flat as possible (successive MUX/DMUX stages otherwise narrow the passband);

Crosstalk: the overlap between adjacent channels, as defined by the transmission where the passbands overlap, typically from -20 dB (poor) to -40 dB (ideal).

In addition, the passband wavelengths of MUX/DMUX devices must be relatively temperature-insensitive, as characterized by a *temperature coefficient* η (typically $\eta \leq 10^{-2}$ nm/°C for AWG and $\eta \leq 5.10^{-4}$ nm/°C for TFF).

We finally focus on *polarization-based devices* which ensure a certain number of key functions in optical transmission systems. These essentially include: *polarization splitters*, *optical isolators*, and *optical circulators*. We shall overlook here the case of *polarization controllers*, which make it possible to change the SOP by

2.4 Passive Optical Components

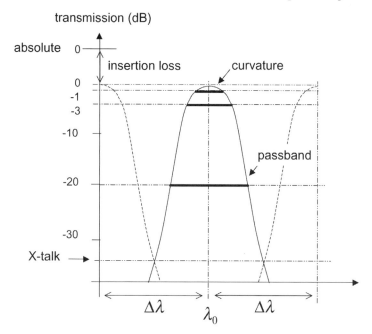

FIGURE 2.46 Key characteristics of MUX/DMUX transfer function in the frequency domain, showing insertion loss, passband shape and top curvature, passband widths at −1 dB/−3 dB/−20 dB (thick lines) and crosstalk point due to overlap with neighboring channels.

mechanically twisting or squeezing the fiber. Since SOP is sensitive to environment and fiber manipulation, these controllers only permit *ad hoc* optimization of SOP conditions by temporarily minimizing any PDL-related effects.

Polarization splitters are based upon the principle of separating the two linear-polarization components (referred to here as parallel // or TE, and perpendicular ⊥ or TM) of an input signal into two different paths. The new paths followed by the TE/TM components can be either orthogonal (*polarization beam-splitter*, or *PBS*), or parallel (*spatial walk-off polarizer*, or SWP). These two effects can be realized by using transparent crystal materials having a high intrinsic birefringence, such as calcite ($CaCO_3$) or rutile (TiO_2). The birefringence causes the effect called *double refraction*, where the refraction angle is different for the two polarizations. One can then realize 45° prisms which provide 100% internal reflection for TE and a finite refraction effect for TM. The TM ray deviation due to refraction can be minimized by placing a second upside-down prism on the TM path, which forms a cube with a thin interface. This is the case of the PBS: the TE component is reflected at 90°, while the TM component follows a straight-through path, as illustrated in Figure 2.47. The explanation concerning SWP splitters is a bit more complex. In this case, the material is cut so that its birefringence axes are not in the plane separating the air/material interface. As a result, Descartes or Snell's law ($\sin \theta_1 = n \sin \theta_2$) which was described in Section 2.2, is no longer applicable!

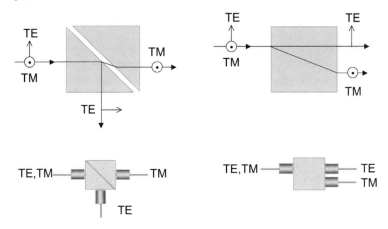

FIGURE 2.47 Two types of polarization-splitting devices: polarization beam-splitter (left), and spatial walk-off polarizer (right). The TE polarization component is parallel to the page plane (arrow), while the TM component is perpendicular to this plane (dotted circle). The second set of drawings (below) show fiber-connected implementations with beam-expanding GRIN lenses.

This is because the k-vector, which defines the initial direction of the light ray, has two finite and different projections on the material's birefringence axes. The phase-matching condition for the ray is then $k_0 \sin \theta_1 = k_2 \sin \theta_2$, or equivalently, $\sin \theta_1 = n_2(\theta) \sin \theta_2$. After analysis, the consequence is that the TM ray propagates in a direction forming some angle with the k-vector, while the TE ray propagates in the same direction. As a result, light rays *with normal incidence* split into TE and TM components, as illustrated in Figure 2.47. The figure also shows equivalent fiber-connected versions using GRIN lenses as beam expanders.

The PBS and the SWP can now be used to build two major optical components: the *Faraday isolator* and the *circulator*. Both are based upon combining polarization path-splitting and the Faraday effect. Recall from Section 2.2 that the Faraday effect consists in the rotation of the E-field linear polarization in the presence of a longitudinal magnetic field. It can be obtained in certain crystal and even amorphous materials having a large *Verdet constant*. Efficient Faraday materials are the *yttrium–iron garnet* (YIG) and other garnets based upon *terbium*. Such materials make it possible to generate Faraday rotations as high as 90° using permanent magnets such as cobalt-iron alloys with small or even miniature sizes. The magnet is then made in the form of a hollow cylinder whose axis, parallel to the magnetic field, is used for the signal path. As previously described, a key aspect of the Faraday effect is that the rotation angle is independent of the E-field propagation direction. Thus, if the polarization rotates by an angle θ in a certain direction and the signal is reflected back on its own path, the total rotation angle will be 2θ (as opposed to zero). An *optical isolator* can then be realized by placing a Faraday rotator between two PBS oriented at 45° from each other, as explained in Figure 2.48. As its name indicates, this isolator prevents any reflected signal from returning to its path, hence the name of "*nonreciprocal*" device. The quality of

2.4 Passive Optical Components 215

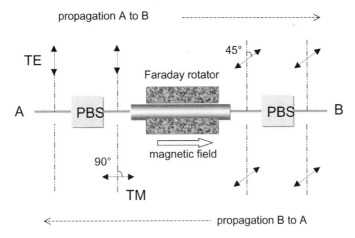

FIGURE 2.48 Principle of the optical Faraday isolator: the upper plane represents the polarization evolution when the signal propagates from left to right (AB), while the lower plane corresponds to the opposite propagation direction (BA). Propagation AB: the first polarization beam-splitter (PBS) transmits a TE input signal; upon passing through the Faraday rotator, the polarization is rotated by exactly 45°; the second PBS is oriented at 45°, parallel to the polarization, which transmits the signal. Propagation BA: the signal is reflected back from point B into the apparatus and transmitted by the first PBS; the Faraday rotator brings the polarization angle to another 45°, corresponding to a TM signal; this signal is blocked by the second PBS and cannot reach point A.

the isolation is measured by the net transmission loss experienced by the reflected signal after traversing the input PBS. Depending upon the PBS crosstalk (fraction of residual TM power passing through the device), this loss is typically -20 dB to -40 dB. Optical isolators are essential in the realization of laser-sources modules, because very small signal reflection into the laser causes large power and spectral instabilities.

The above implementation of a Faraday isolator requires that the input signal polarization be linear and the apparatus (first PBS) exactly aligned with it. Since transmission fibers do not maintain polarization, a new scheme had to be developed for which the isolation works indifferently for TE and TM signals (or by extension, any arbitrary SOP). Polarization-independent devices where the TE and TM paths are split and processed independently are referred to as being of the *polarization-diversity* type. Figure 2.49 shows how to realize a *polarization independent isolator* with two SWP, one Faraday rotator and one *half-wave plate*. The half-wave plate is a slab of birefringent medium whose thickness corresponds to a phase delay between the two axes of π, equivalent to a path difference of $\lambda/2$. If a signal traverses the plate with polarization parallel to any of the axes, there is no change. However, if the incident polarization is at 45° or 22.5° of the axes, the polarization rotates by 90° or 45°, respectively. This second configuration is exploited in the arrangement shown in Figure 2.48. Following the SOP evolution from port A to port B, and from port B towards port A shows that the arrangement

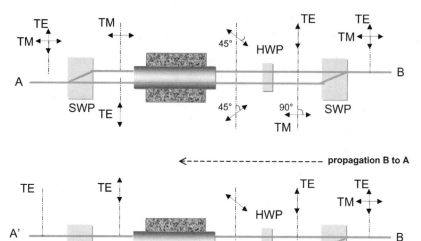

FIGURE 2.49 Principle of polarization-independent Faraday isolator, with propagation from left to right (AB, top) and right to left (bottom), as based upon two space walk-off polarizers (SWP), a Faraday rotator, and a half-wave plate (HWP). The HWP axis is oriented in such a way that the input polarizations are rotated by 45°. The return path (BA) is seen to be blocked by the input SWP, which splits the TE and TM components to different ports A' and A", respectively.

works as a polarization-independent isolator: the return signal splits into ports A' and A", which isolates port A from any signal back-reflections.

An *optical circulator* can be realized from the previous arrangement. In its basic configuration, the circulator can be seen as a three-port (A,B,C), polarization-independent device for which the segments AB and BC act as Faraday isolators. The path AB is already shown in Figure 2.49. To introduce port C, the idea is to retrieve and recombine the two polarized signals initially heading to points A' and A". As Figure 2.50 shows, this can be done by using a PBS and a reflecting 45° prism. The recombined signals then define port C. It is an easy exercise to analyze the reverse path from C to B and to conclude that the output polarizations split into two points B', B", thus isolating port B from any incoming signals. We see that *a circulator is basically a three-port, polarization-independent Faraday isolator*. With the addition of other elements, it is possible to make the circulator symmetrical by creating another isolated path between C and A, completing the path circle. For most applications, only the asymmetrical version is required, as some examples in the next subsection illustrate. More complex implementations permit the realization of *four-port circulators*, usually symmetrical. Figure 2.51 shows the different concepts. A remarkable feature is that in spite of the complexity of these bulk-optics devices, the typical port-to-port loss is only 0.5–1 dB. Such

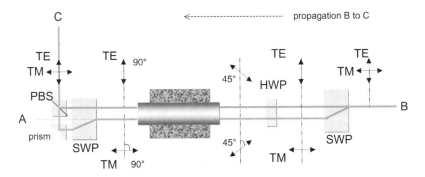

FIGURE 2.50 Principle of optical circulator, as based upon the same apparatus as previous figure for the polarization-independent Faraday isolator, with the inclusion of a PBS and a 45° reflection prism near the input end. The path AB is strictly the same as in the previous case, but the reverse path from port B leads to new port C.

a performance level requires very high optics quality and precision alignment, in addition to anti-reflection coatings, which for symmetric and 4-port devices make the price increase markedly. Three-port asymmetrical circulators make it possible to realize *optical add-drop multiplexers* (OADM), as a versatile alternative to static wavelength routers based upon AWG. They also make it possible to introduce polarization-diversity operation to suppress device PDL, as described in the next subsection.

2.4.2 Optical Filters

As in electronics, *optical filters* permit the accurate control of the signal frequency/wavelength while removing unwanted frequency/wavelength components, either as neighboring channels or as broadband noise. Their spectral selectivity can be of the

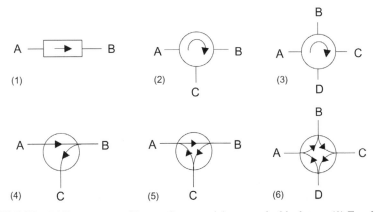

FIGURE 2.51 Different types of 2-port, 3-port and 4-port optical isolators: (1) Faraday isolator, (2)-(3) optical circulators, (4)-(5) 3-port circulators defining paths AB/BC (asymmetric) and AB/BC/CA (symmetric), and (6) 4-port symmetric circulator.

low-pass, *high-pass* or *bandpass* type, as characterized by step-like or window-like passband shapes, defining peak transmission regions. The bandpass type may have a single, narrow transmission window (e.g., TTF, IFBG) or be periodic (e.g., MZI, FP), as described in the following text. The filter's characteristics can be *fixed* (permanent) or tunable. This last feature enables dynamic control, noting that the broader the tuning range, the slower the tuning cycle. More generally, filters are characterized according to the same parameters as previously defined for MUX/DMUX devices and illustrated in Figure 2.46.

In WDM systems, optical filters have a wide variety of functions. In the previous subsection, we saw that *thin-film filters* (TTF) and planar *Mach–Zehnder interferometers* (MZI) can be used for the realization of MUX/DMUX devices. We shall describe here, in detail, another filter type, called the *in-fiber Bragg grating* (IFBG). The IFBG can be regarded as probably one of the most important (and recent) technology developments in WDM communications next to the optical fiber amplifier. In particular, IFBG are used to realize the function of *optical add-drop multiplexing* (OADM) whereby WDM channels can be inserted or extracted from any point in the aggregate signal path. Such a technique completes that of the *static wavelength routers* based upon planar AWG. As another most important application, both AWG and IFBG (and to some extent any filter) can be used as *gain equalizers* or *gain-flattening filters* (GFF) in order to produce a uniform response from optical amplifiers. Here, we shall provide a brief review of the different fiber types and their WDM applications.

We consider again the MZI, which was analyzed in Section 2.4.1 as a periodic wavelength-selective filter. To recall, the MZI bandpass shape is of the type $f(\omega) = \cos^2(\pi\omega/\Omega)$, where Ω defines the cyclic-frequency periodicity, also called the *free spectral range* or FSR with the conversion FSR $= \Omega/2\pi$. This bandpass function is quite broad, since $f(\Omega/4) = 1/2$ yields a 3-dB bandwidth of $\Omega/2$ or half the FSR. The solution to narrow this bandpass is to cascade several MZI whose FSR increase by multiples of two. The net transfer function is given by the product $f(\omega) = f(\omega, \Omega) \times f(\omega, 2\Omega) \times f(\omega, 4\Omega) \times \cdots$, which is illustrated in Figure 2.52 for up to $N = 3$ stages. As the figure shows, the ratio of 3-dB bandwidth to FSR rapidly decreases, approximately as $1/2^N$, which provides increased channel selectivity. The log scale shows that the filter passband has multiple sidelobes, whose transmission maxima also decrease with N. In the case $N = 3$, for instance, the maximum sidelobe extinction is about 5% or -13 dB. Such a filter could be applied to select WDM channels according to a rule of $1:2$, $1:4$, $1:8$, etc., as analyzed earlier in the $1:2$ case. But some applications may require a substantially higher extinction or narrower channel selectivity, in which case other types of periodic filters must be considered.

The second type of periodic filter is the *Fabry–Pérot (FP) etalon*. The principle of a FP filter is analogous to that of the *FP cavities* used for lasers, which were described in Section 2.2 (Figure 2.45). The cavity is realized by placing two partially transmitting mirrors at some distance L. A signal entering such a cavity is internally reflected an infinite number of times. If the cavity length corresponds to an integer multiple k, of the signal's half wavelength, i.e., $L = k\lambda/2$, then a standing E-field wave builds inside the cavity. Note that this resonance condition applies to any

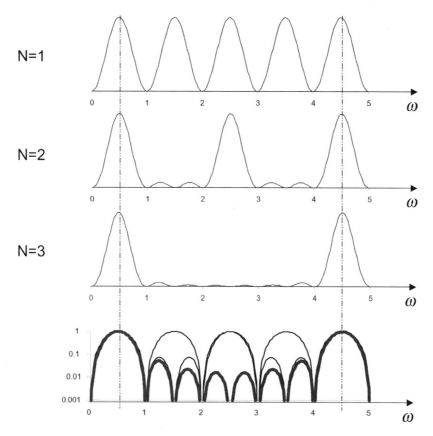

FIGURE 2.52 Filter spectral transmission response obtained by cascading N Mach–Zehnder interferometers with increasing free-spectral ranges, from top to bottom: $N=1$ (FSR = 1), $N=2$ (FSR = 2) and $N=3$ (FSR = 4). The curves at the bottom show the corresponding plots in logarithm scale ($N=3$ in bold), which provides the relative extinction ratios of the unwanted side lobes.

wavelength verifying $\lambda = 2L/k$, which means that it is periodically repeated. Because the mirrors are partially transmitting, the signals exits the cavity from the other end, with maximum transmission under the exact resonance condition. Predictably, if this condition is only approximately verified, some amount of destructive interference occurs because of the mismatched reflections inside the cavity, resulting in signal throughput loss. Thus the FP cavity acts as a wavelength-selective filter. It can be shown that the FP transmission or transfer function, is of the form:

$$f(\omega) = \frac{\left(1 - \dfrac{A}{1-R}\right)^2}{1 + \dfrac{4R}{(1-R)^2}\sin^2\left(\dfrac{nL}{c}\omega\right)} \quad (2.127)$$

where R is the mirrors' (power) reflection coefficient or reflectivity, A their absorption and n is the refractive index of the medium inside the FP cavity ($L/(c/n)$ thus represents the cavity transit time). We observe from this result that the FP transfer function is periodic with a cyclic-frequency period $\Omega = \pi c/(nL)$, or a frequency periodicity $FSR = \Omega/2\pi = c/(2nL)$. The function is plotted in Figure 2.53 for different values of the reflectivity R, assuming a loss-less device ($A = 0$) and $F = 1$. We observe that as the reflectivity increases, the filter response rapidly sharpens, exhibiting single spikes without side lobes. The merit factor to characterize this filter selectivity is the FP *finesse*, which is defined as the ratio of the FSR to the 3-dB passband. Its analytical definition is simply:

$$\text{Finesse} = \frac{\pi\sqrt{R}}{1 - R} \tag{2.128}$$

The examples plotted in Figure 2.53 ($R = 0.05$–0.95) thus have finesses of 0.73–61. It is easily verified from this definition that FP filters with $R \geq 0.99$ mirrors have finesses in excess of 300. For WDM systems, such a high selectivity is an "overkill," since the 3-dB channel bandwidth ($\Delta\nu$) is generally close to the FSR, which corresponds to the concept of *spectral efficiency* ($\eta = \Delta\nu/FSR$). But low-finesse FP filters can be used for MUX/DMUX applications. One of their key advantages is their *tunability*. Indeed, a free-space, FP, cavity can be made through the air

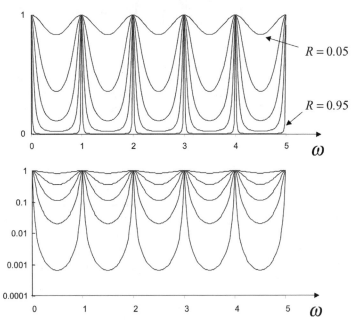

FIGURE 2.53 Fabry–Pérot filter transmission response in linear (top) or logarithmic (bottom) scales obtained for mirror reflectivities of (top to bottom curves) $R(\%) = 5, 25, 50, 75,$ and 95.

gap separating two closely spaced fibers whose ends have been coated with high-reflection TFF. If the fibers are aligned inside a sleeve made of a piezo-electric material, the length of the air gap can be controlled by applying a voltage, making it possible to tune the FSR to the desired channel-spacing value.

We consider next the case of *in-fiber Bragg-grating* (IFBG or FBG) filters, also called for short *fiber gratings*. The discovery and first realization of FBG in the decade 1980–90 has represented a major turn in the history of optical filters and in the field of passive optical devices as well. The underlying principle of FBG is the permanent inscription of a *Bragg reflector* in the core of an optical fiber. Before considering how such inscription can be explained and realized, it is important first to understand the concept of *Bragg gratings*. Basically, a Bragg grating consists of a structure where the refractive index is periodically changed by a small amount δn at every distance d, as illustrated in Figure 2.54. The points where the index locally varies are associated with an effect of E-field reflection, as characterized by a reflection coefficient R and a transmission coefficient $T = A(1 - R)$, where A is the absorption of the segment of length d. The periodic index changes cause the incident light to be multiply reflected. The successive E-field reflections are simply defined by

$$\begin{cases} E_1 = R\cos(\omega t) \\ E_2 = T^2 R \cos(\omega t - 2\beta d) \\ E_3 = T^4 R^2 \cos(\omega t - 4\beta d) \\ E_3 = T^6 R^3 \cos(\omega t - 6\beta d) \\ \cdots \end{cases} \quad (2.129)$$

where $\beta = 2\pi n_{\text{eff}}/\lambda$ is the medium's propagation constant. The infinite sum of these reflected signals is thus

$$E = R \begin{Bmatrix} q^0 \cos(\omega t - 0 \times 2\beta) d + \\ q^1 \cos(\omega t - 1 \times 2\beta d) + \\ q^2 \cos(\omega t - 2 \times 2\beta d) + \\ \cdots \end{Bmatrix} \quad (2.130)$$

where $q = T^2 R = RA^2(1 - R)^2$. It is immediately seen that the condition to achieve maximum reflection is that all cosines are in phase, providing maximum amplitude

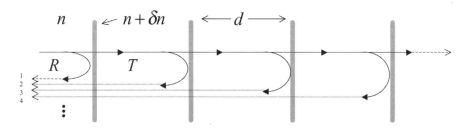

FIGURE 2.54 Principle of Bragg gratings, showing the superposition of signals multiply reflected from mirrors due to periodic index changes.

at each E-field oscillation period. Such a condition is $2\beta d = 2k\pi$, or $2(2\pi n_{\text{eff}}/\lambda)d = 2k\pi$, where k is an integer, or

$$d = k\frac{\lambda_B}{2n_{\text{eff}}} \tag{2.131}$$

which is known as the *Bragg condition*, with λ_B being the *Bragg wavelength*. Sometimes the grading period d is also called Λ_B. We thus observe that the Bragg condition is the same as the resonance condition for a Fabry–Pérot cavity. The cases $k = 1$ (half-wavelength) and $k = 2$ (single wavelength) are said to correspond to *first-order* and *second-order gratings*, respectively. Replacing this condition into equation (2.130) yields

$$E = R\cos\omega t\{q^0 + q^1 + q^2 + \cdots\} = R\cos\omega t\frac{1 - q^\infty}{1 - q} \equiv \frac{R}{1 - q}\cos\omega t$$

$$= \frac{R}{1 - A^2 R(1 - R)^2}\cos\omega t \tag{2.132}$$

which defines the maximum signal which can be reflected by the Bragg grating. If the reflection coefficient R is close to unity (e.g., $>95\%$), the denominator in the above RHS can be approximated by unity, and in this case we have simply $E \approx R\cos\omega t$. In single-mode fibers, it can be shown that for a grating of length L, the maximum reflection coefficient is given by the approximated expression

$$R_{\max} = \tanh^2\left(\eta\frac{\pi L\delta n}{\lambda_B}\right) \tag{2.133}$$

where η is the fraction of optical power contained in the fiber core (as gratings cannot be written in the cladding, see next). We leave it to an exercise at the end of the chapter to determine that with index-changes of $\delta n = 10^{-2}$–10^{-3}, the grating length required to achieve a reflectivity of $R = 99\%$ is of the order of 0.3–3 mm.

The principle of fiber grating inscription is that of *photosensitivity*. The effect consists in a permanent increase of the local refractive index ($\delta n = 10^{-2}$–10^{-3}) after intense UV-light exposure (240–250 nm, 1-W power). It is observed in silica glasses which are lightly doped with germanium (as in the core of SMF/DSF) or other less-conventional dopants such as phosphorus, boron or cerium. The germanium-oxygen (Ge-O) bond can occasionally be broken, forming a structural *defect* in the molecular glass structure. Another type of defect is the Ge-Ge bond, which can be structurally modified under UV exposure. These defects, which are associated with specific electronic absorption bands not found in pure silica glass, are highly absorbing at UV wavelengths. The UV exposure causes complex reactions, with both an effect of electronic excitation of the defects and microscopic compaction/dilation of the glass. The UV absorption, which is accompanied with a refractive-index resonance, is dramatically increased by this process. The evanescent tail of the resonance thus becomes significant at 1.55-μm wavelength, which translates to a measurable index change. This change increases with the duration of UV exposure according to a sublinear power law. The process however saturates with time, leading to maximum refractive-index changes of the order of 5×10^{-4}. It was discovered that submitting the fiber, prior to UV exposure, to a high-pressure hydrogen (or deuterium) environment considerably enhances photosensitivity (a process involving all Ge atoms and not only the defects), leading to maximum index changes of 10^{-2}. As the UV exposure is interrupted, the index change decays over time, with a rapid initial relaxation followed by a very slow decreasing process, leading to a so-called "permanent" index-change. Because such a slow degradation process affects long-term device reliability, it is necessary to stabilize it by a technique of *accelerated aging*. The approach, called *annealing*, consists in exposing the device to high temperatures (300–550°C) for one hour or more, which removes most of the decaying part of the index change and leaves only its stable part. It can be shown that a reflectivity that would normally drop by 10–20% over 1–10 years' lifetime would remain virtually unchanged after annealing.

The photo-inscription of a fiber grating is realized by making two beams of UV light interfere in the fiber core, as illustrated in Figure 2.55. The effect can be produced through various arrangements of bulk-optics interferometers, or through a phase (grating) mask. The phase mask is a surface having a periodic corrugation which, at normal incidence, causes a half-wavelength delay between the peaks and the troughs. This phase grating causes the beams to interfere with a period equal to half that of the mask. Such an arrangement is much more practical that the bulk interferometer, since it requires only one UV beam at normal incidence. The grating length L, which is limited by the size of the interference fringe pattern, can be extended by precise translation of the fiber during the writing process. The basic index-change profile is a raised-sinusoid with a rectangular envelope of length L. The corresponding grating reflectivity as a function of wavelength consists in a main lobe centered at $\lambda = \lambda_B$ and with reflection R_{max}, surrounded by rapidly-decaying side-lobes. The fiber grating thus acts as a reflection filter.

Using fiber translation, other index-profile patterns and filter responses can be obtained. First, the maximum index change can be made to vanish at the grating edges, which is referred to as grating *apodization* (for "cutting feet"). The value of apodization is that the filter side-lobes disappear, leaving only a central lobe at

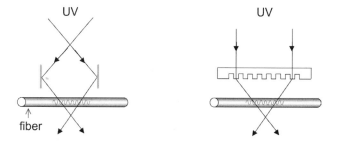

FIGURE 2.55 Two methods of photo-inscription of fiber Bragg gratings: with interferometer (left) and with phase mask (right). In both cases, the interference fringe pattern obtained in the fiber core is made to have the Bragg period Λ_B.

the expense, however, of a broader 3-dB passband. Such a grating filter is highly reflective at this central wavelength and its transmission is about 100% at all wavelengths sufficiently detuned from it. Thus, it can be used to selectively reflect a single wavelength out of a WDM multiplex signal, as shown in Figure 2.56. A most important application of FBG reflectors concerns OADM devices, whose principle is illustrated in Figure 2.57. As seen from the figure, a reflected WDM channel can be routed to a terminal by means of an optical circulator. One can then insert another channel at the same wavelength coming from the terminal, using the two-circulator arrangement shown. Several FBG at center wavelengths (λ_i, λ_j, ...) can be placed in cascade in the center region in order to drop and insert the corresponding channels by blocks. Such an OADM function is particularly useful in both terrestrial and sub-marine networks, where the overall bandwidth can thus be sliced down by WDM channels (or groups of WDM channels) according to the local terminal or subnetwork capacity needs.

It is also possible to introduce one or several discrete phase shifts (quarter- or half-wavelength) at different grating locations. The phase shift introduces a "transmission fringe" in the reflection pattern (changing its "U" shape into a "W" shape), which corresponds to a narrow *transmission* passband. Furthermore, this passband does not have side-lobes in the immediate vicinity, which is advantageous for wavelength-selectivity purposes. The introduction of several phase shifts make it possible to broaden this passband and shape it with a flat top. Thus FBG can also be used as transmission filters, although concatenation at different wavelengths is problematic due to their complex transmission/reflection sidelobe patterns away from their center windows.

In the realization of FBG, another possibility is to continuously vary the grating period along the fiber axis, which is referred to as a *chirped grating* (named after the effect of "chirp" in bird songs, which contain frequency sweeps). As Figure 2.56

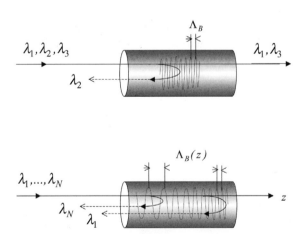

FIGURE 2.56 Two basic applications of fiber Bragg gratings: wavelength-selective reflector, here with λ_2 being the Bragg wavelength (top), and dispersion compensator with chirped-grating structure (bottom).

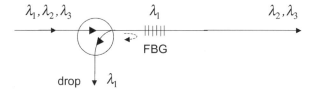

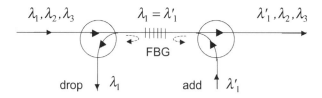

FIGURE 2.57 Principle of channel extraction, or drop function, from WDM multiplex (top), and re-insertion of another channel at same wavelength $\lambda'_1 = \lambda_1$, or add function (bottom), making up the overall function of optical add-and-drop multiplexing (OADM) by means of fiber Bragg gratings (FBG) and optical circulators.

shows, a chirped grating acts as a wavelength-dependent delay line. If the long-period side is placed at the input end, for instance, the longest-wavelength signal is reflected first (until the grating is fully detuned) and the shorter-wavelengths signals are transmitted. The process repeats each time a signal wavelength matches the local Bragg wavelength $\lambda_B = 2\Lambda/n_{\text{eff}}$. Thus each wavelength channel experiences a different time delay. Such an apparatus can be used to compensate fiber dispersion, here in the positive-dispersion region where long-wavelength signals travel slower than short-wavelength ones (the opposite effect, for negative-dispersion compensation, is obtained by reversing the position of the fiber grating ends). See more on this application in the next subsection.

For an economy of size, several FBG can be written inside the same fiber segment, either in cascade or directly on top of each other. This is equivalent to concatenating the gratings, except that there is no splice or connector loss. At least this is true if all FBG can be written into the same fiber types. If the wavelength range is broad (e.g., 1–1.4 μm), different fiber types (with varying Ge-doping and core sizes) must be used in order to achieve gratings of equal performance, which requires fusion splicing with finite coupling loss between the different FBG sections. Such broadband FBG cascades have applications in *multistokes Raman fiber lasers*, which can be used in turn as high-power pumps in fiber amplifiers (see further in subsection below).

The FBG index structure can also be written with an orientation different from the direction normal to the fiber axis, referred to as *blazed FBG* or *tilted FBG*. In this case, the reflected wavelengths are steered to the fiber cladding, corresponding to signal loss or, at maximum reflectivity, to a selective wavelength block. Alternatively, the exiting rays can be collimated onto a detector array, which monitors the intensity changes according to wavelength. With a chirped IFBG, a fraction of the WDM channel power can thus be extracted and monitored, which performs the function of an *in-line spectrometer*. Fiber gratings can also be made *tunable*. This is accomplished by stretching the fiber through a piezo-electric sleeve arrangement (mechanical tuning), or by local heating which dilates the grating (thermal tuning). Note that the FBG is relatively stable with respect to environmental temperature changes, corresponding to 1-nm wavelength shifts over an 80°C range. Such a small effect can also be made negligible by proper FBG packaging.

The above description concerned fiber gratings with periods comparable to the wavelength (i.e., $\Lambda_B = 0.5 - -0.75$ μm). These are referred to as *short-period gratings* or *Bragg gratings*. It is also possible to realize gratings whose period is very large compared to the wavelength, namely from a few hundred times ($n \times 100$ μm) to a few thousand times ($n \times 1$ mm). These are called *long-period gratings*. Their principle is quite different from that of the short-period FBG, because they involve the fiber *cladding modes* (rays propagating between the fiber core and its outer air/glass cladding). If β_{core} and $\beta_{clad,N}$ are the propagation constants associated with the core modes and a given cladding mode labeled by N, the phase-matching condition for constructive interference is simply $(\beta_{core} - \beta_{clad,N})\Lambda = 2\pi$. In the wavelength domain, this condition is equivalent to $(n_{eff}^{core} - n_{eff}^{clad,N})\Lambda = \lambda$. Thus a precise calculation of the propagation constants of the cladding modes makes it possible to determine the period Λ for which a maximum back-reflection effect can be obtained at a desired wavelength λ. This makes possible the realization of *transmission filters* which are dispersive (highly reflective) at some center wavelength and transparent for wavelengths sufficiently detuned. Combining different long-period gratings in the same fiber segment enables one to construct various filter shapes with multiple loss peaks. One can also obtain such a customized, shaping effect by cascading different MZI or FP filters, but the long-period fiber gratings are easier to fabricate and have low intrinsic absorption loss, coupling loss and PDL. One of the major applications of these customized transmission filters is the *gain-equalizing* or *gain-flattening filter* (GEF or GFF), as further described in the next section.

To complete this subsection on optical filters, it is worth describing the case of *acousto-optic tunable filters* (AOTF). The underlying principle of AOTF is the coupling between sound (acoustic) waves and optical waves. An acoustic "surface wave" can be generated by applying a RF drive signal at frequency f_a onto a planar waveguide through a comb of interleaved electrodes. This surface wave creates a perturbation of the refractive index δn with periodicity $\Lambda = V_a/f_a$, where V_a is the wave's velocity (e.g., $V_a = 3.75$ km/s in $LiNbO_3$ substrates). As in long-period gratings, the phase-matching (Bragg) condition is simply $\delta n \Lambda = \lambda_B$. But unlike in the previous case, the signal at the resonant wavelength is not back-reflected. Instead, its polarization is forced to rotate. A fraction of an initial TE (or TM) signal excitation is progressively converted into a TM (or TE) signal according to the phase difference $|(\beta_{TE} - \beta_{TM})\Lambda| = 2\pi$, or $|\delta n \Lambda| = \lambda_B$ with $\delta n = n_{eff}^{TE} - n_{eff}^{TM}$. After a certain interaction length L corresponding to an integer multiple of Bragg wavelengths Λ, the initial TE (or TM) power is completely converted into the other mode, corresponding to a polarization rotation by 90°. Another property of minor incidence is that its optical frequency is up-shifted by the amount f_a. If a PBS is placed at the device output, one obtains a TE/TM *polarization space switch*. For this switch to properly operate, the input signal polarization must be deterministic, which is not amenable to fiber-optics applications. One can however resort to polarization-diversity arrangements whereby (by definition) the process of TE/TM conversion of two polarization components follows independent paths. This can be realized in planar technologies based on a MZI arrangement in which the 3-dB input/output couplers are replaced by TE/TM splitters. It can be shown that the MZI acts as an optical filter which exhibits a center lobe at the resonant wavelength (3-dB width equal to $0.8\lambda_B^2/(L\delta n)$), surrounded by rapidly decaying side-lobes. A first interesting feature is that the filter center wavelength is *tunable*, which is seen by combining the two properties $\delta n \Lambda = \lambda$ and $\Lambda = V_a/f_a$ into $\lambda_B = \delta n V_a/f_a$. Thus increasing or decreasing the RF frequency f_a makes it possible to down-shift or up-shift the AOTF center wavelength, respectively (see illustration Exercise at the end of the chapter). It is also possible to generate a multiwavelength tunable filter by superimposing several RF drive signals. As with FBG, side-lobe suppression for better channel selectivity can be achieved by apodization, which consists in progressively reducing the RF power near the two grating ends. With lithium-niobate ($LiNbO_3$) devices, the best WDM channel resolution is 1 nm, as limited by the available crystal sizes and corresponding interaction length (L). The longer the length, the slower the filter's tuning speed. Considering the sound velocity of $V_a = 3.75$ km/s ($LiNbO_3$), the tuning speed is limited by the sound-wave transit time, i.e. $\tau = L/V_a$, which gives $\tau \approx 10$ μs ($1/\tau 100$ kHz) for a $L = 3.5$-cm interaction region. Note that the absorption and fiber-to-fiber-coupling loss in $LiNbO_3$ devices is of the order of a few dB (e.g., 4–6 dB), depending upon the overall device length, including the input/output waveguide leads.

A second interesting feature of the MZI/AOTF arrangement is that it is a four-port (polarization-independent) device which has the ability to flip the output ports (3, 4) of two signals at input ports (1, 2) having the same resonant wavelength. The condition for this effect is that the signals' wavelengths verify the resonant condition. Since the AOTF can be made resonant at several wavelengths simultaneously (by multiplexing RF signals at different frequencies), it is possible to realize a dynamic wavelength router, in which groups of wavelength can be interchanged between ports 3 and 4.

As a concluding remark for this subsection, we should note that optical filters can also virtually behave as active components, producing dynamic changes in pulse-power. This effect happens in the nonlinear pulse transmision regime previously described as the *soliton* regime. Recall that the peak power of a soliton is given by the relation $P_{peak} = const/\Delta\tau^2$, where the soliton time width ($\Delta\tau$) and frequency width ($\Delta\nu$) verifies the time-bandwidth product TBP $= \Delta\tau \times \Delta\nu = 0.315$. Thus, when a soliton loses a fraction of its power due to fiber absorption (or any other type of discrete loss), it broadens to a new width $\Delta\tau' > \Delta\tau$ corresponding to the lower peak power $P'_{peak} = const/(\Delta\tau')^2 < P_{peak}$. Since the TBP is constant, the pulse broadening is associated with a frequency narrowing to the lower value $\Delta\nu' = 0.315/\Delta\tau' < \Delta\nu$. If we pass the initial soliton (P_{peak}, $\Delta\tau$, $\Delta\nu$) through a narrow filter with width $\delta\nu < \Delta\nu$, such a filter will be partially blocking, causing some power loss with associated transmission $T < 1$. Assume now that the same narrow filter is placed in the path of the soliton with lower power (P'_{peak}, $\Delta\tau'$, $\Delta\nu'$). Since its bandwidth $\Delta\nu'$ is narrower, the filter will have a greater transmission $T' > T$. It is exactly as if the filter transmission were power-dependent! But the remarkable property is in fact that this filter has lower loss at lower powers and higher loss at higher power. If we consider a train of soliton pulses with intensity noise (random peak-power fluctuations), such a filter acts as a power-stabilizing element since upward power deviations correspond to low filter transmissions and the reverse for downward power deviations. This remarkable effect is equivalent to the function of an *automatic power control* (APC) loop, which has near-instantaneous response time. Through this specific illustration example, we see that passive optical filters can serve important functions in the nonlinear transmission regime, for instance as a means of reducing amplitude noise. As described in Section 2.5 on active components this effect is exploited in *all-optical regeneration* (AOR).

2.4.3 Compensation and Power Equalization

In the early-generation perspective, transmission-systems were said to be limited by either loss or dispersion, or both, which would set the maximum transmission distance and bit rate ultimately achievable. The development of *dispersion-compensating devices* (DCF, IFBG) and *optical fiber amplifiers* (EDFA, RFA) has put an end to this early system concept. As the bitrates and unrepeated transmission distances have been increased by these two techniques, two other types of limit were discovered or investigated, namely *polarization-mode dispersion* (PMD) and *fiber nonlinearities*. As described earlier in Section 2.3, the effect of line PMD cannot be compensated but only *mitigated*. Concerning device PMD, it can be canceled by the approach of *polarization diversity*, using Faraday rotators (see below). As for nonlinearities (also described in Section 2.3), they can only be *alleviated* or suppressed. This can be done for instance by lowering the signal powers or locally increasing the fiber dispersion, but not compensated nor reversed. Finally, the *amplified spontaneous emission* (ASE) noise accumulated from the chain of in-line amplifiers represents a last type of limitation as being sometimes the dominant source of signal degradation. In classical systems, such as those based upon passive in-line elements, noise cannot be suppressed or compensated. However, the data can be coded with *forward error-correction* (FEC) algorithms which make it possible to increase the signal-to-noise ratio at the receiving terminal. The principle of FEC is described in Desurvire (2004), Chapter 1, Section 1.6. But it is also possible to compensate ASE and any side-effect related to its accumulation by use of in-line *all-optical regenerators* (AOR). Such devices are described in the section on active optical components. In this subsection, we shall briefly describe the use of passive optical devices for compensation of dispersion, mitigation of PMD and power equalization, while the other above topics are covered in the aforementioned sections.

The topic of dispersion (or GVD) compensation has already been introduced when considering DCF (Section 2.3) and chirped-FBG (previous subsection) devices. Concerning the first approach, we have seen that compensating 100 km of SMF requires one to insert 17–21 km of DCF, which causes extra loss especially in view of their higher attenuation coefficients (0.3–0.6 dB/km). Note that when upgrading an already installed SMF system, the DCF does not contribute to the system transmission length, being just spooled in the optical repeater stations. Two other impairments associated with DCF are nonlinearity, due to their smaller effective area ($A_{\text{eff}} = 18-30$ μm^2) and PMD. A way to cancel PMD from the DCF module (and to reduce two-fold the required fiber length) is shown in Figure 2.58. The apparatus uses a two-pass arrangement with a circulator and a Faraday mirror. This mirror rotates the reflected polarization by 90°, so that each polarized component on the return path experiences the same PMD effect as the other on the input path, which cancels the net PMD. In contrast, chirped FBG do not have loss, nonlinearity or PMD. The corresponding implementation is also shown in Figure 2.58. For SMF dispersion compensation, a drawback is that very long FBG lengths are required. Indeed, 100 km of SMF represents a dispersion of +1,700 ps/nm. A straightforward calculation shows that to generate a path difference of $\delta t = -1{,}700$ ps between signals spaced by 1 nm, a grating length of $L = (c/n)\delta t/2 = 17$ cm is required! Thus a 16-channel WDM system with 1-nm spacing would require a grating of length $16 \times 17 = 2.72$ m, which represents a formidable technology challenge (yet demonstrated in the laboratory). The alternative and far more practical approach is to realize the chirped FBG by discrete segments corresponding to the WDM channels, suppressing the sections corresponding to unused bandwidth and replacing them by mere fiber-length delay. Then the device can be as long as is required, assuming that the system is based upon a standard WDM wavelength comb. A major drawback of FBG is that their GVD is not defined by a smooth line. Instead, the GVD exhibits random local deviations called "ripple." Such a ripple, which is typically of the order of

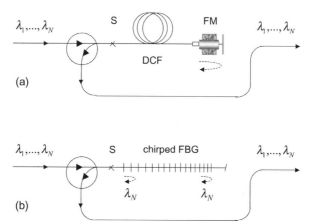

FIGURE 2.58 Two approaches for dispersion compensation through fiber devices: (a) with dispersion-compensating fiber (DCF), using a two-pass arrangement with Faraday mirror (FM) to suppress PMD, and (b) with chirped fiber Bragg grating (FBG), showing difference in delays between short (λ_1) and long (λ_N) wavelengths (S = splice).

one to several picoseconds causes unwanted phase distortion in the compensated signals. It can be only partially suppressed through FBG apodization. Thus DCF and FBG represent two competing approaches for discrete SMF dispersion compensation, each with their inherent advantages and drawbacks. Specialists trust that the ripple problem will eventually be solved, which could give the lead to the FBG approach.

It should be noted that the issue of dispersion compensation is wholly different when considering DSF-based systems. In this case, the dispersion compensation can be implemented by placing short DCF segments along the line, or be managed through the transmission line by alternating fiber segments with different dispersion characteristics. As for SMF, a remaining key issue is to compensate dispersion equally for all channels simultaneously (referred to as *broadband dispersion compensation*). Thus the *dispersion slope* of the compensating means (discrete or alternate line fiber) must be properly managed as well. See more on this issue in Section 2.6 on lightwave system principles.

For future reference on the issue of dispersion compensation, it is useful to mention the case of *photonic-bandgap fibers*, or "*holey*" fibers. Such fibers are made with periodic hollow microstructures in their core. In these microstructures, the periodic air gaps together define a cladding region, given the air/silica index difference. As a result, new types of guided modes can be obtained, and single-mode operation can be brought to the visible-wavelength range. Among the many possible potential applications generated by this new field is dispersion compensation with very short length fibers with high design flexibility. Indeed, the theory shows that negative dispersion as high as -2000 ps/nm/km could be achieved with holey fibers. This high dispersion would provide 100-km SMF dispersion compensation ($-1,700$ ps/nm) with less than one kilometer of holey fiber. By design, it is also possible to achieve very large-area fibers associated with negligible nonlinearity, and ultralow dispersion slope associated with broadband dispersion compensation. A remaining issue to solve is the manufacturing of km-long holey fibers having acceptably low levels of attenuation.

The next issue to consider is that of *PMD mitigation*. To recall, PMD is an effect of pulse distortion or splitting with a randomly varying differential group delay (DGD). Such a DGD also randomly varies between WDM channels with only weak correlation. The mitigation of PMD must be done on a per-channel basis and should be implemented dynamically, due to its spurious and unpredictable time evolution. Note that this evolution has both slow components (scale of days to weeks) and very rapid components (scale of seconds to minutes). The rapid component is due to perturbations such as vibrations from railtracks or highways (along which terrestrial fiber cables are usually deployed) or manipulations of fiber ends which are loosely coiled in the terminals. The two categories of PMD mitigation are electrical or optical. The first involves either linear or nonlinear signal equalization, based upon bit-pattern analysis or the dynamic adaptation of the decision threshold, respectively. The optical approach consists in optimizing the state of polarization (SOP) from the terminals. Based upon a feedback response from the receiver, the transmitter can dynamically adjust the SOP for each WDM channel so that it corresponds to the system's *principal states of polarization* (PSP), see definition in previous section. Such an apparatus is ideal from the standpoint of PMD "compensation," but is heavy to implement and has poor dynamics. The most practical approach consists in minimizing the DGD by passing the received signal through a highly birefringent element. This setup is illustrated in Figure 2.59. First, a polarization controller adjusts the signal SOP. The resulting signal is then launched into the birefringent element, which can be a length of PMF or a series of PMF segments with variable lengths and axis orientation. The idea is to make this element approximately reconstruct the SOP evolution history through the link, while generating the opposite DGD. The PMD monitor provides an error signal which is fed back to a diagnostics/decision circuit. The circuit eventually finds an operating point where the resulting DGD is minimal. The difficulty is to generate the proper PMD-monitoring scheme which ensures rapid convergence and efficient DGD suppression. Another difficulty is the optimization of the control algorithm, which is made complex by the number of degrees of freedom (especially if the birefringent element is made of multiple independent sections). Despite these difficulties, the most simple PMD-mitigation scheme makes it possible to dramatically reduce the bit-error-rate (BER) from 10^{-3} to $10^{-7}/10^{-8}$ or at the same BER to increase the system tolerance to DGD fluctuations by a factor of two (e.g., DGD = 20 ps or 10 ps with or without mitigation, respectively).

Finally, we consider the issue of *power equalization*. This issue is associated with the fact that the gain characteristics of in-line optical amplifiers are far from being perfectly flat. Even in the conditions where the gain spectrum is nearly flat, small oscillations (called *gain ripples*) are still present, in addition to some *curvature* and finite slope (called *gain tilt*). Thus, large or small gain differences (e.g., 3 dB or 0.3 dB) between WDM channels accumulate at each amplification stage, resulting in significant power imbalance (i.e., 30 dB or 6 dB after 20 stages), and possibly channel extinction. More generally, this effect of power-imbalance translates itself into differences of *signal-to-noise ratio* (SNR) between channels, which is associated with a non-uniform BER in the system bandwidth. The first issue is to realize optical amplifiers with near-ideal gain flatness. This gain compensation is achieved by use of *gain-flattening filters* (GFF), also called *gain-equalizing filters*

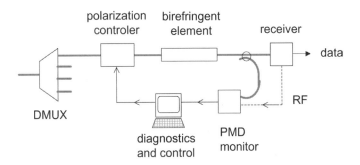

FIGURE 2.59 Apparatus for dynamic PMD mitigation in the terminal (see text for description).

(GEF). But the amplifier gain spectrum is also a function of operating conditions, namely the degree of inversion (or level of pumping power) and signal power load, as described in the next subsection. These conditions slightly evolve along the system because of a variety of complex factors, one of which being the accumulation of gain imperfection. Thus, a second issue is to be able to dynamically compensate the gain. This requires a new type of filter whose transfer function (passband shape) can be controlled and independently optimized for each channel wavelength. Such a function, which is called *equalization*, is performed by either static or dynamic GFF/GEF devices. Note that, properly speaking, equalization is a function applied to a signal, not to a device. Thus, the actions of flattening the amplifier gain through a static GFF, or bringing back all WDM channels to the same power level after some transmission distance, both correspond to this exact definition of signal equalization. In the static case, one needs to know the exact shape of the amplifier gain, or the exact distribution of signal powers at some point along the link. In the first case, the gain shape is known from straightforward device characterization. The effect of gain flattening is then produced by introducing the correct amount of loss in the peak regions having the highest gain G, which lowers the value to some predefined, constant level $\bar{G}$. This can be performed by passing the amplifier output signal through an optical filter having a spectral response of the form $T = \max(0, \bar{G} - G)$, as expressed in decibels. This GFF can be realized by cascading a series of long-period FBG, which exhibit broad and precisely-controllable Gaussian shapes. Figure 2.60 shows a typical amplifier gain spectrum (EDFA) with 5-dB peak-to-peak variations, the ideal GFF reflection $(1 - T)$ which would flatten the gain to $\bar{G} = 10$ dB in the center region, the transmission function of an actual filter made of two long-period FBG (with optimized peak and bandwidth), and the resulting passband function (amplifier + GFF). We observe that the resulting passband is nearly flat, although it exhibits a residual

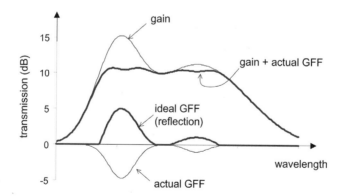

FIGURE 2.60 Typical amplifier gain spectrum (EDFA) with 5-dB peak-to-peak variations (top curve, thin), ideal gain-flattening filter (GFF) reflection which would flatten the gain to $\bar{G} = 10$ dB in the center region (bottom, bold), transmission function of an actual filter made of two long-period FBG (bottom, thin), and resulting passband function combining gain and GFF (top, bold).

ripple of ±0.5-dB amplitude. It is possible to introduce a larger number of FBG to further reduce the ripple within given specifications. In the case where signal equalization must be implemented after a cascade of amplifiers, one can first determine the theoretical power imbalance by numerical simulations. Because of complex higher-order effects, such simulations may not be sufficiently accurate to predict the result within 0.01 dB for instance. Thus the equalization function (GFF shape) must be measured experimentally in a real-life system "testbed" prior to fabricating the corresponding GFF. When the equalization function is *a priori* unknown (or even varies over time or with system utilization), another solution consists in using a *dynamically controllable* GFF/GEF (usually called DGEF). This function can be achieved with tunable MZI or AWG. As previously described, silica-based filters can be tuned by use of *thin-film heaters*, which introduce thermo-optical phase shifts in the MZI branches. Although the AWG is more advantageous that the MZI cascade for realizing complex filter shapes, it is subject to PDL. One possibility is to implement the polarization-diversity scheme shown in Figure 2.61, which uses a PBS and a circulator. Such a scheme uses the fact that the AWG (as a filter) is bi-directional, which enables one to separately filter the TE and TM components via opposite paths. The TE/TM components are launched to and retrieved from the AWG via PMF. At the AWG coupling points, one of the two PMF has its axes rotated by 90°. As seen from the figure, this allows one to convert the clockwise (input) TE signal into an (output) TM signal, and the reverse for the counter-clockwise. This conversion explains why the two counter-propagating signals do recombine in the PBS. With such an arrangement, the polarization of the two signals traversing the AWG is exclusively TE or exclusively TM, which suppresses the device's PDL.

To conclude on the issue of power equalization, it is worth mentioning the technique called *signal pre-emphasis*. The technique simply consists in compensating any power imbalances (or SNR differences) at receiver level by increasing, at transmitter level, the power in the most penalized channels. The idea is to obtain

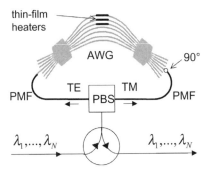

FIGURE 2.61 Polarization-diversity implementation of dynamic gain-equalizing filter (DGEF), using circulator, polarization beam-splitter (PBS), polarization-maintaining fibers (PMF), arrayed-waveguide grating (AWG) filter controled by thin-film heaters. Note the rotation of 90° of the PMF axes at one AWG input.

a near-uniform BER throughout the WDM comb. Increasing the power in those channels is possible only under the limit set by nonlinearities. We thus see that in high-capacity systems operating near performance limits, power equalization is a matter of careful trade-off, with the more or less heuristic determination of an operating point giving best system performance.

2.4.4 Optical Fiber Amplifiers

Optical amplifiers have been known since the invention of the *laser*, back in the '60s. As described in detail in Section 2.2, a laser is basically made from an optically amplifying medium placed inside a FP cavity. After a few round-trip circulations in the cavity, the signal generated by the amplifier's spontaneous noise rapidly builds up into a coherent standing EM-wave, taking the form of a straight and intense light beam. In contrast, one can realize an optical amplifier by removing any possible reflection sources from the amplifying medium, which can be done for instance by placing optical isolators (with high return loss) on each side. As early as the '60s, investigators thought of the potential of optical amplifiers as signal boosters for future telecom applications. The material options were two-fold: *crystalline-based* (semiconductors and garnets) and *amorphous glass* (oxide-based, like silica, and others such as fluoride-based, to be investigated later). We shall leave for the time being the case of semiconductor materials, in order to focus on the glass case, which immediately leads us to the *fiber amplifier* concept.

As bulk devices, *glass lasers* use special dopants called the *rare-earths* (RE). This name comes from the fact that RE appear in certain mining compounds only in the form of very small traces. In the periodic table, the RE form a family of 13 called the *lanthanides*. The favorite lanthanides which exhibit laser transitions of interest at near-IR wavelengths bear the exotic names of: *praseodymium* (Pr), *neodymium* (Nd), *dysprosium* (Dy), *erbium* (Er), *thulium* (Tm) and *ytterbium* (Yb). Since glass fibers offer both long interaction lengths with low background loss and high signal power confinement, it was immediately recognized that they could make efficient signal amplifiers. Because Nd:glass, which has a principal laser transition at 1.06-μm wavelength, turns out to be the most efficient laser material of the RE family, the first investigation of *fiber amplifiers* concerned Nd-doped devices. A major difficulty was to pump the doped fiber core, which at the time could be realized only by coiling the fiber around a multi-kW flashlamp, under pulsed operation. The field of optical telecommunications progressively developed, opening transmission windows at 0.8 μm, then 1.3 μm, and finally in the low-loss region, 1.55 μm. Such a development came with the realization of semiconductor sources at these wavelengths and also of *electronic repeaters*. For about two decades, there was no perceived need for optical amplifiers, since the single-wavelength signals used in fiber trunks at the time could be electronically "repeated" along extended system lengths. However, there was a growing interest in *semiconductor optical amplifiers* (SOA) as an alternative option for 1.3–1.55-μm systems.

By the mid-1980s, the field of *RE-doped fibers* was reactivated, focusing on the RE *erbium* with its broad 1.55-μm laser transition, of potential interest for a future generation of broadband systems. Investigators relentlessly studied and optimized a workable device, *the erbium-doped fiber amplifier* (EDFA). At the time, the EDFA could only be pumped by cumbersome, bulky, high-power, water-cooled gas lasers (argon, krypton), liquid lasers (dye) or garnet lasers (sapphire), which made the approach wholly impractical. By the end of that same decade, however, the identification of the best pumping wavelengths triggered the development of specific semiconductor diodes at 1.48 μm, then at 980 nm, turning the EDFA into a sweeping "killer application" in the field of photonics, as further described below.

The basic EDFA layout is shown in Figure 2.62. It comprises a pump module, a wavelength-selective coupler (called "MUX"), a strand of Er-doped fiber (10–50-m length, typically) and a Faraday isolator at the output (an isolator may also be required at the input if there is a reflection source). A *tap coupler* is usually inserted in the signal path in order to monitor the signal output power and send a feedback to the pump module for *automatic power control* (APC). If the signal is too low or too high, the pump power is then increased or decreased, respectively. The pump module includes a semiconductor laser-diode *pump*, whose back and front facets have been high-reflection and antireflection coated, respectively, to yield maximum output power. In some high-power applications, the diode's temperature must be regulated by means of a *thermo-electric (Peltier) cooler*. The pump output is coupled into the input fiber via a microlens and passed through a Faraday isolator. This isolator is very important to prevent back reflections into the pump, which could cause power instabilities and catastrophic damage. For high-power applications, it is possible to combine the output of two pump modules, using a PBS. In this case, the pump fibers must be of the PMF type, so that the two outputs are combined in the PBS as TE and TM components. As previously mentioned, the pump wavelength can be either 980 nm or 1.48 μm, which is clarified further below.

As illustrated in Figure 2.63, EDFA pumping can be implemented either in the same direction as the signal (*forward pumping*), opposite to the signal (*backward pumping*) or in both directions simultaneously (*bidirectional pumping*). The bidirectional configuration makes it possible to split the pumping power but does not

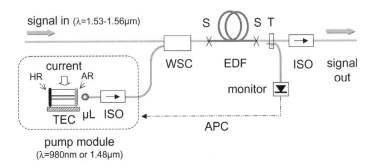

FIGURE 2.62 Basic layout of erbium-doped fiber amplifier (LD = laser diode, HR/AR = high/anti-reflection coatings, μL = fiber microlens, ISO = Faraday isolator, TEC = thermoelectric cooler, WSC = wavelength-selective coupler, S = splice, EDF = erbium-doped fiber, T = tap, APC = automatic power control).

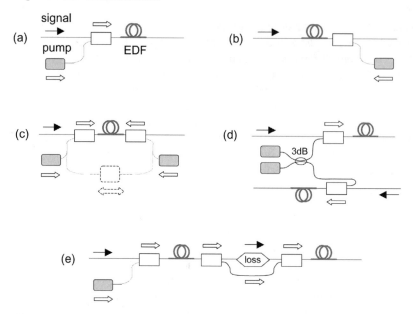

FIGURE 2.63 Different possible configurations of EDFAs (a) forward pumping, (b) backward pumping, (c) bidirectional pumping (single-pump module shown in dashed line), (d) pump-redundancy arrangement shared by trunk fiber-pair, and (e) two-stage implementation with intermediate dispersive element and pump bypass.

necessarily require two pump modules, as seen from the figure (the module output being split by a simple 3-dB fiber coupler). It may also include a 980-nm pump for the forward direction and a 1480-nm pump for the backward direction, referred to as *hybrid pumping*. This approach makes it possible to exploit the best compromise between pumping efficiency and noise figure (see following). The figure also shows a *pump-redundancy* arrangement in which two pump modules are shared by two EDFA in a fiber-pair trunk system, using a 3-dB fiber coupler as a power combiner. In case of pump aging or failure, the power of the other pump can be increased in order to maintain the system in full operation. Finally, the figure shows a *multistage EDFA*, inside which a dispersive element can be inserted on the signal path, while the pump is bypassed. This dispersive element can be a strand of DCF for dispersion compensation, an OADM for extraction/insertion of specific wavelength channels, or a GFF/GEF for power equalization. Thus the loss introduced by any of these elements is intrinsically compensated inside the EDFA "gain block," which provides better performance than if placed outside.

The two wavelengths of 980 nm and 1.48 μm correspond to two out of several absorption bands or energy levels in the Er:glass laser system. To understand the difference, we look at Figure 2.10, which shows *three-level* and *four-level* laser systems. The 980-nm pumping scheme corresponds to the 3-level system shown at the very left: the pump photons cause the erbium atom to be excited from level 1 to level 3, which is followed by a nonradiative (phonon) relaxation to level 2. The signal photons are emitted from level 2, causing the atom to relax to level 1. The 1.48-μm pumping scheme is somewhat different and unusual for laser systems. Assume ideed that levels 1 and 2 are made of a set of closely spaced energy sublevels, say two sublevels each, for simplicity. Looking at the laser system shown in the *far right* of Figure 2.10, we see that the ground level is not "0" but the set "0 + 1," and the excited level is not "2" but the set "2 + 3." Thus it is possible to excite the atom from level 0 to 3 and emit the signal by relaxation from level 2 to 1, consistent with the operation of a 4-level system. But since the energy sublevels 0–1 or 2–3 belong to the same energy-level sets, this laser is actually a *2-level* system. The key difference is that 2-level systems (as split into such multiple sublevels) cannot be fully inverted, unlike 3-level systems. This difference impacts on the EDFA *noise figure* (NF), which is the lowest for 980-nm pumping, as discussed further in following text. Another aspect concerns the *power conversion efficiency* (PCE). The principle of energy conservation states that in a loss-less EDFA, one signal photon (of energy $h\nu_s = hc/\lambda_s$) must be emitted for each pump photon (of energy $h\nu_p = hc/\lambda_p$) absorbed by erbium atoms. The energy difference, $\Delta E = h(\nu_p - \nu_s)$ corresponds to the nonradiative generation of phonons, which is likenable to heat. To a light wave of power P and frequency ν, one can associate a *photon flux* which is the *number of photons per second* carried by the wave. This photon flux is simply given by the ratio $\phi = P/(h\nu)$. If $\phi_s^{in} = P_s^{in}/(h\nu_s)$ and $\phi_p^{in} = P_p^{in}/(h\nu_p)$ represent the input flux for signal and pump, it is clear the total signal-photon flux at the EDFA output is given by the relation

$$\phi_s^{out} \leq \phi_s^{in} + \phi_p^{in} \qquad (2.134)$$

where the equality stands for a perfectly loss-less medium with all pump photons being absorbed by the erbium atoms. In reality, the inequality always stands because of three reasons: (a) doped fibers always have a small amount of background loss because of glass impurities; (b) due to spontaneous emission, a small fraction of the power is radiated sideways into the fiber cladding or in the backward direction; and (c) a fraction of the pump photons escape the fiber without being absorbed by the erbium atoms. Replacing the flux definitions in the above yields:

$$P_s^{out} \leq P_s^{in} + \frac{\lambda_p}{\lambda_s} P_p^{in} \qquad (2.135)$$

which can equivalently be written

$$\text{PCE} = \frac{P_s^{out} - P_s^{in}}{P_p^{in}} \leq \frac{\lambda_p}{\lambda_s} \qquad (2.136)$$

We thus observe that the PCE is bounded by the pump/signal wavelength ratio $\eta = \lambda_p/\lambda_s < 1$. The amplified signal being near $\lambda_s = 1.55$ μm, we find that for 980-nm (0.980-μm) pumping the maximum PCE is 63.2%, and for 1.48-μm pumping it is 95.5%. Thus the two pumping schemes yield quite different performance in terms of the maximum signal power that can be produced at the EDFA output, giving the advantage to 1.48-μm pumping. On the other hand, we have seen that the lowest NF is achieved with 980-nm pumping. Therefore, the important conclusion is that *one can build EDFAs having low NF but limited PCE* (980-nm pumping), or alternatively, *with higher NF but maximum PCE* (1.48-μm pumping).

As detailed in Desurvire (2004), Chapter 1, Section 1.5, the EDFA noise figure is a measure of the signal-to-noise ratio (SNR) degradation caused by the amplifier. In Section 2.2 of the present chapter, we have also established that the amplifier noise is characterized by a parameter called the *spontaneous emission factor* (n_{sp}), which reflects the relative degree of inversion as $n_{sp} = N_2/(N_2 - N_1)$. The link between the NF and the spontaneous emission factor is simply:

$$\text{NF}_{\text{lin}} = \frac{1}{T}\frac{1 + 2n_{sp}(G-1)}{G} \qquad (2.137)$$

where G is the amplifier gain and T ($T < 1$) is the transmission of any lossy element placed in front of the amplifier. The NF is usually expressed in decibels as $\text{NF}_{\text{dB}} = 10\log_{10}\text{NF}_{\text{lin}}$. It is seen that for large gain values ($G \gg 1$), and without input loss ($T = 1$), the NF converges towards the limit of NF $\approx 2n_{sp}$. At maximum inversion, we have $n_{sp} \to 1$ and NF $\to 2$, representing a limit of 3 dB. Such a lower bound for the NF is referred to as the *quantum limit*. If the EDFA inversion is incomplete, we have $n_{sp} > 1$ (since $N_1 \neq 0$) and therefore NF > 2 or NF > 3 dB. This is the case of EDFA pumped at 1.48 μm, for which the lowest spontaneous emission factors are $n_{sp} \approx 1.5$–1.6, typically. This corresponds to noise figures of NF $= 2n_{sp} \approx$ 5.0 dB. In contrast, a 980-nm-pumped EDFA can reach the full-inversion regime whereby NF ≈ 3.0 dB. In either case, however, there is finite amount of input loss caused by optical elements placed on the signal input path, such as the pump MUX (see Figure 2.63), a Faraday isolator, a signal input tap or simply the fiber input splice/connector linking the EDFA to the system. Such losses being typically in the 0.2–1.0-dB range, we find the limits of NF $= 3.2$–4.0 dB for 980-nm pumping and NF $= 5.2$–6 dB for 1.48-μm pumping, respectively.

We focus now on the EDFA gain spectrum. As we have described in detail in Section 2.2, the gain for an amplifier of length L can be expressed as $G = \exp(gL)$, where g is by definition the *gain coefficient*. This gain coefficient is a function of the medium inversion $N_2 - N_1$, i.e., $g = \sigma(N_2 - N_1)$, where σ is the laser-transition *cross-section*, which has the dimension of a surface (cm^2) and is also wavelength-dependent. To recall, N_1, N_2 are atomic (erbium) densities, in units of inverse volume (m^{-3}). The total Er-density being ρ, we have by conservation $N_1 + N_2 = \rho$. By definition, the *absorption coefficient* (α) due to the erbium ions is equal to $\alpha = \rho\sigma_a$, which we will use hereafter.

In basic laser systems with no energy sublevels, the cross-section (σ) has a bell shape with a typical center wavelength λ_s. In Er:glass, the two energy levels involved in the laser transition are split into a finer structure of sublevels called the *Stark levels*. This splitting is due to the static electrical field produced by the surrounding glass molecules. Since the glass structure is random, the electrical field is also random, which makes the Stark levels different from one erbium atom to another. Among other benefits, a key result of this effect is that the emission and absorption cross-sections have wholly different spectral profiles ($\sigma_a(\lambda)$, $\sigma_e(\lambda)$) and center wavelengths. The gain coefficient is thus defined as $g = \sigma_e N_2 - \sigma_a N_1$. We see that the condition $N_2 > N_1$ (inversion) is no longer sufficient to obtain positive gains ($g > 0$) and signal amplification ($G > 1$). The new condition for amplification is indeed: $\sigma_e N_2 > \sigma_a N_1$, or $N_2 > (\sigma_a/\sigma_e)N_1$, which is not necessarily verified for all wavelengths. In order to evaluate the EDFA *gain bandwidth* (i.e., the range over which signals can be amplified), we must plot the gain coefficient as a function of various degrees of inversion. Figure 2.64 shows the corresponding plots obtained for typical EDFAs, with pump wavelength assumed here to be at $\lambda_p = 1.45\,\mu\text{m}$. We immediately observe that the EDFA is a lossy device ($g = -\rho\sigma_a = -\alpha < 0$, $G < 1$) when the pump is turned off, or $N_1 = \rho$ and $N_2 = 0$. As the pump power is turned on, upper-level population N_2 progressively increases, and we observe that the gain coefficient becomes positive for a wider portion of the spectrum, beginning from the longer-wavelength region. At full inversion ($N_1 = 0$, $N_2 = \rho$), the coefficient is positive ($g = +\rho\sigma_e$) over the entire region where $\lambda > \lambda_p$. If we chose the gain coefficient to be at least equal to the value g_{min} (see figure), then the EDFA gain bandwidth is fully determined and is see to be maximum at full inversion. We note that the gain coefficient is proportional to the decibel value, since $G_{dB} = 10\log_{10}[\exp(gL)] \equiv gL \times 10\log_{10}(e) = 4.34 \times gL$. Thus the curves in Figure 2.64 truly provide the EDFA gain bandwidth when expressed in decibel scale. The peak-to-peak gain variations observed within the EDFA bandwidth defined by $G_{dB}^{min} = 10\log_{10}[\exp(g_{min}L)] = 4.34 \times g_{min}L$ can be suppressed by means of gain-flattening/equalizing filters (GFF/GEF), as previously illustrated in Figure 2.60.

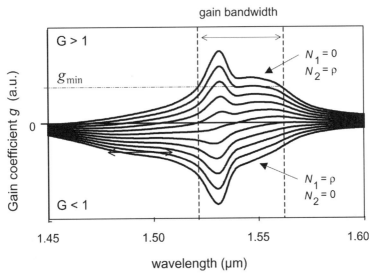

FIGURE 2.64 Spectrum of EDFA gain coefficient obtained for different values of medium inversion, from $(N_1, N_2) = (1, 0)$, corresponding to the unpumped conditions, to $(N_1, N_2) = (0, 1)$, corresponding to fully-inverted conditions, by increments of 0.1. The upper and lower half planes determine the regions where the gain coefficient is positive (amplification) and negative (absorption), respectively. The dot-dashed line, for which the coefficient is greater or equal to a desired value, g_{min}, determines the operating EDFA bandwidth.

Since the pump power is progressively absorbed along the doped-fiber length, the populations N_1, N_2 vary with fiber coordinate. The previous description (and the curves in Figure 2.64 defining the gain coefficient) remains valid if the parameters N_1, N_2 correspond to "path-averaged" quantities as opposed to the local values. Thus, given the amount of pump power launched into the fiber, there exists an optimum fiber length L for which we can obtain a gain of G at a given reference signal wavelength λ_0. But if the input signal power is increased, the population distributions change (N_1 increases and N_2 decreases), due to the competition of stimulated emission rate and pumping rate (see subsection concerning optical amplification in Section 2.2). As the result, the signal gain decreases or *saturates*, corresponding to the *gain saturation regime*. At low signal powers, the gain remains independent of the signal power, which is referred to as the *small-signal gain/amplification regime*. Therefore, there exists a transition from small-signal to saturated regimes which needs to be clarified and formalized. So far, most specialists have relied upon various numerical models (and commercial software) to characterize the EDFA gain spectrum under different pump and signal launching conditions, because no simple formulae were available for the saturated regime. But this picture has changed with the recent development of a practical and exact model (see Desurvire et al., 2002). It is worth providing this model's results, as they are so simple and practical to handle, as well as an illustrative example for WDM-system calculation.

We need first to define several EDFA parameters:

- η = ratio of emission to absorption cross-sections, or $\eta = \sigma_e/\sigma_a$, which is wavelength-dependent; the corresponding parameters for the pump (λ_p) and for the signal (λ_s) are called η_p and η_s, respectively; by convention, the case of 3-level pumping ($\lambda_p = 980$ nm) corresponds to $\eta_p = 0$, while $\eta_p \neq 0$ in the 1.48-μm pumping case, which is important to note;

- μ = a parameter linking the pump and signal absorption coefficients (due to the Er atoms) and the corresponding η-parameters according to $\mu = (\alpha_p/\alpha_s)(1 + \eta_p)/(1 + \eta_s)$; in the foregoing, we use the index k to specify at which wavelength λ_k such a parameter is measured, i.e., $\mu_k = (\alpha_p/\alpha_k)(1 + \eta_p)/(1 + \eta_k)$;

- G_∞ = gain at "infinite" pump power or at maximum inversion, defined as

$$G_\infty = \exp\left(\alpha_s L \frac{\eta_s - \eta_p}{1 + \eta_p}\right)$$

Such a definition makes sense when we set $\eta_p = 0$, corresponding to pumping at $\lambda_p = 980$ nm. Using $\eta_s \alpha_s = (\sigma_{e,s}/\sigma_{a,s}) \times \rho\sigma_{a,s} \equiv \rho\sigma_{e,s}$, we obtain $G_\infty \equiv \exp(\rho\sigma_{e,s})$, which is the maximum gain at full inversion with $N_2 = \rho$; the reduction factor $(\eta_s - \eta_p)/(1 + \eta_p)$ reflects the fact that inversion is not maximum in the other case where $\eta_p \neq 0$ (1.48-μm pumping);

- G_0 = the target EDFA gain at a reference wavelength λ_0 (for instance, 1.55 μm), which we want to achieve, regardless of the degree of saturation; given an EDFA of length L, and knowing the WDM power load at different channel wavelengths, there is a formula which exactly predicts the amount

2.4 Passive Optical Components

of pump power P_p^{in} required to obtain the gain G_0; but it is beyond the scope of this book to go through such details; it is simply nice to know that such a formula exists!

- G_k = the EDFA gain at any signal wavelength λ_k.

After having gone through the previous painstaking definitions, comes the reward! This is a simple formula which provides the gain G_k (at any wavelength λ_k), given the target gain G_0 (at reference wavelength λ_0). Such a formula is

$$G_k = \frac{P_k^{out}}{P_k^{in}} = \{G_0^{\mu_0} \exp[(\mu_0 \alpha_0 - \mu_k \alpha_k)L]\}^{1/\mu_k} \quad (2.138)$$

As an important and remarkable property, this formula applies regardless of the mutual propagation direction of the pump(s) and the signal(s), so it is valid for forward pumping, backward pumping, bidirectional pumping and bidirectional amplifiers as well. To be applied to a practical case, the formula only requires the knowledge of the absorption spectrum, $\alpha_k = \alpha(\lambda_k)$, and the spectrum of the parameter $\mu_k = \mu(\lambda_k)$, which can both be characterized experimentally through loss and cross-section measurements. An illustrative example is provided in Figure 2.65. It concerns the amplification of 10 WDM channels spaced by 0.2 nm, with an input power per channel of 1 μW or −30 dBm and a pump

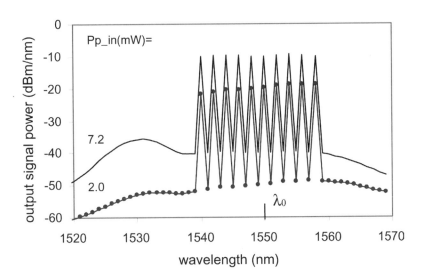

FIGURE 2.65 Output power spectrum in 10-channel WDM amplification with 1480-nm pumped EDFA, as calculated from formula (2.138). The reference wavelength is $\lambda_0 = 1550$ nm, and corresponding target gains are $G_0 = +10$ dB and +20 dB. Each channel is input with 1 μW or −30 dBm. Other EDFA parameters are: $\alpha_p = 0.9$ dBm, $\alpha_0 = 1$ dBm, $L = 37.6$ m. The input pump powers corresponding to the two target gains, $P_p^{in} = 2.0$ and 7.2 mW, were calculated using another exact formula in the model. (After E. Desurvire et al., *Erbium-Doped Fiber Amplifiers: Device and System Developments*, Wiley, New York, 2003).

wavelength of 1.48 μm. The target gains are $G_0 = +10$ dB and $+20$ dB at the reference wavelength $\lambda_0 = 1.55$ μm, to which are associated different inversion conditions (low and high, respectively). Using the parameters (α_k, μ_k) of a typical EDFA, a straightforward application of the above formula [equation (2.138)] yields the spectra shown. The corresponding (or required) pump powers, namely 2.0 and 7.2 mW, were also calculated using another exact formula, which takes into account the WDM power load and channel locations. We see that in the high-pump case (7.2 mW), the output powers are very nearly uniform. This is because in the spectral region chosen here, and in the corresponding high-inversion conditions, the gain spectrum is nearly flat with only a small ripple effect (<0.5 dB). In the low-power case, however, the short-wavelength signals experience less gain than the long-wavelength ones, yielding a 5-dB difference across the WDM comb. The continuous gain curves appearing in the two plots (as a baseline for the WDM comb) correspond to fictitious channels with -60-dBm input power. These have been introduced to show the gain-spectrum shape across the entire 1.52–1.57-μm band. We can conclude from this example that the EDFA gain spectrum is highly dependent upon the inversion conditions, regardless of considerations of gain flatness. Increasing the WDM signal power also results in gain saturation or a reduction of inversion which produces the same power imbalance. It is therefore important to know that in EDFA, gain flatness is not a fixed passband characteristic; rather, it is a function of the WDM power load, and it can be optimized by use of GFF/GEF filtering techniques. Concerning the determination of noise characteristics (i.e., the spontaneous emission factor as a function of wavelength, $n_{sp}(\lambda)$), it can be performed as well through either numerical or analytical/formula models.

How valid is the above analysis in terms of *signal polarization* and *time-varying signal power*? The good news is that *the EDFA gain is completely polarization-independent and insensitive to rapid (≥ 100-kHz) signal-power changes.*

The first property of *EDFA polarization-independence* is explained by two factors: the dipoles associated with erbium atoms are randomly oriented in the glass matrix, so the macroscopic cross-section is 100% insensitive to the signal SOP; the second factor is that Er-doped fibers have a cylindrical symmetry, so the pump/signal mode sizes (hence the intensities) are the same for the TE or TM modes, regardless of the effect of polarization-scrambling. We shall note however that strong signal saturation can cause *polarization-dependent gain* (PDG). This PDG is experienced by weak signals whose SOP is orthogonal to the saturating signal. The effect is associated with a relatively small gain increase (e.g., 0.05–0.1 dB). In WDM systems, however, PDG is not an issue since all channels have nearly equal powers and exert onto each other the same amount of saturation.

The second property of *slow EDFA gain dynamics* is due to the fact that the upper-level (fluorescence) lifetime of the erbium atoms is relatively long, approximately $\tau = 10$ ms. This long lifetime does not correspond to an absolute time-constant characteristic for power transients. Recalling from section 6.2 that the atomic populations are driven by the absorption and stimulated-emission rates, W_{abs}, W_{stim} (in units of s^{-1}), it is possible to show that the dimensionless quantity $W\tau$ corresponds to a power ratio P/P_{sat}, where P_{sat} is a saturation power (and P the signal or pump power). The local time-constant characteristics (τ') of the

EDFA is then power-dependent, and is defined according to $\tau' = \tau/(P/P_{sat})$. Thus, for an average power level (pump or signal) such that $P = 10P_{sat}$, or $P = 100P_{sat}$, the EDFA time constant is $\tau' = \tau/10 \equiv 1$ ms or $\tau' = \tau/100 \equiv 100$ µs. The key conclusion is that the EDFA inversion cannot follow or track any signal (or pump) power changes occurring at frequencies substantially greater than $1/100$ µs = 10 kHz, namely 100 kHz. Since WDM signals have bit rates of at least 100 Mbit/s (and more usually 2.5–40 Gbit/s), the bit-to-bit power changes are four to six orders of magnitude faster than the most dynamic EDFA (10 kHz). As a fundamental result, the EDFA is insensitive to *interchannel crosstalk*, and just responds to average powers, which is essential for WDM applications. Note that average-power changes can affect the channel BER performance, but this is not due to crosstalk or bit-pattern interference. Rather, it is due to the change in inversion conditions, which affects both noise (spontaneous emission factor and noise figure) and output signal power, hence the channel SNR. The slow EDFA gain dynamics are in sharp contrast to the case of semiconductor optical amplifiers (SOA). Indeed, the time constants associated with the SOA gain dynamics are of the order of 200 ps–1 ns, depending upon the drive current, which is *six orders of magnitude* faster! As a result, SOA are not adapted to WDM amplification unless constant-envelope (e.g., FSK; see Vol. 1, Chapter 1) formats are used. Another problem is their residual polarization sensitivity (e.g., 0.1–0.2 dB), which makes concatenation impractical, and higher noise figures (e.g., NF = 6–7 dB). Note that part of this NF penalty comes from the input coupling loss (e.g., 0.5–1 dB) between the fiber and the semiconductor chip. For these different reasons, SOAs are not suitable for in-line amplification applications as "WDM repeaters." But they offer a key advantage which EDFAs do not have: their fast gain dynamics (1–10 GHz) makes it possible to realize bit-to-bit optical signal processing with a variety of new device applications. Thus SOA can be seen as another family of optical amplifiers that work in the ultrafast regime, in contrast to EDFA. Another key advantage of SOA is the possibility of integrating them on single planar circuits (e.g., MZI, AWG, Y-branch splitters, etc.). Consistently with our earlier definition of *passive* and *active devices* (see introduction of this section), we shall therefore place SOA in Section 2.5 concerning "active devices."

Combined with the amplifier fundamentals described in an earlier section in this chapter, the above represents a fast and to-the-point introduction to the intriguing field of EDFAs. We believe that such knowledge is sufficient to master the associated concepts, even if much more interesting features and complex effects could be described. As commercial products, EDFAs are now fully defined according to the three following characteristics (examples only indicative):

1. *Total output power* (e.g., +10 dBm to +30 dBm);
2. *A 3-dB bandwidth* (e.g., 20–30 nm) with associated *gain ripple* (e.g., ±0.1–0.5 dB);
3. *Noise figure* (e.g., NF = 3.5–6 dB), according to 980-nm/1480-nm pumping or hybrids;

and secondarily:

1. *Electrical power consumption* (e.g., 0.5–10 W);
2. *Operating temperature range* (e.g., −20 to +60°C);
3. *Module size* (e.g., 1 × 10 × 20 cm).

Commercial EDFA modules also offer a variety of essential features such as APC and "mid-point" access for introducing GEF/GFF or OADM devices. Additionally, the controlling electronic circuit can be addressed via the Internet (i.e., via SNMP in the TCP/IP protocol suite) for remote monitoring and dynamic optimization under changing system conditions. For future reference, it is also worth mentioning that rapid progress is being made in the field of high-power pump sources. Two leading approaches are *cladding-pumped EDFAs*, and *high-power pump fiber lasers*. In the first case, the EDFA is pumped through the fiber cladding, via a highly-multimode source which can deliver several Watts of output power. The fiber core can be codoped with the element *ytterbium* (another RE), which by dipole-dipole atomic interaction transfers its excitation to the erbium atoms. In the second approach, the goal is to realize high-power fiber lasers to be used for 1.48-μm EDFA pumping. For maximum efficiency, these fiber lasers can also be cladding-pumped. One of the possible approaches is the *multi-Stokes fiber Raman laser*, based upon the principle of multiple SRS signal generation from short wavelengths to long wavelengths (see Section 2.3 on fiber nonlinearities). All these different approaches for EDFA pumping provide alternative solutions for exploiting compact and efficient high-power diodes and yielding the highest EDFA output power. We should however note that progress in 980-nm and 1.48-μm semiconductor diodes has been so far adequate to meet the needs of WDM systems, with numbers of channels as high as 160–320, representing total output powers which can reach up to +25/+30 dBm.

The natural EDFA gain bandwidth, centered near 1.55 μm and having a natural width of 20–30 nm (typically, depending upon GEF and maximum gain specifications), has been called "C" for *conventional band*. Using longer amplifier lengths and higher pump powers, the operating band can be shifted to longer wavelengths, referred to as "L" band, for *long-wavelength band*, with a typical 25–40-nm width (typically, depending upon GEF and maximum gain specifications), and centered near 1.585 μm. Such a L-band is not a new band but rather represents a different way to design and operate the EDFA. Based upon the curves shown in Figure 2.64, we observe that in this long-wavelength region, it is possible to obtain a flatter gain spectrum, which is basically the concept behind the L-band operation. Because the L-band is located in the trailing edge of the EDFA gain spectrum, the pump power requirement is substantially higher, which can be met by bi-directional pumping with polarization-multiplexed pump modules, for instance. Thus EDFAs can be designed and manufactured as either C-band or L-band devices. In a third type of EDFA implementation, the WDM signals are initially split into the C-band and L-band groups via a wavelength-selective (high-pass/low-pass) coupler. The signals are then coupled into L-band and C-band EDFAs, to be eventually recombined via the same type of coupler. Such a complex configuration is referred to as a "C + L" EDFA. The total bandwidth of a C + L EDFA is therefore near 20/30 nm(C) + 25/40 nm(L) = 45–70 nm(C + L), corresponding to $45/70 \times 125$ GHz = 5.6–8.7 THz or over 500–700 channels at 10 Gbit/s (assuming an ideal 0.8 bit s^{-1}Hz^{-1} spectral efficiency). Several demonstration systems have made use of the concept, leading to ground-breaking transmission records for both terrestrial and transoceanic system distances. However, the doubling of EDFA bandwidth through the C + L approach is not without a toll on the other system aspects, compared to the conventional "C" system:

> The WDM signal power is doubled, which impacts on the nonlinearity limitations;
>
> The fiber dispersion must be managed over a bandwidth range twice as large;
>
> The implementation requires a significant amount of pump modules and driving power, which impacts upon both system cost (all systems) and terminal-equipment capability (submarine systems).

These constraints make some experts believe that, in certain long-haul applications, it could be more effective to *implement two C-band systems in parallel* than one single C + L system. In the first case, the advantage would be to use a transmission fiber (and related dispersion-management techniques) that is optimized and mass-produced for the C-band, in addition to using mass-produced C-band repeaters. The drawback is that the "C-only" system requires twice as much fiber length as the "C + L" version. The choice between solutions is dictated by issues of (BER) performance, ($) cost and (long-term) reliability. Such a three-parameter equation may be different for terrestrial and for undersea systems, which do not have the same criteria. Suffice it to state that both approaches are possible and can be either advantageous or disadvantageous concerning the specific systems under consideration.

We shall continue the description of optical fiber amplifiers with the most interesting case of *Raman fiber amplifiers* (RFA).

The popularity of EDFA also triggered a renewed interest for fiber amplifiers based upon the effect of *stimulated Raman scattering (SRS)*, known as *Raman fiber amplifiers* (RFA). As previously described in Section 2.3, SRS is one of the fiber *nonlinearities*. When two signals having frequencies spaced by an amount close to $\Delta \nu_R = 10$–15 THz (or $\Delta \lambda_R = 80$–120 nm at 1.55 μm), the SRS effect causes the power to be progressively transferred from the short-wavelength signal to the long-wavelength signal. Such an effect can be seen as an intrinsic limitation of fiber bandwidth, when a system is operated close to the SRS power threshold. But it is also possible to exploit SRS to realize an RFA device. A nice feature is that SRS occurs on *any* type of silica fiber, provided the effective area (A_{eff}) is sufficiently small (e.g., 20–50 μm^2) and the interaction length sufficiently long (namely $L > L_{\text{eff}} \approx 1/\alpha = 22$ km). Raman gain can therefore be generated directly into the transmission fiber (SMF or DSF), which is achieved by launching a strong pump signal at a wavelength $\lambda_p = \lambda_s - \Delta \lambda_R$, where λ_s is the wavelength of the signal to be amplified.

The two key interests of the RFA are the following:

1. It is a *tunable* amplifier, since it can be implemented at any signal wavelength, provided the adequate pump source at the Raman-shifted wavelength can be realized;
2. It can be used to cancel (or reduce) the signal propagation loss between two repeater stations, whether electronic or optical, as a *distributed amplifier*. This allows one to reduce the *signal excursion* between the stations and hence, to reduce the *nonlinearity limitations* (see Figure 2.32 and description); note that distributed amplification is not practical to implement with EDFA, since low gain coefficients are associated with very low inversion levels, resulting in prohibitively high noise figures.

Other important aspects of the RFA include:

- It is a *broadband* amplifier with 20–30-nm bandwidth, depending upon gain-flatness specification;
- It is a virtually *polarization-independent* amplifier, since relatively long (1 km–50 km) fiber lengths are required, which scramble all polarization-dependent effects;
- It is a *low-noise* amplifier, with NF = 3.4–3.9 dB (n_{sp} = 1.1–1.23; see Section 2.3), which includes the fact that there is negligible coupling loss between the RFA and the transmission fiber;
- It is an *ultra-fast* amplifier, with gain dynamic constant near the *femtosecond* (fs) regime; note that in the absence of saturation, these ultrafast dynamics do not cause any WDM crosstalk; rather, the effect of crosstalk is associated with *self-induced SRS* (SI-SRS) described in Section 2.3, but not with RFA; any pump-power intensity noise (regardless of frequency) is however reflected in the signal gain; this effect can be alleviated (or more accurately, time-averaged to zero) by use of backward pumping and high local dispersion, which make the pump and the signal slide through each other.

The net gain coefficient (g) due to SRS is both a function of the fiber's intrinsic *Raman gain coefficient* (g_R) and the signal attenuation coefficient (α_s). Although the analysis is already made in Section 2.3, we shall reproduce here, for clarity purposes, the Raman gain definition:

$$G = \exp\left(g_R \frac{P_p}{A_{\text{eff}}} L_{\text{eff}} - \alpha_s L \right) \qquad (2.139)$$

where P_p is the launched pump power. Note that such a definition corresponds to the small-signal, or unsaturated regime, whereby the pump power absorption is essentially due to fiber absorption and not to the power transfer to the signal. We leave it as an exercise at the end of the chapter to determine the pump-power requirements to cancel the loss of a single 50–100-km strand of DSF or SMF. The result shows that for RFA, the pump power requirement is somewhat higher than with EDFA (100–300 mW vs. 10–50 mW), but not substantially higher if the application only concerns *in-line loss compensation*. Following that of the EDFA, the industrial exploitation of RFA has only been a matter of pushing further the development of high-power 1.48-μm (InGaAsP) pump diodes, while shifting their center wavelength (i.e., the semiconductor bandgap) to the 1.43–1.47-μm region.

Since RFA makes it possible to reduce the signal power excursion in the fiber trunk, they can be implemented in conjunction with EDFA for reducing somewhat the effect of nonlinearity, while the more efficient EDFA ensure the overall power equalization and loss compensation through the WDM wavelength comb. The basic layout of a hybrid Raman/EDFA optical repeater is shown in Figure 2.66. The principle is to use the RFA as an optical *preamplifier* (the input fiber trunk serving as gain medium) and the EDFA as a post-amplifier with adequate gain-equalization features. Such a configuration also makes it possible to somewhat broaden the usable bandwidth on each tail of the gain passband, although this only represents a secondary benefit.

To conclude the description of optical fiber amplifiers, we must finally consider the various gain bands offered by the other possibilities of RE doping and the Raman effect. Figure 2.67 shows the attenuation spectrum of silica fibers with different regions that were identified according to potential RE-doping/Raman bands. As the figure shows, these different bands have been called (probably to imitate the radio practice!): *original* (O), *extended* (E), *short-wavelength* (S), *conventional* (C), *long-wavelength* (L) and *ultra/extra-long-wavelength* (U/L). The corresponding wavelength/nanometer ranges and the RE-doping possibilities are indicated as well. At first glance, we see that there is plenty of bandwidth available (summing up all the bands yields about 300 nm or 40 THz, approximately). But such an optimistic picture hides several facts:

- Practical fiber amplifiers based on RE other than erbium have not been fully developed and qualified; the glass host cannot be silica-based, but *fluoride-based* (as imposed by constraints of nonradiative relaxation processes); the power requirement for fluoride-based RE-doped fiber amplifiers are one order of magnitude higher; the pump wavelengths are also different, which requires new high-power pump-laser technologies to be developed; for Raman-based solution, the higher fiber loss (0.25–0.35 dB/km) reduces the

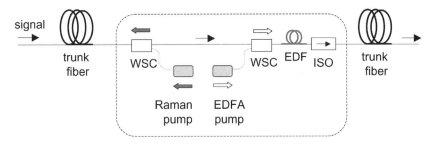

FIGURE 2.66 Basic layout of hybrid Raman/EDFA optical-repeater module.

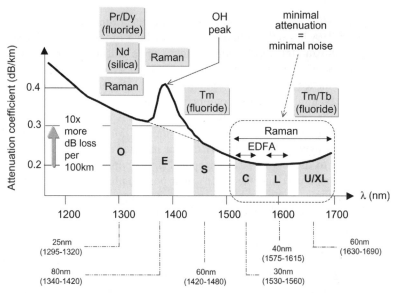

FIGURE 2.67 Ultimate transmission window of silica fibers, as defined by their lowest attenuation coefficient and possible amplification bands called original (O), extended (E), short-wavelength (S), conventional (C), long-wavelength (L) and ultra/extra-long-wavelength (U/L), whose corresponding wavelength/nanometer ranges are indicated below. The low-loss-region, which provides minimal noise, is delineated by a dashed-line. The possible rare-earth doping elements (Pr, Dy, Nd, Tm, Yb) with their glass hosts, or Raman-fiber implementations, are shown for each of the amplification bands.

effective length to a range of 12–17 km, which requires substantially higher pump powers to achieve loss compensation;

- Away from the minimal region, the fiber loss coefficient rapidly increases, namely from 0.25 to 0.35 dB, to compare with 0.2 dB at the minimal (EDFA C/L/C + L) point; for a 100-km fiber trunk, such an increase translates into 25/35 dB gain-compensation requirement, to compare with 20 dB, which is more difficult to implement with practical pumping devices and causes excess noise penalty.

We observe from this figure that in spite of the versatility of the RE-doping possibilities (Pr, Dy, Nd, Tm, Yb), and of the ubiquitous (tunable) Raman-based solution for new amplifier bands, the fiber bandwidth is intrinsically limited by considerations of fiber loss. This issue is not so much of a concern for the current WDM system generation, since the C and C + L bands (and furthermore completed with Raman) have been so far largely under-exploited in actual commercial systems. Since material attenuation, rather than the amplification windows, is the limiting factor, it can be inferred that specialists will investigate new fiber materials having lower loss, a prospect however beyond the current decade's needs, resources and concerns.

■ 2.5 ACTIVE OPTICAL COMPONENTS

The field of *active optical components* concerns *laser sources, modulators, photodetectors, photonic switches* (space and wavelength) and other signal-processing devices such as *wavelength-converting SOA gates* and *all-optical regenerators*. This section provides a brief introductory tour of these different devices with a description of the associated technologies, and their functional blocks.

2.5.1 Laser Sources and Transmitters

In the long chain of devices making up an optical transmission system, the first element is the laser source. The most practical source is the *semiconductor laser diode* (SLD, or LD for short), which takes the form of a miniature chip (e.g., 250-μm length, 50 × 400-μm cross-section). Because of its crystalline nature, the material can be cleaved so that the opposite chip facets are strictly parallel within a few atomic rows. At the air/material interface (30%), the facet's reflectivity causes the chip to be a natural Fabry-Pérot (FP) cavity. The chip's internal structure (active region) acts both as a *laser gain medium* and a *single-mode waveguide*. As we shall see next, when driven by a continuous (DC) electrical current, the LD generates laser light in the TEM_{00} mode. This laser signal can then be intensity- or phase-modulated using an opto-electronic (O/E) conversion device, which is referred to as *external modulation*. Another possibility is to drive the laser directly with the electrical data, which is referred to as *direct modulation*. In another implementation, the modulator can be integrated with the laser on the same semiconductor waveguide, as in the case of *integrated laser modulators* (ILM). In this subsection, we shall review some basics of semiconductor properties, LD structures, and corresponding laser types' associated characteristics. These basics will also be useful for understanding the properties of photodetectors, which are covered in the following subsection.

The principle of *semiconductor materials* and light/current interaction was previously introduced in Section 2.2. Referring to Figure 2.11, recall that in such materials, electrons can be located is either the (lower-energy) *valence band* or the (upper-energy) *conduction band*. The valence band is not 100% filled up by electrons, which leaves *holes*; the same way, the conduction band is not 100% empty (or "filled by holes") which leaves *free or conduction electrons*. Under the influence of an electric field, conduction electrons and valence holes move inside their respective bands, generating a current, hence their names of (charge) "carriers." When a free electron relaxes into the valence band to "occupy" a hole, the corresponding energy difference is released in the form of a photon. Such a process is called *electron-hole recombination*. Conversely, when the energy of an incident photon is absorbed by a valence electron, the electron moves into the conduction band, leaving a hole in the valence band. These are the two fundamental processes underlying photo-emission and photo-detection, respectively.

The energy separation between the valence and conduction bands is called the *bandgap energy*. It is customary to express it in units of *electron-volts* (eV), with the conversion factor $1 \text{ eV} = 1.6 \times 10^{-19}$ Joules. The bandgap energy is determined by the compound of elements forming the semiconductor crystal. Single-element semiconductors such as *silicon* (Si) and *germanium* (Ge) have bandgap energies of $E = 1.11$ eV and 0.66 eV, respectively. Binary-element compounds such as *gallium-arsenide* (GaAs) and *indium-phosphide* (InP) have bandgap energies of $E = 1.42$ eV and 1.35 eV, respectively. By mixing together different proportions of three to four elements, according to the formula $(A_x B_{1-x})(C_y D_{1-y})$, one can realize semiconductors having specific bandgap energies, which fall between that of the AC and BD types. One of the quaternary-material possibilities is (InGa)(AsP), named *InGaAsP* for short, for which the bandgap can continuously vary from $E = 0.36$ eV (InAs) to $E = 2.26$ eV (GaP). The composition yielding $E = 0.800$ eV is of special interest for optical communications, since it corresponds to a wavelength of 1.55 μm ($\lambda = hc/E \equiv 1.24_{\mu m}/E_{eV}$).

In perfect or defect-free semiconductor materials, the numbers of electrons and holes are identical (one electron in the conduction band corresponding to exactly one hole in the valence band). The semiconductor is then called *intrinsic*. But it is also possible to introduce *dopants* (or defects or impurities) in the material. In the crystal lattice, these dopant atoms create a default of an excess of charge, corresponding to a supplementary hole or electron in the valence band, respectively. The materials with excess holes are called *p-type*, and those with excess electrons are called *n-type*. The energy diagrams of p-type and n-type materials, with their electron or hole population differences, are shown in Figure 2.68.

The next concept to grasp is that of a *p-n junction*. A p-n junction can be realized by growing two crystals on top of each other, one from the p-type and the other from the n-type. If the materials are the same, one refers to this as a (p-n) *homojunction* (or *homostructure*). In the contrary case, this is a (p-n) *heterojunction*. What happens in the contact region between the p-type and the n-type? The answer is that the excess electrons from the n-side are attracted by the excess holes of the

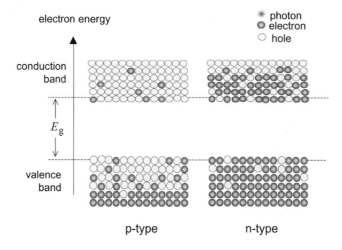

FIGURE 2.68 Energy diagrams of p-type (left) and n-type (right) semiconductor materials.

p-side, and the reverse, and the two carrier types diffuse into the opposite region. This charge motion or carrier diffusion effect stabilizes itself, creating an intermediate region with a built-in potential difference. Such a potential difference shifts the energy levels of the p-side and n-side away from each other by some amount ΔE, as shown in Figure 2.69. Such a energy difference, called a *potential barrier*, prevents the carriers moving from one side to the other. If one then connects the p-n junction to form a closed circuit with a voltage source, the p-side being connected to the positive potential (referred to as *forward bias*), a static E-field is generated. The E-field reduces the potential barrier by an amount $\delta E = eV$ and forces the electrons and holes to move into the junction region with a current proportional to $\exp(eV/k_B T)$. The density of carriers is increased in the junction region, which dramatically enhances the effect of electron-hole recombination, and hence the emission of photons. Because of this effect of photo-emission enhancement, the p-n junction is at the basis of the semiconductor LED (light-emitting diodes) and LD (laser diode).

The above description corresponds to the basic principle of LED. Photons are generated with random k-vector directions and phase, which corresponds to incoherent light. The associated photon flux (number of spontaneously emitted photons per second) is simply $\phi = \eta i/e$, where the ratio i/e is the number of charges injected per second (e = electronic charge, i = drive current), and η the *quantum efficiency* of the charge-photon conversion process. This quantum efficiency also takes into account the fraction of light power that is re-absorbed or experiences total internal reflection in the p-n chip. The LED optical power is given by the number of emitted energy quanta per second, i.e., $P_o = h\nu\phi = \eta h\nu i/e$. One can finally define an overall quantum efficiency, also called *wall-plug efficiency*, as the ratio of optical

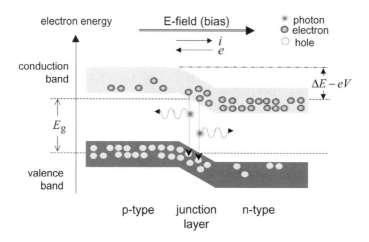

FIGURE 2.69 Energy diagram of a p-n junction, showing (a) the energy difference ΔE between the two regions, (b) the application of a static E-field, which reduces the energy difference by eV and generates a current i (with opposite electron motion e, and (c) the effect of electron and hole diffusion from the p- and n-sides into the junction layer, associated with an increased rate of electron-hole recombination.

power (P_o) to applied electrical power ($P_e = Vi$), i.e. $\eta' = P_o/P_e = \eta h\nu/(eV)$. Another useful LED characteristic is the *L-I responsivity*. It is defined as the ratio of optical power to current, namely $R = P_o/i$, and is expressed in watt/ampere (W/A). From the previous definition, we obtain $R = \eta h\nu i/(ei) = \eta h\nu/e$. Since the ratio $h\nu/e$ represents the photon energy in electron-volts, it also corresponds to the bandgap energy. Using the previous relation $\lambda_{\mu m} = 1.24/E_{eV}$, we finally obtain

$$R_{W/A} = \eta \frac{1.24}{\lambda_{\mu m}} \qquad (2.140)$$

This fundamental result shows that for an LED operating at $\lambda = 1.24$ μm the maximum theoretical L-I responsivity ($\eta = 1$) is $R = 1$ W/A, corresponding to 1 mW of output light for 1 mA of injected current. In practice, LED responses are substantially smaller (e.g., $R = 0.025$ mW/mA or 25 μW/mA). Also note that the linear L-I response only holds for a limited range of drive current. At higher currents, the LED responsivity saturates (until reaching a catastrophic burn-out point), which is observed with LD devices.

Light-emitting diodes are characterized by relatively broad emission spectra. The 3-dB linewidth is given by the formula:

$$\Delta\lambda = 1.45(\lambda')^2 k_B T \qquad (2.141)$$

where $k_B = 1.38 \times 10^{-23}$ J/K is Boltzmann's constant, T the absolute temperature ($k_B T$ being expressed here in eV), and λ' (μm) the effective peak emission wavelength corresponding to the frequency $\nu' = (E_g + k_B T/2)/h$. We leave it as an exercise at the end of this chapter to show that at room temperatures ($T = 300$ K) and for visible to near-infrared wavelengths, the LED linewidth is in the 10-nm range. Note the LED linewidth broadening effect which increases as the square of the operating wavelength. We are very familiar with red, yellow and green LEDs, which illuminate the front panels or keyboards, of practically any electronic appliance we use. These are made with p-n homojunctions based on binary and ternary compounds such as GaP (green) and GaAsP (yellow-red), for instance. Infrared LEDs are also used for short-range, wireless communications between appliances such as printers, mobile telephones or computers (see IrDA in Vol. 1, Chapter 4, Section 4.4). The corresponding materials for 0.9–1.2-μm wavelengths are GaAs and InGaAsP. Due to a carrier lifetime in the order of 1ns, LED can be modulated at frequencies up 1GHz. Because they are quite inexpensive (compared to LD), they can be used to implement short-range optical networks, based upon multimode fiber links. Where does the name "diode" come from? Simply from the fact that if the voltage bias is reversed and increases (p-side on the negative potential), the device blocks the current to a small and constant-intensity level, which is the current/voltage behavior of an electronic diode.

We consider next the case of the *laser diode* (LD). As previously stated, a semiconductor LD is made of a FP-resonator with an active-waveguide structure. Shifting from LED to LD only requires the feedback mechanism provided by the chip facet reflectivities. If we cleave the material in the plane perpendicular to the p-n junction, the spontaneously emitted light is reflected back and forth between the two chip facets, which generates *stimulated emission*, in competition with loss due to the material's absorption and facet transmission loss. Above a certain current threshold, the rate of photon creation exceeds that of photon absorption, leading to a regime of light amplification. A standing, coherent EM-wave then builds up inside this FP cavity, which is the basic laser effect. Simple p-n homojunctions (or homostructures) however work poorly as lasers. The reason is that relatively large threshold currents (>5–10 A) are necessary, due to the lack of light/carrier confinement in the junction. A second reason is that the lack of light confinement and guiding generates a multimode laser beam. The solution to this limitation is brought by the principle of the *double heterostructure* (DH). The idea is to grow three different, crystal-compatible semiconductors (1, 2, 3) on top of each other, each having different bandgaps $E_{g1} > E_{g2} > E_{g3}$. Using both p- and n-types materials, a p-n junction can be realized for instance between material (2) and material (3), forming a *p-p-n heterojunction*, as illustrated in Figure 2.70. The first two materials (1, 2) create a potential energy barrier (approximately of $E_{g1} - E_{g2} - eV$) which confines the carriers into region (2). The two others (2, 3) form an intermediate confinement layer due to the p-n junction, the active region. With the appropriate choice of refractive-index difference $\delta n = n_2 - n_1$ between the center (n_2) and the outer regions (n_1), a planar waveguide can be realized, which confines the light emitted within the active region into a planar-cavity laser mode. In fact, the first lasers used the property that the refractive index is naturally higher in the active region, which produced an intrinsic waveguiding effect. Such lasers were called

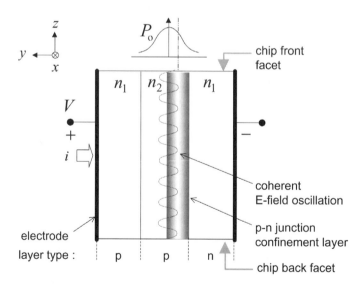

FIGURE 2.70 Principle of a double heterostructure (p-p-n) under forward bias (side view in y, z plane), showing effects of: (a) carrier confinement and electron-hole recombination in the center (shaded) region; (b) light waveguiding due to the difference in materials' refractive index ($n_2 > n_1$); and (c) coherent E-field build-up due to multiple reflections on the chip facets (Fabry-Pérot cavity). The resulting light-power envelope (TEM$_{00}$ mode) is shown on top.

gain-guided DH lasers, as opposed to the later approach of *index-guided DH lasers*, where the index structure is precisely defined by material design. With a two-dimensional index structure, the same index difference can be produced in the perpendicular direction (*y*), as illustrated by possible structures shown in Figure 2.71. The two designs shown are called *ridge-waveguide DH* and *etched-mesa-buried DH*. The *mesa* consists in the active region strip sitting on top of a layer whose sides have been chemically removed or *etched*, hence the name. Other similar structures have the active region buried into a single V-groove channel, or as a mesa surrounded by two channels, hence the fancy appellations of *channel-substrate BH* (CSBH) or *double-channel planar BH* (DCPBH). These various approaches make possible a rectangular waveguide which only supports the fundamental mode, TEM_{00}. Note that the active region and the waveguide need not match exactly, as shown in the examples of Figure 2.71. Suffice it for the waveguide to efficiently "capture" the photons generated in the active region.

The DH-LD concept has made it possible to reduce the laser current thresholds by a hundred-fold factor, bringing them first into the range of 50–200 mA. The rest of the history is a long succession of (index-guided) buried-structure improvements, where the goal is to achieve ever-greater carrier/light confinements and, hence, bring the current thresholds into the 10–20-mW range or lower. For advanced reference, *multi-quantum well* (MQW) lasers are based upon the principle of confining the carriers into periodic p-n structures having thicknesses corresponding to a few atomic layers (50–100 Å). A possible MQW structure could be defined (for instance), as ...p-AlGaAs/GaAs/n-AlGaAs/GaAs/... repeated *N* times. The advantage of the high resulting confinement is lower current thresholds (down to the sub-milliampere regime), greater L-I responsivities (close to 1 mW/mA) and narrower mode linewidths.

The above description concerned basic semiconductor LD called *Fabry–Pérot (FP) lasers*. As the name indicates, the only feedback mechanism is that of the FP cavity provided by the cleaved facets (and additionally, some AR/HR coating to optimize the power output). Consistently, the LD output spectrum contains several "lasing" FP modes (called *longitudinal modes*), an effect previously

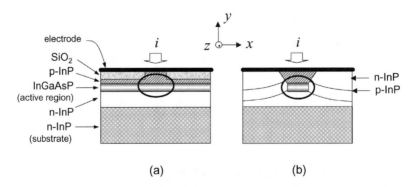

FIGURE 2.71 Alternative designs for index-guided, double-heterostructure laser diodes (facet view in *x, y* plane): (a) ridge-waveguide; and (b) etched-buried-mesa. The ellipse shown in each stucture center corresponds to the lasing spot (TEM_{00} mode).

explained in Section 2.2 and in Figure 2.13. Therefore, the FP-LD are not single-frequency, but *multimode* in the frequency domain. The frequency spacing between adjacent FP modes (also called *free spectral range* or FSR) is given by $\delta v = c/(2nL)$, with the refractive index being $n \approx 3$ and the cavity length $L = 250\,\mu$m, typically. These numbers give $\delta v \approx 0.20$ THz, corresponding to a wavelength spacing of $\delta \lambda = \lambda^2 \delta v/c \approx 1.0$ nm at $\lambda = 1.55$-μm wavelength. If several FP modes are excited, therefore, the net linewidth can be anything between 2–5 nm and 5–10 nm, depending upon the driving current, the number of modes and their relative intensities. A first problem is that the FP mode excitation (wavelength order and intensity) is stochastic, an effect known as *modal partition noise*. Furthermore, the average mode-excitation pattern strongly depends upon the drive current (continuous and modulational transients) and the device temperature conditions, which makes it wholly unpredictable. When a FP-LD is used to transmit signals into a (single-mode) fiber, the longitudinal FP components of the signal spectrum travel at different velocities, resulting in an effect of pulse spreading and envelope distortion. The effect could be easily compensated if the FP-mode patterns were deterministic instead of random. The result is that signal pulses arrive at random times with unpredictable shapes! As long as the pulse spreading, arrival times and envelope oscillations are small compared to the bit period, the impact on the BER is relatively insignificant. As a matter of fact, the first fiber-optic transoceanic cables (e.g., TAT-8) used FP-LD. The specifications for low BER were met because the signals were electronically repeated every 50–60 km, corresponding to the maximum fiber distance traversed at each stage.

In order to carry signals over longer fiber distances and better control their transmission quality, a new family of LD devices with "single-frequency" spectrum had to be developed. In order to achieve single-frequency operation, the principle is to introduce a high loss for all FP modes but one, putting the first group under the lasing threshold. Such a mode selection can be realized through different approaches. One earlier solution, but only of experimental interest, has been the *cleaved-couple-cavity* (c^3) laser, where two chips of different FSR are placed in series; the only lasing mode is the one which belongs to the two FSR sets. A second solution is *the external-cavity* (EC) laser, or ECL. The back facet of the chip is AR-coated, the output is collimated via a lens and a frequency-selective device, namely a *Bragg reflection grating* (see Section 2.4), is placed in the signal path. This Bragg grating can be fiber-based as a FBG (see previous section). The grating selects a single frequency, which is fed back into the laser cavity and yields single-mode operation. One key advantage is the possibility of tuning the laser frequency according to the grating position angle (bulk) or intrinsic tunability (FBG). This is one of the underlying principles of *frequency-tunable lasers* (see following text). Two drawbacks of the approach is the need to precisely align and package the different optical elements (chip, grating) and the overall device size. Another drawback is the effect of *mode-hopping* (jump from one FP mode to the next) as the grating sweeps the FSR, which results in frequency discontinuities. Note that the effect can be completely suppressed through complex mechanical angle-tuning, which is nowadays widely used in laboratory instrumentation. Finally, another possibility is to realize a Bragg grating directly on the laser chip itself, with periodicity $\Lambda_B = k\lambda/(2n_{\text{eff}})$, where k is an integer. The Bragg grating can be written either directly onto the active region, or further away on top of the laser waveguide, as a *distributed* reflection/feedback mechanism. For this feedback to take over, the chip facets must be AR-coated to suppress the unwanted FP cavity effect. The two approaches, called the *distributed-feedback* (DFB) laser and the (three-section) *distributed Bragg-reflector* (DBR) laser, respectively, are illustrated in Figure 2.72. In the DFB case, the Bragg grating covers the entire active region. In the DBR, the grating is either placed on both sides of the active region, as two separate sections, or on a single side, referred to as *external-cavity* (EC-DBR), with an (optional) intermediate region left for phase control. The advantage of the DBR over the DFB is that the control of wavelength and gain are functionally separated. By use of independent electrodes with DC currents (i_1, i_2), which modify the refractive index of the grating regions or the intermediate region (EC-DBR), it is possible to tune the wavelength (3-nm continuously, 8–10-nm with some discontinuities) regardless of the laser drive current (i). The distribution of currents (i, i_1, i_2) which yields a target wavelength is not predictable and must be precisely computer-controlled, which makes the device complex to operate in spite of its tunability advantage. Another drawback of the DBR structure is the loss of

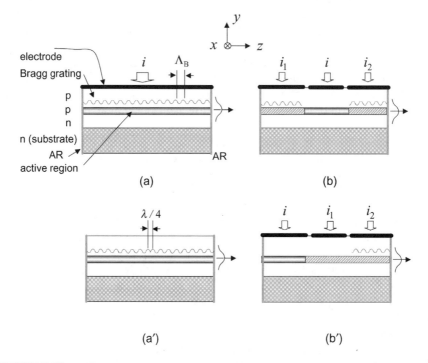

FIGURE 2.72 Different approaches for single-frequency operation of semiconductor lasers (side view in y, z plane: (a) distributed-feedback (DFB) laser; (a′) DFB with $\lambda/4$ phase shift; (b) 3-section distributed Bragg-reflector (DBR) laser, with active region located at mid-point of waveguide; and (b′) external-cavity DBR with intermediate phase section (AR = anti-reflection coating).

output power due to the effect of material absorption in the passive-waveguide regions. This absorption can be somewhat alleviated by generating gain at threshold level (transparency), but this affects the modulation bandwidth and introduces further complexity in the definition of the device's operating point. Having a single electrode, the DFB is far more straightforward to operate. While the DFB wavelength is determined by design (Bragg wavelength), it can also be tuned to a limited extent (1–2 nm) by temperature regulation within a 5–30 K range. However, a limitation of the DFB comes from the fact that an ideal structure would select two longitudinal modes located symmetrically from the Bragg wavelength. Due to the randomness associated with the cleaving process, the phases on the two facets are different and only one mode is actually supported. But this gives two equal possibilities for the lasing wavelength! A way to suppress this unwanted effect is to create an asymmetry by applying an HR coating on the back facet and an AR coating on the front facet, a configuration known as *"hi-lo" DFB*. Another solution for single-frequency operation, which is more practical to implement in terms of process and has the advantage of a substantially lower current threshold, is to introduce a *phase shift* at the center of the Bragg grating (see Figure 2.72). The optimal shift is $\pi/2$, corresponding to one-quarter of a wavelength. Recall from the description of fiber Bragg gratings (Section 2.4) that a $\lambda/4$ shift introduces a narrow "transmission fringe" in the filter's reflection pattern. The effect of such a transmission fringe is to maximize the threshold difference between the lasing mode and the adjacent modes, to which corresponds a highly-efficient side-mode suppression factor. Compared to a non-shifted grating, the laser linewidth is also made significantly narrower. Under cw drive current, the DFB linewidth is in the range 3–10 MHz, compared to 100 MHz for a FP mode in a FP-LD. Under direct modulation at frequency f, this intrinsic linewidth broadens, due to the effect of transient refractive-index changes in the gain medium. Such a broadening is negligible compared to the inherent effect of side-tones generation at $\pm f/2$, which still represents a small fraction of a nanometer (1 GHz = 8×10^{-3} nm at 1.55-µm wavelength). Under such high-frequency modulation conditions, the DFB remains "single-mode," with a FP side-mode suppression ratio typically exceeding 30–40 dB. Because of this superior performance, the DFB is to date the most commonly used laser source in WDM systems. But tunable lasers (EC-DBR or bulk-optics EC) are also used as back-up sources to be activated in case of accidental channel-source failure. Wavelength combs can also be generated with multiple-section DBR structures, as based upon the principle of *"sampled" Bragg grating*. Such gratings are periodically interrupted by blank regions, which produces reflection peaks on both sides of the Bragg wavelength. The wavelength spacing of the side peaks is equal to the period with which the gratings are blanked. By writing two such gratings on each DBR side, each having a different blanking period, it is possible to select a single wavelength (the Bragg wavelength common to the two sampled-gratings) and thus, to achieve single-frequency lasing. By varying the DC current in both grating regions, DBR tunability can be realized over ranges as wide as 100 nm. It is not clear at present whether these monolithic grating-sampled DBR may compete with bulk EC lasers, due to considerations of fabrication cost and operating complexity, but they surely represent an important potential.

The packaging of DFB lasers is essentially the same as that previously described for pump LD sources, as shown in Figure 2.62. The module includes a thermo-electric cooler for temperature regulation and tuning and a chip-to-fiber alignment/coupling apparatus with a Faraday isolator. Additionally, a photodetector (see Section 2.5.2) is placed in the back facet in order to monitor the laser average power. An electronic feedback loop makes it then possible to regulate the laser output power to the specified level under changing conditions (APC).

The above description concerned LD structures which emit from the cleaved crystal facet, the waveguide being in the plane of the p-n junction. Such structures correspond to *edge-emitting* lasers. Another possibility is to make the laser emit in the direction *orthogonal* to the junction's plane. This corresponds to the family of *surface-emitting laser diodes* (SELD), also known as *vertical-cavity surface-emitting lasers* (VCSEL, pronounce vee-xel). The same DBR principle for single-frequency control also applies to VCSEL structures. In this case, the Bragg gratings are realized by superimposing several quarter-wavelength n-type (or p-type) layers to form a vertical, multilayer stack. Unlike with edge-emitting lasers, the active region size (thickness) of surface-emitting lasers is inherently limited, namely by the maximum number of atomic layers that can be grown during a given fabrication process, corresponding to significantly shorter cavity lengths. This shortcoming can be alleviated by using high-reflectivity mirrors, for instance based upon multilayer dielectric coatings. A drawback is the relatively high electric resistance caused by these mirrors, which is also associated with poor thermal conductivity. The key advantage of the VCSEL approach is the possibility of integrating multiple lasers forming a monolithic $N \times N$ array. Within the array, the laser frequencies can be made to vary continuously by locally modifying the DBR parameters during the growth process. Thus large numbers of single-frequency lasers (e.g., 64–100) could be integrated into a single chip for massive WDM applications. This advantage should be moderated by the fact that two-dimensional fiber connectors remain be developed, although VCSEL-to-fiber coupling is comparatively easy to implement when considering a single source point. Another problem is that a single laser failure or built-in defect in the VCSEL array causes the entire chip to be discarded. This is contrast with edge-emitting lasers, which are qualified on a single-device/single-wavelength basis and thus are more practical to replace in case of failure. So far the VCSEL technology has been limited to GaAs structures with operation in the 800–980-nm wavelength range, which limits its application to LAN and broadband optical access, as a competitor for LED-based solutions.

It is worth mentioning the case of *fiber lasers*. A single-frequency fiber laser can be simply realized by putting together a strand of RE-doped fiber (possibly of PMF type), a feedback mechanism (such as FBG) and a laser diode pump. Alternatively, SRS in SMF (or small-area fibers) can be used as the gain process. The main limitation of fiber lasers is that they cannot be directly modulated (owing to the slow gain dynamics in the RE case, and km-long cavity lengths for the Raman case), unlike semiconductor LD. This argument however fades when considering the fact that for best system performance, most applications use external rather than direct modulation (see below). The key advantage of fiber lasers is the ability to scale up their output power to several watts by means of multimode-LD pumping via double-cladding structures. As described in the previous section, this feature is exploited in high-power fiber amplifiers with fiber lasers as pumping sources. But fiber lasers are also attractive for non-telecom applications requiring a high power dynamic range and specific wavelengths unavailable with semiconductor-based sources (e.g., range-finding, weapon guidance, surgery, material processing).

In a single-wavelength transmitter, the two main functions are the signal generation (and its frequency control) and the signal modulation. As previously stated, modulation can be *direct*, i.e., by combining the electrical data with the (cw) laser drive current, or *external*, i.e., by passing the (cw) optical signal through a amplitude/phase modulator. External modulation provides the best signal quality in terms of spectral purity and extinction ratio (see below), although it introduces excess/coupling loss on the signal path (4–12 dB, typically), thus reducing the available transmitter power. The external modulator can be integrated together with the laser chip, as in the case of the *integrated-laser-modulator* (ILM), or be a separate fiber-coupled device as with *lithium-niobate* (LiNbO$_3$) Mach–Zehnder interferometer (MZI) modulators. We describe next these two modulator types.

The ILM approach is based upon the effect of *electro-absorption* (EA). The principle of EA is based on the fact that p-n junctions are strongly absorbing when reverse-biased (as opposed to forward-biased, as in a LD). The condition of this absorption is also that the signal photon energy be higher than the material's bandgap. Thus, turning on and off the reverse bias voltage corresponds to alternating the transmission between high loss and transparency (or minimum loss), which is the effect used to modulate light, the device being called an *electro-absorption modulator* (EAM). It is then possible to integrate a DFB laser and an EAM on a common semiconductor substrate using two different p-n junctions and bandgap materials, and separate electrodes, as illustrated in Figure 2.73. The DFB electrode is driven by an "always on" cw current above threshold, while the EAM electrode is driven by the electrical data voltages (or their logical inverse, since a "1" electrical pulse turns on the absorption, thus generating a "0" light pulse). Note from the figure the insulating barrier which must be grown between the two materials in order for the two electrode circuits to leak into each other. The resulting ILM is a compact device with high-frequency (≥ 10 GHz) modulation capability.

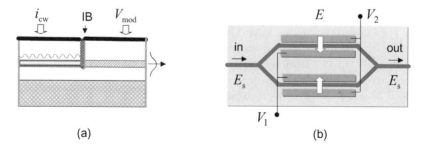

FIGURE 2.73 Two types of modulators: (a) electro-absorption, as integrated with DFB laser (IB = isolation barrier), and (b) electro-optic as based upon Mach–Zehnder interferometer waveguide integrated on lithium-niobate crystal.

The second (and external) modulation approach is based upon the principle of a voltage-controlled MZI. In the description of optical filters made in the previous section, we showed that a MZI can be built as a 4-port planar device with two 3-dB couplers defining either identical or mis-matched optical paths. The same type of interferometer can also be built as a 2-port planar device using two Y-branch waveguides, as illustrated in Figure 2.73. If the two paths have strictly identical lengths (L), the input signal splits up in the first Y-branch and recombines in the second. The recombination is complete or without loss because the two signal components are perfectly in phase. If one can perturb the refractive index by some amount δn on either signal path, for instance by applying a static electrical field (E), then the recombining signals exhibit a phase difference ($\delta\phi = \delta\beta L = \delta n k_0 L$, $k_0 = 2\pi/\lambda$), which leads to destructive interference and signal loss. An intensity modulator can thus be realized using electrically-induced phase modulation in the MZI. Such an interaction between a static E-field and a light-signal E-field, also called the *Pockels effect*, is due to *second-order nonlinearity* ($P_{NL} = \chi_2 E E_s$, $E_s =$ signal field) in the material, see general description in Section 2.3. The Pockels effect provides an index change proportional to the static E-field, as defined by $\delta n = \pm n_{\text{eff}}^3 r |E|/2$, where r is the *electro-optic or Pockels coefficient*. The static E-field can be generated by applying a voltage difference $V = V_2 - V_1$ over a pair of electrodes separated by a small distance d, which gives $E = V/d$, as expressed in Volt/meter (V/m) units. Figure 2.73 shows how the electrodes can be placed in the upper and lower arms of the MZI in order to generate E-fields of opposite directions (this is referred to as a *push-pull* configuration). The total index difference is then equal to $\delta n = n_{\text{eff}}^3 r |E| \equiv n_{\text{eff}}^3 r |V|/d$, corresponding to the net phase difference

$$\delta\phi \equiv n_{\text{eff}}^3 r \frac{2\pi}{\lambda} |V| \frac{L}{d} \qquad (2.142)$$

In order to realize a workable intensity modulator, the phase difference must be at least equal to $\delta\phi = \pi$. There are a variety of electro-optics crystals to choose from (e.g., ADP, KDP, KTP, LiXO$_3$ [X = Nb, Ta], CdTe, InGaAsP) for which the Pockels coefficient ranges from 10^{-12} to 10^{-10} m/V. The first technology constraint is the possibility to integrate single-mode planar waveguides with these types of substrates. So far the technology has essentially focused on lithium-niobate devices (the core index being raised by in-diffusing a titanium strip deposited onto the crystal surface), although semiconductors (such as InGaAsP) would a priori seem to be the best candidates. The reason for choosing LiNbO$_3$ is based on the figure of merit represented by the factor $n_{\text{eff}}^3 r$. Using the crystal parameters $r = 30.8 \times 10^{-12}$ m/V and a refractive index of $n_{\text{eff}} \approx 2.1$ (1.3 µm), we get $n_{\text{eff}}^3 r \approx 2.8 \times 10^{-10}$ m/V. In contrast, semiconductors have for typical parameters $r = 1.3 \times 10^{-12}$ m/V and $n_{\text{eff}} \approx 3.1$, which yields $n_{\text{eff}}^3 r \approx 0.38 \times 10^{-10}$ m/V, which is ten times smaller. Regardless of the material, the second concern is to be able to achieve $\delta\phi = \pi$ with relatively small applied voltages (V_π). We observe that the index modulation is proportional to the geometrical factor L/d. With an electrode spacing of $d = 1$–2 µm and a path length of 5 mm to 1 cm (typically), this ratio is between 5,000 and 10,000. Replacing the different parameters into the definition in equation (2.142), we obtain for 1.55 µm signals the range $V_\pi = 0.6$–1.2 V. Such an estimate of V_π corresponds to ideal conditions with a uniform static field inside the waveguide. In reality, the E-field is only locally uniform, namely at the waveguide's air/crystal interface, causing the electro-optic effect to be less than 100% efficient. The correction due to this nonuniformity takes the form of an overlap factor $\Gamma \approx 0.3$ to include in the RHS of equation (2.142), which brings the extinction voltage to $V_\pi = 2$–4 V. Finally, we must consider the device bandwidth issue. The electrodes cannot be made too closely spaced and too long, because this increases the intrinsic circuit capacitance and hence the RC (time) constant, which defines the modulation speed. Optimized devices for 1.3–1.5 μm operation and having lengths of $L = 2.5$ mm, typically exhibit a modulation bandwidth in excess of 20 GHz (-3 dB) and 40 GHz (-10 dB). The corresponding V_π is close to 8 V (peak-to-peak AC modulation), corresponding to an electrical drive power of $P_e = V_\pi^2/R \approx 1.5$ W for a circuit with impedance $R = 50\,\Omega$. Such a performance can be further improved, namely reducing V_π and extending the 3-dB bandwith towards the 40-GHz region, through more complex electrode designs. The net loss of $LiNbO_3$ modulators (including fiber-to-fiber coupling) is 4–6 dB. Since the Pockels coefficient is polarization-sensitive, the input fiber lead, which connects to the DFB/DBR laser source must be of the PMF type. Finally, we should note the possibility of realizing $LiNbO_3$ strip waveguides for applying *phase modulation*, as opposed to intensity modulation (IM). The approach corresponds to phase-shift-keyed (PSK) transmission systems, in contrast with the *amplitude-shift-keying* (ASK) approach, equivalent to IM when using *direct-detection* (DD) receivers. Although PSK systems can also be used with DD (see next subsection) they are more sensitive to fiber nonlinearities (SPM, XPM, FWM). One of the main reasons for this is the fact that, unlike in the ASK case, PSK signals have a constant E-field envelope. With the combined effects of nonlinearity, dispersion and amplifier noise, constant-envelope signals are subject to large power instabilities and complex deleterious effects.

We have thus described the two main technologies used for external (ASK) signal modulation and their performance parameters. To complete such a description, it is important to mention the criteria of *extinction ratio* (ER). The ER also defines the modulator quality and has a strong impact on the BER performance. As further described in Chapter 1 of Desurvire (2004), ASK signals are made of marks (binary value = 1 or "ON") and space (binary value = 0 or "OFF") symbols which take the form of *return-to-zero* (RZ) or nonreturn-to-zero (NRZ) ON/OFF light pulses. For an average optical signal power P, we should ideally have zero power ($P_0 = 0$) in the OFF state and maximum power in the ON state ($P_1 = 2P$). The ER can be defined as the factor ER $= P_1/P_0$, which is "infinite" in the ideal case. The worst possible value, for which ER $= 1$, would correspond to a meaningless transmission where all symbols would be of equal value. Alternatively, one may define the mark and space powers according to the following:

$$\begin{cases} P_0 = \dfrac{2P}{ER + 1} \\ P_1 = \dfrac{2P}{ER + 1} ER \end{cases} \qquad (2.143)$$

With realistic modulators, it is not possible to bring the ER to arbitrary high values. In the case of EA modulators, this is clearly due to the fact that the difference in absorption between the On and OFF states is limited (e.g., 15–20 dB = −ER). For LiNbO$_3$ devices, the finite ER (e.g., −30 to −40 dB) is due to MZI waveguide imperfections, temperature effects making V_π drift, or uncertainty in the actually applied voltage ($V = V_\pi \pm \delta V$). As described in Section 2.5.2, the bit-error-rate (BER) is determined by the so-called *Q-factor*. Such a Q-factor is proportional to the power *difference* $\Delta = P_1 - P_0$, or $\Delta = 2P(\text{ER} - 1)/(\text{ER} + 1)$. The factor $(\text{ER} - 1)/(\text{ER} + 1)$ thus defines a BER penalty with respect to the ideal system where $\Delta = 2P$. For instance a modulator with ER = −20 dB (10^{-2}) yields a power penalty of $10\log_{10}(199/101) = -0.08$ dB ≈ -0.1 dB, which may be considered acceptable.

We conclude this subsection by summarizing the key elements in the system transmitter: a *LD source* (typically DFB, DBR or ILM, packaged into a module with PMF output) followed by a *modulator* (typically LiNbO$_3$) expect in the ILM case. The transmitter includes *electronic drivers* for both laser and modulation, a *monitoring circuit* for APC and diagnostics, and a *thermo-electric cooler* for temperature regulation and wavelength control. The electronic data may be coupled to the transmitter directly at the nominal bit rate (e.g., 10 Gbit/s) or in the form of four tributaries (e.g., 4 × 2.5 Gbit/s). In the second case, the transmitter must include an electronic *time-domain multiplexing* (TDM) module which comprises clock extraction and buffering circuits. Additionally, the TDM module may introduce *forward error-correction coding* (FEC) as a data overhead for improved BER performance. The TDM output is then electronically amplified and fed to the laser or modulator driver. Finally, the transmitter can include a *post-amplifier*, which boosts the modulated signal (typically 1–10 mW or 0 to +10 dBm) to a nominal output power level (typically 50–100 mW or +17 to +20 dBm. The whole apparatus is completed by a power supply and by intelligent supervision/alarm circuits (on-board micro-processors forming a control network. A full WDM transmitter (e.g., 32 × 10 Gbit/s) is realized by wavelength-multiplexing the different channel outputs into a single fiber. Since most systems must be protected against laser/channel failures, all channels are made redundant with a 1 × 2 switch for electrical data (TDM input) and a 2 × 1 switch for optical data (WDM input). In case of channel failure, the two switches are flipped so that transmission remains virtually uninterrupted. It is left to the reader as an exercise to draw these different elements into a set of functional blocks and control paths.

2.5.2 Photodetectors and Receivers

At the end of the transmission line, the optical data must be converted back into the electrical domain, in order to restitute the original signals. For a single optical channel (i.e., after optical demultiplexing), such an operation includes two basic steps:

1. The conversion of photons into electrons, which is the function of *photodetection*, as associated with a *photoreceiver* device;
2. The integration of the received electrical power over one bit period and the comparison of the result with some reference decision level, which is the function of bit *interpretation*.

A third step which can also be seen as a receiver function is that of *error correction*. This last function makes use of the (logical) data overhead or coding algorithms (FEC). After correction, but under the right input power conditions, the BER can be significantly increased (up to several orders of magnitude), which alleviates a multitude of system limitations and provides extra performance margin over the system's lifetime.

As a whole, the above apparatus is called a *receiver*. From the networking viewpoint, a receiver is merely a *regenerator* or a *repeater*, i.e., a functional block which brings back the electrical signals into their nominal and original characteristics. Since optical links are bidirectional (duplex channel), receivers and the transmitters are usually assembled by pairs, referred to as *transceivers* or *transponders*, which make the interface between optical links and electrical switches/terminals. A complete P2P WDM link with its channel transponders is shown in Chapter 1 of Desurvire (2004), and in Section 2.6.2 of this book.

The main receiver characteristics can be listed as follows:

Operating *bit rate* (e.g., 2.5/10/40 Gbit/s);

Sensitivity, the photodetector input power required to achieve BER = 10^{-9} (one bit mistaken out of one billion received);

Dynamic range, the range of received power where the receiver can operate at nominal performance;

Bit-error-rate (BER), as defined by actual operating conditions and defines prior or after error correction (e.g., BER = 10^{-5} as input, BER = 10^{-12} as output).

The *photodetector* and its associated electronic circuitry is the key element after which the receiver performance is essentially determined, regardless of improvements introduced by FEC. We shall only describe here the two main types of photodetectors, namely the *p-i-n (PIN) diode* and the *avalanche photodiode* (APD). For futher reference, one should note that there are other photodetector types which are called *photoconductors, phototransistors* and *metal-photodiodes*.

Referring to the previous description in Section 2.2 on semiconductors, the conversion of an incident photon into an electron corresponds to the capture of the photon's energy by a valence electron, changing it into a conduction electron (see Figure 2.11). The condition for a photoelectron generation is that the bandgap of the semiconductor material (E_g) be lower than the photon energy ($h\nu$). This condition can be met by the appropriate choice of semiconductor compound, for instance Ge, InP, InGaAs or InGaAsP, which can all be concentration-customized to work for 1.55-μm signal detection. The conduction electron, also called a *photoelectron*, thus brings exactly one elementary charge (per second) to the device current. In the aforementioned section, we have shown that the current *i* resulting from this process (the *photocurrent*) is simply defined by the formula

$$i = \frac{\eta e P_o}{h\nu} \qquad (2.144)$$

where $\eta \leq 1$ is the detector's *quantum efficiency*, P_o the incident or received optical power, and $e = 1.6 \times 10^{-19} C$ is the electric charge. The signal photocurrent is thus proportional to the received power. The efficiency factor reflects the energy loss due to several effects such as electron–hole recombination, material excess loss, finite absorption thickness of the active region and imperfect coupling between the photodiode's active area and the system output fiber. In the case of lasers (previous subsection), we defined the responsivity as the ratio $R = P_o/i$, expressed in watt/ampere (W/A). In the case of photodetectors, we can define likewise the responsivity as the ratio $R = i/P_o$, as expressed in ampere/watt (A/W). From the previous definition, we obtain $R = \eta e/(h\nu)$. Recalling that the ratio $h\nu/e$ represents the photon energy in electron-volts and using the relation $\lambda_{\mu m} = 1.24/E_{eV}$, we finally obtain

$$R_{A/W} = \eta \frac{\lambda_{\mu m}}{1.24} \tag{2.145}$$

This fundamental result shows that for an ideal detector operating at $\lambda = 1.24$ μm the maximum theoretical responsivity ($\eta = 1$) is $R = 1$ A/W, corresponding to 1 mA of photocurrent for 1 mW of incident light. In practice, photodiode responses are somewhat smaller (e.g., $R = 0.5$–0.8 mA/mW), depending upon the material, the depletion region thickness and the signal coupling efficiency. The responsivity is seen to increase linearly with wavelength, but the formula is valid only for photon energies (hc/λ) greater than the (intrinsic) material bandgap energy (E_g). When the wavelength approaches the critical value $\lambda = hc/E_g$, the material absorption drops by orders of magnitude and the responsivity vanishes. Such a critical value is called the photodiode's *cut-off wavelength*. As an example the bandgap energy of $In_{0.53}Ga_{0.47}As$ is $E_g = 0.73$ eV, which corresponds to a cut-off wavelength of $\lambda = 1.70$ μm, which is sufficient to cover the 1.55-μm transmission window.

The first type of semiconductor detector is based upon the principle of a *reverse-biased p-i-n (or PIN) junction*. By definition, reverse-biasing is the action of forming a closed circuit with a p-n junction while connecting the p-side to the negative potential. Recall that in the absence of bias, a fraction of the electrons from the n-type side move into the p-type side and the reverse, which creates a static E-field in the junction region, called the "depletion" layer. Reverse-biasing the junction creates a new static field which is aligned in the same direction and enhances the local carrier density (more electrons in the p-type, more holes in the n-type) as well as the thickness of the depletion layer. The concept of the p-i-n junction consists in introducing an intermediate slice of *intrinsic* (undoped) semiconductor material. The device consists in a double heterostructure (such as InP/InGaAsP/InP) with two different bandgaps and junctions (p-i and i-n). the two junctions are subject to the same carrier motion effect which, with the reverse bias, makes the depletion layer cover the entire thickness of the intrinsic region. The advantages of this thicker depletion region are three-fold: (a) the photon-absorption probability ($1 - e^{-\alpha L}$, $\alpha =$ absorption coefficient, $L =$ thickness) is maximized; (b) the electron-hole recombination rate is decreased because the carriers are rapidly swept towards the other side of the junction, thus increasing the quantum efficiency; and (c) the junction capacitance, C, is decreased, corresponding to shorter RC time constants, to faster dynamics and to greater photodetection bandwidths. By choosing materials for the n-type and p-type side that are transparent to the incident light (i.e., having greater energy bandgaps), such as InP for 1.55-μm wavelength, the PIN junction absorbs photons only in the intrinsic region. Thus the device can be illuminated from either junction side, in the direction orthogonal to the junction plane. When the signal light is incident from the top or from the bottom (substrate), this is referred to as *front (top-entry)* or *back (substrate-entry) illumination*, respectively, as shown in Figure 2.74. Note that in either approach, the metallic electrode of the illuminated facet is annular, which allows an even/symmetric current flow through the heterojunction. The top-entry is the simplest, but the top ring electrode limits the reduction of device size, which imposes excess capacitance. The back illumination with InP for substrate is preferable for two reasons: InP is fully transparent, and the device size/capacitance can be significantly reduced because the top electrode is just a wire contact. Since the semiconductor has 30% reflectivity on the air/material interface, anti-reflection coatings are required for maximum quantum efficiency. Side-illumination (in the direction parallel to the heterojunction plane) is also possible, which allows longer absorption regions. In this case, the top (input) and back facets are AR-coated and HR-coated, respectively.

The second type of semiconductor detector is based upon the principle of *avalanche photodiodes* (APD). The APD structure consists in an *i-p-n* junction with very strong reverse-bias voltages (50–400 V). In this regime, the high energy acquired by photoelectrons and holes from acceleration by the E-field causes an effect called *impact ionization*. Each impact ionization produces a new e-h pair, which multiplies the number of electrons and holes in an effect similar to an avalanche. The average number of carrier multiplications (M) is called the *avalanche gain*. The APD structure is designed so that the multiplication region (p-n junction) and the absorption region (intrinsic layer) are separated for considerations of regime stability and gain optimization. For simplicity, assume first that holes do not contribute to the multiplication process. If α_e is the ionization rate of electrons and L the thickness of the multiplication region, the gain is then

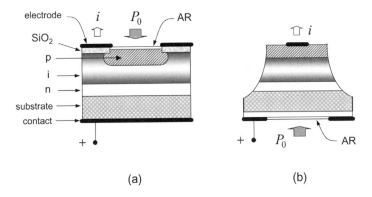

FIGURE 2.74 Basic structure of PIN photodiode with front or top-entry illumination (a), and back or substrate-entry illumination (b).

$M = \exp(\alpha_e L)$. The resulting APD photocurrent is

$$i = M\frac{\eta e P_o}{h\nu} \equiv M R_{A/W} P_o \tag{2.146}$$

showing that the effective response of the APD ($R'_{A/W} \equiv M R_{A/W}$) is M times that of a PIN photodiode. When holes also contribute to carrier multiplication (rate α_h), the gain is enhanced and has a more complex exponential-growth definition, which is a function of the *ionization coefficient* $k = \alpha_h/\alpha_e$. This coefficient largely varies in semiconductor materials. In Si and AlGaAs, for instance, the values can be as low as $k = 0.05$ and as high as $k = 0.85$, respectively.

While there are no physical limits to the avalanche gain, when $k \neq 0$ the associated multiplication process is more noisy as the gain increases. One must therefore find a compromise between gain and noise. Such a noise can be characterized through the parameter $F = \langle M^2 \rangle / \langle M \rangle^2$, called *excess noise factor*. This excess noise factor is defined by the following:

$$F = kM + (1-k)\left(2 - \frac{1}{M}\right) \tag{2.147}$$

We see from this result that the minimum noise-factor value is $F_{min} = 2 - 1/M$, as obtained for $k = 0$. At unity gain ($M = 1$) the noise factor is also unity, which corresponds to the case of a noiseless PIN photodiode. At high gains ($M \gg 1$) the minimum value is $F_{min} \approx 2$. In the general case ($k \neq 0$), and at high gains, the noise factor increases linearly with the gain with a slope of k. To take an example, silicon APD have an ionization ratio of $k = 0.1$. For a gain of $M = 100$, the noise factor is, from the above formula, $F \approx 11.8$. For simplicity, one also defines the noise factor in the form $F = M^x$, with $x = \log F / \log M$. In the last example, we have $x \approx 0.53$ (since $100^{0.53} \approx 11.8$). In the case $k = 0$ and $M = 100$, we have $F \approx 2$ and $x \approx 0.15$, which defines a minimum (high-gain) value for this parameter. It can then be shown that the photocurrent variance $\sigma^2 = \langle i^2 \rangle - \langle i \rangle^2$ is proportional to the factor M^{2+x}, which is generally greater than $M^{2.15}$. It is beyond the scope of this book to analyze the consequence of a finite excess noise factor on the photocurrent signal-to-noise ratio (SNR). The key conclusion is that in comparison to a PIN diode, the APD gain always makes it possible to improve the SNR of the received photocurrent. The improvement corresponds to a factor of 100 to 1,000 (20–30 dB), depending upon the magnitude of the parameters x or k which introduce the M^x excess noise. The improvement rapidly saturates with gain increase, so gain values of $M = 10$–100 are sufficient to reach such optimal conditions.

To complete this basic description of photodetectors, it is worth mentioning the possibility of *optical preamplification* (OPA). The principle of OPA is to optically amplify the signal prior to photodetection with a PIN diode, using for instance a high-gain EDFA or SOA. The optical power is thus amplified with a gain G. As in the APD case, the OPA + PIN system yields a photocurrent which is enhanced by the same factor G, corresponding to an effective responsivity $R'_{A/W} \equiv G R_{A/W}$. And as with the APD, the photocurrent is also characterized by a noise enhancement factor NF, which is the amplifier *noise figure* (see Section 2.4). Thus the parameter NF plays for the OPA + PIN system the same role as the excess-noise factor F for the APD. The difference with the OPA approach is that it is possible to achieve NF = 2 (3 dB), at least with the EDFA approach, whereas the same result in the APD case would require an ideal device having zero ionization ratio ($k = 0$). At identical responsivities, the OPA + D photodetector is therefore less noisy that the APD one, at the price of the complexity introduced by the optical preamplifier. In most systems, however, the OPA + D remains the preferred approach, because of the need to minimize detector noise and optimize the received-signal BER (see further below concerning photon/bit sensitivity). Another advantage of having an optical preamplifier is the possibility of *automatic power control* (APC). This APC, which varies the pump power through a feedback loop, keeps the power incident upon the PIN detector at a constant level. The detector performance and responsivity are thus fixed and specified within some signal dynamic range.

The above concerned the process of converting the optical signal power into a signal photocurrent. The next stage of the receiver chain consists in electronically amplifying the photocurrent to the level required by the decision circuit. This function corresponds to that of the so-called *front-end amplifier*. All electronic circuits with resistance R are characterized by thermal current fluctuations called *Johnson noise* (also called *thermal noise*). The noise electrical power is defined as $R\langle i^2 \rangle = 4k_B T B_e$ (k_B = Boltzmann's constant, T = absolute temperature, B_e = electronic bandwidth). The corresponding noise photocurrent is thus $\langle i^2 \rangle = 4k_B T B_e / R$. Thus the noise photocurrent can be reduced by choosing a high load resistance. But the circuit also has a finite capacitance, C, to which corresponds a bandwidth determined by the time constant, RC. Therefore, a compromise must be found between load resistance (thermal noise) and bandwidth. This situation can however be improved by the way one implements the front-end amplifier. Indeed, two approaches are possible: the *high-impedance* circuit and the *transimpedance* circuit, as illustrated in Figure 2.75. In the first case, the receiver bandwidth is limited indeed by the RC time constant, and we want low load-resistance values

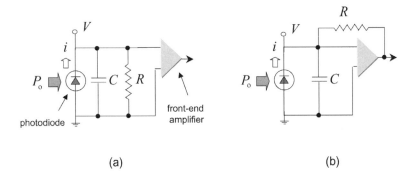

FIGURE 2.75 Two basic implementations of front-end amplifiers in photoreceivers: high-impedance circuit (a) and transimpedance circuit (b). In both cases, R is the load resistance, C is the circuit capacitance, and i is the photocurrent.

(e.g., $R = 50\,\Omega$) for obtaining a maximum bandwidth, at the expense of a higher noise. The transimpedance approach solves this dilemma. Indeed, the effective circuit resistance becomes $R' \approx R/G$ where G is the amplifier gain. Thus, the circuit bandwidth should theoretically be improved by the factor G (namely, $B \leq G/(2\pi RC)$). But the transimpedance circuit introduces more noise photocurrent than the high-impedance circuit. The resulting compromise is a two-fold bandwidth enhancement factor, which still justifies the approach. Another advantage of the transimpedance circuit is its higher dynamic range: large photocurrent fluctuations correspond to reduced output-current fluctuations (i.e., by the factor $1/(G+1)$), unlike in the high-impedance case, where they are proportional. Finally, front-end amplifiers and PIN photodiodes can be integrated onto the same semiconductor substrate. The lowest-noise amplifiers are made with *field-effect transistors* (FET). Their integration with a PIN makes a compact photoreceiver device called *pin-FET* (or *PIN-FET*). The FET can be realized in silicon or GaAs. Compared to the APD, the pin-FET has lower noise and bias voltage. The APD-FET combination is also possible. Advanced receiver designs also include a second-stage, low-noise GaAs-FET *post-amplifier,* using PIN-FET or APD-FET as photoreceivers. The photoreceiver circuit is also completed by an electronic *equalizer,* which compensates frequency-dependent distortion and provides a flat bandwidth response.

We continue the description of the optical receiver by considering the stages following the photoreceiver. For clarity, we shall reproduce this figure here with more functional components, as shown in Figure 2.76. The chain of receiver functions includes an optical preamplifier with pump control via signal feedback for APC. This is followed by photodetection (including front-end amplification) and post-amplification with signal equalization. Between the photoreceiver and the

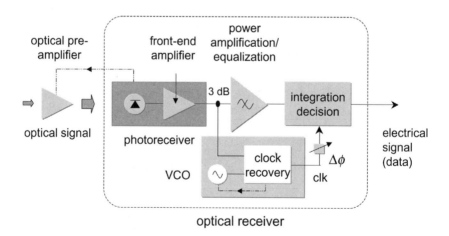

FIGURE 2.76 Functional diagram of complete direct-detection receiver, including (from left to right) optical preamplifier with pump feedback loop, photoreceiver with photodiode and front-end amplifier, 3-dB signal tap, post-amplifier with equalizer, clock-recovery circuit with RF voltage-controlled oscillator (VCO) and phase-locked loop, clock phase adjust ($\Delta\phi$) and integration/decision circuit.

post-amplifier, a 3-dB signal tap is inserted. This signal is fed to a *clock-recovery* circuit. The function of this circuit is to extract the baseband clock reference (f_{GHz}) from the modulated electrical data ($B_{\text{Gbit/s}}$). One of the possibilities is to generate this clock signal through a *voltage-controlled oscillator* (VCO). The circuit compares the baseband signal frequency (f) with that of the VCO (f'). A *phase-locked loop* (PLL) automatically tunes the VCO frequency to $f' = f$. The resulting clock signal (sinusoid at f) is fed to the decision/integration circuit via a tunable phase-adjust device. As its name indicates, the integration/decision circuit integrates the signal power over exactly one bit period, which yields a voltage V_{bit}. This voltage is compared to a decision level V_{D}. If $V_{\text{bit}} < V_{\text{D}}$, the circuit output remains in the OFF position, which is equivalent to a decision that a "0" symbol has been detected. In the opposite case ($V_{\text{bit}} \geq V_{\text{D}}$), the circuit "flips" to the ON state, corresponding to the decision that a "1" symbol has been detected. The output of the receiver is thus a (noise-free) ON/OFF electric signal which replicates the bit sequence of the (noisy) input optical signal. The key difference is that this optical-to-electrical data replication is not 100% accurate, because some bits with too much intrinsic noise, have been mistaken for the other symbols, which is the effect of *bit error*. In standard receivers, the *bit-error-rate* (BER) should be extremely low, namely BER = 10^{-9} to 10^{-12}, corresponding to one bit mistaken out of *1 billion* or *1 trillion* bits. With the development of *forward-error correction* (FEC), the receiver BER can be increased to values as high as 10^{-4} to 10^{-6}, which after processing revert to the previous low-BER levels.

By definition, the *receiver sensitivity* is the average optical signal power required to achieve an output BER of 10^{-9} at some specified bit rate (e.g., $P_0 = -40$ dBm at $B = 10$ Gbit/s, $P_0 = -30$ dBm at $B = 40$ Gbit/s, etc.). The sensitivity is also expressed in terms of required (average) *photons/bit*, $\bar{n}$, which is given by the conversion $\bar{n} = P_0/(h\nu B)$. The theoretical sensitivity depends upon the shape of the *probability-density function* (PDF) corresponding to the "0" and the "1" symbols. It is also strongly dependent upon the receiver's thermal noise ($\langle i^2 \rangle = 4k_{\text{B}}TB_{\text{e}}/R$) which increases linearly with the electrical bandwidth. In the ideal case of a receiver having no thermal noise and an un-distorted signal with an infinite extinction ratio (see Section 2.5.1), the PDF is a Poisson distribution for the "1" and a delta (noise-free) distribution for the "0" ($p_0(0) = 1$ and $p_0(x > 0) = 0$). It can be shown that in this case, the sensitivity is *10 photons/bit*. The physical reality is far different from such ideal conditions! If an optical preamplifier is used in front of the photodiode, it can be shown that the contribution of thermal noise is negligible. If the 1/0 PDF is approximated by a Gaussian distribution, the theoretical sensitivity for (optically pre-amplified) IM-DD receivers is *43–44 photons/bit* (for further reference see, for instance, Desurvire, 1994). If the exact 1/0 PDF of optically amplified signals are used, the sensitivity is $\bar{n} = $ *38.5–39.5 photons/bit*. Thus we can consider $\bar{n} = $ 38–44 photons/bit to represent the theoretical limit for ideal IM-DD receivers. At a rate of $B = 10$ Gbit/s, and 1.55-μm wavelength, such a limit corresponds to a power requirement of $P_0 = \bar{n}h\nu B = 48.5$ nW to 56.5 nW, or -43 dBm to -42.5 dBm. Note that the true receiver sensitivity should include the effect of signal *input coupling loss*, for instance caused by any passive optical device in the signal path (prior to preamplification), such as a fiber connector or a pump/signal

MUX in the EDFA. For instance, if this overall input coupling loss is 1 dB, the net receiver sensitivity is or -42 dBm to -41.5 dBm. In practice, even the most sophisticated receiver electronics always introduces extra penalties, from circuit noise to distortion effects. In the case of optically preamplified IM-DD receivers, the best sensitivities reported are in the 100–150-photons/bit range, representing a 3–5 dB penalty from this theoretical sensitivity limit. Further penalties are introduced by the effect of finite extinction ratio (ER), as discussed in the previous subsection. We have seen that the BER (or sensitivity) penalty is given by the factor $(ER - 1)/(ER + 1)$. For instance, the value $ER = -20$ dB (10^{-2}) yields a power penalty of $10\log_{10}(199/101) = -0.08$ dB ≈ -0.1 dB, which corresponds to $+0.1$ dB signal power requirement with respect to the ideal case of an infinite ER. Other penalties arise from fiber nonlinearity, accumulated amplifier noise, polarization-dependent loss (PDL), polarization-mode dispersion (PMD), WDM crosstalk from passive devices, and many other possible factors. It is important therefore to conceptually separate the receiver sensitivity as an isolated subsystem block (input with ideal signals) and as part of the system (input with nonideal signals). The receiver sensitivity (as specified in the subsystem) however remains a key parameter in the system operation and performance (see more in the next section concerning *power budget* and *required SNR*).

The preceding discussion on sensitivity limits concerned IM-DD systems. The case of *coherent systems* is wholly different. For ideal coherent receivers with modulation format such as ASK, FSK and PSK, the theory predicts sensitivities as lows as *9–20 photons/bit* (PSK), *36–40 photons/bit* (FSK) and *18–40 photons/bit* (ASK), the value depending upon the type of detection namely *homodyne* or *synchronous/ asynchronous heterodyne*, (see, for instance, Senior, 1992, for further reference). Such limits have been approached only in laboratory demonstrations and are very hard to achieve in actual commercial systems, especially at high bit rates (>1 Gbit/s). This is because of the difficulty in realizing narrow-linewidth lasers (used as local-oscillators), high-frequency receiver circuits with low noise, and overall the requirement of polarization control (or diversity) in the reception. Furthermore, coherent signals are more impaired by fiber nonlinearities than non-coherent or IM signals, which introduce at least as much penalty as the improvement in receiver sensitivity of coherent detection. With the recent (1995) introduction and development of FEC in long-haul transmission systems, the issues of "record" sensitivities and comparisons between direct-detection and coherent receivers have become somewhat irrelevant from the industry viewpoint, notwithstanding their academic interest and importance for future potential.

To conclude this subsection on photodetectors and receivers it is useful to recall the concept of *3R regeneration*. The term comes from the three regenerative functions of signal *repowering*, *retiming* and *reshaping*. As we have seen, an optical receiver is a complex device which precisely performs this entire function. Since the input is optical (O) and the output is electrical (E), one may refer to this device as an *O/E regenerator*. If the output is fed to an optical transmitter, the RF/microwave signal (the electrical data) is converted back into a lightwave signal (or optical data). One thus refers to this receiver/transmitter functional block as a *O/E/O regenerator*, or *O/E/O repeater*. One such O/E/O repeater is required for each WDM channel to be transmitted. It is clear that the O/E/O repeater characteristics are primarily determined by the achievable electronic speed (operational bit rate) and electrical-SNR degradation (circuit thermal noise, equalization). In spite of the rapid and continuous progress in the field of opto-electronics, O/E/O repeaters always constitute a bandwidth bottleneck. This explains the recent (1995–2000) technology evolution towards broadband optical amplifiers, to be used as WDM *optical repeaters*. But optical repeaters (such as EDFA and RFA) only provide a "*1R*" function, namely repowering (also called *reamplification*). Thus, one may view optical transmission systems as having 3R O/E/O repeaters in the two terminals for network interface, and 1R optical repeaters and in-line components for long-haul signal transport. A future possibility is the development of *all-optical 3R regenerators* (AOR), which perform O/O regeneration without any electronic interface on the signal path, as discussed in the last subsection of this chapter.

2.5.3 Photonic Switching and Optical Cross-Connects

The concept of *photonic switching* covers several meanings and applications. The term photonic suggests that there should be no O/E/O conversion in the signal path. Most basically, the action of *switching* consists in swapping the paths of two signal channels (2 × 2 space switching), or their wavelengths (wavelength or λ switching with O/O conversion), or their time position (time-slot interchanging). Such a swap may concern individual *bits*, or *packets*, or *frames*. It may also be *connection-oriented*, with the purpose of establishing temporary end-to-end circuit connections using a given group/subgroup of wavelength channels according to slowly changing traffic needs. Therefore, wavelength switching or path switching are closely related functions, as further described below. The time scales corresponding to these different switching applications can be described as follows:

> The function of *bit switching* operates at the same clock frequency as the bit rate B, but with rise/fall times (τ) much shorter than the bit period T_{bit} (e.g., $\tau \leq$ 10 ps for $B = 10$ Gbit/s, corresponding to $\tau \leq T_{bit}/10$;
>
> The function of *packet switching* operates at the packet rate, which has for upper bound $R = B/(8N)$ where N is the number of bytes (8 bits) and B the bit rate inside the packet; the packet arrival time may be deterministic or random; the switch must be able to recognize packets as they arrive, which requires a faster turn-on capability at single-bit level (e.g., at rate $2B$) and in some case the identification of the header information (at rate B);
>
> The function of *frame switching* is similar to packet switching, except that the frame length is generally variable and its arrival time is also randomly defined; the speed requirements are the same as in the previous case, possibly with more sophisticated overhead-processing capability;
>
> The function of *wavelength switching* (also called *wavelength routing*) is essentially connection-oriented; there is no time limit for wavelength channels to be connected according to a given pattern, other than those imposed by

efficient traffic optimization and network use; the switch is then constantly reconfigured according to traffic needs, with a time scale in the 1–10-ms range, if not (for long-haul transport and undersea networks) in the second/hour/days range.

Most generally, the function of *path switching* (also called traffic/service *provisioning*) consists in allocating (or reallocating) a physical signal path to any communication channel to be established (or already established) between two network terminals. The old concept of the "telephone switch" (Desurvire, 2004, Chapter 2) is related to establishing and terminating a circuit connection, according to port and line availability in the operator's central office. The higher-level concept of switching, i.e., the reallocation of physical-channel paths is linked to considerations of *traffic management* and *optimization* (e.g., alleviating network congestion, minimizing the number of network nodes and transit time, optimizing the quality of service and channel bandwidth). Secondarily, path switching provides *network protection* in the event of accidental link breaks or other node-congestion/failure issues. Such a clarification illustrates how complex and fuzzy the concept of "switching" may have become with the development of packet networks, and should bring back *photonic switching* (*per se*) into the right conceptual perspective.

A *photonic switch* should theoretically be able to perform all of the aforementioned functions with N channels for inputs and outputs, referred to as an $N \times N$ switch (or $N \times N$ *cross-connect* from the network terminology). There is some potential confusion between photonic switching and electronic switching. Because it is generally void of advanced signal-processing capabilities, a photonic switch can only perform port-to-port connections, while being "unaware" of the data type and contents. This is in contrast with the electronic switch (e.g., ATM), which is an "intelligent" device which reads and interprets the output-port destination of input packets or frames and operates via O/E/O conversion. It generally has buffering capabilities to handle random packet/frame arrival times and solve contention or congestion issues. Photonic and electronic switches have in common the capability to be dynamically re-configured in order to optimize the traffic flow in the network. The function of switching is also often confused with that of *routing*, although the two concepts have some overlap. To summarize, the function of switching (creating and changing signal paths) belongs to the network "layer 2." This is in contrast with the function of "layer 1," which concerns the physical transport of data from one point to another (possibly including basic 2×2 switching for protection). Routers operate at a functional level immediately higher than that of the switch (layer 3), with enhanced capabilities. For instance, routers can interrogate the network and their router neighbors in order to dynamically optimize their "routing table" according to traffic patterns and achieve optimal use of the network resource. They are also able to identify the frame/packet characteristics (frame/packet length, origin/destination, sequence number, time stamp, etc.), to include specific information in the overheads, to re-fragment or re-assemble payloads and many other sophisticated functions, as in the case of the *IP-router*. There is no conceptual link with such routers (layer 3) and photonic switches (layers 1–2). But some confusion comes from the fact that the market has proposed several prototypes of *photonic IP routers*. These actually represent advanced versions of either *layer-2 photonic switches* (called "wavelength routers"), or actual *layer-3 routers*, with optical ports and internal O/E/O conversion, both being used to handle IP traffic over WDM. Another potentially misleading concept is that of the *all-optical cross-connect* (see below). The term "all-optical" is sometimes indifferently used to designate a photonic cross-connect (space and wavelength switch) with or without internal O/E/O conversion, like a "black box" with optical input/output ports. To be accurate, the term "all optical" should apply only to the second case, i.e., when the switch is exclusively of the photonic type. However, one should not confuse photonic switching with *all-optical switching*. The true concept of all-optical switching only applies to devices that are switched/controlled by optical signals rather than electrical signals. The first and the second are based upon O/O switching and E/O switching, respectively. Some examples of all-optical wavelength switches and other signal-processing devices are described further below and in the next subsection.

We concentrate first on E/O *photonic switching*, with focus on the device and subsystem integration/architecture aspects. The building block is the 4-port, 2×2 E/O coupler. Recall from Section 2.4 that a passive 2×2 coupler supports two normal modes, namely symmetric and anti-symmetric (Figure 2.36, right). Since these modes are polarization-degenerate, the number of modes in the structure is actually four, but this does not change the following description and conclusions. Any input signal from port 1 (or port 2) equally excites the symmetric and anti-symmetric modes of the structure. Since their propagation constants are different, they interfere either constructively or destructively, causing the optical power to oscillate back and forth between the two parallel waveguides over a characteristic period called the *coupling length* (L'). A 2×2 coupler of length L' can be integrated onto an electro-optic substrate (such as $LiNbO_3$), with metallic electrodes overlaid onto each of the two waveguides, as illustrated in Figure 2.77. The application of a voltage difference ($V = V_2 - V_1$) between these electrodes creates a static E-field, which in turn causes a refractive-index change through the *Pockels effect* (see Section 2.5.1 on laser sources and transmitters). In the case of a two-waveguide structure, the refractive-index change is different for the symmetric

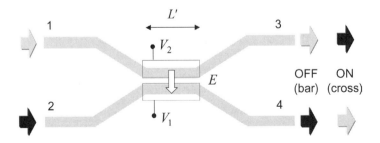

FIGURE 2.77 Elementary cross-bar switch based on planar 2×2 waveguide coupler on electro-optic material and overlaid control electrodes setting the switch in either ON (cross) or OFF (bar) states with respect to input signals from ports 1–2.

and antisymmetric modes. Thus a net phase mis-match $\Delta\beta L'$ is created over the coupling region of length L'. We assume that $\Delta\beta L' = \pi$ for some voltage V_π. When no voltage is applied ($V = 0$), the signals or ports 1 or 2 exit from ports 3 or 4 (respectively), which corresponds to the switch "OFF" "or 'bar' state." When one applies voltage of value $V = V_\pi$, the phase mis-match causes the signals to exit from the opposite ports, which corresponds to the "ON" or "cross" state, as indicated in the figure. Alternatively, the switch can be designed to be in the cross state when no voltage is applied, i.e., by making the coupling region an odd multiple of $L'/2$. In any case, the switch's state can be flipped by applying the voltage V_π. Such a device is called COBRA, after the French *"commutation optique binaire rapide"* which readily translates into English. The COBRA's switching voltage V_π is given by the formula

$$V_\pi = \sqrt{3} \frac{\lambda d}{n_{\text{eff}}^3 r L'} \qquad (2.148)$$

where the parameters n_{eff}, d, r have the same meaning than in equation (2.142) in the previous subsection which corresponds to the case of single-mode waveguides. The same analysis as previously done for the Pockels effect shows that for LiNbO$_3$ devices the switching voltage is below 10 V. Note that alternate versions of the COBRA exist (i.e., "alternate-$\Delta\beta$") for which the control electrodes are split in order to more precisely control the switching efficiency, considering uncertainties in the device realization.

An $N \times N$ crossbar switch can be realized by integrating several 2×2 E/O couplers on the same substrate. The most straightforward implementation is shown in Figure 2.78 in the case of a 4×4 device. Such an implementation requires N^2 elementary 2×2 E/O couplers. There is a large number of possibilities for connecting any input port to any output port. We leave it as an exercise for the reader to determine from the figure that there are 22 different possible configurations for realizing a $1 \rightarrow 4$ port connection (or $4 \rightarrow 1$). But some of these configurations may suppress any connection possibility between other ports, in which case the switch configuration is said to be *blocking*. In contrast, a *nonblocking switch* permits all possible input/output port permutations without blocking any path. The switch is said to be *strictly nonblocking* if there exists at least one single rule

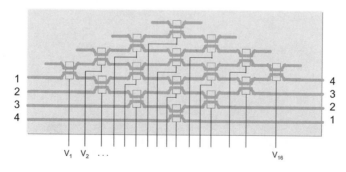

FIGURE 2.78 Implementation of 4×4 crossbar switch on planar substrate through elementary 2×2 electro-optic switches.

for which any input/output path configuration can be defined simultaneously. In contrast, a switch can be *rearrangeably nonblocking*, if the already connected paths must be reconfigured to allow changes in the other channel connections.

A possible architecture of a strictly nonblocking switch is illustrated in Figure 2.79 in the case of a 4×4 device. We observe that in the simultaneous connection $1 \to 3$ and $3 \to 4$, the paths determined by the different switch states are mutually compatible. One can view the switch as an $N \times N$ matrix with rows indexed by the input port numbers (i) and columns indexed by the output port numbers (j). As can be easily verified from the figure, the algorithm for a non-blocking configuration which realizes any $i \to j$ connection is as follows: (a) all switches from row i and column j are in cross-state, except for the intersection switch (i, j) which is in bar-state. With such an algorithm, a complete switching configuration (all input ports connected to an output port) only requires N switches in the bar state, the $N(N-1)$ other ones being left in the cross state. A first drawback of this crossbar switch is that the path length (and corresponding transmission loss) differs according to the switching configuration and port number: it is seen from the figure to vary from l to $2l-1$, where l corresponds to the full length of a single 2×2 coupler. Secondly, the switch dimension, or number of ports (N) is intrinsically limited by the number of elementary 2×2 couplers that can be cascaded onto a single crystal substrate (for LiNbO$_3$, the limit is 4 to 8 couplers, typically). Finally, the number of required couplers grows as the square of the number of ports, which makes impractical the realization of large switches. In order to reduce the path length, to minimize the number of required 2×2 couplers and providing throughput loss to all possible paths, many alternative architectures have been developed. It is also possible to reduce the complexity by use of three-port 1×2 and 2×1 couplers. The principle of 3-port E/O switches is the same

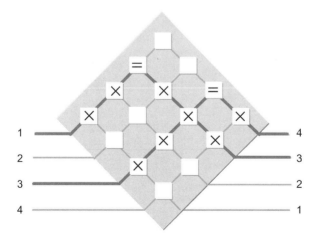

FIGURE 2.79 Architecture of a strictly non-blocking crossbar switch (here 4×4), showing switch configuration connecting ports 1 to 3 and ports 3 to 4. The cross and bar states of each 2×2 switch are shown by the symbols $\times$ and $=$, respectively.

as previously described in the 4-port case, except that 2 × 2 couplers are replaced by Y-branch couplers. Two possibilities for optimized strictly nonblocking switch architectures, in 4 × 4 implementation are provided in Figures 2.80 and 2.81. The first case is called a *Spanke switch*. Both switch types only require 1 × 2 or 2 × 1 couplers, with a total of 24 switching elements. Figure 2.82 shows another possible implementation made with 1 × 2, 2 × 2 and 2 × 1 couplers. The number of switching elements is now reduced to 12. In the three examples provided (Figures 2.80, 2.81, and 2.82) it can be seen that, neglecting the loss due to coupler-to-coupler connections, all paths have the same loss equal to $4l$ or $3l$. A drawback of these optimized architectures is that certain connection paths cross each other with relatively shallow angles, which correspond to many unwanted "Y" or "X" waveguide branches. This fact limits the possibility of massive integration with planar technologies. It is yet possible to implement large switches of these types while using smaller subgroups of integrated 2 × 1/1 × 2/2 × 2 couplers and optical fibers to interconnect them according to the desired architecture, however complex. In any case, such optimized switch architectures can be implemented by free-space or bulk optics technologies. Note that in path switching, the switching function should be bidirectional since optical communications links are generally two-way or duplex. Two implementations are then possible. One consists in duplicating the switch as two separate and parallel elements sharing the same control circuit, which is always the case for electronic switches. The other consists in using a bi-directional switch, as illustrated in Figure 2.83. The bidirectional signals coming from each fiber pair (one direction per fiber) can be fed to the switch via optical circulators. Thus any bidirectional pair can be switched to any

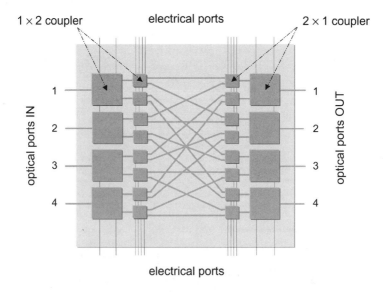

FIGURE 2.80 Strictly non-blocking 4 × 4 switch, based upon Spanke architecture with 1×2 and 2×1 E/O couplers, showing identical number of coupler crossings or path loss for any possible switching configuration.

2.5 Active Optical Components 277

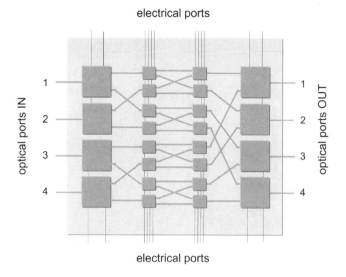

FIGURE 2.81 Another example of strictly nonblocking 4 × 4 switch with 1 × 2/2 × 1 couplers, based upon an architecture different from the previous figure.

port, with an excess loss equal to twice that of a circulator (e.g., 2 × 3 dB = 6 dB). As shown in the figure, the circulators can also be replaced by 1 × 2 passive fiber splitters, which involves a (minimum) 3-dB loss per splitting, or a 6-dB total excess loss. In addition, a Faraday isolator is required to avoid mixing the bi-directional signals in the fibers, which increases this excess loss by a minimum

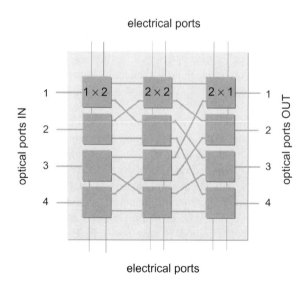

FIGURE 2.82 Another example of strictly nonblocking 4 × 4 switch with 1 × 2/2 × 2/ 2 × 1 couplers, based upon an architecture similar to the previous figure.

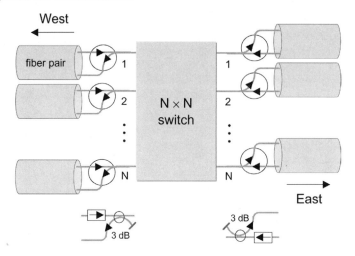

FIGURE 2.83 Implementation of a bidirectional $N \times N$ photonic switch with optical circulators. Circulators may also be replaced by 1×2 passive splitters (bottom of figure).

of 4–6 dB. For large switches, this second configuration may be substantially more economical to implement. In any case, a bi-directional photonic switch is far more advantageous than its electrical counterpart, which requires an O/E/O conversion for each channel and must be duplicated for each tranmission direction. In addition, the photonic switch has a virtually unlimited bandwidth and is transparent to both bit rate and modulation format. However, photonic switches do not have any *buffering* or *retiming* capabilities, unlike with electronic switching. Note that all-optical memories and dynamically controlled optical delay lines exist but their practical implementation for buffering is still at the research level.

All of the preceding concerned optical *path switching*, also called *optical space-division switching*, for which the signal wavelength is unspecified. The second type of switch is the *wavelength switch*, also called *wavelength router*, or *wavelength-selective cross-connect* (WSC), or *wavelength cross-connect* (WXC), which ensures the function of (optical) *wavelength-division switching*. The basic architecture of a 4×4 wavelength switch was previously shown in Section 2.4, Figure 2.45. Referring to this figure, we observe that a wavelength switch is a space switch that can simultaneously support multiple switching configurations, one for each wavelength. To clarify this statement, look only at the four channels (A, B, C, D) with wavelength λ_1. The corresponding switch configuration of the figure, according to the switching table shown, is (A → E, B → H, C → G, D → F). Consider next the channels at wavelength λ_2. The switch configuration is now (A → F, B → E, C → H, D → G), which is wholly different to the previous one, and so on for the other channels.

Just like the space switch, the WXC is inherently bidirectional, and the same comparison with electronic switching as made before fully applies. The key difference is that wavelength multiplexing introduces a phenomenal increase in the number of switching configurations: an $N \times N$ wavelength switch with M

wavelength channels ($M \geq N$) have $N! \times M!$ possible configurations. For instance, a 4×4 switch with 8 wavelengths has $4! = 24$ possible port permutations and $8! = 40{,}230$ wavelength permutations, representing a total of $24 \times 40{,}230 = 967{,}680$ or nearly one million different switching configurations! As described in previous Section 2.4.1, a space/wavelength switch can be integrated at once in a planar AWG device. However, not all switching configurations (or wavelength/port permutations) are possible. The most general implementation architecture which can realize all configurations is shown in Figure 2.84, in the case $N = M = 4$. As seen from the figure, it consists in M independent (space-division) $N \times N$ switches working in parallel, one being attributed to each wavelength layer and to its N possible output ports.

In the previously described implementation of WXC (or wavelength router), the wavelength of each WDM channel is conserved, while the device function is only to shuffle the WDM channels between input and output ports. Consider next the case where each input port is only single wavelength. We can then introduce another switching functionality which is *wavelength conversion*. The principle of a *wavelength-converting switch* is that a signal input from any port at a given wavelength can be switched to any output port with a different wavelength, as illustrated in Figure 2.85 in the case $N = M = 4$. The figure shows that each input port/wavelength is dynamically switched to any output port with an intermediate wavelength conversion (except for the matching ports A–E, B–F, C–G and D–H). Each output port thus receives N signals at the same wavelength, out of which only one must be selected. The proper channel selection is then effected by a $M \times 1$ switch.

As described in the next section, the double functionality of wavelength-conversion with shuffling of ports is central to the network concepts of *wavelength routing* and *optical add-drop multiplexing* (OADM). One can further increase the switch degrees of freedom by including wavelength-conversion in the WXC as well. The principle of a *wavelength-converting WXC*, specifically called *wavelength-interchanging cross-connect* (WIC) is illustrated in Figure 2.86 ($N = M = 4$). In

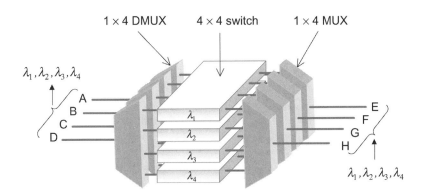

FIGURE 2.84 Principle of a $N \times N/M \times M$ wavelength/optical cross-connect (WXC/OXC), shown here in $4 \times 4/4 \times 4$ implementation. See example of port/wavelength routing table in Figure 2.45.

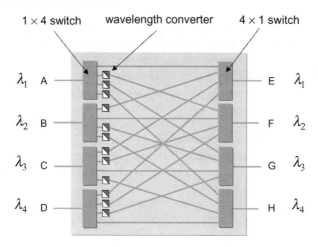

FIGURE 2.85 Basic layout of a wavelength-converting switch.

the general case, each input and output port has M wavelengths. The input channels are all de-multiplexed and fed to a $NM \times NM$ photonic switch. The switch output ports are divided into N functional sub groups of M different wavelengths each. This subgroup includes a wavelength-conversion array which can effect all possible wavelength permutations within this subgroup. Finally, the M resulting signals are multiplexed into a single output port. Thus a WDM channel at any wavelength and any input port can be switched to any output port at any different wavelength, which represents the full functionality of a WXC. If the WIC's switch is photonic, one may refer to it as an *all-optical WIC* (equivalently named *optical cross-connect*,

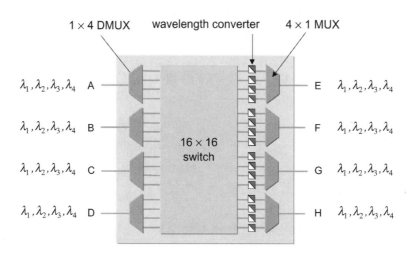

FIGURE 2.86 Basic layout of a wavelength-selective cross-connect (WSC or WXC) or wavelength-interchanging optical cross-connect (WIC or OXC) including wavelength conversion.

or OXC), even if the switching operation is based upon E/O control. Alternatively, the WIC switch can be electrical. In this case, each input and output port of the switch must include O/E and E/O converters, or receivers and transmitters, respectively. It is clear that the electronic implementation of a WIC is very complex compared to the all-optical version. Additionally, it cannot be made bidirectional and thus must be duplicated, unlike the all-optical WIC. Finally, the WDM channel bit rate of electrical WIC is inherently limited, both for consideration of circuit speed and massive-integration cost (a compromise being 2.5 Gbit/s). In contrast, all-optical WIC can operate at WDM bit rates of 10–40 Gbit/s, and possibly higher (160–320 Gbit/s) in future generations. However, the advantage of the electrical WIC is the possibility of recognizing and processing data in the electronic interfaces, which facilitates the functions of monitoring and management. In addition, wavelength conversion is easier to implement since tunable lasers can be used in the output-port transmitters. One may view all-optical WIC as *"transparent"* network nodes, which work for any signal modulation format (NRZ, RZ, etc.), encapsulation type (packet, frames, datagrams) and protocol (SDH/SONET, Ethernet, ATM, TCP/IP, etc.). Electronic WIC do not have this key advantage of transparency, and are inherently more complex to implement, but they provide an additional feature which is *3R signal regeneration*. It is possible that future all-optical WIC will include 3R regenerators as well, but so far the corresponding technology is only at the laboratory stage (see Section 2.5.4). Note that wavelength conversion can be implemented together with the function of retiming (2R regeneration), as discussed further below.

At this stage, we can summarize the various characteristics of WXC and WIC/OXC according to the following list of functions, features and performance criteria:

Architecture: Strictly or rearrangeably nonblocking, number of elementary E/O switches and wavelength converters to integrate and to dynamically control;

Dimension: Number of input/output ports, i.e., $N \times N$ (ports) and $M \times M$ (wavelength);

Number of cross-overs, which may cause significant integration problems for large N;

Operating bit rate and throughput rate (e.g., 10 Tbit/s for $N = 10$ ports and $M = 10$ WDM channels at $B = 10$ Gbit/s);

Switching speed: Rise and fall time of ON/OFF states, including signal propagation delay (3 cm = 100 ps in electrical circuits);

Configuration/reconfiguration latency: processor and algorithm speed;

Switching power per E/O coupler and per configuration;

Extinction ratio: ratio of output powers of the OFF to the ON states in elementary E/O switches;

Crosstalk: noise background from other same- and different-wavelength-channels in E/O switches, wavelength converters and WXC/OXC output ports;

Port-to-port insertion loss, spectral/wavelength transmission flatness and PDL;

Upscalability: possibility for increasing/upgrading at will the number of ports with the same technology and for massive-integration manufacturing;

Packaged size (number of circuit boards), volume (number of shelves) and footprint (total number of bays);

Cost per wavelength or per GHz.

Commercially available WXC/OXC exist in both electrical and all-optical versions with either 256 × 256 or 512 × 512 port configurations. Electrical WXC have throughput capacities of 640 Gbit/s (256 × 2.5 Gbit/s) to 1.28 Tbit/s (512 × 2.5 Gbit/s). In the case of optical WXC, the capacities are 5.12 Tbit/s (512 × 10 Gbit/s) and 10.24 Tbit/s (256 × 40 Gbit/s). These examples well illustrate the advantage of the optical implementation from the standpoint of channel rate and throughput capacity. To the author's knowledge, and at the time of writing, the function of wavelength-conversion (after O/E/O processing) is a potential additional feature available only to electrical WXC/OXC having tunable transmitters.

We consider next the different technologies used in the realization of photonic switches and wavelength routers which constitute the WXC/OXC building blocks. The case of photonic switching has already been extensively described in this subsection with its planar implementation based on E/O switching elements or ON/OFF 2 × 2 gates. We have seen that *lithium-niobate* is one of the possible E/O materials with 10–40-GHz bandwidth, but with the drawbacks of being limited in number of gates, relatively high insertion loss and inherent polarization sensitivity. Another solution is based on *silica-on-silicon* or *polymer* waveguides, which can be controlled by a *thermo-optic effect* (change of refractive index induced by temperature). The advantages are the potential for denser gate integration, lower insertion loss, and low PDL, balanced against higher crosstalk and very slow dynamics (ms scale). Another solution is based on *semiconductor-amplifier* (SOA) gates. A 1 × 2 and a 2 × 1 switch can be realized by using SOA as ON/OFF gates, the ON state corresponding to gain (or transparency) and the OFF state to high loss. The two main advantages of SOA-based switches are the fast dynamics (≤ 1 ns) and loss-less (or amplifying) port-to-port transmission. The main drawbacks are their finite PDL, low extinction ratio (10–15 dB) and cost in massive integration. Finally, switches can be implemented by using mechanical elements with electrically controlled motion, analogous the ancestral telephone switches. For instance, a mirror can be rotated back and forth to couple a collimated light signal from one fiber into any of N output fibers. Mirrors can also be switched between $0°$ and $45°$ incidence in a $N \times N$ array to form a dynamically controlled cross-bar switch (such as shown in Figure 2.79). Over the recent years, considerable progress has been made in the field of silicon-based, *micro-electro-mechanical systems* (MEMS). These MEMS, which are based upon the principle of electrically controlled micromirrors, can be implemented as 2D or 3D devices. The 2D approach essentially consists of the aforementioned crossbar switches. The 3D approach uses micromirror arrays facing a fiber bundle with vertical (or near-vertical) signal incidence. With two analog electrical controls, the mirrors can be made to continuously rock with two independent angles (2D rotation). This 2D angle control defines a unique transmission/reflection path between a given pair of fibers within the bundle. A key advantage of 3D MEMS is their scalability: there is virtually no limit for massive integration (i.e., several hundred to over one thousand ports!), except for the maximum size of the integrated 2D mirror arrays, the path-alignment problems associated with large fiber bundles, the need to achieve a uniform path loss between all possible ports and the complexity of managing the large number of (analog) control signals. To date, MEMS with as many as 1,300 ports have been demonstrated. The MEMS switching and reconfiguration speed is of the order of milliseconds and the port-to-port loss is 2–6 dB, depending upon the device's architecture. Other alternatives for 2D and 3D switching, which have been demonstrated in 64 × 64 and 32 × 32 configurations, use *liquid crystals* (as based upon a polarization-dependent E/O effect) or the principle of thermo-optic reflection in *bubble cells* (hence the figurative name of "bubble switch"), respectively. All these space-switching technologies have ther own advantages and drawbacks, and therefore the choice is ultimately guided by the inventory of functions, features and performance criteria which we have previously listed.

We consider next the function of *wavelength conversion*, which is central to the realization of fully versatile WXC/OXC. Unlike with space switching, which can be slowly reconfigured at millisecond or second scales, wavelength conversion should be as fast as the data rate, namely with rise and fall times of the order of 10–100 ps. A wide variety of ultrafast nonlinear effects can be used for direct wavelength-to-wavelength conversion (such as three- and four-wave mixing in χ_2 and χ_3 nonlinear materials), but with the usual suite of issues: phase-matching requirement, polarization dependence, control/addressing of output wavelength, limited tunability or dynamic range, high switching power and poor conversion efficiency. The alternative solution is to find a fast-responding material which could transfer the modulation from one incident signal at a given wavelength λ_s to another signal, called the "*probe*," at a different wavelength λ_p. Such a modulation transfer can be realized by using the incident signal data to control the gain or the refractive index of an optically amplifying device. A cw probe being passed through this device thus experiences intensity or phase modulation, which (if sufficiently fast) represents a logical "copycat" of the original data. So far, the most promising approach is that based on semiconductor optical amplifiers (SOA). The SOA can be operated as a *cross-gain modulator* (CGM, also called XGM) or *cross-phase modulator* (CPM, also called XPM). The principle of a SOA-based CGM is shown in Figure 2.87 for two different implementations (signal and probe being co-directional or contra-directional, respectively). Each input signal bit, as coded in RZ or NRZ format, saturates the SOA gain with a difference (gain compression) of ΔG, which should at least be 10 dB for efficient modulation. For each input signal bit, the cw input probe thus experiences a decibel transmission change of $\Delta T = -\Delta G$, which corresponds to an envelope modulation equal to the logical inverse of the signal data. Note that in the binary (or Boolean) world, data and their logical complement/inverse represent strictly equivalent information. The data have thus been transferred from one wavelength to another, which is the desired wavelength-conversion effect. The main drawback of the SOA-CGM approach is the relatively low output extinction ratio, since the SOA gain can be saturated only within a limited range (e.g., 10–15 dB) given the low optical power of the input signal (e.g., 0.1–1 mW).

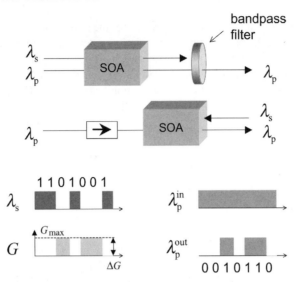

FIGURE 2.87 Principle of cross-gain modulation (CGM) in semiconductor optical amplifier (SOA), showing data transfer from modulated input signal (wavelength λ_s) to probe signal (wavelength λ_p), with two possible device implementation (top = codirectional, bottom = contradirectional). The four diagrams at bottom represent the input signal data (pattern 1101001), the corresponding gain modulation with saturation ΔG from maximum gain G_{max}, the signal probe at input (cw), and the signal probe at output with modulation corresponding to inverse logical data (0010110).

A more complex implementation of SOA-based wavelength conversion utilizes the effect of CPM. To change the CPM into intensity modulation (IM), some kind of interferometer structure is required. Two possible structures are the *Mach–Zehnder* interferometer (MZI) and the *Michelson* interferometer, as illustrated in Figure 2.88. The MZI structure can be made from either 2×2, wavelength-selective couplers or Y-branches. Through a forward-bias current (i), the SOA is set to a certain gain level for which the MZI is balanced, meaning that the output probe is maximized in the absence of any input signal. For most efficient operation, the wavelength-selective couplers should steer the signal exclusively into the SOA path, while having a 3-dB splitting ratio for the probe. An alternative configuration (not shown in the figure) includes one SOA on each of the MZI paths, with un-balanced $\eta/(1-\eta)$ couplers. Note that the relative propagation directions of the signal and the probe are indifferent. The Y-branch approach for MZI implementation is also possible, but 50% of the signal is lost for CGM, unless one uses wavelength-selective Y-branch designs. The Michelson approach represents a clever simplification of the previous structure. In the ideal balanced case (no input signal and SOA properly biased), the structure is 100% reflective for the probe, owing to the HR coating on the right side. The modulated signal can be input from the right side, which facilitates the device connection. In this reflective, Michelson-interferometer implementation, an optical circulator is also required to split the input and output probes.

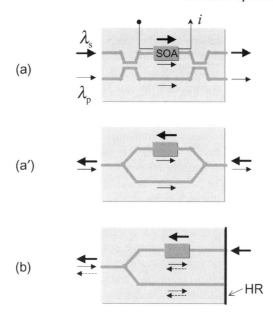

FIGURE 2.88 Different implementations of semiconductor optical amplifier (SOA) based wavelength converters using principles of Mach–Zehnder (a), (a′) and Michelson (b) interferometers (λ_s = modulated input signal wavelength, λ_p = probe wavelength, HR = high-reflection coating at probe wavelength λ_p).

The SOA-based MZI or Michelson devices are more generally referred to as *SOA gates*. Because of the simplicity of operation, switching speed and compatibility with planar-waveguide circuits (e.g., silica on silicon), SOA gates have a promising future for applications in *all-optical signal processing*. The function of wavelength conversion only constitutes one, though significant, example of the potential possibilities. In the next section, we will describe how SOA gates can be applied to even more advanced applications, namely *all-optical 2R and 3R regeneration*.

2.5.4 All-Optical Signal Regeneration

The principle of *3R regeneration* is described in this chapter in Section 2.5.2 dedicated to detectors and receivers. To recall, the term "3R" comes from the three regenerative functions of signal *repowering*, *retiming* and *reshaping*. In current transmission systems, these three functions are performed electronically, namely in the link receivers (one at each fiber end, as the optical links are bidirectional). If the receiver output is fed to a transmitter, one obtains a *3R electronic repeater*. The function of signal *repowering* is also performed along the transmission line by means of lumped or distributed optical amplifiers (i.e., EDFA or RFA). Thus, optical amplifiers can be conceived as being *"1R" optical repeaters*. The key advantage of optical amplification is that it repowers all WDM channels at once, without any signal conversion and any other electronic-circuit bottleneck.

It is also transparent to the data type and the bit rate. Although there is theoretically no limit to the number of in-line amplifiers that can be cascaded to extend the link distance, noise accumulation from amplified spontaneous emission (ASE) is a source of signal-to-noise-ratio (SNR) degradation. Such an SNR degradation is reverted at the receiver stage through the full 3R signal processing, with a certain amount of bit errors. A key question is whether 3R regeneration could also be implemented *all optically*. The interest of all-optical regeneration (AOR) would be to alleviate SNR degradation *along* the transmission line, rather than at the end of the line as with 3R electronic repeaters. It suffices that these regenerators are placed at periodic locations to fully reabsorb (or clean-up) the noise generated in the previous segment. The receiver would thus have the same bit-error-rate (BER), regardless of the distance traversed by the signal, which could represent millions of a kilometers in a fully transparent optical network! One might therefore conceive of the current 1R in-line optical amplification as representing the technology precursor of fully developed 3R in-line optical regeneration. To make such a vision come true, a first issue to address is whether it is possible to realize all-optical devices which could perform *retiming* and *reshaping*. Such devices should inherently be *polarization-insensitive* and *wavelength-independent,* in addition to operating at *multigigabit rates* or over (e.g., 160–320 Gbit/s). Ideally, such devices should also be able to *process all WDM channels simultaneously*, just like an optical amplifier does as a 1R repeater. Other highly desirable features would be to have a low power consumption, to be compact, reliable and cost-effective. The technical answers to some of the above questions already exist. Here, we shall briefly describe various solutions that have been demonstrated for 2R optical regeneration (repowering and reshaping only) and 3R optical WDM regeneration (simultaneous to all channels).

We consider first the simpler case of 2R regeneration. A possible implementation is based upon *saturable absorbers* (SA). The property of saturable absorption corresponds to materials for which the loss is power-dependent. At low powers, the material is characterized by some finite loss. As the power increases, this loss decreases until the material becomes fully transparent or loss-less. Such a property can be exploited to remove the background noise associated with "0" symbols, as illustrated in Figure 2.89 in the case of an ideal, ultrafast SA device. The first stage of the SA-based 2R regenerator consists of an optical amplifier (repowering) followed by a SA. As the figure shows, the SA transmission is increased (say up to $T_{\max} \approx 1$, corresponding to transparency) for the "1" bits and remains low for the "0" symbols (namely $T_{\max} - \Delta T$). Thus the "0" bits have been re-shaped by the SA, meaning that the bit energy is brought close to a zero-power level. In contrast, we observe that the intensity noise in the "1" bits remains essentially unaffected, since the SA is transparent or close to transparent at high power. At this stage, the SNR is improved but only from the contribution of "0"-bit re-shaping. We can then further improve the SNR by suppressing the noise in the "1" bits. This can be done by placing a narrowband optical filter in the signal path. The condition for the filter to act as a dynamic power equalizer is that the signal pulse be a *soliton*. In a previous section concerning passive optical devices, and in the subsection concerning optical filters, we described this principle, which the reader may refer to.

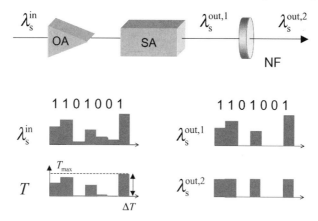

FIGURE 2.89 Principle of all-optical 2R regeneration (re-powering and re-shaping) with optical Amplifier (OA), saturable absorber (SA) and narrowband filter (NF). The four diagrams at bottom represent the input signal data (pattern 1101001) at λ_s with amplitude noise in both "0" and "1" bits, the corresponding transmission changes with saturation ΔT bringing transmission to maximum T_{max}, the signal output from SA before ($\lambda_s^{out,1}$) and after ($\lambda_s^{out,2}$) the NF. Note the removal of noise background in the "0" bits in signal $\lambda_s^{out,1}$, and the removal of intensity noise in the "1" bits in signal $\lambda_s^{out,2}$.

To summarize, the frequency spectrum of solitons broadens at high power and narrows at low power. Thus, if the filter bandpass is chosen sufficiently narrow to block a small portion of the soliton spectrum, any power increase (spectral broadening) will make the filter more dispersive, and any power decrease (spectral narrowing) will make the filter more transparent. The passive filter thus acts like the reverse of a saturable absorber, removing the intensity noise of the "1" bits! To be accurate, the filter is a linear device. The nonlinearity of the response is self-contained in the soliton behavior. The origin of this nonlinearity is *self-phase modulation* (SPM), as described in this chapter, Section 2.3. We can therefore conclude that the combination of an optical amplifier, a SA and a narrowband filter acts as a 2R regenerator providing only *repowering* and *reshaping*. Should a system with these in-line 2R regenerators absolutely require the data to be transmitted as solitons? The answer is no, the only requirement is that signals are RZ (return-to-zero) pulses. The trick is to convert the RZ pulses into soliton pulses prior to passing the 2R regenerator. The conversion is easily realized by placing a short strand of highly dispersive (positive dispersion) fiber next to the optical amplifier. The combination of high power and positive dispersion transforms the linear RZ pulse into a nonlinear soliton pulse, it is as simple as that. At the 2R regenerator output, the effect of transmission loss brings back the soliton into a linear RZ pulse. Thus the link can transmit any type of RZ signal as long as the 2R regenerator includes this internal function. Such an approach has been called the *"black-box" optical regenerator*, or BBOR. The name comes from the fact that the regenerator is designed to work on its own, which leaves the issue of point-to-point signal transmission as a separate optimization problem, just as in any nonlinear/dispersive link. The BBOR principle, based on polarization-independent InGaAlAs MQW, has been demonstrated in the laboratory and proved its benefits (see, for instance, Kaminow and Li, 2002), even as a 2R regenerator without retiming function. See discussion below concerning WDM implementation.

We focus next on all-optical 3R regeneration, which on top of the previous functions introduces signal *retiming*. This regenerator architecture must then include a clock-extraction circuit, corresponding to the function of *clock recovery* (CR). The CR principle can be based on electronics circuitry or be all-optical. In the first case, this is the same principle as in a receiver (see description in Section 2.5.2), but possibly very simplified, since signals reaching an in-line optical regenerator are not supposed to be severely degraded, especially in terms of timing noise (called "jitter"). Alternatively, solutions for *all-optical CR* exist, for instance based upon *synchronous ring fiber lasers* or *self-pulsing semiconductor LD*. When modulated by *random* signal data at bit rate B, both of these devices produce a uniform train of pulses (or a raised sinusoid) at frequency $f = B$, which is the extracted clock. Further, this clock is free from timing jitter and amplitude noise, because the laser pulses experience multiple cavity round-trips, which averages these two effects. In comparison with the well-established electronic solution, the advantage of an all-optical CR circuit is that no high-speed photodiode (and possibly front-end amplifier) is required, which may have a substantial impact on the overall optical-regenerator cost. The background idea includes the possibility of integrating the optical CR (e.g., self-pulsing LD) with the other regenerator functions. Finally, some all-optical regenerator approaches require the clock to be optical. With electronic CR, one would then need an E/O conversion circuit with an LD to transform the electronic clock into an optical clock, which is more complex and expensive. Having now a optimal solution for the CR issue (whether electronic or optical), the next step is to use the clock for signal re-timing and re-shaping. We analyze next two different approaches for all-optical 3R regeneration, as based upon *wavelength-conversion in SOA gates* and *synchronous modulation* (SM), respectively.

The principle of all-optical 3R regeneration with SOA gates is illustrated in Figure 2.90. It is strictly the same principle as previously described with SOA-based wavelength conversion. Here, the difference is that the signal probe is an *optical clock*, which is synchronized with the input data and is ideally *noise-free, both in intensity and timing*. Such a clock signal is generated through a CR circuit, which can be electronic (as shown in Figure 2.90) or all-optical. The electronic CR circuit drives a laser diode which provides a pulsed probe. The all-optical CR circuit directly provides this optical-clock signal. The input (RZ or NRZ) data modulate the SOA gain through CGM, which corresponds to the function of an intensity modulator: clock pulses which are synchronous with a "1" signal bit are blocked (or transmitted with lower level $G_{max} - \Delta G$), while those that are synchronous with a "0" signal bit pass through the SOA (at higher level G_{max}). Note that due to the nonlinear gain saturation (ΔG) in the SOA, large intensity fluctuations of the signal correspond to small gain fluctuations for the probe, which inherently suppresses the initial noise. The modulated clock obtained at the SOA output thus carries the same information as the input signal (as logical complement/inverse). Unlike in the pure wavelength-conversion scheme, this output signal is noise free,

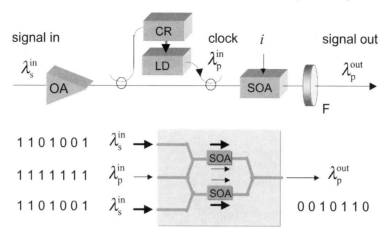

FIGURE 2.90 Top: principle of all-optical regeneration with clock recovery (CR), pulsed clock probe from laser diode (LD), semiconductor optical amplifier (SOA), and bandpass filter (F). Bottom: planar implementation with Mach–Zehnder interferometer (MZI). The input data are at wavelength λ_s and the output regenerated data at wavelength λ_p. Corresponding input and output data patterns are shown for example.

because its bit-timing is based upon a jitter-free clock and its bit-shape upon a pulsed signal source which is void of any intensity noise. The conversion is likenable to a *Boolean/logical AND* operation with the noisy modulated signal and the noise-free clock as the two input ports. The output port is a noise-free modulated signal. Figure 2.90 also shows the actual implementation based upon a MZI structure with two SOA and two optical ports for the driving signal (previously split into two halves). In this interferometric structure, the exploited effect is cross-phase modulation (CPM) as opposed to CGM. Many other device implementations are possible, for instance with counter-propagating signal and clock. Note that the clock wavelength must inherently be different from that of the input signal. For wavelength "transparency," a possibility is to cascade two such SOA-MZI structures (ideally integrated onto the same substrate for loss and compactness considerations), while using a second clock probe having the same wavelength as that of the original signal. Two other advantages of this twin configuration are the possibility of obtaining regenerated signals in NRZ format (as opposed to RZ), and more importantly, of increasing the regenerator bandwidth (namely 40 GHz to date).

A second approach for all-optical 3R regeneration utilizes the principle of *synchronous modulation* (SM). The regenerator includes the same CR function as previously, i.e., electronic or optical. The principle of retiming through a nonlinear interferometer/gate is however different: the signal to be regenerated passes through the device without any wavelength conversion. If one could attribute a "physical identity" to an input data pulse, such an identity remains unchanged through the regeneration process, unlike in the wavelength-converting regenerator case. One could also view the SM regenerator as processing individual photon packets with a reduction of the photon-number and arrival time uncertainty. How does this work? Three SM approaches are possible: the first concerns pure *intensity modulation* (IM), the second pure *phase modulation* (PM) and the third a combination of both (IM/PM), with a sinusoidal modulation shape. It is beyond the scope of this book to describe the fine details of the effects of IM, PM and IM/PM synchronous modulation. Suffice it to state that when applied to soliton pulses, both IM and PM suppress timing jitter, which is the function of signal retiming. The drawback of IM is that timing jitter is converted into intensity noise. This is because pulses that have random arrival times (δt) into an intensity modulator experience random transmission loss: the transmission is maximal when $\delta t = 0$ (peak of IM transmission) and minimal when $\delta t = B/2$ (lowest point of IM transmission). But recall that solitons can get rid of intensity noise when passing through narrowband optical filters (see earlier description of 2R regeneration), but noise suppression is not 100% efficient. One additional advantage of IM is to remove the nonsoliton background noise, such as amplified spontaneous emission (ASE). Since ASE is not a constituent part of the soliton pulse, but rather a light background which is constantly dispersed by the fiber, the effect of IM (as sinusoidal) is to remove 50% of the associated power at each modulation stage. It is clear that dividing the ASE power background by two, periodically along the transmission line, prevents the ASE accumulating or building up, which stabilizes it to some constant level. The price to pay is the timing-jitter to intensity noise conversion, which is only partially compensated by narrowband filtering. In contrast, PM does effectively suppress timing jitter, without intensity-noise conversion. The principle is that PM induces a *frequency chirp* (or time-dependent frequency shift) proportional to the pulse arrival time. The chirp is negative (red shift) if the pulse arrives *before* the modulation peak, and positive (blue shift) if the pulse arrives *after* the modulation peak. A fundamental property of the soliton is that it can redistribute this chirp across the entire pulse envelope. It is as if the modulation red-shifts the early solitons and blue-shifts the late ones. In the soliton-propagation regime where the fiber dispersion is positive, the red solitons have slower velocities than the blue solitons. Thus we conclude that an early soliton, being red-shifted by PM, will automatically have its velocity decreased, while a late soliton, being blue-shifted, will have its velocity increased, which is the desired retiming effect! Because the timing jitter is a random effect, several PM stages disposed at periodic intervals along the link rapidly reduce any timing jitter to a minimized value, which corresponds to a regenerative function of "distributed retiming." But PM does not do anything to suppress intensity noise. This explains why the adequate combination of PM and IM (plus filtering) achieves the complete 3R regeneration function. Electronically controlled MZI can realize both IM and PM simultaneously. To date, the concept has been fully demonstrated with polarization-independent InP-based devices at clock rates up to 40 GHz. A drawback of these devices is their relatively high insertion loss ($>10–12$ dB). Such a loss can be compensated by optical amplification, at the expense of additional noise. With the SM regenerator cleaning up its own noise, the net result is a significant improvement of SNR, namely a complete stabilization over virtually "infinite" transmission distances. As for 2R regeneration (see earlier discussion), there is no actual need for the system to be based upon soliton propagation. With the *black-box optical regenerator* (BBOR) approach, any RZ pulse is eligible for SM-based regeneration. To summarize, the basic components of a SM-based all-optical 3R regenerator are: (1) a CR circuit, which can also be all-optical; (2) a synchronous IM/PM modulator (which should be polarization-independent); (3) a RZ-to-soliton conversion mechanism, which is basically an optical amplifier followed by a strand of dispersive fiber (BBOR); and (4) a narrowband optical filter. If one compares this short-list requirement to that of a fully-fledged electronic repeater (referring to Figure 2.76), the comparison is, beyond any doubts, highly favorable to the optical solution. Such a conclusion however only applies to the function of *in-line regeneration*, as opposed to that of *regeneration at terminals*, for which more complex features are involved (e.g., electronic time-domain multiplexing, error correction, switching, protection and monitoring). Therefore, one should not confuse the function of in-line regeneration, which extends the system haul and improves the signal quality (BER), with that of the regenerators in the network terminals. To summarize, in-line regeneration could accomplish for system transmission performance a task even greater than 1R amplification does for the current technology generation.

The above description concerned the all-optical (2R and 3R) regeneration of single wavelength channels. How about its implementation in WDM systems? Three solutions are possible, as illustrated in Figure 2.91. The first one, the *parallel* approach, consists in demultiplexing the WDM channels and regenerating each one separately, on a channel-by-channel basis, followed by remultiplexing. This would be the same approach for electronic 3R regeneration, except that the optical regenerator devices are considerably simpler and have higher processing bandwidth.

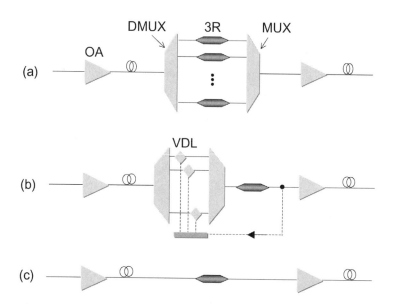

FIGURE 2.91 Three possible implementations of all-optical in-line regeneration: (a) parallel, (b) presynchronized, and (c) self-synchronized (OA = optical amplifier, 3R = all-optical 3R regenerator, VDL = variable delay line).

The second one is based upon the principle of WDM channel *presynchronization* with a demultiplexing/remultiplexing operation. Such a presynchonization function, based upon a feedback loop, aligns all WDM-channel bits in the same clock window. This makes it possible to use a single SM regenerator, working simultaneously for all WDM channels. The third approach is based upon the principle of *self-synchronization*. Depending upon the trunk dispersion, it is possible to make the WDM channels periodically bit-synchronous, which removes any need for a MUX/DMUX re-synchronization loop. Such an ultimate configuration for 3R processing is similar to optical amplification for 1R processing: the regenerator is a single, transparent, two-port device able to handle all WDM channels simultaneously. The experimental investigation of these advanced concepts is only at the laboratory stage. Yet simultaneous all-optical regeneration of 160 Gbit/s signals (4 × 40 Gbit/s) leading to "infinite" transmission at constant BER has already been demonstrated.

Whether the potential of all-optical regeneration remains only speculative at this stage of the market and technology, is only a pragmatic question. Suffice it to say that it could represent one of the last frontiers in the mastery of lightwave communications, even at the border of a new engineering philosophy where *signal degradation (noise)* and *transmission distance* no longer represent any limiting issues.

■ 2.6 WDM NETWORKS

This section provides an overview of the basic principles of *WDM networks*, which form the "optical backbone" of the world's global telecommunications infrastructure. The emergence of "WDM" (as it is familiarly called) has represented a true revolution in the telecom field, to be barely overshadowed by the subsequent Internet. The massive deployment of WDM networks is in fact quite recent, namely less than a decade. It traces back to the development of practical optical amplifiers (EDFA). As the enabling technology of WDM, optical amplification had two major impacts. First, it made it possible to extend signal transmission distances without electronic regeneration by two to three orders of magnitude. Second, by combining channels at different wavelengths in an amplified link, a dramatic increase of capacity could be obtained, corresponding to a fuller use of the fiber bandwidth in the 1.5-μm window. But there is more in WDM than a mere exploitation of bandwidth! Rather, it is its intelligent use and management, which introduces new dimensions, functionalities and unprecedented possibilities in broadband optical networking. The field of WDM networks covers a wide range of issues, to which this section can only be an introduction. What makes this field somewhat difficult to apprehend at first is that it is based upon many conceptual levels: the enabling technologies (passive and active components), the wavelength standards and digital hierarchy (SDH/SONET), the principles and operation of point-to-point WDM links, the network topology and connectivity, and the issues concerning network management, protection, optimization, convergence and global dimensioning. Since the enabling technologies (active and passive), were described in detail in Sections 2.4 and 2.5, this section will focus on their implementation in the network environment. The topic of *passive optical networks* (PON) and their WDM implementation, has been already addressed in the chapter concerning broadband access and will therefore not be discussed here. We will concentrate on point-to-point (P2P) WDM transmission and networking at metropolitan (metro) and continental/transoceanic (backbone) scales. Finally, the issues of *network protection, virtual-topology design, network evolution* and *convergence* (IP over WDM) are described in the last two subsections.

2.6.1 Wavelength Standards and Digital Hierarchy

The early WDM systems of the 1980s consisted in mixing two signals from any of the 0.850-μm, 1.31-μm and 1.55-μm transmission windows, which doubled the system capacity. The later development of 1.5-μm WDM systems (1990) used channels some 3–5 nm apart by groups of two or four, limited by the EDFA gain flatness, the laser tunable range and the resolution possibilities of WDM filters. As the device technologies progressed, the number of channels was increased to 16–32 (and later 64–128–320) and the spacing narrowed to the nanometer scale. At this point came the need to define absolute wavelength and spacing references, which is the task of *standardization*. Wavelength standardization, which has been addressed by the International Telecom Union (ITU-T for telecommunications systems) is indeed central to the *interoperability* of WDM networks. It is also based upon the need for *component compatibility* between different manufacturers and technology sources. In the spectral domain, a WDM signal aggregate basically consists in a set of tones forming a *wavelength comb* or *grid*, where the channels must be equally spaced. But should the channels be equally spaced in frequency or in wavelength? While the issue is indifferent for continuously tunable lasers, it is not obvious for periodic filters. Indeed, filters based upon Fabry-Pérot cavities and Bragg reflectors are periodic in wavelength, while those based on interferometers (MZI, AWG) are periodic in frequency. As it turns out, the channel spacing (25–100 GHz) remains very small compared to the absolute wavelength ($\lambda = 1.55$ μm) or frequency ($f = c/\lambda = 3 \times 10^8$ m/s/(1.55×10^{-6} m) $= 193.548$ THz). An exercise at the end of the chapter shows that the relative position difference between a limited set of channels equally-spaced in frequency or in wavelength can indeed be neglected. The ITU-T has chosen frequency for reference, with a comb center frequency located at $f_{ref} = 193.1$ THz ($\lambda = 1,5536.00$ nm), and which is positioned exactly in the middle of the 1.55-μm transmission window (see Figure 2.26). The main ITU-T standard defines $\Delta f = 100$ GHz as the frequency spacing. At wavelengths near 1.55 μm, such a spacing corresponds to wavelength spacing of $\Delta \lambda = \lambda^2 \Delta f / c = 0.800$ nm. The WDM network engineer knows this number by heart, as well as the other approximation of 1 nm $\equiv 125$ GHz.

The ITU-T standard also recognizes smaller (25 or 50-GHz) and greater (200, 300...600-GHz) spacings, corresponding to finer or coarser WDM grids, still using f_{ref} for center reference. The initial appellation of *WDM* concerned nonstandard systems where channels could be spaced as widely as 1–5 nm apart, or even more. The recent expression of *dense WDM* (DWDM) is more subjective. It principally means WDM systems with less than 1-nm spacing ($\Delta f \leq 100$ GHz), but is now also indifferently used to contrast with the first 0.8/1.3/1.5-μm generation of WDM systems or as a mere commercial argument. One also refers to anything that is not DWDM as *coarse* (CWDM) or *sparse-WDM*, which is also used with commercial overtones! Indeed, systems with fewer channels and broad spacings can be seen as more cost-effective solutions for broadband optical access, in particular in the field of *metropolitan-area networks* (MAN).

In Section 2.4 concerning passive optical devices, we saw that the EDFA defines a set of two windows called C (conventional) and L (long-wavelength), respectively. The full C + L implementation covers a window of 45–70 nm (depending upon gain flatness requirements). Using the approximation 1 nm ≡ 125 GHz, this overall window corresponds to a bandwidth of 45–70 × 125 GHz = 5.6–8.7 THz, representing 55–90 channels with $\Delta f = 100$-GHz spacing. This is where we can observe that the concept of dense WDM could be misleading, because it only tells how channel frequencies are allocated, not how efficiently the bandwidth is used. Indeed, assume that the channel bit rate is $B = 10$ Gbit/s. The *spectral efficiency* can be defined as the merit factor $\eta = B/\Delta f$, namely $\eta = 0.1$ or 10% in this example (and $\eta = 0.025$ or 2.5% for $B = 2.5$ Gbit/s!). If the 10-Gbit/s channels were brought to a narrower 25-GHz spacing, the efficiency would rise to $\eta = 40\%$, which is still not "efficient" from engineering standards, and also from the view that $\eta = 100\%$ is the theoretical limit for IM-DD links. But we must accept this efficiency limitation since some tolerance should be given to the absolute channel frequencies, due to manufacturing specifications and temperature effects in lasers. Also the channel spacing cannot be narrowed to the 100% efficiency limit because passive WDM components (MUX/DMUX, optical filters, OADM) have a finite channel resolution power and the extinction ratio between channels needs to be as high as possible. As a general convention, the comb spacings of 25 GHz, 50 GHz and 100 GHz should correspond to bit rates of 2.5–10 Gbit/s, 10–40 Gbit/s and 40 Gbit/s, respectively, which have efficiencies in the range $\eta = 0.1$–0.8.

Referring to Figure 2.67 in Section 2.4, we observe that the fiber bandwidth can be decomposed into several discrete bands named *original* (O), *extended* (E), *short-wavelength* (S), C, L and *ultra/extra-long-wavelength* (U/XL) which spreads over the 1,300–1,700-nm (500-nm or >60-THz) range. As previously discussed, these different windows correspond to the currently-known possibilities for optical amplification in fibers, using either rare-earth dopants or Raman scattering, or both. An important observation is that only the 1.5-μm band is optimal for WDM systems. Indeed, the lowest loss achieved in this wavelength region corresponds to the lowest gain, and the lowest noise penalty. But should 3R optical regenerators (see previous section) be introduced in WDM systems, such a consideration would no longer be applicable, and the overall operating bandwidth may be expanded as well to 1,200–1,900 nm, provided low-loss *single-mode* fibers can be designed to operate over such a huge wavelength range.

Another issue in wavelength standardization is the *choice* of wavelengths within the previously described ITU-T grid. What should a 16- or 32-channel, 100-GHz-spacing WDM system use as nominal wavelengths? So far there is no answer to this question since no specific agreement has been reached. The choice is left to the network operator, with a concern for maximum wavelength compatibility and reuse within the different network segments. The case of undersea systems (see below) is relatively free of such considerations. Indeed, the wavelength channels are only used for point-to-point transport, with O/E conversion at the terminal ends. Therefore, a great flexibility exists in the choice of these wavelengths, at least when the comb only involves a limited number of channels. This freedom is however restricted by considerations of performance uniformity between channels. The complexity of very long-haul transport (6,500–12,000 km unrepeatered) is such that large BER (Q-factor) differences may be observed between channels, and the grid must naturally be positioned for the set of wavelengths where this difference is minimized. Clearly, the question of wavelength choice does not arise if the system carries *all* possible channels offered by the EDFA windows.

A second and most important level of appreciation of WDM networks concerns the *digital hierarchy*. In Desurvire (2004), Chapter 3, Section 3.1, we have extensively described the two protocols consisting in the so-called *synchronous digital hierarchy* (SDH) and its companion, *synchronous optical network* (SONET). These two protocols, which are only partially compatible, are used in Europe/Asia and North America, respectively. As it was shown, SONET and SDH define a suite of allowed bit rates which are multiples of 64-kbit/s (voice) channels and also include extra bit/s capacity for framing overhead purpose. Here, we shall not recall the intricacies of the SDH and SONET framing (and alternatives), which are fully detailed in the aforementioned reference. Suffice it to refer to Table 3.1 and observe that SDH and SONET have the following compatible bit-rates:

STM-1 or OC-3: 155 Mbit/s

STM-4 or OC-12: 622 Mbit/s

STM-16 or OC-48: 2.5 Gbit/s

STM-64 or OC-192: 10 Gbit/s

STM-256 or OC-768: 40 Gbit/s

with STM standing for *synchronous transport module* (SDH) and OC for *optical circuit* (SONET). The WDM-network engineer knows all these bit-rate definitions and values by heart, although they represent handy approximations (for instance, the *real* bit rate of STM-16 is 2.48832 Gbit/s. We should also note that such a common hierarchy goes by factors of four. Thus a WDM system with $4 \times$ OC-48 channels has the same capacity as a single OC-92 channel (10 Gbit/s). Likewise for $16 \times$ OC-48 WDM with respect to OC-768 (40 Gbit/s). As shown in the table, the number of equivalent voice circuits is roughly given by dividing the bit rate by 64 kbit/s, for instance $10_{\text{Gbit/s}}/64_{\text{kbit/s}} \approx 156{,}250$ circuits. In reality, the number is somewhat less (because of the framing overhead) and varies according to the way the payloads are fragmented into so-called E1-E4 or T1/J1-T3/J3 "containers," which differ according to the aforementioned geographical zones. Concerning pure voice communications, it can also be higher because of voice-compression algorithms.

The STM/SDH and OC/SONET bit-rate hierarchies thus define different standard possibilities for WDM traffic (as multiplied by the number of WDM channels) and equivalent voice circuits (as divided by 64 kbit/s). However one should bear in mind that the traffic not only concerns *voice*, but *video* and (more or less essentially) *data*. By data, one means payload information which computers may exchange, for instance fixed images, text and message files, or any number of records. With the development of the Internet, *e-mail* and *Web pages* constitute the most common and popular type of data. With the addition of 3G mobile telephony and other multimedia services the trend for interactive video will increase, leaving voice as a "commodity," itself encoded in the form of a separate data channel. The lesson to retain is that WDM systems are not concerned with the type of payload (voice, video, or data), which is the matter of higher network layers (Ethernet, ATM, TCP/IP). The SDH/SONET protocols just provide a "layer-1" standard reference to transport payloads between network terminals, while aggregating/deaggregating them according to considerations of destination, optimal route, payload efficiency, traffic patterns and network-use optimization at this just-physical, layer 1 level. Note that the SDH/SONET standards are not unique in WDM networks, since Ethernet, ATM, IP and GPON (for instance) can also make direct use of WDM with different packet/frame encapsulation schemes. The simplest view is to consider that SDH/SONET can simply, reliably and efficiently transport and route *any* payload of *any* higher network-layer format. To understand such functions, it is necessary to consider first the issue of point-to-point transport then that of networking topology and connectivity, which are described in the next two subsections.

2.6.2 Point-to-Point WDM Transport

A *point-to-point (P2P) WDM link* is characterized by a bidirectional transmission line connecting two terminals with transponders (transmitter/receiver pair). In the case of WDM systems, the line is made of an optically amplified fiber pair, and the terminals transmit/receive a certain number of wavelengths (N). The complete layout of such a system, with its two terminals and its optically repeated trunk, is illustrated in Figure 2.92. The figure does not show the rest of the electronic circuitry in the terminals, which concerns (electrical) *time-domain multiplexing/demultiplexing* (TDM), *error-correction processing* and SDH/SONET frame assembly/reassembly. Thus a wavelength-channel rate of $B = 40$ Gbit/s in the optical domain may represent the TDM aggregate of 4×10 Gbit/s (four STM-64) in the electrical domain, but such detailed features remain outside the P2P WDM link characteristics. The WDM link only recognizes bit streams at the wavelength-channel rate. From the network perspective, this defines an "optical layer" which lies just below (or within) the standard "physical" or "transport" layer.

If the channel transponders operate at a wavelength-channel rate B, the total link capacity is $B' = N \times B$. The three other parameters of import in P2P links are the total trunk length (L), the spacing between two optical amplifiers (z_a), and total the amplifier output power, $P_{out} = N \times P_{ch}$, where P_{ch} is the power per channel. The amplifier spacing determines the *amplifier gain*, namely $G = \exp(+\alpha z_a)$, where α is the fiber attenuation coefficient at the channel wavelength, assumed to be uniform. In turn, the gain determines the *amplified spontaneous emission* (ASE) power, which accumulates along the link and represents a first and unavoidable cause of *signal-to-noise ratio* (SNR) degradation. After demultiplexing, the SNR

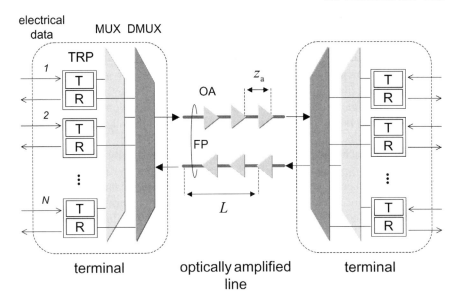

FIGURE 2.92 Layout of a point-to-point WDM link, showing terminals (TRP = transponder, T = transmitter, R = receiver, MUX/DMUX = multiplexer/demultiplexer for N wavelength channels) and amplified line (FP = fiber pair, OA = optical amplifier, z_a = amplifier spacing, L = trunk length).

obtained at the receiver end, as determined by the passband ASE and channel power, corresponds to a certain level of *bit-error-rate* (BER). See more on this issue in subsequent text. One may measure the simplification introduced by the WDM approach when considering the systems of the previous generation. These systems operated at a single wavelength, the optical amplifiers being replaced by electronic repeaters. Looking at Figure 2.92, one may mentally remove the MUX/DMUX and draw N parallel transmission lines instead of a single one. A P2P WDM link is basically a single, bidirectional transmission line which can carry an arbitrary number of channels. The number is determined by the standard channel spacing (ITU-T grid), the optical amplifier bandwidth (region where the gain is specified within some flatness tolerance), and the possible limits imposed by fiber nonlinearities (namely four-wave mixing, or FWM). Since optical fiber amplifiers have a multiterahertz bandwidth (see Section 2.4), the WDM bit rate $B' = N \times B$ can indeed reach the *terabit/s*, with $B = 10$–40 Gbit/s as the channel rate.

Several fiber pairs are generally included in the trunk cable, which constitutes as many independent and parallel P2P links. While such extra fibers may be present in the cable, the fiber-pair links and their line amplifiers are not always activated and installed. One refers to these inactivated links, which may represent 90–95% of the potential capacity, as *"dark fiber."* When a new link must be activated, new line amplifiers are installed in the repeater stations and new transponders are connected in the terminals. In the WDM-engineering jargon, one refers to such an operation as "lighting the dark fiber." Generally, two fiber pairs are activated (or "lit") in a given trunk. This is for reasons of *network protection*, as explained in a dedicated subsection below. This description applies to the case of *terrestrial systems*. In terrestrial systems, the repeater stations are always accessible in small buildings or cabinets called "huts," which belong to the system operator. Upgrading the system link by lighting the dark fiber in the terminals and in the huts is straightforward and can be done at any time. In the case of *undersea systems*, the situation is quite different. The cable with its in-line optical repeaters is buried in the ground (down to 2,000 m) or rests upon the seafloor at depths which can reach 9,500 m. The repeaters thus remain out of reach! While for repair purposes it is technically possible to bring the cable or any specific repeater to the surface, even from such incredible sea depths, the number of optical amplifiers in the repeaters cannot be increased. Thus the fiber pairs are always lit from the first day of system turn-on and powering. The system capacity upgrade is made only from the two terminals. This upgrade simply consists in adding new WDM channels by plugging in transponders. Additionally, the channel bit rate can be increased (two-fold for 2.5–10 Gbit/s or fourfold for 2.5 Gbit/s), but this approach has now become obsolete since sub-marine systems are designed to operate within their own capacity limits. The correct view is that an undersea system has a nominal capacity of $B' = N \times B$, but not all the N channels are initially activated. Instead, the channels are progressively *provisioned* according to the increase of traffic demand. Terrestrial systems do operate the same way, except that they can leave their dark fiber plant without any optical repeater until it becomes necessary to lit the fiber pairs.

The P2P-WDM links are designed to provide "error-free" transmission over any wavelength channel. In the early days of lightwave systems, this expression precisely meant that the BER should be at or under 10^{-9}. Until the end of the 1990s, record transmission distances and/or WDM capacities were acknowledged with the condition that all channels have BER $\leq 10^{-9}$. The introduction of *forward-error correction* (FEC) in recent years, has made it possible to considerably extend the capacities and transmission distances of P2P-WDM links. The new record transmission performances were obtained with receiver BERs as high as 10^{-4} (or even in the lower 10^{-3}) for certain WDM channels before FEC processing. Note that FEC requires *redundancy* (or parity) bits, to which correspond a certain amount of bandwidth *overhead*. For a 10-Gbit/s channel rate, for instance, a 5% overhead means that the signal is transmitted at 10.5 Gbit/s. Desurvire (2004) provides a detailed introduction to FEC processing.

How are P2P-WDM links ever designed? This is a vast question which addresses different levels of fundamental and practical issues. In WDM link design, the way to proceed can be decomposed as follows:

> *Link parameter constraints:* The system length L is usually fixed by geography considerations, as determined by intercity to continental distances in terrestrial networks, and transoceanic distances in submarine networks; in terrestrial systems, there is no flexibility in the amplifier spacing z_a, which is determined by the possible "hut" locations (typically 90–120 km apart); in submarine systems, there is inherent flexibility, but for transoceanic

distances (6,500–12,000 km), the spacing must be reduced to 40–60 km to reduce the ASE noise; for terrestrial systems, another constraint is the use of pre-installed dark fiber, which has specific characteristics (loss, dispersion, PMD);

Basic transmission performance: For a prespecified channel bitrate B (e.g., 2.5/10/40 Gbit/s) and modulation format (e.g., NRZ, CRZ, RZ, PSK,...), and given the link length and amplifier spacing, the channel power P_{ch} required to achieve a target BER is straightforward to calculate; the result corresponds to a "linear" transmission system which is only impaired by ASE; regardless of other factors, the system cannot be operated with a better performance than that determined by the ideal, linear case (unless 2R/3R in-line regeneration schemes be implemented); this level of analysis replaces the old-fashioned "power budget," which is still in use in wireless (mobile and satellite) links; the reason is that for optically amplified links, power is not an issue since the signal is periodically repowered;

Numerical simulations: The effect of fiber nonlinearities (and in some cases, PMD) must then be taken into account; an optimization strategy must be found between channel power and dispersion management (periodicity, strength of dispersion jumps, amounts of pre- and post-compensation, effect of dispersion slope); the goal is to obtain an ideally uniform BER performance for all channels, since it is generally wavelength-dependent and there is no unique optimization solution; the simulations are performed by extensive computer calculations using long pseudo-random bit sequences (PRBS) for truly accurate statistical predictions of the channel BER; as an over-riding rule the simulation result should include the effect of component imperfections or random deviations from nominal "specs"; optimal operating points which depend upon some critical parameter value (e.g., dispersion to be accurate within 0.1 ps/nm/km); this may represent a difficult, multi-dimensional optimization problem;

Advanced simulations: To be complete and dependable, numerical simulations should also provide two scenarios: one at the system's beginning of life (BOL), where components are ideally specified, and one at the system's end of life (EOL) where the performance has been degraded by a variety of component-aging effects, causing the BER to increase to an unacceptable level; the BER difference between best BOL and EOL performance is called the system margin; simulations should finally address various issues of accidental component failure, which may not be catastrophic to the system's operation but temporarily impact (until repair or replacement) the signal quality in some of the channels or all of them;

Subsystem design: This level of analysis concerns three separate aspects:
- Terminals: electrical/optical packaging and volume integration, architecture, signaling/monitoring, redundancy, connection to TDM/FEC and SDH/SONET subsystems;
- Transmission fiber: dispersion management with alternative negative-dispersion segments, special reduced-dispersion-slope and/or large-effective-area fibers, upgrade of fiber plant with high loss or PMD; and

- Line components: amplifier gain and noise characteristics, pump modules, equalizing filters, dispersion-compensating elements, and in some cases OADM, with electrical powering and monitoring/signaling/control requirements;

Network and operations analysis: Integration of the link into the larger (local) network, traffic estimates with channel provisioning scenarios, link protection schemes, supervision and management, reliability and maintenance issues, options for future link upgrade (e.g., transition from STM-16 to STM-64 from terminals).

This preceding description illustrates the high complexity of P2P-WDM link design, for which nothing should be left to chance. The fact that links are inherently protected against accidental failures (cable breaks, accelerated aging, natural catastrophes, manipulation errors or terrorism) should not preclude the fact that they must (in theory) be absolutely reliable. This is particularly true for submarine systems where cable repairs involve highly expensive marine operations, which may take days to complete. From the raw economic standpoint, one may calculate the loss of operator profit caused by one single day of system shut-down having a normal average of 1,000–100,000 telephone connections at 1–3 €/U.S.$ per minute! This is the reason why all high-capacity WDM systems are built redundant, i.e., with at least two active fiber pairs per link. In case of such accidents, the system automatically switches to the spare link while maintaining the P2P connection. In case of cable break (or any event which interrupts the traffic on this specific P2P link), the network activates a higher form of protection, based upon a ring topology, as further described in Section 2.6.4.

These P2P-WDM links define yet another layer for the purposes of network supervision and management, which has been called, by ITU-T, the *optical layer*. Such an optical layer operates "under" the client layer which here is SDH/SONET. See more on this issue in the next subsection, which clarifies the definitions of the SDH/SONET P2P link and its different segments defined by intermediate ADM/OADM stages.

It is beyond the scope of this book to describe the state of the art of P2P terrestrial and sub-marine WDM transmission, in terms of best capacity × distance (C × D) figures. For advanced and detailed descriptions of the state of the art in terrestrial and undersea systems, see Chesnoy (2002), Desurvire et al. (2002), and Kaminow and Li (2002). Outside such a context, even a brief description of the state of the art would be complex, subjective, and potentially misleading, for two main reasons:

1. The transmission performance (as expressed in terms of C × D or of maximum WDM capacity at given distance) is defined in three different types of system: (a) laboratory (amplified recirculating loop, typically 250–500 km long); (b) straight-line testbed (typically 1,000–5,000 km); and (c) commercial/installed systems (6,500–12,000 km); these systems do not have the same requirements for stability/reliability and BER margin, although the performance gap between installed systems and laboratory systems has so far been bridged within 1–2 years; finally, commercial systems use multiple fiber pairs (for both working and

restoration purposes), which tends to blur realistic performance comparisons with the experimental/research domain;

2. Most experimental results utilize very different transmission technologies and strategies (modulation format, channel bit rate, amplifier spacing, amplification bands, linear or nonlinear propagation, dispersion compensation, forward error correction). This diversity of approach defines a best-performance "*cloud*" rather than a so-called "*Moore's law*" for optical systems, for this reason, it is hard to predict any future trends in commercial P2P systems when considering only record experimental performance, since the issues of technology cost, system margin and reliability have gained increased importance.

These reservations being stated, we can say that the current status of P2P-WDM transmission corresponds to a C × D performance slightly above *10 Petabit.km/s*, representing the transmission of 100 channels at 10 Gbit/s (1 Tbit/s) over 10,000 km (10,000 Tbit = 10 Pbit), or 250 such channels over 4,000 km. Interestingly (but not surprisingly), the two distances of 4,000 km and 10,000 km cover transcontinental (e.g., United States) and transoceanic (e.g., Pacific) scales. Such a level of achievement nearly represents a *thousand-fold performance increase over 10 years*. This 10-year evolution exactly corresponds to the introduction of practical optical amplifiers (i.e., EDFA). Popular belief has it that such a "bandwidth explosion" has been caused by the "advent of WDM." In reality, there has been no more C × D growth than over the past 30 years! Rather, it has remained the same over this whole period and is equal to a *ten-fold increase every four years*, as shown from the data in Figure 2.93. The data in the figure represent the record C × D performance according to the succession of key technologies, from 0.8-μm multimode-fiber to 1.3-μm SMF, 1.5-μm DSF, coherent transmission, EDFA, WDM, dispersion management, FEC after 1999 and Raman. The round curves define technology envelopes, better known in linear scale as "*S curves*," which are observed in practically all technology fields. The introduction of EDFA corresponds to a "*fifth generation*" in the telecom field, as we called it at the time (*Scientific American*, January 1992). The immediate exploitation of the EDFA bandwidth, largely oversized for single-channel transmission, corresponds to the famous "WDM revolution." The other fiber-based and electronic-based technologies of dispersion management and error correction just came along serendipitously to keep the C × D pace of growth to the same constant rate of × 10/4 year or $10^{Y/4}$ (Y = year difference between two performance quotes). Over the scale of 30 years, which means practically since the birth of optical communications in 1970, such a growth has represented an increase of seven orders of magnitude, as shown in the figure. The recent WDM "bandwidth explosion" only reflects the smooth continuation of this exponential growth.

The observation of an exponential law for the growth of WDM performance has led analysts to draw a parallel with the famous *Moore's law* of the electronics (IC) industry (for an authoritative description of the history of Moore's law, see paper from Tuomi, which can be accessed from the URL listed in the Bibliography). In order to discuss any parallel between Moore's law and the exponential growth of WDM, it is useful to provide some background and figures for the former, which have seemingly received little attention.

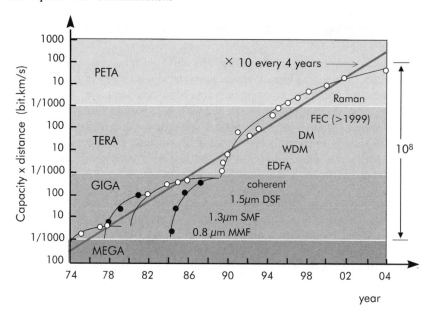

FIGURE 2.93 Increase of capacity × distance performance over the past 30 years, highlighting the different technology generations (from left to right: 0.8-μm multimode fiber, 1.3-μm SMF, 1.55-μM DSF, coherent transmission, EDFA, WDM, dispersion management, FEC and Raman). The rate of growth (straight line), corresponds to a ten-fold increase every 4 years (after Desurvire, 1994, with data complements from Desurvire et al., 2002 and end 2003 status).

In a 1965 paper, G. Moore (cofounder of Intel) reportedly stated that the number of transistors that could be integrated per square inch was *doubling every 1.5 years* (18 months) since the IC was invented. In 1975, the figure was revised to a *doubling time of 2.0 years (24 months)*, taking into account and balancing out new considerations of chip complexity, cost and packaging issues. As a self-fulfilling prophecy, Moore's law was absolutely and remarkably followed by the IC industry, at least between 1975 and 1990. From then on to the 2000s, the trend has lowered to a *doubling time of 2.5 years* (30 months). The current doubling-time target (*International Technology Roadmap for Semiconductors*, ITRS) is 3.0 years (36 months) while the actual performance is 4.5 years (54 months). To further complicate the analysis, Moore's law was later interpreted in terms of *"doubling chip processing or computing power every 18 months"* or *"halving the cost"* of said power in said period, which was not observed with any reasonable accuracy. Although these different corrections to Moore's law do not appear at first sight so important, they actually represent a tremendous difference in performance outcome when accumulated over four to five *decades* (the currently accepted validity range of Moore's law, i.e., 1965–2015). Consider indeed that the factors $2^{40/1.5}/2^{40/4.5}$ and $2^{50/1.5}/2^{50/4.5}$ represent performance-estimate differences between 4 and 5 orders of magnitude, respectively! Our own conclusion is that

Moore's law remains applicable and valid as long as the different *technology cycles* and their physical limits are taken into account, which means that the exponent must be periodically revised and lowered in order to fit the technology trend, which (to recall) may yet concern five years to a full decade of industry progress.

Consider now the bandwidth (or C × D) growth in WDM systems. Our $10^{Y/4}$ heuristic figure, which covered *two decades* (1974–1994), corresponds to a *doubling every 1.2 years*, or 14.5 months ($10^{Y/4} = 2^{Y/1.2}$). Since then, this law has been tentatively named by computer-industry scientists *"teraKIPS"* (terabit-kilometer per second merit figure). In view of current progress, this law should be now called "petaKIPS," but this is not the most important issue. Based upon the accurate data shown in Figure 2.92, which now cover *three decades*, the teraKIPS/petaKIPS law can be put into either of the following two formulations (reference points being 1994 ≡ 0.1 Pbit · km/s, 2001 ≡ 10 Pbit · km/s):

$$C \times D_{\text{Pbit·km/s}} = 5.178 \times (1.9306)^{Y-2000} \qquad (2.149)$$

where Y is the absolute year (e.g., 2003), or

$$C \times D_{\text{Pbit·km/s}} = 5.178 \times 2^{\frac{Y}{1.053}} \qquad (2.150)$$

where Y represents the difference between the prediction year and the reference year 2000.

We observe from the last formula that the doubling occurs at an interval of 1.05 years (12.5 months), representing a small correction from our initial 1994 prediction (1.2 years). The magic feature is that the corrected number (in comparable evolution times) happens to champion the early Moore's law, with a doubling every year, as opposed to every two years! Thus the argument that WDM growth follows more or less exactly Moore's law principles seems to be strongly consolidated by this factual analysis. Yet, replacing Moore's law into the previous perspective, we ought to expect the same type of exponent corrections as technology cycles unfold, taking into account increasing complexity, cost and integration issues. Under this reference for guidance, the rate of growth of WDM system capacitiy should roll off over the next two decades according to the same trend, namely that performance will only double by two ($2^{Y/2}$) to four ($2^{Y/4}$) year increments. Starting from 2003 (the date of this book's preparation) with an approximate performance of 40 Pbit · km/s, the prediction for 2015 ("end of Moore's law") is calculated to be $40 \times 2^{(2015-2003)/4} = 320$ Pbit · km/s. This eight-fold improvement factor over the next 12 years, seems to be reasonably conservative in view of the possibilities for expanding the amplification bands (2× to 4×) and the number of working fiber pairs in transmission cables (2× to 8×). Even being so conservatively corrected, the teraKIPS or petaKIPS law predicts a bright future to P2P-WDM transmission capacities.

2.6.3 WDM Network Topology and Wavelength Connectivity

The first generation of optical networks used WDM as a practical means to upgrade the P2P capacity of optical-fiber trunks. The real beginning of *WDM networks* came with the introduction of *wavelength routing* and *optical add-drop multiplexing* (optical ADM or OADM). The principles of these two functions are illustrated in Figure 2.94 and explained hereafter. In the description, we will use the word "terminal" to designate the endpoints or communicating stations. The term "node" will be reserved to designate the point where the terminal connects to the network. For simplicity, we will also use throughout this subsection the general term OXC to differentially designate WSC (wavelength-selective XC) and WIC (wavelength-interchanging XC, which includes wavelength conversion), except when this second feature is specifically discussed.

In *wavelength routing*, the channel wavelength defines a specific network path and a bi-directional connection between two given terminal endpoints. The network nodes are configured to passively route the wavelengths according to a preassigned *routing table*. For instance, it is seen from the figure that the routing tables make the channel at λ_1 connect terminal 1 to terminal 2, while the channel λ_2 connects terminal 1 to terminal 4, and so on. We observe that the same wavelength (λ_1) can be *reused* to create other paths between other terminals (e.g., connecting terminals 3 and 4). Given the number of network nodes and terminals, all possible paths can be created in a nonblocking way while using a minimum set of wavelength channels through this principle of wavelength reuse. The network nodes thus act as passive *wavelength routers*. If their routing table is fixed, the wavelengths define a permanent connection path between terminals, forming many independent networks overlaid on top of each other within a unique physical-path mesh. Each terminal can be

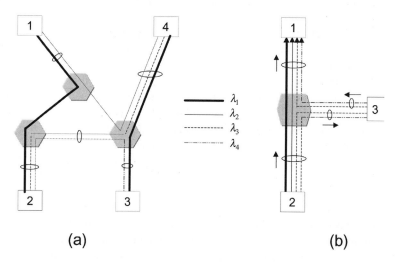

FIGURE 2.94 Principle of wavelength routing (a) and optical add-drop multiplexing (b). The ellipses designate the physical fiber paths.

equipped with a tunable transmitter and receiver in order to be able to set connections with any other network terminal. But this approach is technically expensive and not efficient in bandwidth use. It also becomes complex (or intractable) to manage when the number of network nodes and terminals becomes large. The most efficient use of the network resource is obtained with dynamically *reconfigurable* wavelength routers. Such a reconfigurability makes it possible to create new paths according to evolving demand, the network acting like a giant, nonblocking wavelength switch. It also makes it possible to increase the traffic between a subset of terminals by assigning to them additional wavelength channels, according to momentary demand. Networks based upon either fixed or reconfigurable wavelength connectivity are called *wavelength-routed networks* or *wavelength-switched networks*, respectively.

Optical ADM is conceptually close to wavelength routing, but serves a different purpose. As seen from Figure 2.94, it consists in extracting (or dropping) a subgroup of wavelength channels from the multiplex to route them to a different or local endpoint. The same subgroup of wavelength is reassigned to insert (or add) traffic to the multiplex from this endpoint. In the example of the figure, the traffic carried by channels at λ_3, λ_4 from terminal 2 is dropped and routed *to* terminal 3. Conversely, the traffic *from* terminal 3 at the same wavelengths λ_3, λ_4 is added to the multiplex. Note that the OADM function is also *bidirectional*. Namely, traffic from terminal 3 to terminal 2 is added, and traffic from terminal 1 to terminal 3 is dropped. One may better visualize such a bidirectional OADM operation by removing all arrows from the figure. The wavelength channels thus become bidirectional, but as with wavelength routing, the corresponding physical connections are ensured by *fiber pairs*. To make a basic comparison, an OADM works the same way as a highway exit: small local traffic comes in and out, while the larger, long-distance traffic remains unaffected, both in terms of route and volume.

One of the most effective applications of OADM concerns *wavelength-routed ring networks*, whether local-area (LAN), metropolitan-area (MAN) or wide-area (WAN). As illustrated in Figure 2.95, each of the ring terminals is assigned one unique wavelength. For instance, terminal 1 is assigned wavelength λ_1, terminal 2 is assigned wavelength λ_2, and so on. The links are fully bi-directional. Due to the ring topology, only one single optical fiber is needed, as illustrated in Figure 2.96 in the case of a "counter-clockwise" traffic implementation. All terminals can thus individually and bidirectionally communicate with the central terminal (0), which connects the ring to the outside network. Usually, a second fiber is also installed so as to form a second overlaid ring (conceptually equivalent to reversing all arrows in Figure 2.96). This second ring is used only for redundancy and protection purposes in case of link or OADM-node failure (see Section 2.6.4). It can also be used to carry nonpriority traffic, to be preempted in case the protection is activated and the priority-traffic switched onto it. The wavelength-routed ring-network implementation shown in Figures 2.95 and 2.96 does not allow the terminals to communicate with each other (e.g., 1 with 2). But such a functionality can be introduced by at least four possible approaches:

Including an extra wavelength channel (λ_0) to be shared by all terminals; the bandwidth or channel use can be allocated for instance using "token-ring," Ethernet and other varieties of LAN protocols;

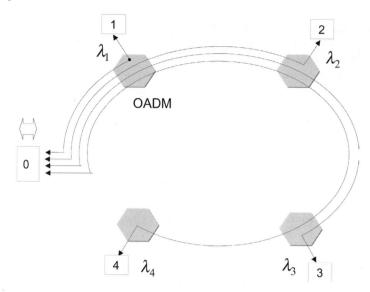

FIGURE 2.95 Implementation of optical add-drop multiplexing (OADM) in ring WDM networks, with central station (0) connected to four terminals (1-2-3-4). The bidirectional arrows correspond to two different fiber paths (see next figure).

- Adding more wavelengths: terminal 1 is then assigned four new wavelengths ($\lambda_5 - \lambda_8$) to communicate to all ring terminals in exactly the same way as 0 does with wavelengths $\lambda_1 - \lambda_4$, and so on, for each of the terminals;
- Connecting the OXC of terminal 0 with a *digital cross-connect* (DXC or DCS, see below); the terminal recognizes whether the received traffic at any wavelength is destined for internal use (1–4 communications) or to the outside network; in the first case, the DXC sends back the data to the OXC, which, after wavelength conversion, routes them towards the designated ring terminal;
- Adding more fibers and OADM into the ring, making many independent and overlaid OADM rings which re-use the same basic wavelength set (e.g., $\lambda_1 - \lambda_4$); each terminal thus requires multiwavelength transponders and an OXC to switch traffic to the different rings (possibly with wavelength conversion but not necessarily).

In order to adapt to changing traffic demand or requirements, the OADM function can be made *dynamically reconfigurable*. This should not be confused with the function of the optical cross-connect (OXC), which selectively switches wavelengths into different *physical input/output paths* (see previous section). Concerning a dynamic 2×2 OXC, the two principles yet become conceptually identical. Figure 2.97 illustrates how a 2×2 OXC can be implemented for providing OADM functionality, using simple 2×2 cross-bar switches. Such a configuration makes it possible to switch *any* wavelength channel from/to the mainstream path to/from a local terminal. For a given wavelength channel, such an operation is

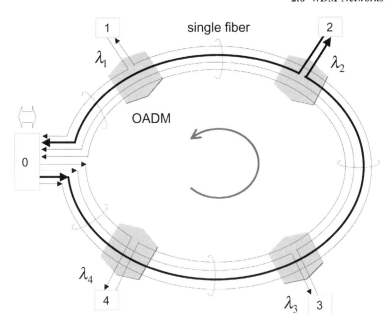

FIGURE 2.96 Physical fiber path followed by bidirectional WDM signals in OADM ring network, with counter-clockwise implementation. To clarify a given path, the channel at wavelength λ_2 is emphasized with a bold line. The ellipses designate single-fiber trunk connections.

performed by turning the corresponding switch to the ON (or cross) state. Since there is no reason why OADM should be limited to a single (bidirectional) entry/exit path, one may see dynamic OADM as performing the same function as an OXC. Referring to Figure 2.84, one may attribute the mainstream traffic to the paths A–E and B–F and the local add/drop traffic to the paths C–G and D–H.

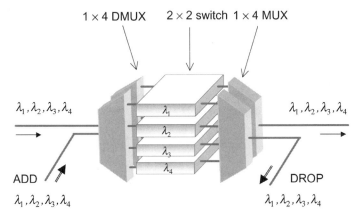

FIGURE 2.97 Implementation of dynamic optical add-drop multiplexing (OADM) with a 2×2 optical cross-connect (OXC) with four wavelengths.

As we have seen, OXC can also be equipped with the function of *wavelength conversion*. One can thus connect two WDM networks (e.g., A and B) having independent wavelength-routing or wavelength-switching configurations, as illustrated in Figure 2.98. The combination of passive/fixed OADM, dynamically reconfigurable OADM and (wavelength-converting) OXC provides a new tool to manage multipoint traffic without any electronic conversion of the optical data. If the OXC is of the WSC type (no wavelength conversion), the channel path and the network are said to be *transparent*. One also refers to the approach as *transparent routing*. Transparent routing represents the simplest and least expensive approach, with a virtually unlimited bandwidth and bit-rate/format insensitivity. The transmission distance (or number of segments/nodes traversed) is however limited by the effect of SNR degradation. In large networks having multiple operators and vendors, network transparency is in fact difficult to implement since the same wavelength must be attributed to the communicating end-to-end circuit, regardless of considerations of local traffic management and wavelength-channel allocation. This approach also causes problems of "wavelength blocking," where the network may be temporarily unable to find a free wavelength path to assign to a channel. In contrast, the OXC of WIC type (with wavelength conversion or interchange capability) offers the required wavelength flexibility, with 2R/3R optical regeneration as an additional feature (see Section 2.5). Wavelength blocking is no longer an issue, and due to signal regeneration the transmission distance is virtually unlimited. As previously described, the OXC/WIC can be either all-optical or electronic (internal O/E and

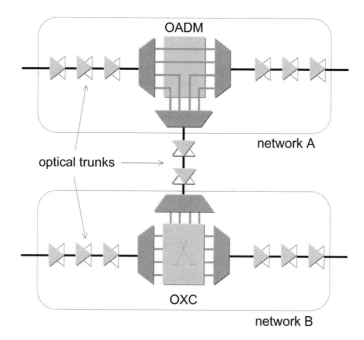

FIGURE 2.98 Connection of two networks (A and B) via optical add-drop multiplexer (OADM) and optical cross-connect (OXC) as node interfaces.

E/O conversion). In the second case, the term OXC refers to the "black-box" functionality of the approach, which from the outside works the same as an all-optical OXC. When wavelength conversion occurs in either case, the network is said to be *opaque*, as opposed to "transparent." Such an opacity is advantageous for considerations of network optimization (maximum reuse of wavelength, shortest path assignment) and interoperability (nonblocking operation in multivendor environment). The bandwidth is however limited by the speed of the electronic transceivers (or all-optical wavelength conversion), the routing function is sensitive to the bit rate (as not upgradable) and the implementation cost is substantially higher. We see that *transparent* and *opaque WDM networks* have both intrinsic advantages and limitations. One could summarize the picture by stating that fully transparent networks are not practical and fully opaque networks are inefficient and costly. The choice between the two approaches must therefore be guided by trade-offs which take into account the progress in OXC technologies, the "legacy" of the already-existing network infrastructure and interfaces, and finally, the need for interoperability at massive scale. See more on this issue in the following text addressing the "digital wrapper."

The unique capabilities of wavelength-switched networks with fixed or reconfigurable OXC is managed as a new *optical layer*. From Section 2.6.2, this optical layer is within the physical "layer 1" of the (OSI) network model, and lies under the SDH/SONET layer. A P2P SDH/SONET link is actually managed according to a suite of sublayers. We shall clarify here their functions, from top to bottom, as follows:

> *"Path" sublayer*: Manages the full end-to-end path connection between two SDH/SONET terminals;
>
> *"Line" sublayer*: Manages the multiplexing of several path-layer connections into a link separating two intermediate SDH/SONET nodes;
>
> *"Section" sublayer*: Manages the line-layer connection between a SDH/SONET node and any ADM/OADM or between two OADM;
>
> *"Physical" sublayer*: Manages the section connection between two link regenerators.

The bottom physical sublayer of this first generation of SDH/SONET systems has now become the *optical layer*. To further refine the network management, ITU-T has then divided this optical layer into three more sublayers which are called, from top to bottom:

1. *Optical channel (OC or OCH/OCh)*;
2. *Optical multiplex section* (OMS);
3. *Optical transmission section* (OTS), sometimes called *optical amplifier section (OAS)*.

As a result of this overall decomposition, the OSI network layer-1 with SDH/SONET as the protocol now forms a stack of six different sublayers, which is illustrated in Figure 2.99. The fine structure of the OCH sublayer, also shown in Figure 2.99, will be discussed in relation to the "digital wrapper" concept (see following description).

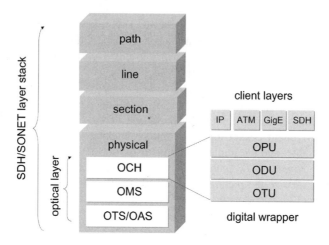

FIGURE 2.99 Network-management stack of point-to-point SDH/SONET links, showing different sublayers with its "optical layer" and own layered structure at bottom (OCH = optical channel, OMS = optical multiplex section (OMS), OTS = optical transmission section, OAS = optical amplifier section). The right side shows the "digital wrapper" hierarchy In side the optical layer, as discussed further down in text (OPU = optical channel payload unit, ODU = optical channel data unit, OTU = optical channel transport unit).

The optical layer management does not recognize the data contents, payloads, or bit rates of the SDH/SONET frames, but only the channel wavelengths and corresponding paths, which ITU-T called *lightpaths*. The function of recognizing and processing the data contents belongs to the upper three SDH/SONET layers previously listed. It is performed by *digital cross-connects* (DXC, also called DCS or EXC). These DXC are what we have called the ring "terminals" in the previous description of WDM ring networks. We may now refer to Figure 2.100A in order to visualize how the first generation of optical SDH/SONET rings operated with DXC and electronic ADM. This figure shows that the network is actually built from different optical rings that are interconnected by electrical ADM nodes (also called EADM). The electrical ADM node extracts and reinserts VC payloads from SDH/SONET frames according to destination and traffic needs. The "drop" data of ADM are reconstructed SDH/SONET frames of the immediately lower hierarchy level (e.g., STM-4 formed from STM-16). From the example of Figure 2.100A, it is seen that the top ring supports STM-16/OC-48 (2.5-Gbit/s) traffic. The ADM in this ring extract/insert STM-4/OC-12 (622-Mbit/s) tributaries from/into this higher level of traffic. Smaller local rings perform the same function down to the STM-1/OC-3 (155-Mbit/s) level. The function of the DXC is to switch the data from/to local networks (e.g., PSTN, CO, LAN) or from/to larger external networks (MAN, WAN), establishing SDH/SONET *connections*. The DXC also performs *traffic grooming*, the action of splitting, combining or re-arranging a large number of low-speed connections (e.g., DS3/T3/J3 down to DS1/T1/J1).

We can redraw a figure similar to Figure 2.100A, but corresponding now to the second generation of SDH/SONET rings, namely based upon WDM, OADM, and

2.6 WDM Networks 311

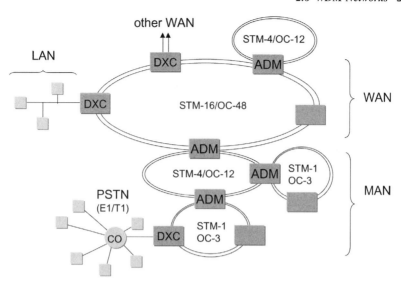

FIGURE 2.100A SDH/SONET ring topology, showing different subnetworks and traffic levels. The sub-network rings that handle similar traffic types are linked through add-drop multiplexers (ADM), while the others use digital cross-connects (DXC) as interfaces.

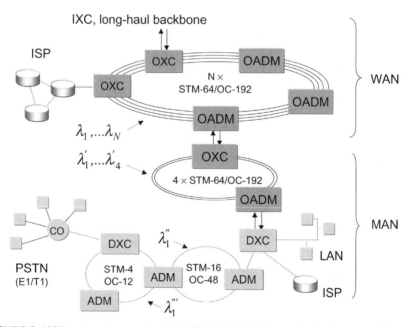

FIGURE 2.100B Second-generation SDH/SONET WDM network, showing central ring and connections to external networks (ISP, IXC, long-haul backbone) forming a wide-area network (WAN), and secondary ring with connections to local networks (PSTN, LAN, ISP) forming a metropolitan-area network (MAN). Acronyms: ISP = Internet service provider, IXC = inter-exchange carrier, OXC = optical cross-connect, OADM = optical add-drop multiplexer, DXC = digital cross-connect, ADM = electronic add-drop multiplexer, PSTN = public-switched telephone network, CO = central office, LAN = local-area network.

OXC technologies at metropolitan-area (MAN) and wide-area (WAN) network levels. Figure 2.100B shows such a network with its different hierarchy levels. The wavelength channels $(\lambda_1, \ldots, \lambda_N)$ from the main ring at $N \times 10$ Gbit/s are extracted/inserted via OADM, possibly with wavelength conversion as shown in the example of the figure. This conversion enables wavelength reuse from the ITU-T grid $(\lambda'_1, \ldots, \lambda'_4$ being any wavelengths common to the previous set) and dynamic wavelength allocation according to channel loading in the main ring. Here, the secondary ring operates at 4×10 G bit/s, meaning no change in the SDH/SONET hierarchy. Thus OADM plays the same role as OADM for wavelength channels, except that the hierarchy level is conserved and the SDH/SONET frames are unchanged. Alternatively, a single 10-Gbit/s wavelength channel can be extracted from the main ring with an OADM and then transformed into 4×2.5-Gbit/s WDM signals via a DXC (see Figure 2.100B at bottom right). The rest of the network is similar to that of the previous generation of SDH/SONET rings with traffic grooming all the way down to low-level payloads (e.g., E1/T1).

Any group of neighboring network nodes which are connected through optical fibers can be upgraded to operate as a WDM ring. The concept of "ring" thus applies to any set of WDM nodes forming a triangle, square or any irregular polygon. The OXC/OADM technique, which makes it possible to connect different subnetworks together and create wavelength paths forming many overlaid networks is applicable to optical networks of all possible sizes, namely from LAN and MAN to WAN, including backbone/terrestrial and undersea systems. Figure 2.101 illustrates the concept of a global network with its different coverage areas, represented in the form of network planes operating on top of each other. Contrary to what this picture suggests, such planes do not represent "network layers" in the sense of the OSI model or in the sense of the "optical layer" concept. Instead, these planes represent different levels of aggregated traffic and inter-node/repeaterless distances. The top plane, which concerns the long-haul/backbone optical network represents any continental/terrestrial or undersea system. The middle plane is that of the metropolitan-area network (MAN or "metro"), which can be accessed by a variety of carriers and service providers for high-bandwidth services, such as interoffice or private networks and internet transport. The network interface connecting a long-haul OXC (top plane) to a metropolitan OXC (middle plane) is sometimes called *optical gateway*. But in view of the "optical layer" management, such a distinction is superfluous and more a commercial overtone. The bottom plane concerns the network *access*, which provides limited bandwidth (still under the name of "broadband" services such as xDSL, FTTx, or PON) to a variety of residential and enterprise customers. The global WDM network can thus be interpreted as the superimposition of a high-speed, transparent *backbone* plane, a *metro* plane that does the traffic and multi-vendor/multi-protocol payload aggregation, and at the bottom, the enterprise/residential *access* plane, from which both telephone and broadband services (e.g., xDSL, FTTx) can be accessed. One may also decompose the metro plane into an upper *"metro core"* plane, a middle *"metro edge"* plane, and a lower *"metro access"* plane, according to terms of bandwidth aggregation (10–40 Gbit/s down to 2.5 Gbit/s per WDM channel), network topologies

2.6 WDM Networks   313

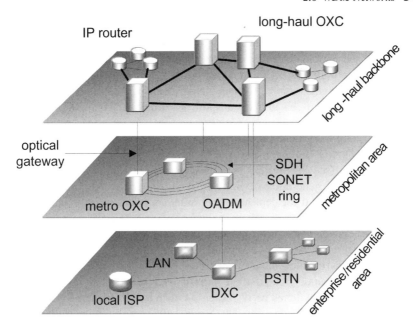

FIGURE 2.101 Vertical representation of global WDM network according to three planes: long-haul/backbone (top), metropolitan-area with SDH/SONET WDM rings (middle) and local-area, enterprise/residential (bottom). The thick lines on the top plane indicate long-haul, optically amplified WDM links with highest aggregate capacities.

(mesh to ring to tree and P2P) and jurisdiction domain (50–250 km to 20–50 km to 1–20 km). However, such a decomposition into *metro*, *edge* and *access* planes is not always so clear-cut. This is because WDM networking is migrating from the metro-core to the metro edge, in order to reach towards the LAN/enterprise access level and provide seamless, high-capacity *lightpath* connections (namely, 2.5 Gbit/s per WDM channel). One refers to this last frontier of metro networks as "*metro enterprise*." The global network hierarchy is thus usually conceived according to the plane hierarchy *backbone/metro/edge/enterprise/access*, but market considerations and technology advances often tend to blur the differences (if any!) at the plane interfaces.

This previous description concerned WDM networks with SDH/SONET as the unique physical-layer (or layer-1) protocol operating at MAN and WAN levels. In reality, nothing forbids other protocols such as Ethernet or ATM (layer-2) or TCP/IP (layer-3) being transported by wavelength channels with OADM/OXC as *protocol-transparent* interfaces. Figure 3.27 in Desurvire (2004), illustrates how these protocols can relate to the WDM "optical layer" either directly (e.g., IP over WDM) or as a convoluted suite (e.g., IP over ATM over SDH over WDM). The reader may refer to that section for detailed discussion concerning these different possibilities. The driving concept behind using WDM as a direct transport medium for such higher-level protocols is to reduce the number of DXC interfaces

(O/E/O conversions) and the corresponding overhead or bandwidth "tax." It is also a matter of simplifying network operations/management, of reducing capital-investment and operations costs, of accelerating the time to deployment (getting more rapidly into the market), and of lowering service/access fees at equal bandwidth/QoS for higher competitiveness. Note that such considerations so far do *not* apply to long-haul/backbone transport, for which SDH/SONET remains the most efficient and dependable encapsulation protocol. Another reason for this difference is that long-haul (continental/transoceanic) systems are essentially P2P links for which SDH/SONET represents the most practical interface for traffic grooming/aggregation (see Section 2.6.4). Focusing on terrestrial (MAN) systems, it is clear that the first generation of SDH/SONET rings with ADM/DXC provided the natural background for deployment of WDM networks with OADM/OXC, as a second generation. The dilemma is that ATM and TCP/IP networks are also rapidly developing, jointly or separately, coming ever closer to the edge and the access boundaries of the global network picture (see in particular Chapter 1 on broadband access). It then makes every sense to introduce these protocols in the WDM layer, being understood that the purpose is to reduce the number of interfaces between two ATM switches or IP routers. Two possible implementations of this concept are illustrated in Figures 2.102 and 2.103, respectively. In the first case (Figure 2.102), a specific wavelength (or subgroup of wavelengths) is assigned to each type of traffic data. The WDM node is just an aggregator which links together the different network types (SDH/SONET, Ethernet, FDDI, ATM, IP ...) without any function other than routing and managing the traffic through the preassigned wavelength channels. The second approach (Figure 2.103) is based upon the principle of the so-called *digital wrapper* (DW), which is based upon the principle of a *protocol-independent WDM transmission* where wavelength channels are

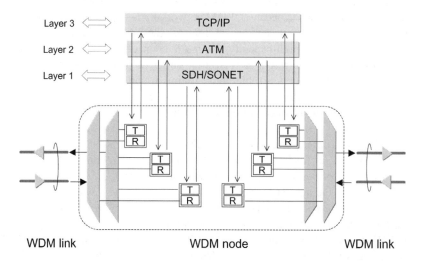

FIGURE 2.102 Implementation of protocol-transparent, bidirectional WDM node (e.g. TCP/IP, ATM and SDH/SONET): a specific wavelength channel is assigned for each type of traffic.

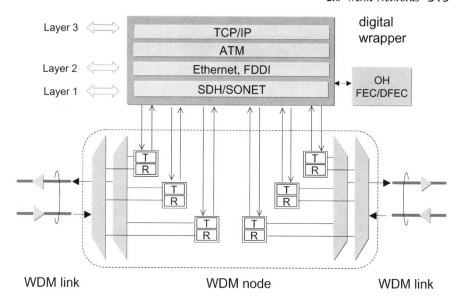

FIGURE 2.103 Implementation of protocol-transparent, bidirectional WDM node with digital wrapper: frames containing any mix of data (e.g., TCP/IP, ATM, Ethernet, FDDI, SDH/SONET) are allocated any of the wavelength channels according to changing traffic needs.

assigned according to the overall traffic demand in the node, regardless of the data-protocol type. In order to achieve such a "tour de force," the DW assembles (and disassembles) a special frame that encapsulates whatever payloads (e.g., frame-relay, SDH/SONET, Gigabit-Ethernet, FDDI, ATM, TCP/IP) may come in as "client signals." The basic DW frame, also called an *OCH container*, has been standardized as *ITU G.709*. The frame structure is shown in Figure 2.104. It consists in a 255-byte × 16-row (4,080-byte) arrangement with three fields: a 1-byte/row overhead field (OCH-OH), a 238-byte/row payload envelope (OCH-PE), and a 16-byte/row FEC trailer. The aggregation of four such frames in a 4 × 4,080-byte × 4-row arrangement constitutes a *DW superframe*.

The need for "superframing" comes from the fact that ITU-T G.872 distinguishes (yet another) three sublayers within the OCH layer, as previously illustrated in the diagram of Figure 2.99. These three sublayers are called: OPU (*optical channel payload unit*), ODU (*optical channel data unit*) and OTU (*optical channel transport unit*). Basically, the DW superframe is an OTU frame which encapsulates ODU and OPU subframes. The result is shown in Figure 2.105. It is seen that the OPU is the actual payload subframe: it carries the client signals into a 4 × 3,808-byte payload field and includes a 4 × 2-byte overhead (OPU-OH). The ODU framing appends an overhead (OPU-OH) to this OPU subframe for the purpose of optical-channel maintenance/operation, path/connection monitoring and fault alarms. The OTU framing appends an overhead for frame alignment (FA-OH) and another overhead (OTU-OH) to perform the same functions as ODU-OH at OTU-level. Finally, it completes the superframe with a 4 × 256-byte FEC trailer. We

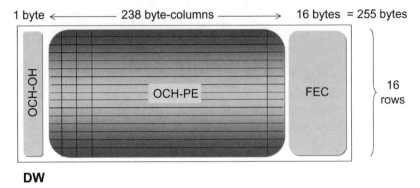

FIGURE 2.104 Basic frame structure of digital wrapper (DW) container (16 × 255 = 4,080 bytes), showing overhead field (OCh-OH), payload envelope (OCh-PE) and forward-error-correction (FEC) trailer. The payload envelope contains any client signal (e.g., frame-relay, SDH/SONET, Gigabit Ethernet, ATM or TCP/IP) by variable-size blocks.

notice that the total overhead cost of the DW, including FEC, represents only $(256 + 16)/3{,}808 = 7.1\%$ of the bandwidth, corresponding to a payload efficiency of about 93%. The remarkable feature is that such an efficiency applies to *all* types of client signals, regardless of their built-in protocol. The *"protocol-agnostic"* DW approach, whether proprietary or standardized under G.709, may represent one of the most promising solutions to fully exploit WDM networks with multivendor/multiprotocol environments, with maximum interoperability and efficient bandwidth use. The standard version G.709 also makes it possible to simplify the interfaces between different network types (WDM and non-WDM) and their management systems, thus reducing capital and operations costs. As a conclusion, one may view DW as performing the same OAM&P (*operations, administration, maintenance and provisioning*) functions for WDM networks as SDH/SONET does for single-wavelength network connections.

So far, our description has exclusively concerned *optical cross-connects* (WSC/WIC) which act as reconfigurable wavelength switches/routers. What about

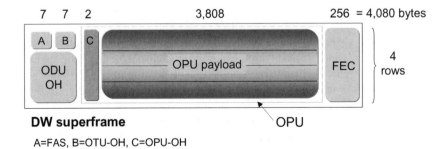

FIGURE 2.105 Structure of 4 × 4,080 bytes digital wrapper, superframe (OPU = optical channel payload unit, ODU = optical channel data unit, OTU = optical channel transport unit, FA = frame alignment, OH = overhead, FEC = forward error correction).

switching and routing *packets*? The requirements for a packet switch are substantially more complex than for a wavelength switch. In addition to being strictly nonblocking, the packet switch should include two other functions:

1. *Buffering*, since packets have random arrival times and may have conflicting destination ports, and
2. *Scheduling*, which requires the identification of destination and priorities through processing header data.

Since all-optical memories and all-optical signal processing is still quite difficult (if not impractical) to implement, these functions of buffering and scheduling must all be performed electronically. Such a feature introduces a bandwidth bottleneck, which as well as considerations of price and complexity, reduces the number of input/output ports that can be processed. This explains that the best commercial packet switches for WDM networks have throughput capabilities that are about one order of magnitude (i.e., ≤ 160 Gbit/s) lower than the OXC. Optics can play a role as a means to simplify the switch backplane connections and upscale the number of ports. The internal electronic switch (between transceivers and buffers) can also be replaced by a photonic switch in order to simplify the signal-processing stages, to reduce the number of signal connections and the overall power consumption. The realization of a truly *optical packet switch* (ideally channel- and protocol-transparent) still remains at the research stage.

2.6.4 Network Protection and Virtual-Topology Design

In this subsection, we shall describe two important aspects of WDM networks which concern the issues of *protection* (service continuity in the event of link or node failures) and *optimization* (managing and optimizing bandwidth use). Since the second issue represents an entire field in itself, rich in complex solution algorithms and intensive numerical treatment, our description will only reduce to a brief, but hopefully useful, introduction for further reader reference.

Network protection represent built-in mechanisms which guarantee the continuation of service in the event of accidental failures. These may concern *link breaks* (fiber or cable cuts due to civil engineering), or *in-line component damage* (e.g., amplifier-pump burnout from electrical/optical transients), or *node outage* (e.g., accidental power shutdown, electronic-circuit contact problems, IC burnout, laser/photodiode failure, software bugs, mistaken plug-in removal, fire, terrorism, etc...). In any of these events, the solution is to immediately re-route the traffic away from the broken link or the failed node. One refers to the action of activating the protection as *restoration*. There are two main types of protection scheme: *linear* and *ring*.

The different approaches for *linear protection* are called $1+1$, $1:1$, $1:N$ and $N:N$. The first three are illustrated in Figure 2.106 for the transmitter to receiver path (the same exists in the reverse direction). The $1+1$ approach consists in passively splitting the traffic into two fiber routes. At receiver level, a 2×1 switch selects one out of the two signals and switches in the other position in case of power failure (or fiber break). The 1:1 approach uses the second fiber as a

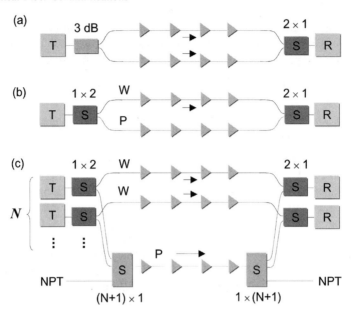

FIGURE 2.106 Possible types of linear protection from transmitter (T) to receiver (R): 1 + 1 (a); 1 : 1 (b); and 1 : N (c). Fibers carrying traffic are shown in full lines, protection fibers are shown in dashed lines (S = switch, W = working fiber, P = protection fiber, NPT = non priority traffic).

"protection fiber." If the "working fiber" breaks, two switches route the traffic through the protection fiber. The 1 : N approach is the generalization of 1 : 1 protection with N transmitters/receiver links, but with a single protection fiber common to all links. Additionally, this protection fiber can carry *nonpriority traffic*. Any link break (assumedly occurring only in one working fiber) causes the corresponding traffic to be switched through the unique protection fiber. The nonpriority traffic is thus interrupted in favor of the traffic to be maintained in priority. It is easy to visualize from Figure 2.106 and the 1 : N case what a N : N protection looks like. The two protection switches are of the $N \times N$ type (plus an extra port for non-priority traffic) and the protection path is made of N independent fibers. Thus the entire traffic of the N transmission lines can be switched into the protection-fiber path. For both 1 : N and N : N, it is safer (when possible) to have the protection fiber(s) located in a separate cable, however following the same physical route or duct. In this case, any accidental break of the *working-fibers cable* would not also disable the protection cable, unless the two cables were cut at once (a less likely event however). Note that in P2P undersea links, linear protection does not protect so much against fiber breaks as against line-component failures (such as optical amplifiers in the submerged repeaters). This is because when the cable happens to break (under the rough treatment of supertanker anchors and fishing-trailer ploughs), it usually cuts all fibers contained in the cable! For these systems, linear protection is therefore necessary but highly insufficient, as further discussed in the following text.

Clearly, linear protection is maximum with the $N:N$ approach, but it is also the most expensive to implement. In contrast, the $1:N$ protection is cost-effective, since only one protection fiber is shared by the N links, especially if N is large (fiber-optic cables can contain from 4–12 to over 100 fibers). For links with $1+1$ protection, the switching operation is automatically triggered when a receiver detects any optical power failure. For $1:N$ and $N:N$ protections, the receiving end must inform the transmitting end of which line has failed and which transmitter to switch to the protection fiber. For uni-directional systems, this requires a network signaling algorithm called *automatic protection switching* (APS). For bidirectional systems, the failure is immediately sensed by the two end transponders. Depending upon the system complexity and switch technology (usually mechanical), the response time of the protection loop is of the order of several tens of milliseconds at the minimum and 300 ms as a maximum. At the human scale, this keeps voice and video traffic seemingly uninterrupted, despite the fractional loss of signal payload during the whole operation.

The linear protection approach also applies to node failures such as previously mentioned. In high-capacity systems, DXC and OXC nodes may just be duplicated with $1:1$ protection. In WDM O/E/O interfaces, both transmitters and receivers should also use $1:1$ protection. This works the same way as shown in Figure 2.106 (b), expect that the two end switches are 4×4 and there are two independent transmitters and receivers ready to be activated from each side. To reduce the WDM terminal cost, a $1:N$ protection may be implemented instead. From the transmitter end, such a protection requires a tunable laser which automatically locks onto the failing channel wavelength. For maintenance purposes, and to be on the safe side when there are tens to hundreds of transmitting WDM channels, more than one tunable laser may be included in the terminals, corresponding to a $M:N$ protection approach (e.g., $M \leq N/10$).

We consider next the *ring protection* scheme. In Chapter 2, Section 2.5 of Desurvire (2004), it is shown that multifiber rings have an intrinsic capability for protection. This capability is illustrated in Figure 2.27, which shows that a ring containing one working fiber pair and one protection fiber pair can ensure continuation of traffic by the same node-switching principle as in the $1:1$ linear approach, hence the name of "*self-healing*" (SH) rings. The SH principle can be applied with unidirectional or bidirectional protection, and the corresponding rings are called USHR and BSHR, respectively. Similarly to the previous description concerning linear protection, one distinguishes *path-switched* protection (analogous to $1+1$) and *line-switched* protection (analogous to $1:1$). In path-switched protection, traffic is always split into two paths; failure of one fiber causes the receiving node to switch to the other fiber. In line-switched protection, the traffic is carried by a "working fiber"; in case of failure, it is switched to the "protection fiber," previously unused, or used for nonpriority traffic. In a ring, both protection schemes can be implemented either unidirectionally or bidirectionally. These are named as follows:

UPSR, for unidirectional path-switched ring (also called by ITU-T *subnetwork connection protection*, or SNCP;

BPSR, for bidirectional path-switched ring;

ULSR, for unidirectional line-switched ring;

BLSR, for bidirectional line-switched ring.

Since both schemes can be implemented with either two or four fibers, this makes eight basic possibilities, which are named UPSR/2, UPSR/4, BPSR/2, BPSR/4, ULSR/2, ULSR/4, BLSR/2 and BLSR/4. Figures 2.107 and 2.108 illustrate the 2-fiber implementation of UPSR and BPSR protection principles, respectively, before and after fiber break between two nodes. Concerning UPSR and ULSR, we observe that there is no difference with a P2P linear protection between two given nodes, as based upon the $1+1$ or $1:1$ schemes, respectively. For BPSR and BLSR, the protection is seen from the two figures to involve traffic re-routing though the entire ring. The scheme BPSR/2 involves changing the receiver switch configurations in all nodes. The scheme BLSR/2 is more advantageous, since only two switches need to be reconfigured. The same principles can be genera lized with 4 fibers (including either $1:1$ or $1:2$ protections for BLSR/4), which leads to identical observations. We can thus conclude that regardless of the number of fibers involved, UPSR, ULSR and BLSR are the most practical protection schemes (namely, eliminating BPSR). Note that similarly to the linear-protection case, the USLR/BSLR approaches offer the potential to carry nonpriority traffic in the protection fibers, unlike with UPSR. It is important to note that these protection schemes are fully compatible with WDM signals. It is sufficient to connect each node with an OADM interface, and assign it a fixed wavelength (or several fixed wavelengths). The same applies to rings with OXC/WXC, namely using optical cross-connect nodes with or without wavelength conversion.

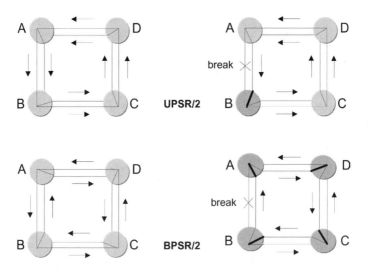

FIGURE 2.107 Principle of unidirectional (UPSR) and bidirectional (BPSR) *path-switched* ring protection (2-fiber ring): normal operation (left) and after fiber break between nodes A–B (right). Arrows indicate live traffic. Bold lines indicate new configuration of a node receiver switch after protection is activated.

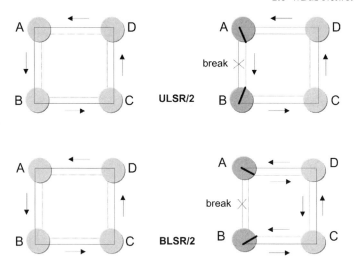

FIGURE 2.108 Principle of unidirectional (ULSR) and bidirectional (BLSR) *line-switched* ring protection (2-fiber ring): normal operation (left) and after fiber break between nodes A–B (right). Arrows indicate live traffic. Full and dashed lines indicate working fiber and protection fiber, respectively. Bold lines indicate new configuration of a node receiver switch after protection is activated.

If one uses the full capabilities of node switching and SDH/SONET, there are even more possibilities for ring protection. A first possibility, which protects *single fibers*, consists in using an outer ring for working traffic and an inner ring for protection. In the event of fiber break, the traffic is switched from the outer ring to the inner ring, with the contribution of all nodes, as Figure 2.109 (top) illustrates. Such a scheme, which operates at the OMS layer (i.e., concerning multiplexed aggregates), is known as MS-DPRING for *multiplex-section dedicated protection ring*. A second possibility, which protects *fiber pairs*, exploits the framing properties of SDH/SONET signals. Consider indeed that SDH/SONET frames carry a certain number of unit payloads. Using SDH terminology, each STM-4n ($n = 1, 4, 16 \ldots$) frame carries exactly four VC-4 payloads. Thus we may allocate two VC-4 for working traffic and leave empty two VC-4 for protection. The four working VC-4 payloads are then split between the inner and outer rings. In the event of a fiber break, the nodes connect the outer ring to the inner ring, while the two working VC-4 of one channel are loaded into the two protection VC-4 of the other channel, as illustrated in Figure 2.109 (bottom). When implemented at the OMS layer, such a scheme is known as MS-SPRING for *multiplex-section shared protection ring*. When implemented at the higher OCH layer (i.e., concerning individual optical channels), the protection is then called OCH-SPRING.

The above description concerned the protection against failures associated with fiber or span breaks. The same algorithms strictly apply to the case of *node failures*, which are even less likely to occur. However, the restoration mechanism is more complex. This is because a single node failure is detected by all the immediately

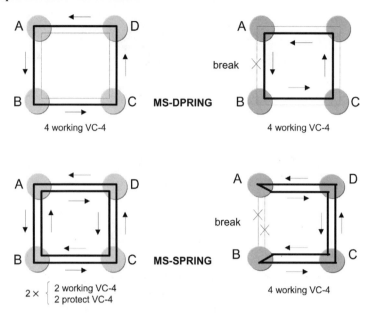

FIGURE 2.109 Two principles of ring protection with SDH/SONET signals, before (left) and after (right) fiber break (arrows and bold lines indicate live traffic direction and path, respectively). Top: dedicated protection (MS-DPRING) for unidirectional traffic with outer working ring and inner protection ring. Bottom: shared protection (MS-SPRING) for bidirectional traffic, with each signal initially having two working VC-4 and two protection VC-4.

adjacent nodes, in contrast to a fiber/span break, which is detected by only two nodes. If each adjacent node was to attempt to restore the network on its own, this would unmistakably cause routing errors and other blocking situations. Thus the node must interrogate the network and apply a common restoration solution which avoids such loopholes. The local traffic is then optimally rerouted in order to avoid the failed node.

Comparing ULSR with BLSR in Figure 2.108, or MS-DPRING with MS-SPRING in Figure 2.109, we notice that these schemes reflect two different protection strategies: the first consists in directly restoring traffic in the failed span (A–B in Figure 2.108), while the second consists in rerouting the traffic through the other spans of the ring (A–D, D–C and C–B in Figure 2.108). One refers to the first strategy as *span protection,* and to the second as *link protection* (also called *line protection*). Here, "span" means the physical route existing between two nodes (to be re-utilized upon restoration), while "link" means the shortest alternative route connecting the nodes (to be used upon restoration). While both protection schemes are applicable in the event of a *fiber* break, only link protection works in the event of a *cable* break, where the entire span is severed.

Being now familiar with the concepts of span or link protection/restoration for ring networks, we can look at the larger picture of *mesh networks*. A mesh network represents the most general topology where all nodes may be connected to all others with many by-pass paths. At geographical scales, a mesh network rather resembles a highway map, which connects the big towns through a honeycomb structure of distorted rings. This extended topology makes it possible to introduce a third network-protection concept, which is *path protection*. Figure 2.110 illustrates the difference between *span*, *link* and *path* protections. The initial and final traffic routes between two nodes are called the *working path* and the *restoration path*, respectively. It is seen that the two are identical in the case of span protection. With link protection, the restoration path is the shortest route available between the two nodes concerned. With path restoration, the traffic is made to follow a completely different route. In both link and path restorations, the choice of the restoration path can be either preplanned (*static restoration*), or dynamically attributed (*dynamic restoration*). Static restoration, which is the most rapid to implement, relies upon a fixed database of optimal alternative routes (fibers and wavelength channels) which have been spared for protection (as $1:N$) and for which the number of node hops is minimized. For optimal network-resource use, such routes must also have a certain number of segments in common. The corresponding strategies are thus called *shared static link restoration* (SSLR) and *shared static path restoration* (SSPR). Dynamic restoration is slower and more complex, because it must take into account the current network status, check the availability of alternate links/paths and decide on the optimum solution under these constraints. It is however more able to handle multiple failures. It is also best suited to operating in expanding networks, which constantly add new possibilities for restoration paths. Table 2.4 summarizes the general attributes of static and dynamic restoration strategies concerning span link and path restoration. Two most important criteria are the time required for completing the restoration and the capacity to handle more dramatic events such as multiple fiber/span breaks and node failures. Ideally, the restoration should be sufficiently rapid to cause no interruption of service and no loss of frames or packets. It should also be sufficiently robust to prevent any possibility of *network outage* due to unforeseen events. Other important criteria concern the "tax" that the protection strategy imposes on the use of network resources (number of hops, spare-channel requirements, algorithm and signaling). Table 2.4 shows that in fact, no solution is intrinsically or entirely ideal for handling link and path protection. As usual in all network optimization

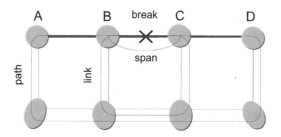

FIGURE 2.110 Elementary network-protection schemes: *span*, *link* and *path*. The dark line corresponds to the initial traffic route, or working path, between nodes A–D and B–C. The dashed lines correspond to alternative traffic routes, or restoration paths, after the different restoration schemes.

TABLE 2.4 General attributes of span, link and path restoration

Attribute	Span Static	Span Dynamic	Link Static	Link Dynamic	Path Static	Path Dynamic
time required to completion	short		short	long	short	long
number of hops and segments of restoration path	1		large	small	small	large
requirement for spare fiber/wavelength channels	yes		yes	no	yes	no
algorithms and signaling	simple		simple	complex	simple	complex
capacity to handle multiple fiber or span breaks	none		poor	good	poor	good
capacity to handle node failure	none		none		simple	complex

issues, the best solution is a matter of trade-off, which itself depends upon the network topology, the type of service, the reliability of the technologies and the evolving network use.

The second issue to address in this subsection concerns *traffic optimization*, namely how to make the best use of network bandwidth at minimal resource, deployment, effort and operation/service cost. As previously stated, this issue is a field in itself which could fill up an entire library section. For the scope of this book, we can only partially cover it and even so in the form of a brief introduction to provide the reader with a start for future investigation and study. This introduction will only concern the optimization of the *wavelength-routed transport layer*, without consideration of types of signal and service. It is clear however that the best network use requires *service-differentiated* OCH connections which must take into account the specificities of the various client layers (SDH/SONET, ATM, TCP/IP) and their circuit/packet modes of operation. Also, we will be only concerned with the issue of *static* optimization, bearing in mind that real networks must be dynamically reconfigurable on a scale varying from minutes to months to years in order to meet rapidly to evolving service demands.

Analyzing and optimizing the traffic flow in a wavelength-routed network requires a minimum of analytical notions and tools, for which we shall provide here some of the basics. Consider first the 5-node network shown in Figure 2.111. All nodes are connected by physical links (optical fibers). The links are traversed by signals having different wavelengths and propagation directions. The nodes are also logically connected by *lightpaths*. The lightpath is another term for the *logical link* established at a given wavelength between two distant nodes. This lightpath is also made of a sequence of physical links. Such a sequence is called a *link lightpath*: for instance, the link lightpaths at wavelengths λ_1 and λ_2 are $A \to B \to C$ and $A \to B \to C \to D$, respectively. These correspond to the lightpaths (or logical links) $A \to C$ and $A \to D$ respectively. One can thus distinguish the *physical network*, which is the set of nodes and their mesh of physical links, and the *virtual network*, which is the set of nodes and lightpaths

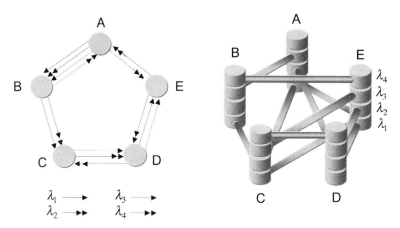

FIGURE 2.111 Left: graph representing a *physical network* with five nodes (A–D) and directed links with their wavelength assignments. Right: corresponding *virtual network* of logical links formed by each lightpath (directions not shown for clarity).

(or logical links). The figure illustrates the difference between the physical and the virtual network topologies.

The physical and virtual topologies are completely defined by what the theory calls a *graph*, referred to as $G(V, E)$. The letter V stands for *vertex* (node), and represents the set of all vertices ($V_1, V_2 \ldots$) present in the network. The letter E stands for *edge* (link). It is the set of either physical links or logical links. These links, which are always directed from a node source to a node destination, are noted $\{V_i, V_j\}_k$ for all possibilities $i \neq j$ and all wavelengths k. By rule, there cannot be more than one directed link between V_i and V_j at a given wavelength. The graphs of the physical and virtual topologies are thus noted $G_p(V, E_p)$ and $G_v(V, E_v)$, respectively. We leave it as an exercise at the end of the chapter to define the graphs corresponding to the physical and virtual topologies shown in Figure 2.111.

The following analysis of the traffic-optimization problem (also called *virtual-topology design*, or VTD) may look very complicated at first sight, but it is in fact relatively easy to grasp if one carefully follows all the steps. To simplify the problem, we shall assume that all physical links are unidirectional, based upon a single fiber. The logical links corresponding to lightpaths are thus unidirectional. In the following definitions, we will use the indices (m, n) to refer to physical links from node m to node n, the indices (i, j) to refer to lightpaths (or logical links) from node i to node j, and the indices (s, d) to refer to the source and destination nodes of a given end-to-end logical connection or circuit. These notations are illustrated in the partial graph of Figure 2.112. As this figure suggests, a single logical connection between nodes (s, d) may require the concatenation of successive lightpaths (i, j), with wavelength conversions from one to the next. Note that only one wavelength is allowed per logical connection. The source/destination nodes can have several logical connections using different wavelengths and many bifurcated lightpaths. But two lightpaths having the same source/destination nodes cannot share the same wavelength!

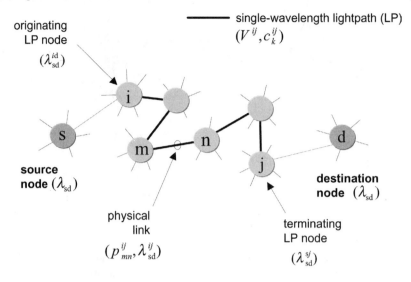

FIGURE 2.112 Partial network graph for virtual-topology design, showing source (s) and destination (d) nodes for a given logical connection, originating (i) and terminating (j) nodes for a segment of single-wavelength lightpath, and a physical link between two adjacent nodes (m, n) belonging to the lightpath. The various parameters shown in parenthesis are defined in the text.

In order to formalize the VTD-optimization problem we first introduce the fixed (or given) network parameters:

N: total number of network nodes;

M: maximum number of wavelengths per fiber;

T_i: number of transmitters at node i, which is at least equal to one;

R_i: number of receivers at node i, which is at least equal to one;

λ_{sd}: total *traffic flow* required between a given source and a given destination; it can be expressed either in average number of packets transmitted per second, or in multiplex bit rate (not to exceed $M \times B$, where B is the wavelength-channel bit rate); by definition, $\lambda_{ss} = \lambda_{dd} = 0$;

$P_{mn} = P_{nm}$ (binary): *link indicator*, equal to 1 if there is a physical connection between nodes m and n, or equal to 0 otherwise ($m, n = 1, \ldots, N$);

h_{ij} (integer): maximum allowed number of physical hops over any lightpath (i, j); this number puts a limit on both signal-to-noise ratio degradation and propagation delay associated with lightpaths.

We introduce next the *variable* network parameters, which can be listed as follows

V_{ij} (binary): *Lightpath indicator*, equal to 1 if there is a logical connection (lightpath) between nodes node i and j, otherwise equal to 0;

p_{mn}^{ij} (binary): *Physical-route indicator*, equal to 1 if the physical link (m, n) is present in the lightpath (i, j), otherwise equal to 0;

c_k^{ij} (binary): *Lightpath wavelength indicator* equal to 1 if the wavelength k ($k = 1, \ldots, M$) has been assigned to the lightpath, otherwise equal to 0;

λ_{sd}^{ij} (real positive): Amount of *partial traffic flow* over the lightpath (i, j) which concern source/destination nodes (s, d);

The previously defined variables are subject to a certain number of constraints, which we list below with their justifications (referring to Figure 2.112, or making drawings with a small number of nodes defining different s, i, m, n, j, d paths might in some cases be of help):

1. $\sum_i V_{ij} \leq R_j$ and $\sum_j V_{ij} \leq T_i$
   There should be no more lightpaths on any terminal node j than its number of receivers R_j; there should be no more lightpaths from any originating node i than its number of transmitters T_i; the equalities stand for full utilization of the node's transponders;

2. $\sum_k c_k^{ij} = V_{ij}$
   Only one wavelength is allowed per lightpath and each lightpath must be attributed one wavelength (recalling that $V_{ij} = 0$ or 1);

3. $\sum_k c_k^{ij} p_{mn}^{ij} \leq 1$
   Only one wavelength is allowed per physical link (m, n) belonging to a given lightpath;

4. $p_{mn}^{ij} \leq V_{ij} P_{mn}$
   As required by the existence/non-existence of logical ($V_{ij} = 1/0$) and physical-link ($P_{mn} = 1/0$) connections: the product in the RHS summarizes the two inequalities $p_{mn}^{ij} \leq V_{ij}$ and $p_{mn}^{ij} \leq P_{mn}$;

5. $\sum_n p_{in}^{ij} = \sum_m p_{mj}^{ij} = V_{ij}$
   Concerning a given logical connection (i, j), the sum of all physical-route indicators starting from node i (p_{in}^{ij}) or ending at node j (p_{mj}^{ij}), cannot exceed V_{ij} (namely, 1 or 0);

6. $\sum_n p_{ln}^{ij} = \sum_m p_{ml}^{ij}$ for $l \neq i, j$
   Concerning a given logical connection (i, j), the sums of all physical-route indicators starting from any intermediate lightpath node l (p_{ln}^{ij}), or ending at it (p_{ml}^{ij}), give the same result (namely, 1 or 0);

7. $\sum_i \lambda_{sd}^{id} = \sum_j \lambda_{sd}^{sj} = \lambda_{sd}$
   The sum of all partial-traffic flows originating from all possible lightpath nodes i and terminating at the destination node d is equal to the source-destination traffic flow (λ_{sd}); same statement for the sum of partial traffic flows originating from source node s and terminating at lightpath node j;

8. $\sum_i \lambda_{sd}^{il} = \sum_j \lambda_{sd}^{lj}$ for $l \neq$ s, d
   The total (s, d) traffic flow reaching or leaving an intermediate node l is conserved;

9. $\sum_{s \text{ or } d} \lambda_{sd}^{ij} \leq V_{ij} \lambda_{sd}$
   The sum of all partial-traffic flows from multiple sources/destinations

which share a given lightpath (i, j) cannot exceed the total allowed traffic in the fiber (λ_{sd}); it is zero if the lightpath is idle $(V_{ij} = 0)$;

10. $\sum_n \sum_m c_k^{ij} p_{mn}^{ij} \leq h_{ij}$

The argument in this double sum is equal to unity whenever a physical link (m, n) belongs to the lightpath (i, j) with wavelength k being assigned; it thus corresponds to the total number of physical hops in the lightpath of this wavelength channel, and should therefore not exceed the maximum number of hops allowed (h_{ij}).

These constraints ensure consistency between virtual and physical topologies, attribution and conservation of traffic flow, wavelength allocation and reuse, and lightpath/logical-link exclusive correspondence. The goal of the problem is to find a virtual topology matrix $V = [V_{ij}]$ defining all the network lightpaths with their wavelength allocation (c_k^{ij}), physical route (p_{mn}^{ij}) and total traffic flow. This total traffic flow is defined by the quantity

$$F_{ij} = \sum_{s\ or\ d} \lambda_{sd}^{ij} \quad (2.151)$$

We define the *congestion F* as the maximum value of F_{ij}, namely:

$$F = \max_{ij}\{F_{ij}\} \quad (2.152)$$

(the braces designating the set of all possible values for F_{ij}). The best solution $(V = \lfloor V_{ij} \rfloor)$ is the one for which the congestion F is *minimized*. The underlying idea is not only that one should make maximum use of the available network routes for all nodes but rather that none of these routes or intermediate nodes experience traffic saturation. In such conditions, the network offers a bandwidth reserve for future upgrades and has the capability to handle momentary traffic increases or bursts between any two ends. One refers to such an optimization strategy as "*maximizing the load offer*," which corresponds to minimizing the average traffic flow in any of the network's physical links.

For packet-switched networks, a second optimization criteria is *packet delay*. To simplify the problem, one possibility is to introduce the constraint (j) limiting the number of hops in a given lightpath to a predefined maximum. But this does not take into account the actual delays incurred by packet queuing in the two switching nodes (i, j) and actual transmission delays due to variable physical distances between nodes. Concerning the first issue, if λ_{sd}^{ij} is the packet rate (number of packets per second) over the lightpath and $1/\mu$ the mean packet delay due to queuing, it can be shown that the mean queuing delay with traffic flow λ_{sd}^{ij} is

$$t_{ij} = \frac{1}{\mu - \lambda_{ij}} \quad (2.153)$$

This delay approaches infinity (full congestion) when $\lambda_{ij} \to \mu$ (hence the need to minimize the congestion function F!). The second issue concerns the propagation delay due to the physical length of the lightpath. For every physical link between adjacent nodes (m, n), one can associate a *propagation delay* d_{mn}. The complete solving of the delay-minimization problem thus consists in finding the minimum value of the quantity

$$G = \sum_{ij}\left[\sum_{sd}\lambda_{sd}^{ij}\left(\sum_{mn}p_{mn}^{ij}d_{mn} + t_{ij}\right)\right] \quad (2.154)$$

The two quantities F and G to be minimized are called *objective functions*. The first objective function with its constraints is *linear* with both integer and real (positive) values. The computer program which can perform the minimization is called a *mixed-integer linear program* (MILP), and in this case the program is thus called VTD-MILP. The second objective function is *nonlinear* since it contains the terms $\lambda_{ij}/(\mu - \lambda_{ij})$. The reality is that there is no known algorithm to solve such linear and nonlinear problems. Therefore, the search for a converging and optimal solution must be performed heuristically, with the objective to find not only a workable solution but a solution close to optimal. The approach first consists in decomposing the problem into main subproblems. There are essentially four of these subproblems, which can be described as follows:

1. *Topology subproblem*: determine a "good" virtual topology ($\lfloor V_{ij} \rfloor$) which offers an acceptably large number of lightpaths;
2. *Lightpath-routing subproblem*: determine those lightpaths that yield a minimal number of hops ($c_{sd}^{ij}(m, n)$);
3. *Wavelength-allocation subproblem*: optimally allocate wavelengths to lightpaths (c_k^{ij});
4. *Traffic-routing subproblem*: optimize packet routing by minimizing the (lightpath) traffic flow over the virtual topology (λ_{sd}^{ij}).

Even as simply stated, each of these subproblems is quite complex to solve, and they also require a fair amount of heuristics. There is even a risk that such a decomposition into subproblems and their sequential optimization leads to the absence of a global (full-problem) solution! The VTD can be made more complex by taking into account the notion of *quality of service* (QoS), which introduces various additional constraints such as path reliability, transmission quality, cost of use, and restoration times. As previously mentioned, the network can also be *service-differentiated*, where optimization must take into account the client layer and the type of payload, all coming with their own priorities and QoS (service class) requirements. We shall leave the discussion on this note, hoping that the above presentation provided the reader with an adequate, first-understanding and flavor of VTD. For examples of optimization results in realistic WDM networks (e.g., NSFNET in the United States) and advanced analysis, the reader may refer to Jukan (2001), Mukherjee (1997), Ramaswami and Sivarajan (1998), and Sivalingam and Subramanian (2000) (for instance), which are listed in the Bibliography.

2.6.5 Network Evolution and Convergence

The issue of *traffic optimization* (maximizing the offered load) in WDM networks should be placed in the broader perspective of network *evolution* and *convergence*.

The network *evolution* begins with the legacy of an initial, fully opaque network generation (DXC/DCS switching) and moves towards the next, fully transparent network generation (OXC switching). Our current intermediary stage is based on an overlay of both network generations, with a mix of opacity and transparency (hybrid DXC/OXC switching). As previously discussed, all-optical cross-connect solutions are mature with *wavelength routing*, but not yet with *packet switching* (packet being understood here in the largest sense, i.e., not only ATM cells but also IP datagrams and other formats such as DW). However, the concept of *optical packet switching* (OPS) does not necessarily call for all-optical processing from port to port. Indeed, it is sensible that a minimum electronic edge-interface will always be required for practical OPS, due to the triple need for buffering/queuing the packets, analyzing/processing the headers and recognizing the data/client/service types, and finally, managing contentions/congestion events. Photonics can then be advantageously exploited when it comes to reducing the number of internal switch connections, signal propagation delays, electrical-power consumption and performing bandwidth/protocol-transparent switching in both space and wavelength domains. The concept of a global transparent network with OPS must therefore be used with this understanding.

The network *convergence* is another key issue within the above network evolution scenario. The rationale behind OPS is not only to improve the traffic flow with higher efficiency and lower operations costs, but also to *simplify* and *tighten* the integration between the different client layers (namely GigE, ATM, frame-relay or IP) and the physical layer, namely WDM and its OCH interface (see Section 3.3.10 in Desurvire, 2004). In the previous subsection of this chapter, we described the digital wrapper (DW) approach whereby any client data can be packaged and routed on WDM networks using a single transport format and "protocol-agnostic" switching. Focusing specifically on networks dominated by IP traffic (as the trend goes at global scale), the DW approach does not provide a truly optimal solution. This is because the number of IP packets transferred between two endpoints during the same internet session can be quite large. Both IP routers and OPS (at OCH layer) must then manage the routing and switching of this dedicated traffic flow on a *per-packet* basis, which represents a waste of time and resource. The same observation can be made for other packet types which may be exchanged during a session (e.g., IP over ATM, IP-over Ethernet).

The simplification of the preceding problem is introduced by *multiprotocol label switching* (MPLS), a new function which performs *"flow aggregation."* Such a function consists in defining a virtual *end-to-end path* for a flow of packets having the same destination addresses. This path is called a *label-switched path* (LSP). The LSP is initiated and controlled by a succession of *label-switched routers* (LSR), which can be any layer-2 switch (ATM, Ethernet). But as configured through MPLS, these LSR are able to immediately recognize the aggregated flow and thus directly forward all matching packets without looking into their inside address fields and routing information. In order to initiate a MPLS path (LSP set up procedure), two adjacent LSR must agree on a *label* (4-byte word) to be used for flow identification. The upstream LSR first sends a request to the downstream LSR to specify the bandwidth requirement and service type. The downstream

LSR then allocates a label value meeting the requirements and communicates it to the upstream LSR. This label value is then written into the *protocol field* of the packets to be aggregated. Once two LSR have agreed upon this means of identification, all packets having this label will be switched the same way. The choice of a specific label, which must match one of several predefined network routes, is controlled by the *label distribution protocol* (LDP). For advanced reference, the two higher types of LDP are called CR-LDP (*constraint-based routing* LDP) and RSVP-TE (*traffic engineering resource-reservation protocol*). Instead of just exchanging labels, these protocols also introduce bandwidth allocation and service class into the label definition. The mesh of LSR sharing such an explicit routing and bandwidth/service reservation mechanism forms an MPLS network, as illustrated in Figure 2.113. If the LSR are of the OXC type, the LSP is exclusively a wavelength-switching path, as illustrated in Figure 2.114. In this case, the MPLS network is referred to as an *MPλS* network. The MPLS (or MPλS) networks are accessed by *label edge routers* (LER, also called ELSR), which control the input/output traffic and make the interface with any client IP router and other networks. At the end of the LSP, the destination LER removes the MPLS label from the packets, which are then handled by the next IP router in the path (or by the entry LER of another MPLS network) or the ISP point of presence (POP). Recalling that IP only represents a "best-effort" transmission protocol, we can observe that MPLS introduces a new level of QoS by guaranteeing path and bandwidth availability to the aggregated traffic, in addition to the acceleration of packet/datagram transit times.

As its name suggests, the *generalized multiprotocol label-switching* (GMPLS) comes as a natural extension of MPLS, which can be applied to all technology and switching platforms, such as SDH/SONET EADM, OXC, reconfigurable OADM, and fiber-bundle switches (e.g., MEMS). To stress the key difference, the previous MPLS is often referred to as *nongeneralized MPLS*. The GMPLS labels, whose lengths are no longer restricted to 32 bits, are defined under five categories which

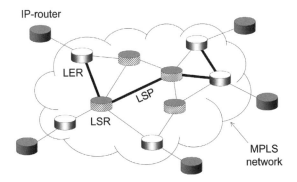

FIGURE 2.113 Layout of multiprotocol label-switched (MPLS) network, with label edge routers (LER), label-switched routers (LSR), and label-switched path (LSP) shown in bold line.

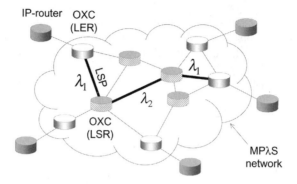

FIGURE 2.114 Layout of multiprotocol lambda-switched (MPλS) network, with optical cross-connects (OXC). See previous figure for other definitions.

concern (a) fibers, (b) wavelengths, (c) wavebands, (d) time slots, and (e) SDH/SONET. *Fiber labels* attribute a unique fiber path (as a single fiber number) to be identified by a fiber switch. *Wavelength* and *waveband labels* attribute a unique wavelength (as a single channel number) or set of wavelengths (as channel numbers with a group identification), to be identified by an OXC. *Time-slot labels* define a bandwidth allocation (as number of reserved slots), to be recognized by a TDM switch, such as based on SDH/SONET. Finally, SDH/SONET labels reserve bandwidth for virtual containers (VC-n in SDH) or virtual tributaries (VT-n in SONET) at specified hierarchy levels (n). Thus different types of label-switched paths can be generated, forming the nested hierarchy shown in Figure 2.115. This hierarchy is defined after its *"switching-capable"* type: FSC (fiber), LSC (wavelength, L being for "lambda"), TDM (time-slot and SDH/SONET) and PSC (packet-switching) to the very top. The corresponding label-switched paths are referred to as FSC-LSP, LSC-LSP, TDM-LSP and PSC-LSP, respectively.

The GMPLS label-request procedure is the same as in MPLS (based on modified CR-LDP and RSVP-TE signaling), except for those extra options and parameters which need to be precisely specified. Another difference between GMPLS and (nongeneralized) MPLS is in the LSP setup procedure, which consists in reaching a mutual agreement of possible label options. The upstream LSR, which first makes the request to the downstream LSR, imposes constraints on the choice of labels. The requested labels must belong to what is called a predefined *label set*. Such a set either explicitly lists the different label possibilities requested, or is absent. In the last case, all valid labels are then implicitly eligible. In response, the downstream LSR returns to the upstream LSR its own label set, in which labels corresponding to impossible options (e.g., wavelength conversion) have been removed. If the returned label set is absent, this means that the downstream LSR is able to handle any label options. The label set thus propagates from one LSR to the next until a complete and valid LSP path has been defined and agreed upon by all. Thus each LSP has a specific set of label options, which depend upon the switching technologies and capabilities involved in the path LSR nodes.

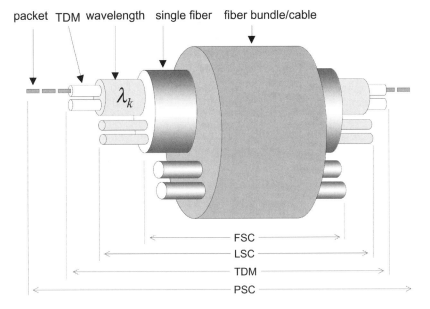

FIGURE 2.115 Nested hierarchy of generalized multiprotocol label-switching (GMPLS) path according to router interfaces: fiber-switching capable (FSC), lambda-switching capable (LSC), time-slot and SDH/SONET-switching capable (TDM) and packet-switching capable (PSC).

Network protection is ensured by the fact that for each primary LSP there is at least one "*maximally disjoint*" secondary LSP backup, which can be automatically activated in case of link failure. The backup LSP can be either re-provisioned (*hot standby*) or activated on demand (*cold standby*).

Another (optional) constraint introduced by GMPLS is the *explicit label control* (ELC). This ELC makes it possible to match the labels for both communication paths, creating a *bidirectional LSP setup*. For instance, this setup can consist in allocating the same wavelength, waveband or SDH/SONET labels to both paths. Such a match establishes consistency in signal quality (SNR, BER) and assignment of virtual container/tributaries in the two transmission directions. Unlike non-generalized MPLS, GMPLS signaling (LSP set up and LSP link management, see the following text) does not borrow the same physical path as the data to be routed. This is to avoid wasting wavelength channels and their optical bandwidth. Such out-of-band signaling thus requires a low bit-rate channel, for instance based on a parallel Ethernet link. The upstream LSR must therefore determine first a data path, then a separate signaling path, linking to the downstream LSR and LER, which complicates the signaling formats. The overall control and supervision (OAM) of a GMPLS network is ensured by the *link-management protocol* (LMP). The LMP controls the link parameters, the physical connectivity between switches (ping signaling) and their link properties. It also detects and locates any faults occurring within the optical layer.

From the above description, we can construct a new *protocol stack* for the different client layers (IP, ATM, GigE, Frame relay, SDH/SONET), described in Figure 2.99, but which now includes LMP/GMPLS as a control plane on top of the IP layer, as shown in Figure 2.116. One recognizes (as key examples) the different suites ranging from IP-over-ATM-over-SDH/SONET to IP-over-WDM, with GMPLS managing the paths and connections at TCP (and UDP) level through LMP and LDP signaling. While GMPLS opens the vistas of the long sought-after "intelligent optical networks," other alternative or complementary approaches are still under constant investigation. For advanced reference, it is worth mentioning two "overlay" reference models. The first is an intuitive client-server node interface, in which the IP router directly commands the OXC at OCH layer. The internetworking messages (called "primitives") basically consist in setting up and releasing OCH connections (connecting, disconnecting, bridging and switching input/output OXC wavelength ports) and receiving alarm OXC messages for failed connections. The second reference-model example is ASON (*automatically switched optical networks*), which has been standardized as *ITU-T G.8080*. Under ASON control, both *permanent* and *switched* network connections can be established. The permanent connections are set up by a specific management system (the *optical connection controller*, or OCC), while the switched connections are dynamically initiated and released according to customer demand, with signaling/routing protocols based on GMPLS or similar.

The deployments of MPLS/MPλS, along with its more versatile "sibling," GMPLS, and the ASON reference model are now considered to represent the best-in-class and fastest approach towards achieving *network convergence*. This

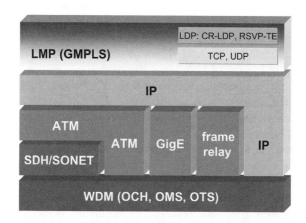

FIGURE 2.116 Protocol stack of GPMLS network (LMP = link-management protocol, RSVP-TE and CR-LDP = label distribution protocols [see definitions in text], TCP = transmission control protocol, UDP = user datagram protocol, IP = Internet protocol, OCH = optical channel layer, OMS = optical multiplex section layer, OTS = optical transmission section layer).

most precise concept actually conveys several and mutually compatible levels of meaning, namely (and not exhaustively):

1. Establishing secure network paths and bandwidth allocation for provisioning increased IP-based traffic and services having enhanced QoS;
2. Making the best use of layer-1/layer-2 switching technologies and their network-legacy platforms (SDH/SONET, ATM, Ethernet, Frame-relay);
3. Bringing IP ever closer to the WDM layer with simplified OAM management and signaling, and with dynamic bandwidth provisioning/path restoration directly from this layer;
4. Bringing IP closer to the network edge and access, while providing multivendor service aggregation;
5. Preparing and optimizing the network (from access to core) towards the likely dominance of IP-based traffic/services over traditional voice and data traffic/services;
6. Reducing operational costs and capital expenditures for operators and service providers, and access costs for private/enterprise customers with greater bandwidth use, enhanced QoS and diversified (multivendor) service provisioning.

In the next generation of transparent networks, as based upon such an "intelligent" optical layer, the *convergence* of different (and emerging) service offers is indeed primordial. The key issues hiding behind this high and sensible cause remain the following: *inter-operability* between remaining network domains, global-scale *management*, *cost effectiveness* and *return on investment* (should new technologies/standards be implemented), and *scalability* of the load offer and emerging services (providing more users with more services). These could be seen as the overarching concepts of any long-term strategy towards global network deployment and dimensioning. Other secondary, but no less important, issues for service provider and network operators, naturally concern *capital growth*, *market competition* (ILEC/CLEC), *partnerships strategies*, *customer satisfaction*, and *expert recruiting, training and retraining*. The last issue is actually one of the main rationales for this book, as developed in the Preface.

✅ EXERCISES

Section 2.2

2.2.1 A manned craft is sent to Jupiter's vicinity. What is the minimum and maximum time for exchanging messages with the Earth? Express the result in hour/min/s (Parameters: mean Earth distance to Sun $d_E = 149.6$ million km, mean Jupiter distance to Sun $d_J = 778.3$ million km.) Conclusion and solutions to create a duplex communication channel?

2.2.2 Consider the 45° glass-prism structure shown below. What should be the minimum value of the glass' refractive index n_1 in order for light rays to be totally reflected in the 90° direction? (Assume $n_2 = 1$ for air).

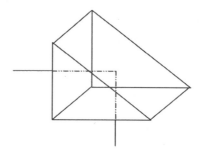

2.2.3 Obtain the time-independent relation of equation (2.7) between the E-field coordinates defined in equation (2.8). Clue: use the formulae $\cos(a+b) = \cos a \cos b - \sin a \sin b$ then weight and sum/subtract the resulting equations in such a way as to eliminate the time-dependent terms through the property $\cos^2 a + \sin^2 a = 1$; also use $\sin(2a) = 2\sin a \cos a$ and $\cos(2a) = \cos^2 a - \sin^2 a$.

2.2.4 Show that elliptical SOP with phase $\Delta\varphi$ comprised between 0 and π corresponds to a right-handed (clockwise) rotation of the E-field polarization, and the opposite rotation direction for $\Delta\varphi$ comprised between π and 2π. (Clue: Evaluate the time change of the angle γ defined by $\tan\gamma = Y'/X'$ from $t=0$, assuming $\Delta\varphi = \pi/2$ and $\Delta\varphi = 3\pi/2$, then generalize the conclusion to all phase-delays $\Delta\varphi$).

2.2.5 Show that the Stokes coordinates are uniquely defined by the two ellipse angles (ψ, χ) and the power (S_0) according to equation (2.13), using the definitions of (E_o, E_e), (α, ψ, χ), and (S_0, S_1, S_2, S_3) from text.

2.2.6 What is the photon flux (rate of incident photons per second) in a light beam of power $P = 1$ W, 1 mW, 1 µW and 1 nW at a wavelength of $\lambda = 1.5$ µm (IR light)? Same question for a wavelength of $\lambda = 100$ nm (UV light). Conclusions?

2.2.7 An optical amplifier has a gain coefficient of $g = 0.7$ cm^{-1}. What are the required amplifier lengths L to achieve gains of $\times 100$ and $\times 1{,}000$?

2.2.8 A semiconductor amplifier chip of length $L = 250$ µm provides a gain of $G = +25$ dB. What is its gain coefficient in dB/m and m^{-1}?

2.2.9 Prove that the two different gain-coefficients definitions in [dB/m] and [m^{-1}] are related through $g_{m^{-1}} \approx 0.2302 \times g_{dB/m}$.

2.2.10 The amount of output power that can be extracted from a laser cavity is given by the approximation formula $P_{out} \approx P_0(1-R)/[2\alpha L + \log(1/R)]$, where P_0 is a constant, R the reflectivity of the output mirror ($R < 1$), and $2\alpha L$ the cavity's round-trip loss. Determine by trial-and-error (within 0.5% accuracy) the optimum reflectivity for which this output power is maximized (assume a loss $2\alpha L = 0.7$, corresponding to $\exp(-0.7) = 0.5$ or 3 dB).

Section 2.3

2.3.1 Determine the maximum number of ray incidence-angles making it possible for light waves to propagate inside an index-layer structure of thickness $d = N\lambda$, with $N = 100$, $N = 10$, $N = 1$ and $N = 1/2$. Conclusion for the lowest number?

2.3.2 We want to realize a planar dielectric waveguide with a thin slab of glass surrounded by air. What should be the slab thickness in order for the waveguide to have a cut-off wavelength at $\lambda_{cutoff} = 10$ µm? (glass index $n_1 = 1.450$). Same question if the slab is surrounded by another glass material with index $n_2 = 1.449$? Conclusion for practical realization of the waveguide?

Exercises 337

2.3.3 We want to realize a planar dielectric waveguide with a thin slab of glass (index $n_1 = 1.45$) with thickness $d = 10\,\mu m$ surrounded by two identical glass layers with a different index n_2. Give the relative index Δ for which the waveguide has a cutoff wavelength of $\lambda_c = 1.4\,\mu m$.

2.3.4 Based upon the figure below, show that the condition for an external ray to be captured (totally internally reflected) into a waveguide of numerical aperture NA is that its incidence angle be $\varphi < \varphi_a$, where φ_a is an acceptance angle defined by $\sin \varphi_a = NA$.

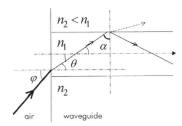

2.3.5 Plot the LP_{01} E-field mode envelope $E(r)$ and corresponding power distributions $E^2(r)$ of a step-index optical fiber assuming a numerical aperture of $NA = 0.1$, a core radius of $a = 5.75\,\mu m$ and three different signal wavelengths $\lambda_1 = 1.3\,\mu m$, $\lambda_2 = 1.4\,\mu m$, $\lambda_3 = 1.5\,\mu m$, $\lambda_4 = 1.6\,\mu m$ and $\lambda_5 = 1.7\,\mu m$. You may use a plotting software that includes Bessel and Hankel functions, or alternatively use the approximations:

$$J_0(x) \approx 1 - 2.25(x/3)^2 + 1.26(x/3)^4 - 0.31(x/3)^6 \text{ for } x \leq 3,$$

and

$$K_0(x) \approx \sqrt{\pi/(2x)}[1 - 1/(8x)]\exp(-x) \text{ for } x > 1.$$

(Clue: first make a table with the values of V, U and W for each wavelength, then use the above definitions to plot the E-field between $r = 0$ and $r = 3a$. See text for final plot result and conclusions.)

2.3.6 How much fiber length L can be manufactured from a single glass preform with dimensions $l = 1$ m (length) and $od = 1.5$ cm (diameter)? Same question with $od = 5$ cm?

2.3.7 An optical-transmission receiver requires a minimum incident signal power of $P_R = 0.1\,\mu W$ to guarantee a standard bit-error-rate of 10^{-9}. What is the maximum possible system length, assuming that the transmitter power is limited to $P_T = 100\,\mu W$, 1 mW, 10 mW, 100 mW or 1 W, respectively (fiber attenuation coefficient $\alpha = 0.2\,dB/km$)? Conclusion from the results?

2.3.8 Show that the bandwidth-length (BL) limitation of fiber-optics systems:

$$BL(\text{THz} \times \text{km}) < \frac{1}{|D|_{\text{ps nm}^{-1}\text{km}^{-1}} \Delta\lambda_{\text{nm}}}$$

where $\Delta\lambda \geq (\lambda^2/c)B$ the signal spectral width can also be put in the B^2L form:

$$B^2L(\text{THz}^2 \times \text{km}) < \frac{0.3}{|D|_{\text{ps nm}^{-1}\text{km}^{-1}} \lambda_{\mu m}^2}.$$

338 Optical Fiber Communications

2.3.9 What should be the zero-dispersion wavelengths of NZDSF$^\pm$ fibers assuming dispersions of $D = \pm 2.5$ ps nm^{-1}km^{-1} and identical slopes of $D' = 0.07$ ps nm^{-2}km^{-1} at $\lambda = 1{,}550$-nm wavelength? (Clue: use linear approximation formula for dispersion).

2.3.10 Calculate the fourth-order dispersion coefficient D'' of the SMF ($\lambda_0 = 1{,}300$ nm), assuming its dispersion and slope at 1,550 nm to be $D = +17$ ps/nm/km and $D' = 0.045$ ps/nm^2/km, respectively.

2.3.11 What is the range of maximum allowed PMD (14%–30% of the bit period) for bit rates of 2.5 Gbit/s, 10 Gbit/s and 40 Gbit/s? Express the result in picoseconds.

2.3.12 Two optical transmission systems A and B are built with fibers with PMD of 0.1 ps/$\sqrt{\text{km}}$ and 0.8 ps/$\sqrt{\text{km}}$, respectively. Make a table showing for systems A, B the maximum achievable transmission distances for bit rates of 2.5 Gbit/s, 10 Gbit/s and 40 Gbit/s (clue: the PMD should not exceed 14% of the bit period).

2.3.13 Calculate the full development in single-frequency tones of the third-order nonlinearity

$$P_{NL} \equiv \chi_3 EEE \equiv \chi_3 \left\{ \begin{array}{l} (E_{01} \cos \omega_1 t + E_{02} \cos \omega_2 t + E_{03} \cos \omega_3 t) \\ \times (E_{01} \cos \omega_1 t + E_{02} \cos \omega_2 t + E_{03} \cos \omega_3 t) \\ \times (E_{01} \cos \omega_1 t + E_{02} \cos \omega_2 t + E_{03} \cos \omega_3 t) \end{array} \right\}$$

using the basic rules $\cos^2 x = (1 + \cos 2x)/2$ and $2 \cos x \cos y = \cos(x+y) + \cos(x-y)$.

2.3.14 Determine the SRS and SBS pump thresholds for single-mode optical fibers such as SMF ($A_{\text{eff}} = 80\ \mu\text{m}^2$) and DSF ($A_{\text{eff}} = 50\ \mu\text{m}^2$), using $L_{\text{eff}} \approx 22$ km and the gain coefficients $g_R \approx 6.5 \times 10^{-14}$ m/W and $g_B \approx 5 \times 10^{-11}$ m/W. Express the results in dBm.

2.3.15 Determine the required pump power to achieve Raman gains of 0 dB, +3 dB, +10 dB and +20 dB at 1.55 μm wavelength, assuming a DSF length of $L = 30$ km with pump/signal attenuation coefficient of $\alpha_p \approx \alpha_s = 0.2$ dB/km (0.046 km^{-1}), effective area $A_{\text{eff}} = 50\ \mu\text{m}^2$ and Raman gain coefficient of $g_R = 6 \times 10^{-14}$ m/W. Same question with SMF ($A_{\text{eff}} = 80\ \mu\text{m}^2$).

2.3.16 Show that the condition of 1-dB power penalty in a DSF-based WDM system having an effective length L_{eff} and M channels and spaced by $\Delta\lambda$ with individual power P_0 is equivalent to the following:

$$M(M-1)\Delta\lambda_{\text{sm}} P_0^{\text{mW}} L_{\text{eff}}^{\text{km}} \leq 42{,}000_{\text{mW.nm.km}}$$

(Clue: use the condition $\dfrac{M(M-1)}{2} \dfrac{\Delta\lambda}{\Delta\lambda_R} g_R^{\text{peak}} P_0 \dfrac{L_{\text{eff}}}{A_{\text{eff}}} \leq 0.205$ with $\Delta\lambda_R \approx 120$ nm, $g_R^{\text{peak}} = 6 \times 10^{-14}$ m/W, and $A_{\text{eff}} = 50\ \mu\text{m}^2$.)

2.3.17 Show that for SMF and DSF transmission fibers at $\lambda = 1.55$-μm wavelength, the nonlinear constant γ associated with the nonlinear phase $\phi_{NL} = \gamma P_i L_{\text{eff}}$ is of the order of $\gamma \approx 1$ km^{-1}W^{-1}. (Clue: use the definition $\phi_{NL} = n_2(2\pi/\lambda)P_i L_{\text{eff}}/A_{\text{eff}}$ with $n_2 = 2.3 \times 10^{-20}$ m^2/W and $A_{\text{eff}} = 50\ \mu\text{m}^2$ (DSF) or $A_{\text{eff}} = 80\ \mu\text{m}^2$ (SMF).)

2.3.18 What is the minimal linewidth of a RZ signal at $\lambda = 1.55$ μm wavelength transmitted at rate $B = 1$ Gbit/s, assuming Gaussian-envelope pulses?

2.3.19 What is the fundamental-soliton peak power in a 10 Gbit/s SMF-based 1.55 μm transmission system using soliton pulses of width $\Delta\tau = 25$ ps? (Use for SMF dispersion and effective area $D = +17$ ps/nm/km, and $A_{\text{eff}} = 80\ \mu\text{m}^2$, respectively.) Same question for a DSF-based system ($D = +0.2$ ps/nm/km, $A_{\text{eff}} = 50\ \mu\text{m}^2$).

Section 2.4

2.4.1 Calculate the net loss experienced by a signal through a 4×4 star coupler according to the two implementations shown in Figure 2.40 (A = based upon 4×4 modules, B = based upon perfect shuffle). Same question for 16-port, 32-port and 64-port star couplers. Application: the 2 × 2 unit couplers are assumed to have a 0.2-dB coupling/excess loss and have a −3-dB splitting loss.

2.4.2 Determine the path imbalance Δl (mm) of Mach–Zehnder interferometers to be used as a MUX/DMUX device for $\lambda_0 = 1.5$ μm WDM systems with $\Delta \nu = 100$ GHz frequency spacing (clues: use $n_{\text{eff}} = 1.5$ for the effective index, $\Delta \omega = 2\pi \Delta \nu$, and $\omega_0 = 2\pi c/\lambda_0$ for the reference frequency).

2.4.3 Determine the fiber Bragg-grating lengths required to achieve a reflectivity of $R = 99\%$, at 1.55-μm wavelength, assuming a power fraction in the fiber core of $\eta = 0.5$ and index-changes of $\delta n = 10^{-2} - 10^{-3}$.

2.4.4 Determine the frequencies of the RF drive signals to realize an acousto-optic tunable filter (AOTF) in LiNbO$_3$ able to select 1.55-μm WDM signals spaced by 10 nm? (Use for the index difference $\delta n = 0.07$ and for the sound velocity $V_a = 3.75$ km/s.)

2.4.5 Determine the pump powers required to cancel the link loss at 1.55-μm wavelength, assuming a DSF/SMF lengths of $L = 50$ km and 100 km with attenuation coefficient of $\alpha = 0.2$ dB/km (0.046 km^{-1}), effective area $A_{\text{eff}} = 50/80$ μm^2 and a Raman gain coefficient of $g_R = 6 \times 10^{-14}$ m/W.

Section 2.5

2.5.1 Calculate the linewidth of LEDs emitting with a peak wavelength in the range $\lambda' = 0.5-1$ μm at room temperature ($k_B = 1.38 \times 10^{-23}$ J/K).

Section 2.6

2.6.1 Assume that the WDM channels form a regular comb with increasing frequencies $f_1, f_2, \ldots, f_N$ such that the frequency spacing $\Delta f = f_{i+1} - f_i$ is uniform or constant, with $N = 32$, $\Delta f = 100$ GHz, $\lambda_1 = 1.550$ μm and $c' = 3 \times 10^8$ m/s. Calculate the wavelength spacing corresponding to the first two (1, 2) and the last two ($N - 1$, N) channels, respectively, and show that they are not strictly identical. What can be concluded?

2.6.2 Determine the graphs of the physical and virtual topologies associated with the network shown in Figure 2.111.

a MY VOCABULARY

Can you briefly explain each of these words or acronyms in the context of optical components and networks?

Absorption (of photons)	regenerator, switching	AOTF	ASON
Access (network, metro)	Amplifier (optical)	APC	Automatic power control
Acousto-optic tunable filter	Amplified spontaneous emission	APD (photodiode)	Automatic protection switching
Add-drop multiplexing (optical)	Anisotropy (medium)	Arrayed-waveguide grating	Automatically switched optical networks
ADM	Annealing	Apodization (grating)	
All-optical regeneration/	Antisymmetric (mode)	Arsenic	Avalanche photodiode
	AOR	As	Avalanche gain
		ASE	
		ASK	

AWG
Attenuation coefficient
Backward pumping
Beat length
BER
Bandgap
Bandwidth (fiber, pulse)
BBOR
Bidirectional pumping
Birefringence (axis)
Bit-error-rate
Black-box optical regenerator
Blazed FBG
Bloch modes
Blocking (switch)
BLSR, BLSR/n
Beginning of life (system)
BOL
Bow-tie fiber
BPSR, BPSR/n
Bragg grating
Bragg condition, wavelength
BSHR
Butt-coupling
C-band (EDFA)
C × D
C^3 laser
C + L (EDFA)
Carrier
CGM
Chirp
Chirped FBG
Chromatic dispersion
Circular polarization
Circulator (optical)
Cladding
Cladding mode
Clock recovery
Coarse WDM
COBRA
Coherence (temporal or spatial)
Conduction band
Comb (WDM)
Compensation (dispersion, PMD)
Congestion
Connector
Core (fiber)
Core (network, metro)
Coupling length
CPM
CR
CR-LDP (MPLS)
Cross-gain/phase modulation
Crossbar (switch)
Cross-connect (optical)
Cross-phase modulation
Cross-section (atomic)
Crosstalk
CSBH
Cut-off wavelength (fiber, photodiode)
CWDM
Dark fiber
dB
DBR
DCF
DCPBH
DCS
DD
De-excitation (atomic)
Decibel
Demultiplexing
Dense WDM
Descartes' law
DFB
DFF
DGD
DGEF
DH
Dielectric medium
Digital cross-connect
Digital wrapper
Dipole
Direct detection
Dispersion (index)
Dispersion compensation, management
Dispersion slope
Dispersion-flattened fiber
Dispersion-shifted fiber
Distributed amplification
Distributed feedback, or Bragg-reflector (laser)
DM
DM soliton
DMUX
Double heterostructure
DW
DWDM
DXC
Dynamic restoration
Dysprosium
E-field
EA, EAM
EC
EC-DBR
EDFA
Edge (network, metro)
Edge (graph)
Edge label-switching router (MPLS)
Effective (refractive/mode) index
Effective length
Effective (mode) area
ELC (GMPLS)
Electrical cross-connect
Electro-absorption
Electro-optic (effect, coefficient)
Electromagnetic
Electron
Electron-hole recombination
Electron-Volt
Elliptical polarization
ELSR
EM
EMI
Emission (of photons)
End of life (system)
Energy level (atomic)
Enterprise (metro)
EOL
Equalization (power, gain)
Equivalent step-index (fiber)
Er
ER
Erbium
Erbium-doped glass/fiber amplifier
ESI
Ev
Evanescent wave
EXC
Excess loss
Excess noise factor (APD)
Excitation (atomic)
Explicit label control (GMPLS)
External cavity
Extinction ratio
Extraordinary (index)
Fabry-Pérot (cavity, interferometer, filter, etalon, laser)
Faraday effect/rotation

My Vocabulary

FA-OH (DW)
FBG
FDM
FEC
FET
Fiber (optical)
Fiber Bragg grating (filter)
Fiber amplifier
Field-effect transistor
Fifth generation (of communication systems)
Finesse
Flow aggregation (MPLS)
Forward bias
Forward error correction
Forward pumping
Four-wave-mixing
Fourier transform
Fourier-transform limited (pulse)
Fourth-order Dispersion
FP (filter)
Free carrier
Free spectral range
Frequency
Frequency mixing
Fresnel lens
Front-end amplifier
FSC-LSP (GMPLS)
FSR
FTL
Fundamental mode
Fundamental soliton
Fusion splicing
FWM
G.709 (DW)
GI-POF
Ga
Gain
Gain coefficient

Gain ripple, curvature, tilt saturation
Gain-equalizing or flattening filter
Gain-guided laser
Gallium
Gaussian (distribution, approximation)
Ge
Geometrical birefringence
GEF
Generalized multiprotocol label switching
Germanium
GFF
Glass optical fibers
GMPLS
GOF
Gordon-Haus effect
Graded-index (profile, fiber)
Graph
Grating (fiber)
Grooming (traffic)
Group index
Group velocity
Group-velocity dispersion
Guided mode
GVD
H-field
Half-wave plate
He-Ne
Heterojunction
Heterostructure
Hi-Bi (fiber)
Hi-lo (DFB)
Higher-order mode
Higher-order PMD
Hole
Holey fibers
Homojunction
Homostructure
Hybrid pumping
IFBG
ILM

IM-DD
Illumination (front, back)
Impact ionization
Index-guided laser
Indium
InGaAsP
In-fiber Bragg grating
Index (refractive, relative, profile)
Infrared
Integrated laser-modulator
Intensity modulation
Inter-modal dispersion
Interaction length (coupler)
Interaction length (nonlinearity)
Intrinsic (semiconductor)
Ionization coefficient
IR
Isolator (optical)
Isotropy (medium)
ITU-T
Jacket (fiber)
Johnson noise
Kerr effect
L-band (EDFA)
L-I responsivity
Label distribution protocol
Label edge router
Label-switched path
Label-switched router
LAN
Lanthanides
Large effective-area fiber
Laser (diode)
Laser cavity
Layers 1, 2, 3 (network)
LDP (MPLS)

LEA (fiber)
LED
Left-handed (SOP)
Legacy (fiber)
LER
Lifetime (fluorescence or spontaneous-emission)
Light-emitting diode
Light-year
Lightpath (WDM, graph)
Lightwave
Line-switched protection (ring)
Linewidth
Linear polarization
Linear response
Link-management protocol
LMP (GMPLS)
Local-area network
Logical link (graph)
Long-period grating
Longitudinal mode
Lorentzian (line shape)
LOS
Loss (fiber, coefficient, insertion, excess, absorption or intrinsic, waveguide, return, coupling, splitting)
LP_{lm} mode
LSC-LSP (GMPLS)
LSR
Lumped amplification
Mach–Zehnder (interferometer)
MAN
MASER
MEMS
Mesh (network)

Metro (network, access, core, edge, enterprise)
Metropolitan-area network
MFD
Michelson interferometer
Microelectromechanical system
Microstructured fibers
MILP
Mitigation (PMD)
Mixing products
MM (fiber)
Modal dispersion
Mode
Mode coupling, overlap, hopping
Mode size, field-diameter
Modulation (external/direct)
Moore's law
MPLS
MPλS
MPOF
MQW
MS-DPRING
MS-SPRING
Multiquantum-well
Multistage EDFA
Multimode (laser)
Multiplexing
Multiprotocol label switching
MUX
MZ or MZI
n-type (semiconductor)
NA
Nd
Neodymium
Net gain
Network convergence
NF
Noise figure
Non-blocking (switch, strictly, rearrangeably)
Nongeneralized MPLS
Nonradiative decay
Nonreciprocal (device)
Nonlinear (susceptibility, response, refractive index)
Nonlinearity (fiber)
Normal modes
Normalized frequency
NRZ
NSFNET
Numerical aperture
NZDSF
NZDSF±
OADM
OAM&P
OAS (SDH/SONET)
OC-n
OCC (ASON)
OCH (SDH/SONET)
OCH container
OCH-OH (DW)
OCH-PE (DW)
OCH-SPRING
OD
ODU (DW)
O/E (modulation, conversion, regeneration)
O/E/O (repeater, regenerator)
OMS (SDH/SONET)
Operations, administration, maintenance and provisioning
OPA
OPS
Optical activity
Optical amplifier section (SDH/SONET)
Optical (fiber) amplifier
Optical channel (SDH/SONET)
Optical connection controller (ASON)
Optical circuit (SONET)
Optical cross-connect
Optical fiber
Optical gateway
Optical layer (network)
Optical multiplex section (SDH/SONET)
Optical packet switching
Optical preamplifier
Optical rectification
Optical repeater
Optical transmission section (SDH/SONET)
Opto-electronic
OPU (DW)
Ordinary (index)
OTS (SDH/SONET)
OTU (DW)
OSI model
Outage (system)
OXC
P
P2P
p-i-n (diode)
p-type (semiconductor)
Panda fiber
Parametric (effect, mixing, gain)
Partition noise (modal)
Passive optical network
Path-switched protection (ring)
PCD
PCE
PDF
PDG
PDL
Perfect shuffle
Phase-locked loop
Phase matching
Phase mismatch
Phase velocity
Phonon
Phosphorus
Photocurrent
Photodetection
Photoelectron
Photoreceiver
Photosensitivity
Photon
Photon flux
Photonic-bandgap fibers
Photonic switching
Physical link (graph)
Physical topology
PIN (diode)
Planar (technologies, waveguides)
PLL
PMD
PMD compensation or mitigation
PMF
PMMA
p-i-n
PIN
pin-FET
p-n junction
Pockels (effect, coefficient)
Point-to-point link
POF
Polarization
Polarization crosstalk

My Vocabulary

Polarization-
  maintaining
  (fiber) dependent
  loss/gain
  diversity (beam)
  splitter controller
Polarization-mode
  dispersion
Polarization-
  dependent
  chromatic
  dispersion
Polarizer
Polymer fiber
PON
POP
Population (atomic)
Population
  (inversion)
Potential barrier
Power conversion
  efficiency
Pr
Praseodymium
PRBS
Pre-emphasis
  (signal)
Preform (fiber)
Principal states of
  polarization
Probability-density
  function
Propagation
  constant
Protection (ring, span,
  line)
Protocol-transparent
PSC-LSP (GMPLS)
Pseudo-random bit
  sequence
PSK
PSP
Pump
Pump multiplexer
Pulse width
Push-pull (MZI)

Q-factor
Quanta, quantum
Quantum limit
Quantum efficiency
QED
Radiative decay
Raman fiber
  amplifier
Raman gain
  coefficient
Raman shift
Rare-earth
Ray (light)
RDF
RE
Receiver
Restoration (network,
  path, link)
Reverse-dispersion
  fiber
RFA
Right-handed (SOP)
Ring (network)
Reflection
Refraction
Refractive index
Relative index
Restoration fiber
Restoration path
Reverse bias
Reverse-dispersion
  fiber
Rotatory power
Router
RZ
SA
Saturable absorber
Saturation (gain)
SBS
SDH
Second-harmonic
  generation
Second-order PMD
SELD
Self-healing (ring)
Self-induced SRS

Self-phase
  modulation
Semiconductor
  optical amplifier
  intrinsic n-type or
  p-type
Sensitivity
  (receiver)
SHG
Si
SI-POF
SI-SRS
Silicon
Single-mode
  (laser,
  waveguide, fiber)
Single-polarization
  fiber
Skewed ray
SLD
Slope (dispersion)
SM (fiber)
Small-signal gain
SMF
SNCP
Snell's law
SOA
SOA gate
Soliton
Soliton interaction,
  collision,
  self-frequency
  shift,
  path-averaged,
  guiding-center
SONET
SOP
Spanke switch
  (architecture)
Spatial walk-off
  polarizer
Speckle (laser/mode)
Spectral efficiency
SPM
Spontaneous
  emission

Spontaneous
  emission factor
Spot size
SRS
SSFS
SSLR
SSPR
Stark levels
State of
  polarization
Static restoration
Step-index (profile,
  fiber)
Stimulated emission
Stimulated Raman
  scattering
STM-n
Stress-induced
  birefringence
Stokes parameters
Stokes wave/signal
Strip waveguide
Substrate-entry
  illumination
Susceptibility
  (linear/nonlinear)
Switch (photonic)
Switching (path,
  wavelength,
  time)
SWP
Symmetric (mode)
Synchronous digital
  hierarchy, optical
  network transport
  module
TBP
TDM
TDM-LSP
  (GMPLS)
TE
Terabit
TeraKIPS law
RSVP-TE (MPLS)
TEM_{lm} mode

Temperature coefficient (MUX/DMUX)
TFF
THG
Thermal noise
Thermo-electric (cooler)
Thin-film filter
Third-harmonic generation
Third-order dispersion
Threshold (laser, pump)
Thulium
Time-bandwidth product
Time width
Timing jitter
Tm
Top-entry illumination
Total internal reflection
Transceiver
Transmission window(s)
Transmitter
Transparent routing, network
Transverse electric/magnetic (mode)
Tree coupler
Tunable laser
ULSR, ULSR/n
Ultraviolet
Un-polarized (light)
UPSR, UPSR/n
USHR
UV
V-number
Vacuum fluctuations
Valence band
VCO
VCSEL
Velocity (light)
Verdet constant
Vertex (graph)
Virtual topology (network, design)
Voltage-controlled oscillator
VTD, VTD-MILP
Wall-plug efficiency
WAN
Wave mixing
Waveguide (planar, slab, ridge, fiber)
Waveguide grating router
Wavelength
Wavelength converter, cross-connect, comb, grid
Wavelength routing
Wavelength-routed network
Wavelength-selective coupler (or cross-connect)
Wavelength-interchanging cross-connect
Wavelength-switched network
Wavevector
WDM
Weakly-guiding fibers
WGR
WIC
Wide-area network
Working fiber
Working path
WSC (coupler, cross-connect)
WXC
XPM
Y-branch (coupler)
Yb
YIG
Ytterbium
2×2 (3dB) coupler
3R regeneration

CHAPTER 3

Cryptography and Communications Security

This chapter first describes the basics of cryptography through history. We then address "modern" cryptography, as based upon binary coding and algorithms. The main encryption standards are reviewed and illustrations of basic and futuristic applications are provided. The field of quantum cryptography and its potential implications are also addressed.

■ 3.1 MESSAGE ENCRYPTION, DECRYPTION AND CRYPTANALYSIS

If some content-sensitive or classified *message* must be transmitted to someone in exclusive secrecy, for fear of the consequences of its being known or revealed to other people, one can chose one out of four basic approaches.

The first approach is to deliver the message yourself. Or you can have it delivered by a trusted party. This happens when ambassadors carry inside their suit pockets a letter addressed to a chief of state, or when documents are sent through a sealed "diplomatic bag." Locally, sensitive documents can be handed out after traveling with a police or army escort. Normal citizens trust the post office, certified mail, or express-delivery carriers to guarantee, at a lower scale of importance, the same type of ultimate confidentiality or privacy. But all these approaches are intrinsically insecure. This is because there is always a finite probability that the documents could be maliciously intercepted, read, duplicated (altered or even substituted!), without the awareness of the sender, the carrier or the destination party.

•

Wiley Survival Guide in Global Telecommunications: Broadband Access, Optical Components and Networks, and Cryptography, by E. Desurvire
ISBN 0-471-67520-2 © 2004 John Wiley & Sons, Inc.

The second approach is similar to the first one, but the message is dissimulated in such a way that no one could suspect its very existence, or would have some difficulty in finding it. This approach is referred to as *steganography*, from the Greek *steganos* (covered) and *graphein* (to write). The early Chinese used invisible ink on silk, early Greeks used clay tablets covered with wax. The spy history abounds in examples of steganographic methods, the most classical one remaining the tiny microfilm, for instance glued on a dot in a mail letter. Yet the same negative conclusion applies to any of these approaches. The messages are in no way protected against third-party interception, since all possible "professional tricks" become eventually known, let alone the good old and classical ones.

The third approach consists in communicating the message *in the open*, under the form of innocuous or even meaningless *words, expressions* or *messages*, whose second-degree significance is intelligible only to the receiving party. In earlier centuries, a young woman wearing a rose (instead of a ribbon) on her hat during the afternoon stroll could indicate to her lover (incidentally passing by) "Sunday right after church service is the right time (or not the right time) to introduce yourself to my father." There are no limits to the variety of open codes, from *visual signs*, to *spoken* or *written* messages. A text version could take the form of an informal letter, where only certain characters carry the secret message, using a reference grid to determine their positions. The letter has then to be re-arranged by the recipient as an array of characters of the same grid size. The recipient uses the same grid to mask unwanted characters and reveal the significant ones (e.g., WILLYOUMARRYME in the sentence "... by now, I'm st*ill* a *you*ng *m*an in the *a*rmy *r*anks, *y*et *me*ditating..."). Such open texts, normally having a large proportion of characters to be discarded, are called *null ciphers*. They are efficient but never safe, since the meaning can be inferred without a grid, provided the interceptor has some intuition of what is going on and looks for specific letters or word patterns (e.g., previous example). The main difficulty for the openly-communicating parties is to first exchange the grid and then keep it hidden in some secret place. There are many other null ciphers which use different rules of re-arrangement by rows and columns, which can then be read vertically or diagonally. The advantage is that once the coding/decoding rule is agreed, it can be memorized without leaving physical clues.

Open-code *voice* messages were broadcast during World War II for the attention of active intelligence and resistance groups. Routine details at the midpoint of a weather forecast could hide meanings (e.g., "gusty winds blowing westwards" = "enemy [from East front] may strike tomorrow"). *Jargon codes* to send/acknowledge information in the form of anonymous "personal communications," for instance "the train will be on time" (supplies coming as planned), "the dice are on the table" (go ahead with certain option of sabotage plan), or "the elephants eat strawberries" (new Russian collaborator authenticated), etc. Such open messages are destined only to parties that have been tipped beforehand on whatever they mean in terms of action/contingency plans, news and other mutually agreed acknowledgment procedures. The most famous example is the announcement of D-Day through the first stanza of Verlaine's poetry "*Chansons d'Automne*." The first half of the stanza (in English, "The long sobs of violins in autumn..." would be periodically broadcast to announce the imminence of the Allied invasion. But if the stanza ever included the second half ("...wound my heart with monotonous langor"), this meant that D-day was scheduled for tomorrow, starting at 00:00 hours! This broadcast took place indeed on the eve of June 5th, 1944. Reportedly, it was even intercepted by the occupiers, who discarded it in ignorance. Here, the key was not to be found in the message contents but in the inclusion of the second-half! Despite these historical successes, jargon codes have inherently limited information power, which is essentially reduced to YES/NO and OK/not-OK timing signals. Communicating the meaning or *key*, of these codes is also a very risky task. Third parties may catch up with the keys through different interception and coercion means in the network. The approach remains secure only if the codes can be regularly and rapidly renewed. But this also means extensive communication of code keys, and therefore increased risks of interception.

The fourth approach is *cryptography*. It consists in making the original message completely unintelligible by means of different *encryption* algorithms. The word comes from the Greek *kryptos* (secret), or the action of making a message a self-contained secret. Encryption thus changes the original text, or *plaintext*, into a *ciphertext*. The reverse operation, changing a ciphertext into the original plaintext, is called *decryption*. A third party trying to make sense of a ciphertext must resolve the enigma presented by the crypted message through more or less extensive work, called *cryptanalysis*. The goal of cryptanalysis, also called *code-breaking*, is to *break* the cipher, meaning recovering the plaintext by practical, inexpensive, and relatively effortless methods. The concept of *code cracking* is somewhat different. It could be defined as breaking a code *for the first time*. Put simply, a code cracker is a champion code breaker. The goal of the code cracker is not so much to retrieve secret messages (for which he cannot care less) as to prove to himself and to history that he is the one who "cracked" the previously reputed "invulnerable" or "unbreakable" cipher. There will be more on this interesting topic later in this chapter.

In the traditional code-breaking job, rapidity of execution is essential, since the plaintext may be more or less highly time-sensitive, for instance in situations of war, diplomatic crisis or threat. This observation stands for both the *receiving* party and the *intercepting* third-party. Decryption is based upon the principle of a *secret key*. As described below, this key can be anything from a method (the step-by-step reverse *algorithm* of encryption), a code word (named *keyword*) or a reference database (correspondence *tables*, *squares* or *grids*). As in the three previous approaches, the secret key must be exchanged by some safe and discreet means prior to communicating the crypted message. This exchange involves some risk, for the same reasons as previously. However, the key can be made very complicated, be held in high secrecy, be communicated by very indirect means, be rapidly and randomly changing, so that little chance is left for the interceptor to figure it out on time and to succeed in decrypting or breaking the code.

The triple problem of securely transmitting the secret key, of being totally sure that no one will able to intercept it and break/decrypt the ciphertext, and of being "absolutely confident" that the message secret remains invulnerable to cryptanalysis, is central to cryptography. Until the discovery of *public-key cryptography* (see later section), the lack of truly absolute confidence has been prevalent and caused rapid progress in the sophistication of encryption algorithms. In the rest of this section, we shall review a certain number of classical encryption algorithms which were used throughout history, all of them having at some turning point proven definitely unsafe or inefficient, long before the computer appeared.

Before going through the technical details of cryptography, it is important to stress an essential requirement. This requirement is that the language of the message to be encrypted and decrypted be known and mastered by the communicating parties. This may seem like an obvious statement but in fact it is nontrivial. Two people speaking a common, but rare dialect (Papua New Guinea having for instance as many as 700) don't need any cryptography! The idea behind using cryptography is the fair supposition that the destinee is fully conversant with the tongue and language used, or has all means to have it interpreted by some native. This brings the issue of the *ancient scripts* and *lost languages* which still strike our imagination. Clearly, the documents whose alphabet and language signification disappeared were never written for any cryptography purposes. Yet, some of these remained until the nineteenth and the twentieth centuries like gigantic enigmas, challenging generations of the most gifted archeologists and linguists. The two famous examples are the Egyptian *Rosetta stone*, and the Greek *Linear B* tablets. A riveting account of the heroic endeavors to crack these enigmas can be found for instance in the famous books by Singh (*The Code Book*) and Kahn (*The Codebreakers*), as listed in the Bibliography. Although the discipline of "breaking" ancient mysteries seems remote from modern crypto-graphy, the investigation methods have many similarities. It is also a little-known fact that the U.S. Army used the complex Indian *Navajo* tongue in order to secure communications during the Pacific war. Other possibilities for native-American languages were discarded because of the possibility that academic linguists from the other camp could have been already exposed to them! Rare or very difficult tongues can thus be used as a second level of encryption in order to fool the most agile code breaker. In the foregoing, we will just assume that the plaintext should not require extra interpretation or translation work. This illustrates that ideally, cryptography should remain simple to learn and to implement, with a single coding/decoding procedure. Such a compromise puts the emphasis on the strength of the cipher (its supposed invulnerability), rather than the complexity of the message (its possible enigmatic, unintelligible, or second-degree meanings).

3.1.1 Mono- and Multialphabetic Encryption

In this subsection, we review the two main types of encryption, which are based on various *alphabetic translation* or *substitution* methods. A straightforward way to encrypt a plaintext message is indeed to substitute the letter alphabet into another one, using a one-to-one symbol (or phonetic group) correspondence. The reader may have played with this in his or her childhood! The substitution alphabet can be made of novel characters or symbols (e.g., Martian-looking scripts), or just the same Roman letters with different forms of one-to-one mapping, translated or random. This is referred to as a *monoalphabetic* cipher algorithm. An example is provided here with the message "How are you doing?" to be encoded as "howareyoudoing" (plaintexts do not have capitals, spaces or punctuation):

Plaintext	h	o	w	a	r	e	y	o	u	d	o	i	n	g
Ciphertext	A	Y	C	X	D	Z	T	Y	U	B	Y	H	W	E

Just considering the Roman-letter substitution as above, the task of deciphering would seem to be easy if one just puts in the effort. But this is far from being the case. Indeed, there are as many as 26! possible permutations between the 26 letters, namely $26! = 1 \times 2 \times 3 \cdots \times 25 \times 26 = 4.0 \times 10^{26}$, which represents the huge number of *four hundred million billion billion* possibilities ($400 \times 10^9 \times 10^9 \times 10^6$). Assuming that each possibility could be checked out at once every second, breaking such a code would take some 13 billion billion years, i.e., about *one billion times the age of our universe*! A paradox is that even if such a requirement is indeed "impossible," such a cipher algorithm could in fact be broken by a five-year old child. This amazing feature will be demonstrated in the next subsection.

In any case, the main drawback of the monoalphabetic random substitution is that the destinee must have the corresponding coding table or key. It would be nicer if there were some secret rule for changing letters into one another. The emperor Julius Caesar used a letter-shifting algorithm, now referred to as the *Caesar shift cipher*. The concept is to shift letters by a certain number of places, from 1 to 25, forward or backward. For instance, a 3-place forward shift changes A to D, B to E, etc., and Z to C. Using this cipher algorithm, the previous example gives:

Plaintext	h	o	w	a	r	e	y	o	u	d	o	i	n	g
Ciphertext	K	R	Z	D	U	H	B	R	X	G	R	L	Q	J

In the above example, the cipher key is simply "+3", meaning that decryption can be done by a "−3" shift in letter places. But such a key is not so secure, considering that natural human weakness in that era (and the urgency of the communication) would normally force one to chose relatively small shifts. The intercepting third party could then guess that the possible key must fall the range −1 to +3, which greatly simplifies the cryptanalysis work. Note that this human weakness is still exemplified today in the way most people choose or modify their computer passwords!

A more sophisticated approach for monoalphabetic substitution is to use a pseudo-random mapping from an easily memorized (or communicable key). Two friends have the same book edition on the works of Leonardo Da Vinci. The key could then be "leonardo, page 535." On top of page 535 of this book, they can both read (for instance) the sentence "...by the turn of the century, Leonardo..." and generate the key "bytheturnofthecenturyleonardo." Removing any repeated letter, the key then becomes "bytheurnofcylad." It is then used for the alphabet mapping, as follows:

Plain alphabet	a	b	c	d	e	f	g	h	i	j	k	l	m	n	o
Cipher alphabet	B	Y	T	H	E	U	R	N	O	F	C	Y	L	A	D

and so on to letter Z. If the key does not contain all letters (which may happen if limited to a single book page), then the first available letter (not already used) is assigned in alphabetic order. Because of the quasi-impossibility for a third party to know in which book and page to find the key, the approach seems relatively safe. However, the two friends should have their own individual books instead of using a common one (e.g., in a public library). For improved security, they should change the book titles and pages often, while mutually agreeing on what they should be, which limits the efficiency and practicality of the key selection and exchange. But weakness might dictate always using the same book and page and bookmark it for quicker reference, which facilitates third-party investigation! Furthermore, the algorithm reduces to yet another type of monoalphabetic substitution, which can be broken by straightforward *frequency analysis* (see next subsection).

Long before monoalphabetic ciphers were broken, the above pitfalls in terms of key secrecy and communication risks had been identified. This led to the development of *polyalphabetic* cyphers, as early as the Renaissance Age. The florentine Alberti first suggested the use of more than one cipher alphabet, as illustrated in the following example:

Plain alphabet	a	b	c	d	e	f	g	h	i	j	k	l	m	n	o
Cipher alphabet 1	L	Y	W	H	A	K	O	C	N	U	R	X	P	T	J
Cipher alphabet 2	B	U	G	M	K	C	W	N	H	T	D	Z	A	E	I

The rule is to use the cipher alphabets alternatively. For instance, the name MICHEL ANGELO would be coded as PHWNAZ LEOKXI. In this example, we observe that the letter succession EL is first coded as AZ then as KX. The fact that the same letters can have different substitutions according to their position in the plaintext makes the ciphertext much more difficult to analyze and to break.

Based upon the works of Alberti and others, the French diplomat Vigenère finalized the development of an even more powerful algorithm, known as the *Vigenère cipher*, which will remain unbreakable for centuries. Hence its name, *le chiffre indéchiffrable* (the unbreakable code), which has long been up to its reputation. It is a polyalphabetic cipher with 26 substitution alphabets and a keyword which defines which alphabets to use. One needs first to define the substitution alphabets, which are shown as an example in Table 3.1. Each of the 26 alphabets appear in the orderly sequence ABC ... Z while their letters are moved by one to 26 positions according to the alphabet's row number, like many Caesar shifts. The first alphabet (row 1) is shifted by one position, the second (row 2) by two positions, and so on down to the last one (row 26), which is shifted by 26 positions and matches the plain alphabet. Anyone can thus remember this construction rule and write down his/her Vigenère square. The secret is how to use it, which requires a key exchange. Assume this key is the word ANGEL. In order to prepare the ciphertext, the originator first writes this key on top of the plaintex, as shown in the table below. Looking then at the second column of the Vigenère square in Table 3.1, the letter A of the key calls for using the substitution alphabet of line 26, which begins with A. Then we use that of line 13, which begins with N, and so on until we are back to A. The plaintext

TABLE 3.1 Example of *Vigenère square*

Plain	a	b	c	d	e	f	g	h	i	j	k	l	m	n	o	p	q	r	s	t	u	v	w	x	y	z
1	B	C	D	E	F	G	H	I	J	K	L	M	N	O	P	Q	R	S	T	U	V	W	X	Y	Z	A
2	C	D	E	F	G	H	I	J	K	L	M	N	O	P	Q	R	S	T	U	V	W	X	Y	Z	A	B
3	D	E	F	G	H	I	J	K	L	M	N	O	P	Q	R	S	T	U	V	W	X	Y	Z	A	B	C
4	E	F	G	H	I	J	K	L	M	N	O	P	Q	R	S	T	U	V	W	X	Y	Z	A	B	C	D
5	F	G	H	I	J	K	L	M	N	O	P	Q	R	S	T	U	V	W	X	Y	Z	A	B	C	D	E
6	G	H	I	J	K	L	M	N	O	P	Q	R	S	T	U	V	W	X	Y	Z	A	B	C	D	E	F
7	H	I	J	K	L	M	N	O	P	Q	R	S	T	U	V	W	X	Y	Z	A	B	C	D	E	F	G
8	I	J	K	L	M	N	O	P	Q	R	S	T	U	V	W	X	Y	Z	A	B	C	D	E	F	G	H
9	J	K	L	M	N	O	P	Q	R	S	T	U	V	W	X	Y	Z	A	B	C	D	E	F	G	H	I
10	K	L	M	N	O	P	Q	R	S	T	U	V	W	X	Y	Z	A	B	C	D	E	F	G	H	I	J
11	L	M	N	O	P	Q	R	S	T	U	V	W	X	Y	Z	A	B	C	D	E	F	G	H	I	J	K
12	M	N	O	P	Q	R	S	T	U	V	W	X	Y	Z	A	B	C	D	E	F	G	H	I	J	K	L
13	N	O	P	Q	R	S	T	U	V	W	X	Y	Z	A	B	C	D	E	F	G	H	I	J	K	L	M
14	O	P	Q	R	S	T	U	V	W	X	Y	Z	A	B	C	D	E	F	G	H	I	J	K	L	M	N
15	P	Q	R	S	T	U	V	W	X	Y	Z	A	B	C	D	E	F	G	H	I	J	K	L	M	N	O
16	Q	R	S	T	U	V	W	X	Y	Z	A	B	C	D	E	F	G	H	I	J	K	L	M	N	O	P
17	R	S	T	U	V	W	X	Y	Z	A	B	C	D	E	F	G	H	I	J	K	L	M	N	O	P	Q
18	S	T	U	V	W	X	Y	Z	A	B	C	D	E	F	G	H	I	J	K	L	M	N	O	P	Q	R
19	T	U	V	W	X	Y	Z	A	B	C	D	E	F	G	H	I	J	K	L	M	N	O	P	Q	R	S
20	U	V	W	X	Y	Z	A	B	C	D	E	F	G	H	I	J	K	L	M	N	O	P	Q	R	S	T
21	V	W	X	Y	Z	A	B	C	D	E	F	G	H	I	J	K	L	M	N	O	P	Q	R	S	T	U
22	W	X	Y	Z	A	B	C	D	E	F	G	H	I	J	K	L	M	N	O	P	Q	R	S	T	U	V
23	X	Y	Z	A	B	C	D	E	F	G	H	I	J	K	L	M	N	O	P	Q	R	S	T	U	V	W
24	Y	Z	A	B	C	D	E	F	G	H	I	J	K	L	M	N	O	P	Q	R	S	T	U	V	W	X
25	Z	A	B	C	D	E	F	G	H	I	J	K	L	M	N	O	P	Q	R	S	T	U	V	W	X	Y
26	A	B	C	D	E	F	G	H	I	J	K	L	M	N	O	P	Q	R	S	T	U	V	W	X	Y	Z

The keyword ANGEL defines which substitution alphabets should be used.

is then coded according to the sequence of substitution alphabets thus defined, which yields the ciphertext shown.

Key	A	N	G	E	L	A	N	G	E	L	A	N	G	E
Plaintext	h	o	w	a	r	e	y	o	u	d	o	i	n	g
Ciphertext	H	B	C	E	C	E	L	U	Y	O	O	V	T	K

It is seen from the above example that letters appear several times in the ciphertext, but have different meanings, as in all polyalphabetic substitution. The difference here is that the choice is not reduced to two or three alternate substitutions according to a cyclic rule, but is virtually unlimited. Indeed, the key can be as long as the message itself, and can have repeated letters (e.g., "bytheturnofthecenturyleonardo") which makes the substitution close to random. In practice (weakness showing again), the two parties will agree on a shorter key, easy to memorize and quick to refer to, for instance a first name like LEONARDO. This brings the period of the substitution cycle to 8, a respectable number for the time, but definitely not out of reach for modern code crackers.

In order to recover the plaintext, the destinee just needs to write the key on top of the ciphertext and do the reverse Vigenère square substitution. The great strength of the Vigenère cipher is that the letters do not show any recognizable frequencies or group patterns (see next subsection). Its main drawback (at the time) was its complexity of use, requiring a minimum of patience and application. But political or military communications often require expediency and simplicity, which excludes the principle of such complicated solutions (especially to the extent that the two communicating parties are usually very important and busy people). For this reason, the monoalphabetic substitution remained the preferred approach for several centuries, in spite of its weakness and the tremendous superiority of the Vigenère cipher. Predictably, but not before the nineteenth century, the Vigenère cipher was eventually "cracked," owing to the work of the British genius C. Babbage. This event is considered to be the second important milestone in the field of cryptography after the earlier discovery of *frequency analysis*, which we describe in the next subsection. During World War I, however, the U.S. Army cryptographers realized that the Vigenère cipher was "absolutely" safe if the keywords were chosen as long as the plaintexts, and furthermore, if they were void of any structure, as based upon a random letter selection. This approach required the mutual communication of long and complex keywords, but to increase security these keywords were used only once and destroyed immediately thereafter. This approach is referred to as *onetime pad cipher*. It can be shown mathematically that onetime pad ciphers with random keys are absolutely unbreakable. But this requires lots of time, effort and resource to generate and securely distribute the keys, which is incompatible with the operations of army command in times of crisis. It is however used nowadays for hotline communications between chiefs of state.

3.1.2 Frequency Analysis

As previously stated, the pitfall of monoalphabetic and polyalphabetic ciphers is that they are cyclic. Therefore, a more or less rapid analysis of the ciphertext may reveal that certain letters, groups of letters or group patterns frequently come back. The technique of searching for such repeated occurrences, which was invented and developed in the ninth century by the Arabian scientist and linguist Al-Kindi, is called *frequency analysis*.

In any language, alphabetic letters do not have the same frequency of occurrence. Some letters are always used more often and some more rarely. In English, for instance, the most three common letters are $E/T/A$ and the least three common are $Q/J/Z$, as further described below. Such a frequency distribution is language-dependent. Considering for instance some of the European tongues, the first three common letters in French, German, Italian, Portuguese and Spanish are $E/A/I$, $E/N/I$, $E/A/I$, $A/E/O$ and $E/A/O$, respectively. The same observation applies to groups of two and three letters, called *digrams* and *trigrams*. In English, the ten most common digrams and trigrams are $TH/IN/ER/RE/AN/HE/AR/EN/TI/TE$ and $THE/ING/AND/ION/ENT/FOR/TIO/ERE/HER/ATE$, respectively (for complete data see for instance the book from Fouché Gaines, *Cryptanalysis* listed in the Bibliography). Digrams and trigrams should not be confused with 2-letter and 3-letter words, which represent isolated sequences. To illustrate this point, here are first ten most-frequent 1–3-letter English words: $THE/OF/AND/TO/A/IN/THAT/IS/I/IT$.

Although appearing as one-, two-, or three-symbol substitutes, these frequency patterns betray any possible meaning. The first task of any cryptanalysis is thus to make the inventory of symbols used and classify them in decreasing order as single-symbols, then digrams and trigrams. The result is a histogram, also called *frequency spectrum*. If the ciphertext is long enough and was made from an English plaintext, it is very likely that the most common three symbols have the meaning of $E/T/A$, the most common symbol digrams have the meaning of $TH/IN/ER$, and so on. One can then try out the different matching possibilities between symbols (then digrams, trigrams, and 2/3-letter words, as required), and see if such a substitution begins to generate a fragmented text that makes any sense. If we know that the message is a military order, we might search for words such as "strike," "attack," "supply," "plan," "tomorrow," etc. Such an initial hunch, along with the lucky matching of frequently used symbols and patterns, is called a *crib*. This notion of crib is essential in code-breaking science, as further discussed through this chapter. Once a crib has lifted a bit of mystery from the ciphertext, code-breaking can proceed extremely rapidly. Then there is no need to spend the age of the universe (or so) to check out the 26! alphabet substitution possibilities: at human-logic scale, the result may come in minutes, and at computer scale, in a few microseconds.

The longer the ciphertext, the most dependable the frequency statistics, and the more rapid the discovery of cribs. Obviously, these statistics depend upon the tongue, the era, and the style and the type of the original plaintext. Figure 3.1 shows the plot of one of the reported probability distributions (relative %) of English letters, including spaces. We observe from this plot that the letters are exponentially distributed, with "space" being the most frequent symbol (19%), followed by $E/T/A/O$ (>6%), and so on. Since spaces are so highly probable, it is sensible to remove them from plaintexts. Otherwise, this property would be reflected in the

FIGURE 3.1 Probability distributions or frequency spectrum of written-English letters (after tabulated data from F. Pratt, *Secret and Urgent*, Bobbs-Merril Book Company, Indianapolis, 1942).

ciphertext, immediately revealing where any of the words begins and ends, with their full lengths. This feature would surely facilitates any code-breaker's task! But even removing spaces does not make the work so much more difficult, since several other frequently-appearing symbols should be revealed by frequency analysis.

It would be too easy if frequency analysis could rely upon a universal probability distribution for each language. As a matter of fact, the distribution varies with the type and contents of the text. Figure 3.2 shows a comparison between the previous distribution and a different one obtained by sampling English newspapers and novels over 100,000 characters. Not surprisingly, we observe a good correlation between the two distributions (with $E/T/A/O$ as the most common letters), but some differences appear here and there after the fifth position. Dependable as such statistics may be, they have a reference value only if the ciphertext is sufficiently long and representative. Let us make a test by taking this very paragraph of 600 alphabetic characters as an example.

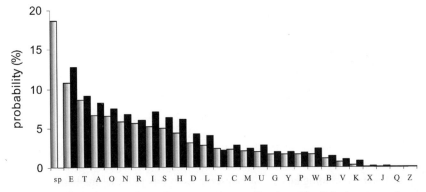

FIGURE 3.2 Comparison of two frequency spectra of written-English characters: in gray, from data of Figure 3.1, and in black from another source (after tabulated data in S. Singh, *The Code Book*, Anchor Books, New York, 1999).

The frequency analysis of the preceding paragraph is easily performed through a personal computer. While it is easy to write up a small program, an even quicker solution for a one-shot test consists in using the command CTRL-H to substitute each of the 26 characters one after another by a space (for instance), and directly read the number of substitutions reported in the command window. The percentage probabilities can then be plotted after normalization of the results by the factor $100/N$ ($N = 600$). Figure 3.3 shows the frequency analysis plot, in comparison with the previous reference distribution. The agreement between the two is quite remarkable (the reader may trust that we only performed a single, one-shot try), except for letters O and H. If this plaintext example was encrypted into a monoalphabetic cipher, it would takes only seconds using the cipher spectrum to make a good match with the original characters.

Let us now perform the same experiment with *digrams*, using the same text sample. Figure 3.4 shows the corresponding frequency spectrum, along with that of the 25 most common English digrams, rearranged according to the decreasing symbol order obtained in the first spectrum. In this second reference spectrum, the distribution has also been normalized so that the two exponential decays more or less coincide. If we overlook the *TH* overshoot in the second position, we observe that for the first 10 digrams, the match and correlation of the two spectra are excellent. Past this point, the reference spectrum becomes increasingly chaotic and uncorrelated. This is expected, since the probability differences between

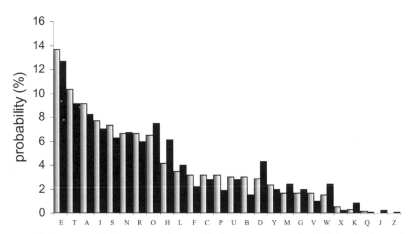

It would be too easy if frequency analysis could rely upon a universal probability distribution for each language as a matter off fact the distribution varies with the type and contents of the text figure shows a comparison between the previous distribution and a different one obtained by sampling English newspapers and novels over characters not surprisingly we observe a good correlation between the two distributions with etao as the most common letters but some differences appear here and there after the fifth position dependable as such statistics may be they have a reference value only if the cipher text is sufficiently long and representative tet us make a test by taking this very paragraph of alphabetic characters as an example

FIGURE 3.3 Frequency analysis (single letters) of preceding 600-character paragraph "It would be too easy... as an example" (text below with only alphabetic characters), corresponding to the gray bars. For comparison, the reference spectrum of the previous figure is shown in black bars.

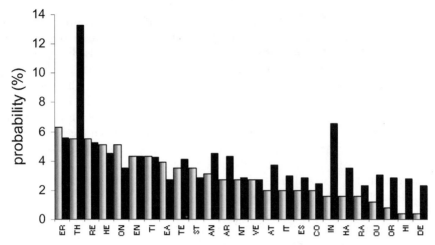

FIGURE 3.4 Frequency analysis (digrams) of same text sample as in previous figure (gray bars), and corresponding histogram (arbitrary scale) of the 25 most common digrams in English (black bars).

digrams progressively decrease, or the digrams become increasingly equiprobable. A longer sample text (e.g., 5,000 characters) however, would surely provide a much better correlation over this 25-digram scale.

While this preceding frequency-analysis experiment with single-letters and digrams may appear fairly conclusive, one should not conclude too rapidly that monoalphabetic code-breaking is so straightforward. The ciphertext may be way too short to make sense of any statistics. Or it may be very specific in terms of less-likely letters (e.g., "a zebra from Zanzibar named Zazie has a bizzarre zest for Jazz"). Or it can be a memo full of technical words and acronyms that are very unusual for the language (e.g., "www," "dichlorodiphenyltrichlorobenzene"). However, if one knows that the plaintext is about internet or organic chemistry, these may represent many potential cribs. Finally, the plaintext may use tricks to fool any frequency analysis. In his book, Singh mentions the intriguing case of a 200-page novel from G. Perec (*"La Disparition"*) which does not contain *any* letter *E*. Even more amusing is that when this book was later printed in English, the translator fully respected the author's choice of never using letter *E*! It would be interesting to make the frequency analysis of these two books and compare them to the French and English references. We can infer that they would look quite different, based upon the fact that avoiding *E* forced both author and translator to pick less usual words. Code breakers can only hope that ordinary plaintext messages make fair use of the most common letters and words!

As mentioned in a previous subsection when describing the Vigenère algorithm, frequency analysis also applies to *multialphabetic* ciphers. It is made more complicated since the repetition cycles are arbitrary. But if the communicating parties use reasonably small keys (e.g., 5–6 letters long), and if the text is sufficiently long, then frequency patterns can be readily identified. Any permutation trick, by line or column and any combination thereof, aiming to further scramble the cipher

3.1 Message Encryption, Decryption and Cryptanalysis

symbols, whether mono- or multialphabetic, remains ineffective in view of this frequency analysis. The analysis will detect symbol patterns anyway.

Since frequently-appearing characters and groups of characters are spotted by frequency analysis no matter how they are scrambled (unless impractical, heavy-duty multialphabetic substitution is used), cryptographers have sought for more effective and straightforward ways to encrypt messages. The idea is now to erase traces of any such character/pattern occurrences. A clever solution to this problem is to assign not *one* symbol, but *several* symbols to each character to be substituted. The number of corresponding symbols should be in proportion to the character frequency in the plaintext language. As a result, characters must be coded by as many symbols, to be chosen at random in a list (known to the destination party), as the character is frequent. This means that more than 26 symbols are required. A simple way to generate extra symbols is to code the characters by two numbers, i.e., from 00 to 99, which is referred to as a *binumeral code*. Table 3.2 shows one possible symbol attribution, which reflects (and also lists) the reference probability distribution of Figure 3.2 or Figure 3.3. We can see from this table that character E and T are attributed 13 and 9 symbols, respectively, and so on down to the least frequent characters (P–Z), so that the relative weights (or plaintext frequencies) are closely respected. Each time a character must be substituted, one of the attributed symbols must be picked at random, avoiding repeats until the list is exhausted. Here is an example:

Plaintext   l   e   o   n   a   r   d   o   a   v   i   n   c   i
Ciphertext 37 63 32 38 74 17 72 87 61 43 33 57 91 80 59

Since the total of symbols is exactly 100, all symbols have a close to 1% equal chance of appearing in the ciphertext. On a 00–99 scale, the frequency spectrum is nearly uniform and essentially featureless, as illustrated in Figure 3.5, unlike in the previous cases of mono/multialphabetic ciphers.

The above approach is referred to as a *homophonic* substitution cipher. The word means "which sounds the same," like all 13 symbols for E, or all 9 symbols for T do "sound" like E or T in the ciphertext, but only to the "ears" of the communicating parties.

Homophonic substitution can be related to the concepts of *coding efficiency* and *information entropy*. While we recommend referring to that section, we shall briefly recall here the concepts. The underlying idea is that a communication channel is most efficient in transmitting information when all symbols (n) forming an alphabet have *equal* probabilities of occurrence in the message (p_n). This means that in this case, there is maximum uncertainty as to which symbols are used, and to that extent, the information content of the message is also maximum. The entropy (H) is the direct *measure of information*, and is expressed in *bits*. Such a measure represents the number of elementary bits of information that can be conveyed (on average) by the alphabet symbols. Formally, the entropy is defined as:

$$H(X) = -<\log_2 p> \equiv -\sum_n p_n \log_2 p_n \qquad (3.1)$$

If all N symbols of the X alphabet have equal probabilities of occurrence (namely, $p_n = 1/N$), the entropy is maximum and equal to $H_{\max} \equiv \log_2 N$ bits. For binary

TABLE 3.2 Example of homophonic substitution alphabet, with number of symbols attributed in proportion of the character frequency

Character	Probability	Character Weight	Symbols												
E	12,7	13	09	12	18	22	31	35	48	63	75	77	81	88	97
T	9,1	9	05	08	27	40	55	67	78	79	86				
A	8,2	8	16	25	28	43	58	74	84	94					
O	7,5	7	32	39	46	54	62	87	89						
I	7	7	00	29	42	57	59	85	90						
N	6,7	7	34	38	50	52	60	91	95						
S	6,3	6	47	49	56	68	82	93							
H	6,1	6	04	10	11	15	53	92							
R	6	6	17	26	36	64	76	99							
D	4,3	4	41	61	72	83									
L	4	4	01	23	37	51									
C	2,8	3	21	30	80										
U	2,8	2	19	96											
M	2,4	2	13	69											
W	2,4	2	24	98											
F	2,2	2	03	70											
G	2	2	14	20											
Y	2	2	33	73											
P	1,9	1	65												
B	1,5	1	07												
V	1	1	44												
K	0,8	1	71												
J	0,2	1	06												
X	0,2	1	02												
Q	0,1	1	45												
Z	0,1	1	66												
		100													

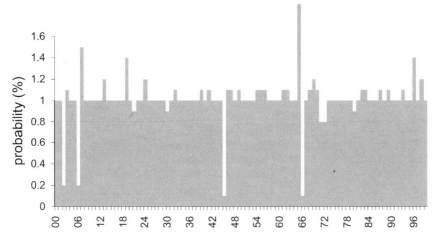

FIGURE 3.5 Frequency spectrum of the 00–99 monoalphabetic cipher defined in Table 3.3 (probability distribution p').

signals ($N = 2$), the maximum entropy is thus $H_{max} = 1$, meaning that there is one bit of information in each symbol transmitted via the communication channel. In the non-optimal case, some symbols have a higher degree of occurrence than others, meaning that their uncertainty is lower, and the entropy is also lower. In summary, if some symbols are more likely to be used, we can learn more of the information content of the message.

How does the above ever relate to cryptography? Simply through the fact that cryptography is meant to transmit a given amount of information while making maximum use of uncertainty in the use of alphabet symbols, so that no single-character or higher-degree patterns may be expected and readily unmasked by interceptors. The entropy of the cipher should then be maximum. For an N-character alphabet, this entropy should be at the very least equal to $H_{max} = \log_2 N$. Thus for 26- and 100-symbol alphabets, the cipher entropy should ideally be $H_{max} = \log_2 26 = 4.700$ and $H_{max} = \log_2 100 = 6.643$, respectively. Let us now consider again the previous data in Table 3.2, and determine the respective entropies of the usual English alphabet (A–Z) in monoalphabetic substitution, and the proposed symbol alphabet (00–99) in homophonic substitution. The corresponding character and symbol probabilities are listed (now in alphabetic order) in Table 3.3. Note that if p_n is the probability of a character having m possible symbols, the symbol probability is $p'_n = p_n/m$. We observe from the data listed in the table that our character/symbol mapping yields symbol probabilities fairly close to 1% ($p'_n \approx 0.01$), except for the least-likely characters which are over-represented in the 00–99 scale. From the table data, we find that the entropy of the monoalphabetic code (A–Z) is $H = 4.192$, while that of the (00–99) code it is $H = 6.616$. Compared to the maximum possible values, the efficiencies defined by $\eta = H/H_{max}$ are $\eta = 89.2\%$ in the first case, and $\eta = 99.6\%$ in the second. We can thus conclude that entropy is a good measure of how uniformly the symbols of the ciphertext are distributed and unpredictable, at least on a one-symbol basis.

TABLE 3.3 Character and symbol probabilities (p, p') and entropies (H, H') associated to the alphabets listed in Table 3.2

Character (Out of 26)	Symbols (Out of 100)	Character Probability p	Character Entropy H	Symbol Probability p'	Symbol Entropy H' (Group)
A	16, 25, 28, 43, 58, 74, 84, 94	0,082	0,296	0,010	0,542
B	07	0,015	0,091	0,015	0,091
C	21, 30, 80	0,028	0,144	0,009	0,189
D	41, 61, 72, 83	0,043	0,195	0,011	0,281
E	09, 12, 18, 22, 31, 35, 48, 63, 75, 77, 81, 88, 97	0,127	0,378	0,010	0,848
F	03, 70	0,022	0,121	0,011	0,143
G	14, 20	0,020	0,113	0,010	0,404
H	04, 10, 11, 15, 53, 92	0,061	0,246	0,010	0,133
I	00, 29, 42, 57, 59, 85, 90	0,070	0,269	0,010	0,465
J	06	0,002	0,018	0,002	0,018
K	71	0,008	0,056	0,008	0,056
L	01, 23, 37, 51	0,040	0,186	0,010	0,266
M	13, 69	0,024	0,129	0,012	0,153
N	34, 38, 50, 52, 60, 91, 95	0,067	0,261	0,010	0,449
O	32, 39, 46, 54, 62, 87, 89	0,075	0,280	0,011	0,491
P	65	0,019	0,109	0,019	0,109
Q	45	0,001	0,010	0,001	0,010
R	17, 26, 36, 64, 76, 99	0,060	0,244	0,010	0,399
S	47, 49, 56, 68, 82, 93	0,063	0,251	0,011	0,414
T	05, 08, 27, 40, 55, 67, 78, 79, 86	0,091	0,315	0,010	0,603
U	19, 96	0,028	0,144	0,014	0,172
V	44	0,010	0,066	0,010	0,066
W	24, 98	0,024	0,129	0,012	0,153
X	02	0,002	0,018	0,002	0,018
Y	33, 73	0,020	0,113	0,010	0,133
Z	66	0,001	0,010	0,001	0,010
		1,00	**4,192**	1,00	**6,616**

The bold numbers at bottom are the total entropies obtained after weighted summation.

How are the above analysis and conclusions applicable to *multialphabetic* ciphers, such as Vigenère's? In this case, ciphertext symbols may represent as many different characters as the key lengths. In our previous example with ANGEL as the key, any character symbol has 5 possible plaintext meanings. It is then straightforward to calculate the corresponding probability distribution and its entropy. The occurrence probability of a given character-symbol in the ciphertext should take into account not only the frequency of the plaintext character it corresponds to, but also of its repetition due to the multialphabetic coding. Considering symbol A, for instance, we see from Table 3.1 that it can mean any of the plaintext characters a/n/p/u/w. If p_n is the probability of the plaintext characters n, the probability of the cipher symbol A is $p'_A = (p_a + p_n + p_p + p_u + p_w)/5$. The complete probability distribution $p'_{A \to Z}$, as calculated though this formula for all A–Z symbols, is shown in Table 3.4. The calculated Vigenère-cipher entropy is now $H' = 4.627$, compared with that of the 26-symbol monoalphabetic cipher $H = 4.192$. Predictably, the

TABLE 3.4 Vigenère-cipher character (p) and symbol (p') probabilities, and associated entropy (H') corresponding the 5-letter key (ANGEL) and the Vigenère square shown in Table 3.1

Character (Out of 26)	Symbols (Out of 5-key)	Character Probability p	Symbol Probability p'	Symbol Entropy H'
A	A, N, P, U, W	0,082	0,044	0,198
B	B, O, Q, V, X	0,015	0,021	0,115
C	C, P, R, W, Y	0,028	0,030	0,152
D	D, Q, S, X, Z	0,043	0,022	0,121
E	E, R, T, Y, A	0,127	0,076	0,283
F	F, S, U, Z, B	0,022	0,026	0,136
G	G, T, V, A, C	0,020	0,046	0,205
H	H, U, W, B, D	0,061	0,034	0,167
I	I, V, X, C, E	0,070	0,047	0,209
J	J, W, Y, D, F	0,002	0,022	0,122
K	K, X, Z, E, G	0,008	0,032	0,157
L	L, Y, A, F, H	0,040	0,045	0,201
M	M, Z, B, G, I	0,024	0,026	0,137
N	N, A, C, H, J	0,067	0,048	0,210
O	O, B, D, I, K	0,075	0,042	0,193
P	P, C, E, J, L	0,019	0,043	0,196
Q	Q, D, F, K, M	0,001	0,020	0,111
R	R, E, G, L, N	0,060	0,063	0,251
S	S, F, H, M, O	0,063	0,049	0,213
T	T, G, I, N, P	0,091	0,053	0,226
U	U, H, J, O, Q	0,028	0,033	0,164
V	V, I, K, P, R	0,010	0,033	0,164
W	W, J, L, Q, S	0,024	0,026	0,137
X	X, K, M, R, T	0,002	0,037	0,176
Y	Y, L, N, S, U	0,020	0,044	0,197
Z	Z, M, O, T, V	0,001	0,040	0,186
		1,00	1,00	**4,627**

The bold number at bottom is the total code entropy obtained after weighted summation.

entropy of Vigenère cipher (as with a 5-letter key) is higher than the monoalphabetic case. The corresponding efficiency is $\eta = H'/H_{max} = 4.627/4.700 = 98.4\%$, compared with 89.2% in the previous case. The Vigenère cipher and any other multi-alphabetic cipher alternatives thus generate more entropy and uncertainty to make the code-breaker's task more difficult. It is interesting to look at the frequency spectrum of our 5-letter key, Vigenère example, which is shown in Figure 3.6. We observe that the spectrum is closer to uniformity, while the most common characters (E/T/A/I) are still somewhat privileged. Yet, when compared to the classical monoalphabetic spectrum (also shown in figure), we can see that single-character frequency analysis should be much more difficult. What about a Vigenère cipher which had a 26-letter key, corresponding to the maximum level of alphabet substitution (according to the Vigenère square)? In this case, *any* symbol in the cipher corresponds to *any* plaintext character. The probability for symbol A is thus $p'_A = (p_a + p_b + p_c + \cdots p_z)/26 \equiv 1/26$ and the same for all other symbols. The resulting entropy is thus simply $H' = \log_2 26 = H_{max} = 4.700$. We can therefore conclude that a Vigenère cipher with a 26-letter key is 100% efficient. Furthermore, its frequency spectrum is 100% uniform ($p'_{A \to Z} = 1/26$). Its entropy is not as high as that of our previous 100-symbol, 00–99 cipher ($H_{max} = 6.643$), but its definite advantage is to have a 100% uniform spectrum. The frequency analysis must then be carried out on higher-order fixed patterns such as digrams, trigrams and words, and anything that looks cyclic. The observed cycle period immediately betrays the length of the key, and the code-breaking task can progress from there.

The same conclusion as above also applies with keys longer than 26 letters or having indefinite lengths, such as a full page text. This would keep the code-breaker a bit more busy finding fixed or cyclic patterns. Yet, the use of Vigenère keys with 26 letters or more is not so attractive a solution. Here again this is because of the

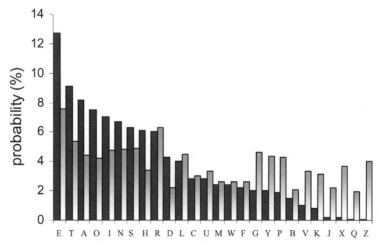

FIGURE 3.6 Gray bars: frequency spectrum of the Vigenère-cipher example defined in Table 3.1 (key = ANGEL) whose symbol probability distribution (p') is listed in Table 3.4. Black bars: frequency spectrum of written English shown for comparison (same as in Figure 3.2 or 3.3, black bars, as rearranged in decreasing order).

human weakness factor. It is not so difficult to generate and memorize a 26-letter key or to use a full-page text as the key. But having to go through the entire Vigenère square to encrypt and decrypt a message, especially if long, might deter the most courageous, unless some mechanically-computing machine with gears and cranks could be devised. Also, when performing encryption by hand, the risk of making mistakes (i.e., accidentally skipping substitution alphabets) is higher with long keys. If the destinee was unable (or did not have the patience) to work out these possible mistakes, the result could be the loss of the message. But nowadays this encryption can be made fault-free through a personal/home computer, which unfortunately is the same machine able to break the code with comparable computing effort!

The above description of frequency analysis leads to a conclusion which takes the form of the following paradox: *in cryptography, "entropy" is both a measure of channel information and a measure of the difficulty (or impossibility) for intercepting parties to retrieve the information.* The maximum single-symbol entropy, for which all symbols are equiprobable, yields a uniform frequency spectrum. But this does not remove any potential from higher-order frequency analysis, such as based upon digrams, trigrams, words and all possible cyclic patterns revealed by this analysis. A more tricky code would be one where all higher-order frequency spectra would be made uniform, which is only obtained by digital cryptography.

3.1.3 Other Classical Ciphers

We conclude this section with a brief (yet nonexhaustive) review of other interesting classical ciphers which have been used through history. The purpose of this review is to give a hint of the rich variety of encryption possibilities, to further develop our cryptology background, and why not, to entertain ourselves before entering the austere domain of modern cryptography. Note that in most of these classical ciphers, frequency analysis is generally of limited help, because the messages are usually kept relatively short. For the code-breaker, collecting several messages from the same source is of considerable help for this analysis, but provided they have the same *key*, which is not always the case as we shall see. Some ciphers use scrambling by rows and columns, rather than alphabet transposition or substitution. In that case, searching for syllables helps one to unscramble the text towards its initial order until well-known words and digrams/trigrams start showing up. The selection of classical ciphers described herewith only provide a flavor of the wide variety of possibilities. In the older times when such ciphers were used, code breaking had to be performed with paper, ink or pencil, methodology and trained intuition. The first task, which could be rapid, consisted in identifying the cipher type and finding its key. If the type was unknown (an unlikely occurrence) one had to try discovering its hidden algorithm through some expert methodology. There was no possibility for a *"brute-force"* approach (trying at random a large number of substitution possibilities), until computers and supercomputers came into the picture (namely by the end of WW II). Such a patient algorithm and plaintext retrieval required a deep familiarity with all possible cipher families, types and variants. Nowadays, hundreds of classical algorithms are available from computer databases, so breaking any of them could just be the matter of milliseconds. Moreover, any attempt to devise a new cipher based upon these classical approaches may prove unsuccessful, because the computer will likely find out what transposition/substitution algorithms are being used, no matter how tricky. And if not, the computer will break the code anyway, based upon sophisticated frequency-analysis methods and an extended database of words and character correlations, no matter the language used. This is why nowadays it is not useful anymore to learn how to break classical ciphers, unless for entertainment or hobby purposes (which we leave the reader to appreciate). This is also the reason why this chapter has no exercises but only illustration examples. The following selection of ciphers is mostly inspired by the aforementioned references (Fouché Gaines, Singh), with our own (original) applications and comments.

Lord Bacon's biliteral cipher. This is a monoalphabetic transposition which uses alphabetic symbols having two font types (hence *biliteral*), such as A/B/C and *A/B/C*, or a/b/c or *a/b/c* (ideally, the font difference should be invisible to the naked eye). Let's call "0" the first font type and "1" the second. Table 3.5 establishes a possible (secret) correspondence between the characters and their code. Shown in this binary form, it looks pretty much like some modified ASCII code restricted to a 5-bit alphabet! Any random mapping instead of the regular progression shown is of course eligible. The regular progression can however be memorized in seconds, thus with no need to keep the correspondence table on a piece of paper. According to the coding, we need a five-symbol word for each character to be encrypted. In fact, these symbols can be any arbitrary letters, as long as the letter fonts change according to the coding rule. Here is an example:

plaintext	s	e	e	y	o	u	a	t
code	10001	00100	00100	10110	01101	10011	00000	10010
plaintext	m	i	d	n	i	g	h	t
code	01011	01000	00011	01100	01000	00110	00111	10010

This is a 16-character plaintext requiring $16 \times 5 = 80$ symbols. How about the following ciphertext (bold font = 1, other = 0) which reproduces the above 5-bit word sequence:

the political situation **is** not without concerns,
as **the** national elections date **is** now **approaching**

TABLE 3.5 *Lord Bacon's* cipher, listing correspondence between alphabetic characters and fonts (0 = type "0", 1 = type "1")

a	00000	i-j	01000	r	10000
b	00001	k	01001	s	10001
c	00010	l	01010	t	10010
d	00011	m	01011	u-v	10011
e	00100	n	01100	w	10100
f	00101	o	01101	x	10101
g	00110	p	01110	y	10110
h	00111	q	01111	z	10111

In this 83-character example, the difference of fonts obeys the coding rule by groups of five characters (overlooking nonalphabetic ones). The last three characters are in excess and just made to complete the statement, and are called *nulls*. In a real cipher, the font difference should be practically invisible by normal reading means. A microscopic dot in the vicinity of any "type 1" character, to be observed with a big magnifying lens, may do the trick. While the coding technique is obvious to the expert's eye, it may remain unnoticed to most people, and for this reason is called a *concealment cipher*.

What is very attractive in this cipher is that because the open statement is fully intelligible, it can be used to distract the mind and attention of any interceptor, or purposefully lead him in the wrong direction! But if the code is known, it is fully transparent. For reasons of easy memorizing, human weakness will choose an alphabet transposition rule as straightforward as that shown in Table 3.5, with changes of 1 to 5 shifts. The key (number of shifts) would have to be periodically communicated, but the interceptor is able to try any of these few possibilities. The intriguing feature of this cipher is that it can be implemented without any text, using any object that has two possible states. The possibilities range from a 54-card deck with card faces flipped up or down according to the bit value, to crayon boxes or book piles! This coding can also be implemented through any document where the first word of each page begins with either a vowel or a consonant. To remove any apparent cycle, the bit polarity change can also be made to follow differential algorithms. This approach still remains quite complicated to implement, especially in comparison to open-jargon techniques (see Section 3.1), also considering that in practice, the message content is limited to a few short sentences. Those techniques have nevertheless been extensively used by WW-II spies, to the point that any object collection having two possible states, such as a folded newpaper, could be suspected to contain a cipher message!

Abbé Thrithème's trinumeral cipher. This represents another type of *concealment* cipher. The principle is first to map the alphabet with a set of numbers based on 1/2/3, namely from 111 to 333, as shown in the example of Table 3.6. Hence the name *trinumeral*. This coding reduces the length of the cipher symbols, and also makes it possible to introduce one to three useful characters (such as &, ? and /, for instance). The corresponding ciphertext consists in an innocuous sentence in which the *count of syllables*

TABLE 3.6 Example of *Abbé Thrithème* cipher, listing correspondence between alphabetic characters and number codewords

a	111	i-j	133	r	232	&	331
b	112	k	211	s	233	?	332
c	113	l	212	t	311	/	333
d	121	m	213	u-v	312		
e	122	n	221	w	313		
f	123	o	222	x	321		
g	131	P	223	y	322		
h	132	q	231	z	323		

represents the coded symbol. Consider the plaintext message "meeting is cancelled", which for expediency can be contracted into "metig is canceld." According to the table, this plaintext must be coded into (spacers * shown for clarity):

213 122 311 133 131*133 233*113 111 221 113 122 212 121

Since the above sequence indicates the number of syllables found in any arbitrary text, how about the following:

> Seeing my bicycle get broken again suddenly is not
> my ideal conception of practicing sports, yet repairing
> dysfunctions after careful diagnostics does not diminish
> my own feel towards getting a good sense of expertise
> for complex repairs, being a person who doesn't quit.

As imperfect it may sound for grammar and style (and within a strictly accurate definition of "syllable"), a ciphertext like the preceding only takes a few minutes to devise and to write down (the reader may try this revealing exercise). A talented cryptographer would even turn such a ciphertext into a perfect statement, including love declarations and poetic rhymes. But it remains quite taxing in the required number of symbols (43 words and 226 alphabetic characters, just to cancel a meeting!). It may take a good sense of secrecy or artistic disposition to go to such creative pains, but the fact is that the approach works if no one suspects the existence of a concealed message within the open text. The code is otherwise easily cracked by doing the reverse operation, especially considering that the most likely character-transposition table is the one we have used.

Knight's tour cipher. This is based on the principle of a monoalphabetic transposition with a difficult key. The message must be a 64-character sequence (possibly including a few nulls at the end). The characters are then assigned a position inside a "magic square" (the key). The strength of this code is that the key is particularly hard to guess. It is defined by the move of a knight on a chessboard, in which every square is touched only once. Table 3.7 shows one possible solution. The 64-character message "We shall discuss our contingency plan at midnight in the usual meeting

TABLE 3.7 Magic square used in the *knight's tour* code

1	52	**3**	16	63	50	**13**	26
4	17	64	51	**14**	27	62	49
53	**2**	15	28	35	40	25	**12**
18	**5**	54	39	32	29	48	61
55	38	31	34	41	36	**11**	24
6	19	42	37	30	33	60	47
43	56	21	**8**	45	58	23	**10**
20	**7**	44	57	22	**9**	46	59

The first 14 moves from the the top left square (1) are shown in boldfont for clarity.

place" can then be crypted with the magic square as follows (characters from first three words "We shall discuss" highlighted for clarity):

W	L	S	U	C	U	S	N
H	R	E	A	S	C	A	S
M	E	O	Y	M	G	E	U
C	A	E	I	N	P	U	L
E	N	A	T	H	I	C	G
L	O	T	D	L	A	P	E
I	T	T	D	T	N	N	S
N	L	N	I	I	I	H	G

One may argue that any random permutation of the 64 letters may be used as a key to crypt and decrypt any 64-character message. But in that case, the (non) magic square must be physically communicated and kept in a secret place for reference, which implies some risk of interception. Instead, the magic square shown in Table 3.7 can be memorized at once (as difficult as it may appear after the first 15 knight hops!). The expert then only has to move the end of his pencil making knight hops onto the 8×8 cipher grid, in a quasi-automatic way, and memorize the letters as they come out. We may bet that a chess champion could do the same without a pencil...

Zanotti transposition cipher. The principle is to rearrange the plaintext inside a fixed-size grid (also called "grille") where the first letter of each row is shifted by a certain number of positions, which can be implemented by some sliding-rule machine. The set of shifts to be assigned to each row is the key for both coding and decoding. Here is an example using a 9×9 grille, the key being 304251011:

–	–	→	w	e	s	h	a	l
l	d	i	s	c	u	s	s	o
–	–	–	→	u	r	c	o	n
–	→	t	i	n	g	e	n	c
–	–	–	–	→	y	p	l	a
→	n	a	t	m	i	d	n	I
g	h	t	i	n	t	h	e	u
→	s	u	a	l	m	e	e	t
→	i	n	g	p	l	a	c	e

The corresponding ciphertext is the character sequence as it can be read column by column, and transmitted for instance by groups of four letters (or three with possibly nulls):

LGDN HSII TATU NWSI TIAG ECUN MNLP SURG
YITM LHSC EPDH EAAS ONLN EECL ONCA IUTE

With a template with sliding rules, such a cipher is very easy to implement and to undo. A first drawback is that the key must be transmitted via a separate channel, and ideally, be changed on a daily, or even one-message basis (*onetime pad cipher*). A second drawback is that the code can easily be cracked. Suffice it to first assume, from the human-weakness argument, that the grille is not gigantic, i.e., its size should be 10×10 at the very most. From the message length (L), one can also assume that at least 25–75% of the grille is used for coding, so the grille size ($N \times N$) must correspond to the integer values N that are closest to the boundaries $x = \sqrt{L/0.75}$ and $y = \sqrt{L/0.25}$. In the present example ($L = 64$), we find that $x = 8.5$ and $x = 16$, which points to 9×9 or 10×10 grille sizes, if assumed square (a rectangle would have less rows and more columns, thus decreasing the key size and the security). Then the key must lie somewhere between 0000000000 and 9999999999, with the constraint that the sum of the key numbers be exactly $81 - L$ or $100 - L$. We would not expect to see too many "9" or "8" in the key (yielding only one or two characters coded per line) or too many "0" either (yielding less line/column transposition potential). Under such constraints, the human-weakness factor will lead to choosing keys that have nothing random or are more or less always the same in a limited catalog of possibilities, like a relative's birthdate (15/08/2003 ≡ 15082003[00]) or even worse, a familiar dog name (Fido = 6/8/4/15 ≡ 68415[00000]). The code cracker may easily construct his sliding machine and try out a few key possibilities out of intuition, based upon the above considerations. If this proves unsuccessful, the second step consists in spreading out individual vowels so that their distribution within the grid becomes more or less uniform. Indeed, juxtapositions of vowels such as AA/EEE/UU in either rows or columns are a measure of disorder, since they do not correspond to (usual) English words. Minimizing these juxtapositions restores some order and might bring luck. One can then attempt to form likely words (e.g., "THE," "OK," "PLAN," "HOLD," etc.) which may rapidly lead to cribs.

Nihilist square cipher. This is based on the principle of transposing either columns or rows, or both, under a specific order of permutations defined by a *keyword*. The *keyword*'s size defines the number of columns and the message length the number of rows. Consider for instance the 38-letter plaintext: "The Queen wants to hear more about our project" with the 7-letter keyword CHATEAU. These numbers define a rectangle of seven columns and 6 rows, in which we write the plaintext as follows:

C_3	H_5	A_1	T_6	E_4	A_2	U_7
t	h	e	q	u	e	e
n	w	a	n	t	s	t
o	h	e	a	r	m	o
r	e	a	b	o	u	t
o	u	r	p	r	o	j
e	c	t	x	x	x	x

the last four positions being filled with nulls. The rule to assign the order into which columns should be permuted is the following. The first letter of highest rank in the keyword gets label one (i.e., A_1). The second letter of same or lower rank gets label two (i.e., A_2), and so on with the rest of the key (i.e., C_3, E_4, H_5, T_6, and U_7). The ciphertext is then generated by writing down the plaintext column after column, while following the column hierarchy, i.e.,

EAEA RTES MUOx TNOR OEUT RQRx HWHE UCQN ABPx ETOT Jx

An alternate possibility is to fill in the plaintext letters in each row according to the column label priority, namely:

C_3	H_5	A_1	T_6	E_4	A_2	U_7	
e	u	t	e	q	h	e	the quee
a	t	n	s	n	w	t	n wants to
e	r	o	m	a	h	o	hear mo
a	o	r	u	b	e	t	re about
r	r	o	o	p	u	j	our proj
t	x	e	x	x	c	x	ect

We note that the resulting columns are the same as in the previous grid, but the order is different. Either coding technique is referred to as the *"nihilist square,"* which is basically an *irregular column transposition* algorithm.

To further complicate the task of the would-be code breaker, we can introduce another permutation step, this time by rows. The same keyword is used for the row permutation, as follows:

C_3	e	r	o	m	a	h	o	o hear mo
H_5	r	r	o	o	p	u	j	our proj
A_1	e	u	t	e	q	h	e	the quee
T_6	t	x	e	x	x	c	x	ect
E_4	a	o	r	u	b	e	t	re about
A_2	a	t	n	s	n	w	t	n wants t
U_7	x	x	x	x	x	x	x	(null)

(noting that the last row is of no use in this example), and the ciphertext now writes as follows:

EROM AHOR ROOP UJEU TEQH ETXE XXCX AORU BETA TNSN WTXX XXXX X

Having knowledge of the keyword (CHATEAU), the recipient of this crypted message can draw the corresponding 7×7 grid and then fill in the squares with the ciphertext characters. Then he must reorder the rows according to the priority labels indicated by the keyword at left (row A_1 comes to the top, followed by row A_2, etc.). Finally, he must reorder the columns according to the priority labels with the keyword now written on the top, as in the preceding grid. It is clear that a "nihilist square" with double transposition makes code breaking significantly harder. As in the previous Zanotti-transposition example, cracking such a code requires testing different permutations by rows and columns until the distribution of vowels becomes more or less uniform, and until bits and pieces of easily recognizable words or cribs appear. There are many other variants of irregular-transposition algorithms whose description and solving methods fills entire books and bookshelves. These previous examples might largely suffice for first-familiarization purposes.

Polybius square cipher. This is a monoalphabetic transposition cipher where plaintext characters are represented by two-number symbols and a keyword number is used as the key. The interesting result is that the same cipher symbol can represent several possible plaintext characters, but to an extent which depends upon the choice of the key and which is not uniformly distributed over all symbols. Let us consider the character-symbol correspondence (called *Polybius square*) shown in Table 3.8, which is easy to memorize. The principle is that character "a" is coded as 11, character "b" as 12, etc. all the way to characters y-z, which are coded as 54-55. With characters i-j having the same meaning, the character alphabet is converted into a two-number cipher alphabet of $5 \times 5 = 25$ symbols. The same table is used to convert the letter keyword into a number key. For instance, the keyword ANGEL corresponds to the key 11 33 22 15 31. The coding consists in adding the key to each of the number-symbols by groups lengths corresponding to the key length (here 5). Here is an example:

plaintext	s	e	n	d	s	u	p	p	l	i	e	s
symbol code	43	15	33	14	43	45	35	35	31	24	15	43
+ key	11	33	22	15	31	11	33	22	15	31	11	33
ciphertext	54	48	55	29	74	56	69	57	46	55	26	76

The inverse operation, i.e., subtracting the key from the ciphertext and converting the number symbols into characters restitutes the plaintext. A nice feature of this coding is that the same cipher symbol can correspond to two plaintext characters, for instance "55" can be "n" or "i." Another possibility is to use a keyword as long as the plaintext, for instance generated by a series of verses known to both originator and destinee, or the first sentence at the top of a mutually agreed book and page (overlooking nonalphabetic characters). This principle of adding a number key to the plaintext code gets us a bit closer to modern cryptography, which is based on digital codewords, as described in the forthcoming sections. Here, the

TABLE 3.8 *Polybius square*, providing correspondence between letters and two-number symbols (e.g., c = 13, z = 55)

	1	2	3	4	5
1	a	b	c	d	e
2	f	g	h	ij	k
3	l	m	n	o	p
4	q	r	s	t	u
5	v	w	x	y	z

3.1 Message Encryption, Decryption and Cryptanalysis 371

approach is extremely basic. But it has a major flaw in the fact that not all numbers are represented, unlike in a conventional 00–99 symbol code. Indeed, the ciphertext does not have symbol with zero in the first or second place, i.e., 00/01/.../09 or 10/20/.../90. The multiplicity of possible character/symbol correspondence is also non-uniform and depends upon the key itself. The lowest-value symbol, which has a single correspondence, is 22 ($=a_{\text{plain}}+a_{\text{key}}$) and the highest-value symbol, which has two correspondence possibilities, is 99 ($=z_{\text{plain}}+u_{\text{key}}$ or $99=u_{\text{plain}}+z_{\text{key}}$). It is easily established that symbol 55 (for instance) may correspond to as many as 28 plaintext characters. These number gaps in the cipher code may first betray the code origin and type. But an unfortunate choice of keyword, in relation to the plaintext may also reveal well-known frequency patterns.

Phillips cipher. This is based upon the principle of *polyalphabetical* substitution by groups, without key. The idea is to use eight different substitution alphabets (called "type-1" to "type-8"), to be used one after another for each groups of five characters to be coded. As illustrated in Table 3.9, type-1 alphabet is a reference 5×5 square whose characters have random locations. It is seen from the table that the seven other alphabets are constructed by successive two-by-two row permutations from one to the next. The substitution rule between a plaintext character and a ciphertext character consists in taking the letter immediately on the right, following a diagonal. For instance, letters "z" and "q" of type 1 square are coded into letters "u" and "x," respectively. For letters located in the bottom row or the last column, one must imagine the square to be repeated underneath or at right, or both. Thus for type 1, letters "y," "m," and "r" are coded into "g," "p," and "w," respectively. The coding rule is to use the type-1 square to code the first group of

TABLE 3.9 Example of eight different substitution alphabets (type 1 to type 8) to be used in *Philips* cipher

type 1

w	g	d	f	k
z	ij	t	o	b
q	u	n	h	m
p	x	c	l	e
y	v	s	a	r

*

type 2

z	ij	t	o	b
w	g	d	f	k
q	u	n	h	m
p	x	c	l	e
y	v	s	a	r

*

type 3

z	ij	t	o	b
q	u	n	h	m
w	g	d	f	k
p	x	c	l	e
y	v	s	a	r

*

type 4

z	ij	t	o	b
q	u	n	h	m
p	x	c	l	e
w	g	d	f	k
y	v	s	a	r

*

type 5

z	ij	t	o	b
q	u	n	h	m
p	x	c	l	e
y	v	s	a	r
w	g	d	f	k

** *

type 6

q	u	n	h	m
z	ij	t	o	b
p	x	c	l	e
y	v	s	a	r
w	g	d	f	k

** *

type 7

q	u	n	h	m
p	x	c	l	e
z	ij	t	o	b
y	v	s	a	r
w	g	d	f	k

** *

type 8

q	u	n	h	m
p	x	c	l	e
y	v	s	a	r
z	ij	t	o	b
w	g	d	f	k

** *

The marks (*,**) indicate rows that are moved down between two successive alphabet types.

five ciphertext letters, then the type-2 square for the next group of five and so on, until $8 \times 5 = 40$ letters have been substituted. The process may be repeated beyond this point for longer messages (e.g., up to 80 or 120 characters), but this would be at the expense of security (see below). Here is an example showing how the first three groups of plaintext characters ("The Queen wants to") are coded under the above principles:

	type 1				type 2				type 3		
Plaintext	t	h	e	q	u	e	e	n	w	a	n	t	s	t	o
Ciphertext	H	E	Y	X	C	Y	Y	L	U	B	F	H	O	H	K

One may intuitively conclude that this cipher, used once in the military, should be quite safe considering the use of eight different substitution alphabets for only 40-character messages. But such an intuition is wrong. Looking at the cipher messages, appearing by groups of five and not exceeding 40 characters, any code breaker immediately suspects that it must be a Philips cipher. Having intercepted a few messages from the same source, he also has plenty of material to work with! All these messages have strictly the same character substitution rules (i.e., type-1 for group 1, and so on). The task is then to find the correct substitutions to apply to each group of five characters. The expert also knows that in Philips cipher, any plaintext letter has up to 4 substitutes at maximum (one may check this property from Table 3.9: for instance "a" has two substitutes [k, b], and "b" has four substitutes [p, q, w, y]). The frequency analysis and search for cribs is far easier than in a 8-alphabet Vigenère, because the substitution by groups is unique. Furthermore, any repeated character within a group has the same meaning (e.g., "YY" for "ee" in group 2, above). A given group may accidentally begin with a frequent word, for instance "the", which is the case of group 1. By putting all the intercepted messages on top of each other, so that all groups with the same number form a block 5 columns wide, one can easily perform frequency analysis for single letters, digrams, trigrams and frequent words (see previous subsection) and rapidly break the code.

Playfair cipher. This is based upon the principle of one-to-one *polygram substitution*. This principle, which we have not considered so far, consists in substituting letters by digrams or even trigrams. In the simplest case, the key could be any $26 \times 26 = 676$-element table such as partially shown in Table 3.10. For easy memorization, the rows and columns can be defined by practical keywords such as "Europe" and "America." To define all possible characters (1 to 26) in each row and column, we have expanded "Europe" and "America" into "europqstvwxyzabcdfghijklmn" and "americdfghjklnopqstuvwxyzb," respectively, thus avoiding character repetitions. The resulting table thus provides all possible digram codes. For instance, the plaintext "The general arrives tomorrow" is coded through the following:

Plaintext	th	eg	en	ra	la	rr	iv	es	to	mo	rr	ow
Ciphertext	JH	IA	NA	AC	AX	DC	UU	RA	OH	OY	DC	VD

While the frequency of single letters is now hidden by the context-dependent digrams, this remains still a single-substitution algorithm. We see from this example that the symbol DC comes twice, betraying the same plaintext digram

TABLE 3.10 Example of square key for digram substitution (upper corner with first 16 rows/columns shown)

	a	m	e	r	i	c	d	f	g	h	j	k	l	n	o	p
e	AA	BA	CA	DA	EA	FA	GA	HA	IA	JA	KA	LA	MA	NA	OA	PA
u	AB	BB	CB	DB	EB	FB	GB	HB	IB	JB	KB	LB	MB	NB	OB	PA
r	AC	BC	CC	DC	EC	FC	GC	HC	IC	JC	KC	LC	MC	NC	OC	PA
o	AD	BD	CD	DD	ED	FD	GD	HD	ID	JD	KD	LD	MD	ND	OD	PD
p	AE	BE	CE	DE	EE	FE	GE	HE	IE	JE	KE	LE	ME	NE	OE	PA
q	AF	BF	CF	DF	EF	FF	GF	HF	IF	JF	KF	LF	MF	NF	OF	PA
s	AG	BG	CG	DG	EG	FG	GG	HG	IG	JG	KG	LG	MG	NG	OG	PA
t	AH	BH	CH	DH	EH	FH	GH	HH	IH	JH	KH	LH	MH	NH	OH	PA
v	AI	BA	CI	DA	EI	FA	GI	HA	II	JI	KI	LI	MI	NI	OI	PI
w	AJ	BJ	CJ	DJ	EJ	FJ	GJ	HJ	IJ	JJ	KJ	LJ	MJ	NJ	OJ	PJ
x	AK	BK	CK	DK	EK	FK	GK	HK	IK	JK	KK	LK	MK	NK	OK	PK
y	AL	BL	CL	DL	EL	FL	GL	HL	IL	JL	KL	LL	ML	NL	OL	PL
z	AM	BM	CM	DM	EM	FM	GM	HM	IM	JM	KM	LM	MM	NM	OM	PM
a	AN	BN	CN	DN	EN	FN	GN	HN	IN	JN	KN	LN	MN	NN	ON	PN
b	AO	BO	CO	DO	EO	FO	GO	HO	IO	JO	KO	LO	MO	NO	OO	PO
c	AP	BP	CP	DP	EP	FP	GP	HP	IP	JP	KP	LP	MP	NP	OP	PP

(here "rr"). Frequency analysis is able to rapidly identify the one-to-one correspondence between the most frequent digrams and fill out the table. Another drawback is that the recipient must either keep the 676-element table in a secret place, or rewrite it again at each decrypting session, which is a hassle. A common solution to both issues is brought by the *Playfair cipher*. It consists into a single 5 × 5 square key with certain digram-coding rules. An example of Playfair square is shown in Table 3.11. The four easy coding rules for any plaintext digram "xy" can be described as follows:

- "x" and "y" belong to a same row within the square: shift letters by one position to the right; for instance, "sv" becomes "FM," "zp" becomes "DZ," and "ly" becomes "YU";
- "x" and "y" belong to a same column: shift down letters by one position; for instance, "sz" becomes "ZU," "fk" becomes "DF," and "oa" becomes "VQ";
- "x" and "y" belong to a same diagonal: replace by the pair of letters that have the same diagonal relationship in the opposite direction; for instance, "sd" becomes "ZF,", and "ds" becomes "FZ"; "fr" becomes "BC," and "rf" becomes "CB";
- "x" and "y" belong to different rows, columns or diagonals: shift down the letter from the upper row by one position, and shift up the letter from the lower row by one positon; for instance, "si" becomes "ZV," "fw" becomes "DL," and "qc" becomes "AP";

Clearly, the above rules must be reversed for decryption. One problem is caused by letter repeats where the digram has the same two letters. In this case, a null (e.g., "x") must be inserted to push the repeated letter into another digram. If this causes another repeat (e.g., "less sure" → "le**sx** **s**sure"), another null must be inserted (i.e., "le**sx** **s**sure" → "le**sx** **sx**sure"). Should the repeat occur more than once in the plain text, the null should also be changed so as to mask these occurrences. The substitution of this arbitrary null will result in a typo which the decryptor will easily figure out. Based on such rules and according to the Playfair square shown in Table 3.11, we can then crypt our previous message as follows:

Plaintext	th	eg	en	ra	la	rr	iv	es	to	mo	rr	ow
Ciphertext	BR	WX	ML	YQ	YL	HG	AI	WZ	BQ	NQ	YG	QE

TABLE 3.11 Example of *Playfair square for digram substitution*

s	f	v	m	c
z	d	ij	n	p
u	t	a	l	y
g	b	q	w	r
x	k	o	e	h

3.1 Message Encryption, Decryption and Cryptanalysis

Note that for the first and second repeat digrams "rr,", we used "rx" and "ru," which yield HG and YG, respectively. The destinee will then read "the general a**rx**ives tomo**ru**ow," which suggests immediate and unambiguous correction.

Beale papers cipher (also called book cipher). This is based upon the principle of using a book text as the key (named the *keytext*) to substitute characters into any-size preudo-random numbers. The text should be sufficiently long to contain all alphabetic characters, and should be relatively unlimited (such as a book) in order to offer more random choices for each possible symbol substitution, as we shall explain below. Here is an example of keytext, based upon a *Hamlet* soliloquy (Act II, Sc. I):

[1]To [2]be [3]or [4]not [5]to [6]be: [7]that [8]is [9]the [10]question:/[11]Whether [12]tis [13]nobler [14]in [15]the [16]mind [17]to [18]suffer/[19]The [20]slings [21]and [22]arrows [23]of [24]outrageous [25]fortune,/[26]Or [27]to [28]take [29]arms [30]against [31]a [32]sea [33]of [34]troubles,/[35]And [36]by [37]opposing [38]end [39]them? [40]To [41]die:, [42]to [43]sleep;/[44]No [45]more; [46]and, [47]by [48]a [49]sleep [50]to [51]say [52]we [53]end/[54]The [55]heart-[56]ache [57]and [58]the [59]thousand [60]natural [61]shocks/[62]That [63]flesh [64]is [65]heir [66]to, [67]tis [68]a [69]consummation/ [70]Devoutly [71]to [72]be [73]wished. [74]To [75]die, [76]to [77]sleep;/[78]To [79]sleep: [80]perchance [81]to [82]dream; [83]ay, [84]there'[85]s [86]the [87]rub;/[88]For [89]in [90]that [91]sleep [92]of [93]death [94]what [95]dreams [96]may [97]come/[98]when [99]we [100]have [101]shuffled [102]off [103]this [104]mortal [105]coil,/[106]Must [107]give [108]us [109]pause: [110]there'[111]s [112]the [113]respect/[114]That [115]makes [116]calamity [117]of [118]so [119]long [120]life/[121]For [122]who [123]would [124]bear [125]the [126]whips [127]and [128]scorns [129]of/[130]The [131]oppressor'[132]s [133]wrong, [134]the [135]proud [136]man's [137]contumely/(...)

Each of the words in the keytext has been numbered, which offers a pseudo-random correspondence between this number and the first letter of the word, as shown in Table 3.12. We observe that interestingly enough, this 137-word keytext contains all alphabetic characters but four (namely, k, x, y, z). We might go further down into Hamlet in order to find these missing letters, although the task is painstaking (letters "y" and "k" appearing only near positions 250 and 600 in the words "you" and "knaves," respectively!). It is not a problem if "x" and "z" were definitely missing in the keytext, because they can always be replaced by nulls. Here is a cipher example for "Safe in two days":

Plaintext	s	a	f	e	i	n	t	w	o	d	a	y	s
Ciphertext	111	35	63	38	64	44	58	126	37	95	83	64	51

which as a ciphertext write

111, 35, 63, 38, 64, 44, 58, 126, 37, 95, 83, 64, 51

The nice feature about *book cipher* is that there is plenty of choice for random character substitution, this choice growing with the keytext size. Any frequency pattern is thus erased under the condition that one never uses the same number/symbol twice and picks it up at random from all the possibilities such as listed in Table 3.12. We still observe that the choice is limited by the frequency of English words beginning by the character to be coded. For instance, our 137-word keytext

TABLE 3.12 Correspondence between plaintext characters and ciphertext number symbols, as based upon the keytext (Hamlet soliloquy) selection shown in text

a	21, 22, 29, 30, 31, 35, 46, 48, 56, 57, 68, 83, 127	n	4, 13, 44, 60
b	2, 6, 36, 47, 72, 124	o	3, 23, 24, 26, 33, 37, 92, 102, 117, 129, 131
c	69, 97, 105, 116, 137	p	80, 109, 135
d	41, 70, 75, 82, 93, 95	q	10
e	38, 52, 53	r	87, 113
f	25, 63, 88, 121	s	18, 20, 32, 43, 49, 51, 61, 77, 79, 85, 91, 101, 111, 118, 128, 132
g	107	t	1, 5, 7, 9, 12, 15, 17, 19, 27, 28, 34, 39, 40, 42, 50, 54, 58, 59, 62, 66, 67, 71, 76, 78, 74, 81, 84, 86, 90, 103, 110, 112, 114, 125, 130, 134
h	55, 65, 100	u	108
i	8, 64	v	11
j	89	w	52, 73, 94, 98, 99, 122, 123, 126, 133
k	?	x	?
l	119, 120	y	?
m	16, 45, 96, 104, 106, 115, 136	z	?

offers only one possibility for characters "g," "i," "q" and "v," which is incidental to this specific Hamlet soliloquy. But any other keytext such as an essay or a thriller would not show any such gaps. Because of the unlimited abundance of number/ character choices, it is then virtually impossible to break such a code, unless the interceptor has a hunch of what keytext could have ever been chosen by the source. The story of the *Beale papers* constitutes a famous application of *book ciphers*, which is fraught with thrilling details and controversies (see Singh, 1999). In short, Beale left to posterity three cipher sheets concerning (1) the exact location, (2) the detailed description and amount, and (3) the would-be heirs of a fabulous hidden treasure. Such a treasure came from the exploitation of a miraculous gold mine situated 250–300 miles north of Santa Fe that he, and his workers, once discovered and exploited. To date, only the second cipher has been broken. It took several years to figure out that this second sheet utilized for keytext the *Declaration of Independence*, starting with word numbered 115, in the sentence:

(...); [108]That [109]to [110]secure [111]these [112]rights, [113]governments [114]are [115]instituted [116]among [117]men, [118]deriving [119]their [120]just [121]powers [122]from [123]the [124]consent [125]of [126]the [127]governed; (...)".

which represents the letter "i" in the opening sentence "I have deposited (...)". Whether or not the story was a hoax, or could not be solved because of other more complex factors, the conclusion to retain is that if the keytext reference is unknown, lost or destroyed, the document definitely remains *indecipherable*. We can just infer (for possible consolation) that the fabulous treasure was eventually spirited away by a person or team who may have found the keytext of the first paper!

ADFGVX cipher. This is based upon the principles of alphabetic transposition followed by substitution with a keyword. Both originator and destinee use the same 6 × 6 square whose rows and columns are indexed by the letters ADFGVX, as shown in Table 3.13. The square contains the 26 alphabet letters and the 10 numbers (a useful new feature) placed in random positions. Coding consists first in attributing to each character the letter pair of the corresponding row and column. Here is an example:

Plaintext   h   o   l   d   p   o   s   i   t   i   o   n
Ciphertext AA AG GG VD DV AG XA XG AF XG AG XD

Plaintext   t   i   l   6   p   m   t   o   d   a   y
Ciphertext AF XG GG GD DV DF AF AG VD FD GF

The second stage consists in rearranging the pairs by columns under a certain keyword, which should be 4–5 letters. Using the keyword ATOM, the result of this operation is the following:

A	T	O	M
AA	AG	GG	VD
DV	AG	XA	XG
AF	XG	AG	XD
AF	XG	GG	GG
GD	DV	DF	AF
AG	VD	FD	GF

The ciphertext is then obtained by writing the message column by column, which gives:

AADVAFAFGDAGAGAGXGXGDVVDGGXAAGGGDFFDVDXGXDGGAFGF

We see that the cipher has only six symbol letters, namely ADFGVX. The reason for choosing such letters is that they are very clearly recognizable when transmitted through *Morse code*. This new cipher was introduced in all confidence by the Germans at the outbreak of WW-I. The code was cracked three months later (and within only a few days) by a French army cryptoanalyst, G. Painvin.

TABLE 3.13 Example of square used in the ADFGVX cipher

	A	D	F	G	V	X
A	h	x	t	o	z	b
D	9	q	m	l	p	0
F	7	a	4	f	w	k
G	e	6	y	1	g	3
V	8	d	r	u	c	j
X	s	n	5	i	2	v

Enigma-machine cipher. This is based on the principle of polyalphabetic substitution with a key (different each day). The substitution is done mechanically through a machine looking like a typewriter with a keyboard and a top A–Z display. Each letter typed on the keyboard causes another letter to light in the display, which designates either the ciphertext symbol (in the process of encryption) or the plaintext character (in the process of decryption). The correspondence is generated by changing electrical contacts between keyboard and screen. These changes are caused by the automatic rotation of gears (or rotors, or scramblers) each time a letter is typed. Figure 3.7 shows a possible arrangement (here limited to 9 keyboard letters, for clarity) with two rotors. The edges of the rotors, with their electrical connections, are shown here in two dimensions like ribbons which otherwise are closed onto themselves. We see from the figure that in this rotor/scrambler configuration, typing letter "W" lits letter "E" on the display. So each alphabetic keyboard letter has a unique correspondence with a letter on the display. But as soon as a letter is typed, the left rotor rotates by 1/9th of a revolution (or 1/26th in the real Enigma machine). Figure 3.8 shows the new arrangement. If we retype letter "W," we then lit letter "R" on the screen. The keyboard and the display now have a completely new alphabetical correspondence, and so on each time a new letter is typed. It is clear that after having typed 26 letters, the first rotor is back to its initial position. But this full cycle causes the second rotor to move up one notch, which generates another new set of alphabetic substitutions. The total number of alphabetic substitutions is therefore $26 \times 26 = 676$, and is completed after typing 676 letters. The second version of the machine used three rotors, which comes to $26 \times 26 \times 26 = 17{,}576$

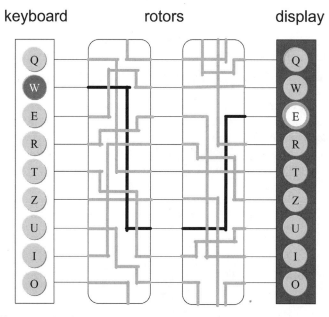

FIGURE 3.7 Principle of the Enigma machine (keyboard and display limited to 9 characters for clarity). Using two rotors, typing the letter "W" lights up letter "E."

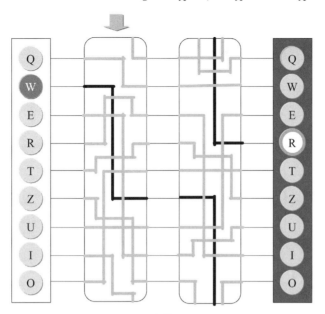

FIGURE 3.8 Same as in Figure 3.7, after $1/9^{th}$ rotation of left rotor, or one-notch transposition. Typing again the letter "W" lights up letter "R."

substitution possibilities. Before typing the message, one must chose a *key* to indicate in which orientation position the rotors must be placed. The key is simply given by the respective orientations of the three rotors, as shown by characters printed on the sides (e.g., ZKQ, BJX, HYE, etc.). The second part of the key is the sequence according to which the three rotors 1, 2 and 3 must be arranged: 123, 132, 312, 213, 231 or 321. This makes up to $3! = 6$ additional possibilities, or a total of $17{,}576 \times 6 = 105{,}456$ different substitutions or keys. This apparatus works for encryption, but how about *decryption*? The answer is that decryption works exactly the same way: it suffices to type the cipher, letter by letter, from the beginning of the message, to see the corresponding plaintext letters appearing in the display. This amazing feature is made possible because of the introduction of a *back reflector*, as illustrated in Figure 3.9. As the figure shows, the reflector connects the keyboard/display letters by pairs: Q-Z, W-E, R-I, T-O and U-A. So typing "R" lits up "I" and typing "I" lits up "R." If the decryptor has set his machine according to the same scrambler key as the encrypter, typing in the ciphertext exactly restitutes the plaintext on the screen. Thus only one machine type is required at both ends of the crypted communication channel.

The Enigma machine was further improved by introducing a *plugboard* between the keyboard and the scramblers. This plugboard makes it possible to swap up to six pairs of letters (leaving 14 other letters unchanged) through manually plugged jumpers. The key should then also specify which pairs must be swapped (e.g., A/K-V/J-D/F-Z/M-T/R-Q/X). With the introduction of this plugboard, the actual number of keys becomes phenomenal. Consider indeed that

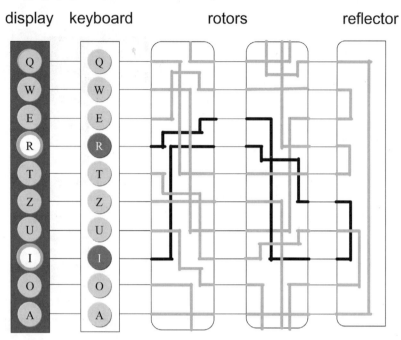

FIGURE 3.9 Enigma machine completed with a back reflector, showing that typing "R" or typing "I" on the keyboard lights up letter "I," or "R" on the display, respectively.

Choosing 12 letters out of 26 gives $C_{26}^{12} = 26!/(12!14!) = 9.6577 \times 10^6$ possibilities;

Choosing 6 pairs among these 12 letters gives $A_{12}^6 = 12!/6! = 665,280$ possibilities;

Since a given pair X/Y gives the same result as the pair Y/X, we must divide the result by $2^6 = 64$.

The total number of combinations introduced by the plugboard is therefore $9.6577 \times 10^6 \times 665,280/64 = 100,391,791,500$. With the scrambler orientations and sequence, the grand total comes to the number (calculation without any truncation):

$$100{,}391{,}791{,}500 \times 17{,}576 \times 6 = 1.058691676 \times 10^{16}$$
$$= 10{,}586{,}916{,}764{,}424{,}000$$

which corresponds to about *10 million billion* (or 10 peta) combinations.

The *Enigma machine* was invented by the German A. Scherbius immediately after the conclusion of WW-I, and used by the German military and private business (in a non-classified version) over two decades, until the outbreak of WW-II. Despite its superior complexity, the Enigma was eventually cracked in 1932 by a Polish cryptoanalyst, M. Rejewski, who worked out how to find the scrambler settings. The starting point for this cracking was a built-in weakness in the message communication protocol. Indeed, the Germans used a *day key* recorded in a monthly reference book (e.g., 321 for the rotor positions, ZKQ for the rotor orientations, and A/K-V/J-D/F-Z/M-T/R-Q/X for the plugboard). But they also used a *message key* which was sent at the beginning of each message. This additional key indicated the new orientation of the rotors to be used for this specific message. It was also repeated twice to avoid risks of errors due to radio interference. Considering the examples in Figures 3.7 and 3.8, the message key QRI typed twice was first coded into TIU then ITE. The message would then start with TIUITE, which would then be decrypted as the message key QRIQRI. The receiving operator would then stop and set his scramblers in the new position QRI in order to decrypt the rest of the message. But for the interceptor, this way to proceed betrays some relationship between letters. Indeed, in TIUITE, letters TI, IT and UE have a one-to-one relationship in the first scrambler. If the interceptor could collect enough messages (with different message keys) during the course of a single day, a full relationship table could be established between these pairs. Using an original Enigma machine, a complete catalog of the $6 \times 17{,}576$ possible scrambler settings was created (this took a full year). From this catalog and the observed pair relationship table, it was possible to exactly determine the original setting of the three scramblers. This represented enormous information as part of the cipher key, but was insufficient, since the plugboard setting, the rest of the key, was unknown. But in spite of the huge number of plugboard setting possibilities (100,391,791,500), typing and retyping the cipher while playing around with the plugboard cables would eventually produce some *crib*s, such as "wxe generzl pilj prrive komorrob" easily reconstructed into "the general will arrive tomorrow." Other possible cribs came from the habit of beginning messages with the same expressions (e.g., "Herr General"), or being short of ideas for choosing random scrambler settings. A tired crypto-operator would indeed likely use successive letters on the Enigma keyboard, such as shown in Figure 3.7 (e.g., QWE, WER, ERT, etc.) which, along with the message key, would betray some of the plugboard settings. Once the complete plugboard setting (of the day in question) is thus identified, the message can be deciphered in full, since its message-specific key is communicated at the beginning (first 6-letter group). And since the plugboard settings remain unchanged for the day, all messages intercepted in this day's course can be successfully decrypted. Finally, this code-breaking "disaster" is more complete considering that the same plugboard settings are used by the entire army for all messages of the day. Many more interesting facts and background stories concerning the fabulous Enigma machine and the development of the first *electrical code-breaking machines* (dubbed "*bombes*") can be found in the book/Web site selection shown in Bibliography.

These previous examples of classical ciphers complete our initiation tour in the domain of cryptology, which provides the reader with some familiarization and indispensable background. The rapid development in the 1940s of powerful computers, which were only available to states and big corporations, of the "personal" computer in the 1980s, and of the Internet in the 1990s, has completely revolutionized the entire field of cryptography, which marks a second historical development phase. The big computers made it possible to introduce phenomenal ciphering complexity (obviously, to be cracked by machine power!), while the widespread use of personal computers for electronic correspondence, commerce and transactions called for popular and cheap applications of "safe" cryptology. Therefore, our next task is to revisit cryptography in the digital domain, which requires a fully different approach and new analytic tools.

3.2 MODERN CRYPTOGRAPHY ALGORITHMS

This section reviews the basics of what could be called "modern" cryptography. As we shall see, the principles used in modern cryptography are not just the same older ones implemented with powerful computers and bigger keys, like super Enigma machines. Rather, they represent a second phase of history with the introduction of radically new algorithms, some leading to ciphers that are extremely hard to crack, even with the latest supercomputers. Although this may look a difficult subject (and it really is!), the elementary basics are quite simple to explain and understand. In the next subsection, we will first look at how encryption and decryption work with *binary numbers*, from simple transposition to more complex algorithms which opened the way to modern standards. In the other subsections, we will then consider the principles of *double-key* encryption, cryptography *without key exchange*, and *public-key* cryptography, while introducing the necessary basics of *modular algebra*.

3.2.1 Encryption With Binary Numbers

Alphanumeric characters, including spaces, punctuation, numbers and other signs which we use as a written language, are seen by computers in the form of standard binary-number codes. The reader not familiar with binary-number representation should refer to Desurvire (2004), Chapter 1, Section 1.2. As mentioned in that specific section, the two main alphanumerical codes used in computer communications are ASCII (*American standard code for information interchange*), and its extension EBCDIC (*extended binary coded decimal interchange code*). These two standard codes use 7-bit and 8-bit (1 byte) words, respectively. Table 3.14 shows an extract of the ASCII table for the most common keyboard characters. For instance, the letter A is coded as 100 0001 and character # is coded 010 0011. Note that the convention is to always to show the bit of lower weight (factor of 2^0) to the right: the binary number $a_3 \times 2^3 + a_2 \times 2^2 + a_1 \times 2^1 + a_0 \times 2^0$ is therefore written as $a_3 a_2 a_1 a_0$, just as we do with decimal numbers. The ASCII and EBCDIC codes have $2^7 = 128$ and $2^8 = 256$ possible alphanumerical character and other keyboard-command possibilities, respectively. For cryptography purposes, this represents a huge extension of the previous 26-letter alphabet from earlier times.

How can one make ciphers from binary numbers? The answer is that there are an infinite number of ways to proceed, even in the most simple cases. Instead of 26 symbol characters, we now have only two symbols, namely "0" and "1." Since each group of 7 bits (ASCII) represents one plaintext character, there are many new possible schemes for transposition, substitution or permutation. The number of ways any single 7-bit word can be modified into anything is $2^7 = 128$, as opposed to only 26 with the ordinary alphabet. But the interesting feature of the binary system is that now we can code the bits not only within a single 7-bit block, but over the entire

TABLE 3.14 Extract of ASCII table showing the most common alphanumeric and keyboard characters

4	3	2	1	7 6 5	0 1 0	0 1 1	1 0 0	1 0 1	1 1 0	1 1 1
0	0	0	0		sp.	0	@	P	\	p
0	0	0	1		!	1	A	Q	a	q
0	0	1	0		"	2	B	R	b	r
0	0	1	1		#	3	C	S	c	s
0	1	0	0		$	4	D	T	d	t
0	1	0	1		%	5	E	U	e	u
0	1	1	0		&	6	F	V	f	v
0	1	1	1		'	7	G	W	g	w
1	0	0	0		(	8	H	X	h	x
1	0	0	1		)	9	I	Y	i	y
1	0	1	0		*	:	J	Z	j	z
1	0	1	1		+	;	K	[	k	{
1	1	0	0		,	<	L	\	l	|
1	1	0	1		−	=	M	]	m	}
1	1	1	0		.	>	N	^	n	~
1	1	1	1		/	?	O	−	o	

message sequence, which scrambles the characters themselves and not only the alphabet. For instance, the (4-bit/character) sequence

$$a_3a_2a_1a_0b_3b_2b_1b_0c_3c_2c_1c_0d_3d_2d_1d_0 \ldots$$

can be transformed by switching the bit positions by pairs, which gives

$$a_2a_3a_0a_1b_2b_3b_0b_1c_2c_3c_0c_1d_2d_3d_0d_1 \ldots$$

Alternatively, we can take characters by pairs and swap their center bits, i.e.,

$$a_4a_3b_2a_1a_0b_4b_3a_2b_1b_0c_4c_3d_2c_1c_0d_4d_3c_2d_1d_0 \ldots$$

Let us see now what these two approaches would produce as cipher texts. Consider the following example (ASCII characters, see Table 3.14 for conversion), noting that now we can use capital characters in the plaintext:

Plaintext	I	m	i	s	s	y	o	u
PT-ASCII	**1**00100**1**	110110**1**	110100**1**	111001**1**	111001**1**	111100**1**	110111**1**	111010**1**
CT-ASCII	**0**110001**1**	111001**1**	111000**1**	110110**1**	110110**1**	111100**1**	111011**1**	110101**1**
Ciphertext	1	s	q	m	m	y	w	k

Plaintext	I	m	i	s	s	y	o	u
PT-ASCII	1001001	1101101	1101001	1110011	1110011	1111001	1101111	1110101
CT-ASCII	1001001	1101101	1100001	1111011	1111011	1110001	1100111	1111101
Ciphertext	I	m	a	{	{	q	g	}

In the first case, we observe that the last bit (a_0) of each character is unchanged (since the number of bits is odd), and that the character "y" is invariant. In the second case, the first two characters, "I" and "m", are invariant. But if we were to swap more bits up to a complete permutation algorithm, we would achieve most perfect scrambling. For an 8-character/56-bit codeword, the number of possible permutations is $56! = 7.1 \times 10^{74}$, which beats any random-key Vigenère for the same character-word length! Although with a sufficiently complex substitution algorithm, this approach would give undecipherable messages, the drawback is that its key is fixed. No matter how complex, a fixed key has many drawbacks. It can be intercepted during its communication (over the Internet, for instance) without the two parties ever being aware. It can be retrieved from the software used to crypt or decrypt, should one of the terminal computers be compromised by dishonest third-parties. The alternative is to use a new random key for each new message sent (onetime-pad cipher). As already discussed in the case of Vigenère ciphers with random keys, this approach is not practical for any routine use, as in the central command of an army. If the keys are to be communicated over the Internet (or any other wide-range transmission medium), it is also unsafe.

The solution to these issues, which is far from trivial, will be addressed in the next two subsections. In the meantime, we will get further prepared by considering how *secret keys* can be used in the binary system.

In the cipher examples provided in Section 3.1, we saw that the key can be anything from a word to one or several magic squares, to algorithms of row-and-column manipulations, to the arrangement of machine scramblers. In the binary system, keys are reduced to a single codeword of any desirable length. Mixing the key with the plaintext provides both encryption and decryption. Such a mixing is performed by simple *Boolean operations*. In the decimal system, the four basic operations are ($+$, $-$, $\times$, $\div$). In the binary system, the four operations, called Boolean, are NO (logical inversion), AND (multiplication), OR (addition or subtraction), and XOR (exclusive OR). The rules of these operations on one (NO) or two (AND, OR, XOR) bits or "operands" are shown in Table 3.15. We note that with binary numbers, addition and subtraction are the same ($1 + 1 = 1 - 1 = 0$, $1 + 0 = 1 - 0 = 0 + 1 = 0 - 1$). We also note that XOR is the same as OR except for the rule $1 + 1 = 0$, which is similar to $9 + 1 = 0$ in the ordinary digital system. For further reference, the operators (NO, AND, OR and XOR) are also noted by mathematicians

TABLE 3.15 The four Boolean operations in the binary system: NO, AND, OR, XOR

a	b	NO a	NO b	a AND b	A OR b	a XOR b	a XOR b XOR b
0	0	1	1	0	0	0	0
0	1	1	0	0	1	1	0
1	0	0	1	0	1	1	1
1	1	0	0	1	1	0	1

The last column sows the repeated operation of XOR with operand "b", which yields operand "a."

($\neg$, $\wedge$, $\vee$ and $\oplus$), respectively. Thus "NO a," "a AND b," and "a XOR b" are noted $\neg a$ and $a \wedge b$ and $a \oplus b$, respectively. The other basic Boolean operators are $\leq$ (noted $\rightarrow$) and $=$ (noted $\leftrightarrow$). Thus $a \rightarrow b$ is identically zero if a $>$ b and equal to one otherwise. Similarly, $a \leftrightarrow b$ is zero if $a \neq b$, and one if a $=$ b. It is easily checked that the binary result of the operation $(a \rightarrow b) \wedge (b \leftrightarrow c)$ is unity if the condition $a \leq b = c$ is fulfilled. In this book, we will not be concerned by Boolean logic other than just using XOR.

As Table 3.15 shows, an interesting property of XOR is that its double application to "a" with the same operand "b" restitutes operand "a". We can thus use this property for the purpose of encryption with a key. Indeed, if we perform XOR between each bit of the plaintext and the key, we obtain a ciphertext. If we do the same operation again between the ciphertext and the key, we obtain the plaintext. Here is an illustrative example, with "sAXOPHON" as a (case-sensitive) key:

Plaintext	I	m	i	s	s	y	o	u
Key	s	A	X	O	P	H	O	N
PT-ASCII	1001001	1101101	1101001	1110011	1110011	1111001	1101111	1110101
K-ASCII	1110011	1000001	1011000	1001111	1010000	1001000	1001111	1001110
CT-ASCII	0111010	0101100	0110001	0111100	0100011	0110001	0100000	0111011
Ciphertext	;	'	1	<	#	1	space	;

We see from this example that the same ciphertext character may correspond to different plaintext characters, namely the first/second ";" correspond to "I"/"u" and the first/second "1" correspond to "i"/"y," respectively. We could therefore conclude that this XOR operation with a key as long as the plaintext (or sufficiently long and repeated) and as many as 128 possible characters would produce quite a strong cipher. But it is completely the opposite case! Let us prove this point right away.

Consider indeed the first codeword, "0111010." Because of the XOR operation, we are absolutely sure that

1. For bits 7,3,1 $=$ 0: operands must be of same parity, namely, 0 XOR 0 or 1 XOR 1;
2. For bit 6,5,4,2 $=$ 1: operands must be opposite parity, namely, 0 XOR 1 or 1 XOR 0.

Since there are two operand choices for each bit, the number of possibilities is $2^7 = 128$. But we can reduce this number by correlating the different conditions (1)–(2) for each bit. To establish such a correlation, we can tabulate all correspondence possibilities between each ciphertext character and each key character under the form of a 128 × 128 = 16,384-element array. To simplify the task (for this description, not for the computer), we can assume that both plaintext and key characters have been selected from the reduced ASCII set shown in Table 3.14. This reduces the field of investigation to a matrix of 6 × 6 main codewords starting with bits $a_7 a_6 a_5 =$ 010 [1], 011 [2], 100 [3], 101 [4], 110 [5], and 111 [6]. Each of these main codewords, which we call [1]–[6] for rows and columns, have four other bits, which have $2^4 = 16$ possibilities. So our real matrix dimension is

(6 × 16) × (6 × 16) = 96 × 96 with 9,216 elements. But we don't need to fill the matrix of possibilities with YES/NO on a element-by-element basis. Indeed consider our first cipher codeword. Condition (1) implies that the first ciphertext bit corresponds to plaintext/key bits with same logical value. Therefore, the 16 (main) matrix elements [1][3], [1][4], [1][5], [1][6], [2][3], [2][4], [2][5], [2][6], [3][1], [4][1], [5][1], [6][1], [3][2], [4][2], [5][2], and [6][2] are excluded because their leading bits are different. This makes 16 × 16 = 256 eliminated possibilities. The same procedure applied to the second ciphertext bit all the way down to the last one, makes it possible to eliminate a certain number of possibilities. Using a simple spreadsheet editor for the 6 × 6 array with its 16 × 16 inner structure, and comparing rows and columns by blocks and subgroups, it took the author less than *10 minutes* to proceed with this systematic elimination. This "by hand" processing, which would have taken microseconds of computing time with a program, made it possible to identify only 4 relevant (main) matrix elements, in which 16 possibilities were left (namely, 4 × 16 = 64 possibilities). Table 3.16 shows the main matrix element [6][3] (row [6] and column [3]), which contains the solution, with its 16 pair-possibilities total. In this table, we recognize the "I-s" match of our previous example, along with 15 other possible pairs. This demonstration thus shows that the 9,216 initial guesses for the first ciphertext character can be reduced to 64, which represents over a hundred-fold reduction factor. We also note from the table that certain solutions for keyword characters are less likely ($\{, |, \}, \sim, n/a$). This fact reduces, in this specific matrix area, the number of possibilities to $64 - 5 = 59$. There is no small gain in cryptanalysis!

What about the other ciphertext characters? We have seen that within microseconds (or 10 minutes for the author's brains), the computer is able to tabulate 64 possible choices for the first plaintext/key character pairs. Let us leave the computer to run for another few microseconds until all possible plaintext/key character choices are listed for each of the cipher codewords. Considering our previous 8-codeword example, we then obtain eight tables of (say) 64–100 valid choices for plaintext/key character pairs. Can one make sense of tables having as many as 8,000 entries? The answer is yes, because the computer (and even a human being), is able find cribs considering either rows and columns separately. Let us write down the data of Table 3.16 (which concern the possible plaintext/key pair possibilities of the first codeword) in the form of a condensed table:

plaintext	@	A	B	C	D	E	F	G	H	I	J	K	L	M	N	O
key	z	{	X	y	$\sim$	n/a	|	}	r	s	p	q	v	w	t	u

In the above table, we framed in bold the combinations that are unlikely, either because a plaintext message is not expected to begin with "@," or because a normal key is not expected to have characters such as ($\{, \sim, n/a, |, \}$). We see that from this simple "by hand" analysis, the number of correspondence possibilities for the first cipher character has been reduced from 128 to 10, which represents over a hundred-fold reduction factor. The same conclusion

TABLE 3.16 Selection of the correlation table showing one of the matrix regions (here row [6], column [3]) where plaintext and keyword ASCII characters match according to ciphertext XOR analysis

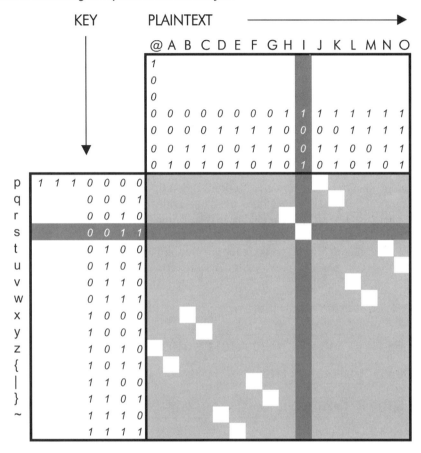

The white squares define the possible solution pairs in this region. The dark lines correspond to the actual solution in the text example (plaintext = I, key = s). The crosses (x) indicate unlikely choices for the key.

applies to the key, which represents important information for the rest of the message decryption. Thus the actual reduction factor is $128 \times 128/10 = 1{,}638$, or over a thousand-fold. The same systematic elimination and prescreening procedure can be performed through a computer on *all* ciphertext codewords. A straightforward frequency analysis of both plaintext and key, searching for the most common letters, digrams, trigrams and other most commonly used words will finish the job. Should a few nulls be present in the beginning or at random places in the plaintext, as in an attempt to fool decryption, this would have no effect on the outcome of the frequency analysis.

This demonstration suggests that in the binary system, encryption on a one-to-one character basis is as unsafe as any Vigenère cipher, unless complex random keys be used. As we have seen, however, the binary system offers a tremendous advantage, which is the ability to mingle the characters together in the coding process, on a bit-by-bit or block-by-block basis (a *block* being a string of bits of any prescribed length). This is the principle used by the first attempt at standard cipher, developed in the 1960s by H. Feistel of IBM and called *Lucifer* (as a rhyme with "cipher"). Figure 3.10 shows the Lucifer coding procedure. In the first step, the plaintext is converted into a string of binary digits (not necessarily ASCII). Then the string is split into 64-bit blocks (step 2) which are individually *encrypted* (step 3). The resulting blocks are *shuffled* (step 4), then split into two 32-bit elements called $left^0$ and $right^0$ (step 5). The next step is called a *round*, and this round will be repeated 16 times. Call these two elements L0 and R0, respectively. The round first passes R0 through a *mangler*, which performs a complex bit substitution according to a *mangling function*. The result is XORed with L0 to give R1. The original R0 becomes L1. The next round does the same with the pair (L1, R1), and so on until one gets (L16, R16). The cipher is then formed with the concatenated string of (L16, R16) elements.

We see that this new approach for encryption is quite complex and definitely harder to break than any previous scheme. Even with the full knowledge of the

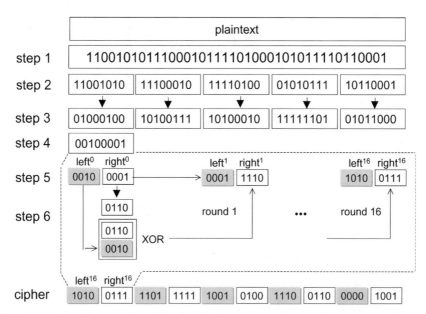

FIGURE 3.10 Encryption principle of the Lucifer system. Step 1: plaintext conversion into binary; step 2: decomposition into 64-bit blocks; step 3: encryption by individual block; step 4: shuffling; Step 5: splitting into two 32-bit blocks, called $left^0$ and $right^0$ (shown only for first block); step 6 (or round 1): mangling $right^0$ and XOR with $left^0$, yielding $right^1$, with $right^0$ becoming $left^1$. The process is iterated 16 times to give $left^{16}$ and $right^{16}$ which form one block element in the final cipher (bottom).

above encryption procedure, the code breaker must know which mangling function (the key) has been used for the 16 rounds, which shuffling algorithm is used in step 4, and which encryption algorithm is used in step 3. The combination of these three unknowns presents a formidable obstacle to code-breaking attempts. But when these different elements are known, decryption is "straightforward" for the machine. The cipher is decomposed into 32-bit block pairs representing as many L16-R16 groups. The element L16 becomes R15. The element R16 is then XORed with L15, then the result is passed through the inverse mangling function (the key), which yields R15. And so on until L0 and R0 are obtained. The L0 and R0 elements are then concatenated into a 64-bit block. The bits are then reordered, according to the inverse of the shuffling algorithm. The resulting block is decrypted. Then all decrypted blocks are concatenated together and converted into plaintext. We can infer that the whole decrypting operation can be performed in a few milliseconds of computer time, depending upon the complexity of the step-3 cipher and the mangling function. As we have stated it, the mangling function represents the key to the encryption and decryption process. Such a key can be defined by a *number*, which represent the different integer parameters to be used for generating the mangling function. Thus the number of possible keys is virtually infinite. To prevent any possibility that this code became *absolutely* unbreakable, it was agreed that the number of keys should be $2^{56} \approx 72 \times 10^{15}$, which corresponds to a 56-bit key. Twenty-five years later, the principle of the Lucifer system with 56-bit key was officially endorsed in the United States to become the *data encryption standard* (DES), which is still in use today. See detailed description of the actual DES algorithm with an illustrative example in the next section.

3.2.2 Double-Key Encryption

In spite of the tremendous level of protection offered by algorithms such as DES/ Lucifer, a major weakness remains, which is the obligation for the originator to communicate the *key* to the destinee. Like the Germans with Enigma, both persons could keep a secret copy of the same book where the keys to be used are indicated for each day of a 4-week period. A new set of books must be created and exchanged every month. But a malicious third party may intercept the book and rapidly copy the contents, while making sure that no one notices anything. As a result, all secret communications based upon this book's keys could be read "in the open" by this third party, the worse being that the two ends are unaware of it.

In cryptography, the tradition is to call the originator *Alice*, and the destinee *Bob*. The malicious interceptor who eavesdrops, is *Eve*. From this point on, we will use these three nicknames. For Alice, the encryption task is like putting a message into a locked steel box and sending it to Bob through the mail. If Bob does not have a copy of Alice's key to open the lock, then he can't retrieve Alice's message. Alice must then provide Bob with a copy of her key, for instance by meeting regularly in some agreed time and place. But this way to proceed amounts to the same thing as passing on and hiding key books, notwithstanding the time and effort required.

This intractable problem of *key exchange* caused cryptographers to search for new directions in the way message "boxes" could be locked and unlocked. They found that one simple solution could be to use *two locks* instead of one! To visualtize how this can work, assume that both Alice and Bob have different keys of their own. Alice locks her message box and sends it to Bob (step 1). Bob cannot open it, since he does not have Alice's key. What he does is to put a second lock onto the box, lock it with his own key, and send the box back to Alice (step 2). Then Alice opens the first lock with her key. She then sends back the box to Bob (step 3). Now Bob can open the box, since he has the key of the remaining lock (step4). Such a process is referred to as *two-key encryption*. The remarkable result is that the message was safely communicated without Alice and Bob exchanging any key whatsoever. Consider now a practical illustration. The following table shows the different steps of double-key encryption, using the principle of adding (XOR) a same-length key (since the purpose is just to prove the point through successive XOR additions, we have chosen random keys for Alice and Bob):

Plaintext	I	m	i	s	s	y	o	u
PT	**1001001**	**1101101**	**1101001**	**1110011**	**1110011**	**1111001**	**1101111**	**1110101**
Alice key	0001001	1010100	0101011	0111101	0111010	0001001	1001001	1100100
Step 1	1000000	0111001	1000010	1001110	1001001	1110000	0100110	0010001
Bob key	1011001	1000111	0111010	1111101	0101110	0011001	1001000	0001011
Step2	0011001	1111110	1111000	0110011	1100111	1101001	1101110	0011010
Alice key	0001001	1010100	0101011	0111101	0111010	0001001	1001001	1100100
Step 3	0010000	0101010	1010011	0001110	1011101	1100000	0100111	1111110
Bob key	1011001	1000111	0111010	1111101	0101110	0011001	1001000	0001011
Step 4	**1001001**	**1101101**	**1101001**	**1110011**	**1110011**	**1111001**	**1101111**	**1110101**
Plaintext	I	m	i	s	s	y	o	u

We see from the above example that at the end of this double-key encryption process, Bob is able to retrieve the plaintext unaltered (step 4). We note again that in the different steps, Alice and Bob only used their own keys. The explanation of this successful retrieval is the following. Call M the plaintext message block and A, B the keys from Alice and Bob. The cipher in step 1 is thus $M\ XOR\ A$, or $M \oplus A$. The cipher in step 2 is $M \oplus A \oplus B$. That in step 3 is $M \oplus A \oplus B \oplus A = M \oplus B$ (using the previously described property of *XOR*). Finally, the result in step 4 is $M \oplus B \oplus B = M$, which is the plaintext.

While the previous double-key encryption surely works without any key exchange, it is absolutely unsafe. Indeed, if Eve is ever able to intercept all three messages (ciphers in step 1, step 2 and step 3 in the above), the only thing she has to do to retrieve the plaintext is just to XOR these together:

Step 1	1000000	0111001	1000010	1001110	1001001	1110000	0100110	0010001
Step 2	0011001	1111110	1111000	0110011	1100111	1101001	1101110	0011010
XOR	1011001	1000111	0111010	1111101	0101110	0011001	1001000	0001011
Step 3	0010000	0101010	1010011	0001110	1011101	1100000	0100111	1111110
XOR	**1001001**	**1101101**	**1101001**	**1110011**	**1110011**	**1111001**	**1101111**	**1110101**
Plaintext	I	m	i	s	s	y	o	u

There is nothing surprising in this result if we consider from the previous definitions that what Eve does is the operation

$$(M \oplus A) \oplus (M \oplus A \oplus B) \oplus (M \oplus B) = M \oplus M \oplus M = M$$

Since Eve can so effortlessly retrieve the plaintext, without even any cryptanalysis, we conclude that this "no key exchange" approach based upon the addition of independent keywords is by and large the worst possible and riskiest way to proceed.

A safer solution for Alice and Bob could consist in using *substitution* algorithms instead. But the problem is that substitution algorithms have no reasons to be commutative. If Alice first encrypts the message with her own substitution-key algorithm, then Bob does the same with his own, then Alice and Bob again, the end result would be nonsense. Some more creative approach had to be devised, which led to the breakthrough of *public-key cryptography* (PKC). This requires some more mathematical (easy) basics, which are described in the next subsection.

3.2.3 Modular-Algebra Basics

In this subsection, we shall briefly review some principles and properties of *modular algebra*, also called in schools "*clock arithmetic*." As the following subsection will show, such a modular algebra is at the root of PKC, and this is why we need to look at it more closely. The following is not meant to be a rigorous introduction, but rather to define the shortest and easiest conceptual path possible.

Let m and x be two integers, with m being non-zero. Define $[x/m]$ as the integer part of the division of x by m. If we consider for instance $m = 3$ and the ratios $2/3 = 0.66$, $4/3 = 1.33$, $6/3 = 2$ and $11.3 = 3.66$, we thus obtain $[2/3] = 0$, $[4/3] = 1$, $[6/3] = 2$ and $[11/3] = 3$. It is then possible to express any integer number x in the form:

$$x = m[x/m] + r \tag{3.2}$$

where r stands for "residue." With the previous examples, we have:

$$\begin{cases} 2 = 3[2/3] + 2 \\ 4 = 3[4/3] + 1 \\ 6 = 3[6/3] + 0 \\ 11 = 3[11/3] + 2 \end{cases} \tag{3.3}$$

We observe that the residues are positive numbers which range from $r = 0$ to $r = m - 1$. We can also define these residues in the form

$$r = x - m[x/m] \tag{3.4}$$

By convention, one designates the residues according to any of the following equivalent notations

$$\begin{cases} r = |x|_m \\ r = x \bmod m \\ x \equiv r[m] \\ x = r(\bmod\ m) \end{cases} \quad (3.5)$$

which reads "r equals x modulo m," or "x equals r modulo m," being understood which one is the residue. Here, we shall mainly use the second definition, with the convention that the last "mod m" carries over the entire RHS expression or even the full line (this to avoid a parenthesis inflation).

Modular arithmetic is strictly the same as our regular algebra (with any integer numbers) when considering addition/subtraction or multiplication, except that the result of the operation is reduced to "modulo m." The following examples illustrate how this algebra works:

$$\begin{cases} (3+4) \bmod 6 = 7 \quad \bmod\ 6 = 1 \\ (8+7) \bmod 15 = 15 \quad \bmod\ 15 = 0 \\ (7-4) \bmod 4 = 3 \quad \bmod\ 4 = 3 \\ (2 \times 3) \bmod 4 = 6 \quad \bmod\ 4 = 2 \\ (7 \times 9) \bmod 11 = 63 \quad \bmod\ 11 = 8 \end{cases} \quad (3.6)$$

In modular arithmetic, the operations of addition and multiplication are *commutative*, *associative* and *distributive*, meaning the following properties:

$$(a \pm b) \bmod m = (a \bmod m) \pm (b \bmod m) \bmod m \quad (3.7)$$

$$(a \times b) \bmod m = (a \bmod m) \times (b \bmod m) \bmod m \quad (3.8)$$

$$[a \times (b+c)] \bmod m = ((a \times b) \bmod m + (a \times c) \bmod m) \bmod m \quad (3.9)$$

Here are two other obvious or less obvious properties (k = integer):

$$(-a) \bmod m = (m-a) \bmod m = [(m-1) \times a] \bmod m \quad (3.10)$$

$$(k \times m) \bmod m = 0 \quad (3.11)$$

$$(k \times a) \bmod (k \times m) = k \times (a \bmod m) \quad (3.12)$$

While addition and multiplication are straightforward in modular arithmetic, *division* is not. For instance, what is the residue r as defined through $r = (11/5) \bmod 3$? In modular arithmetic, r should be the number which verifies $5 \times r = 11 \bmod_3 = 2$. The only solution to this equation is $r = 1$, since $5 \times 1 \bmod 3 = 5 \bmod 3 = 2$. We thus see that $|11/5|_3 = 1$, which was not at all intuitive. One can also define the number $(11/5) \bmod 3$ as the product $(11 \times 1/5) \bmod 3$, or equivalently $(11 \bmod 3) \times (1/5 \bmod 3) = 2 \times (1/5 \bmod 3) \bmod 3$. According to the same procedure, we find that $|1/5|_3 = 2$, since $5 \times 2 = 10 \bmod 3 = 1$. We thus obtain $2 \times (1/5 \bmod 3) \bmod 3 = 2 \times 2 \bmod 3 = 4 \bmod 3 = 1 = (11/5) \bmod_3$, which is consistent with the previous definition of division. Since inverses can be defined in modular arithmetic, there is no need to use the division operation. One restriction however concerns the relation between the number to be inverted and the modulus: they should not be multiples of each other (or have 1 as the greatest common divider). This issue is further discussed in the following text.

What about other functions, such as powers, exponentials, square roots or logarithms? For powers and exponentials, it simply consists in expressing the result of the usual function to its modulo m residue, as illustrated in Table 3.17 in the case of the functions $y = x^2$ and $y = 2^x$. We observe that the residues take periodic values, except that some values are missing (i.e., 2, 3 for $y = x^2$, and 0 for $y = 2^x$). Consider next the square root ($y = \sqrt{x}$) and logarithm ($y = \log_2 x$) functions. In the first case, we notice that in this algebra, number 4 has two possible square roots, namely 2 and 3 (i.e., $4 = 2^2 = 3^2 \mod 5$), while numbers 2, 3 have no square root. In contrast, the $\log_2$ function is seen to present no oddity, but this is due to our specific modulus choice. We conclude that in modular arithmetic, it is possible to define a wide variety of elementary functions (e.g., powers and exponentials) regardless of the modulus m, but their reciprocal functions (e.g., square root and logarithm) do not always exist.

The conversion between x and its residue r (modulo m) is called *modular reduction*. Such a reduction operation corresponds to a *one-way function*. This terms refers to the fact that the correspondence that is established is not reversible, since there is an infinity of numbers which have the same residue. It is like putting a drop of ink in a glass of water, the result cannot be undone to retrieve the clear water. Powers and exponentials of the type $y = x^p$ and $y = a^x$ ($a, p =$ integers) are also one-way functions, since the result y is associated with an infinity of possible values x. Here is an important property concerning the commutation of exponents, which is the same as in ordinary arithmetic, and which we will have the opportunity to exploit:

$$(a^p \bmod m)^q \bmod m = (a^q \bmod m)^p \bmod m = a^{pq} \bmod m \qquad (3.13)$$

Having introduced the basics of modular arithmetic, we can introduce two fundamental theorems which have been derived by Fermat and Euler. Before considering these theorems, it is necessary to recall the concept of *prime number*. A prime number (also called "prime") is an integer number greater than unity which can be divided only by one and by itself. As a consequence, a prime cannot be factorized as a

TABLE 3.17 Squares, square-roots, exponentials and logarithms (base 2) in modular arithmetic ($m = 5$)

x	x^2	2^x	$\sqrt{x}$	$\log_2 x$
0	0	1	0	–
1	1	2	1	0
2	4	4	–	1
3	4	3	–	3
4	1	1	2 or 3	2
5 = 0	0	2	0	–
6 = 1	1	4	1	0
7 = 2	4	3	–	1
8 = 3	4	1	–	3
9 = 4	1	2	2 or 3	2
10 = 0	0	4	0	–

product of smaller integers different from unity. The first ten primes are 2, 3, 5, 7, 11, 13, 17, 19, 23, 29. The number of primes is infinite, and there is no known algorithm to define their increasing sequence. Extensive tables of prime can be found on the Internet, and many web pages are dedicated to this fascinating mathematical subject. The largest known prime (record of year 2001, apparently unchallenged to date) is $2^{13466917} - 1$. This number is made of no less than 4,053,946 digits (this can be checked by using the transformation $2^p = 10^{p \times \log 2/\log 10} = 10^{p \times 0.301}$. Numbers which are defined in the form $M_p = 2^p - 1$ are called *Mersenne numbers*, or *Mersennes*, after the name of the eleventh-century French monk who discovered them. A Mersenne is not always a prime, but $M_2, M_3, M_5, M_7, M_{13}, M_{17}, M_{19}$, $M_{31}, M_{61}, M_{81}, M_{107}$ and M_{127} are. As it turns out, when M_p is a prime, the exponent p is also a prime. The current list of known Mersenne primes is limited to 39 numbers, and stops at $M_{13466917}$. It can also be shown that for given any number N, the number of primes that are lower or equal to N is approximately $N/(\log N - 1)$. Thus, if $N = 10^9$, there exists a maximum of $10^9/(\log 10^9 - 1) \approx 50.7$ millions of primes, and if $N = M_{13466917}$, the number of primes becomes $10^{4,053,939}$. While this last figure is 10^7 times smaller than $M_{13466917}$, it is yet another "monster" number. There is no point in calculating how much memory space would be required to tabulate all primes existing under $M_{13466917}$ if they were all known, even if $M_{13466917}$ only takes about 3.5 Mbytes using ASCII.[1]

Two numbers are said to be *relatively prime* if they do not divide each other, or if their greatest common divisor (gcd) is unity. For instance, (3, 4), (2, 9) or (5, 12) are relatively prime. Clearly, two prime numbers are relatively prime. Primes and relatively prime numbers play a central role in modular arithmetic, as the two following theorems and their application will illustrate.

Fermat's theorem (named after the French mathematician) states that if m is prime and (a, m) are relatively prime, then

$$a^{m-1} \bmod m = 1 \qquad (3.14)$$

which can be put into the equivalent form: $a^m \bmod m = a$. Thus we have $a \bmod 2 = 1$ or $a^2 \bmod 3 = 1$ for any a that is prime with number two or number three, and so on with any prime modulus, for instance $11^{23-1} \bmod 23 = 1$. We see that as a first benefit, Fermat's theorem makes it possible to considerably simplify calculations in modular arithmetic.

Euler's theorem (named after the Swiss mathematician) states that if p and q are two primes and $n = pq$ is their product, then we have for any integer a which is prime with n:

$$a^{(p-1)(q-1)} \bmod n = 1 \qquad (3.15)$$

The factor $(p-1)(q-1)$ with $pq = n$ is also called $\phi(n)$. A consequence of Euler's theorem is that the inverse $1/a$ is readily defined as

$$\frac{1}{a} = a^{-1} = a^{\phi(n)-1} \bmod n \qquad (3.16)$$

For instance, what is the inverse of 5 modulo 21? We have $21 = 3 \times 7$ so $\phi(21) =$

---

[1] Note (2004 update): $M_{24036583} = 2^{24036583} - 1$ is the 41st known Mersenne prime (May 28, 2004, source http://www.utm.edu/research/primes/largest.html).

$(3-1) \times (7-1) = 12$ and $1/5 = 5^{12-1}$ mod $21 = 48,828,125$ mod $21 = 17$. Thus $|1/5|_{21} = 17$ which we verify as (17×5) mod $21 = 85$ mod $21 = 1$. We see that the calculation of an inverse through Euler's theorem is quite straightforward, compared to the method previously described. But the requirement is that the modulus n be defined by the product of two primes, which represents a special or fortuitous case.

It is important to note that if (a, n) are *not* relatively prime, the inverse a^{-1} does not exist. If the two are relatively primes but Euler's theorem cannot apply, the inverse a^{-1} can be found using the *extended Euclidian algorithm*. For a source code of this algorithm, see for instance Schneier (1996). Here, we shall explain the algorithm through two representative examples: for instance, find the inverse of 7 in modulus 39, and the inverse of 5 in modulus 16,272, i.e., the values of 7^{-1} mod 39 and 5^{-1} mod 16,272. Taking the first example, the algorithm proceeds as shown in the following table with the steps as numbered in the first column:

1	39	39
2	7	1
3	$5 \times 7 = 35$	$5 \times 1 = 5$
4	$39 - 35 = \mathbf{4}$	$39 - 5 = \mathbf{34}$
5	$1 \times 4 = 4$	$1 \times 34 = 34$
6	$7 - 4 = \mathbf{3}$	$1 - 34 = \mathbf{-33}$
7	$1 \times 3 = 3$	$1 \times (-33) = -33$
8	$4 - 3 = \mathbf{1}$	$34 - (-33) = 67 = \mathbf{28}$

In steps 1–2, we fill out the table as shown. We first divide 39 by 7, to find that the integer part of the result is $[39/7] = 5$. In step 3, we multiply line 2 by "5," and in step 4 we subtract the result of line 3 from that of line 1. This completes the first round. Consider next the ratio 7/4, whose integer part is $[7/4] = 1$. In step 5, we multiply the results of line 4 by "1," and in step 6, we subtract the result of line 5 from that of line 2. This completes the second round. Consider next the ratio 4/3, whose integer part is $[4/3] = 1$. In step 7, we multiply the results of line 6 by "1," and in step 8, we subtract the result of line 7 from that of line 4. If we have obtained "1" in the first column (line 8), we have finished. When positive, the number to the right is the inverse we have been looking for. Here, we find that the inverse of 7 is 28. Proof: $7 \times 28 = 196 = 1$ (modulo 39).

Consider next the second example, which concerns the search of 5^{-1} mod 16,272.

1	16,272	16,272
2	5	1
3	$3,254 \times 5 = 16,270$	$3,254 \times 1 = 3,254$
4	$16,272 - 16,270 = \mathbf{2}$	$16,272 - 3,254 = 13,018$
5	$2 \times 2 = 4$	$2 \times 13,018 = 26,036$
6	$5 - 4 = \mathbf{1}$	$1 - 26036 = -26035$

In steps 1–2, we fill out the table as shown. We first divide 16,272 by 5, to find that the integer part of the result is $[16,272/5] = 3,254$. In step 3, we multiply line 2 by "3,254," and in step 4 we subtract the result of line 3 from that of line 1. This completes the first round. Consider next the ratio 5/2, whose integer part is $[5/2] = 2$. In step 5, we multiply the results of line 4 by "2," and in step 6, we subtract the result of line 5 from that of line 2. This completes the search, since we have obtained "1" in the first column (line 6). However, the number in the right column, $b = -26,035$" is negative. In this case, the inverse is the complement $km + b$, namely $5^{-1} = 2 \times 16,272 - 26,035 = 6,509$. The proof is that $5 \times 6,509 = 32,545 = (2 \times 16,272) + 1 = 1$ (modulo 16,272). This numerical value will be used in Subsection 3.2.5 below concerning PKC applications.

Although the mathematical toolbox of modular arithmetic may seem somewhat heavy, it is in fact quite easy to use. Most importantly, it makes possible to implement a new type of cryptography in which there is no need to exchange any secret key, as described in the next subsection.

3.2.4 Cryptography Without Key Exchange

The seemingly intractable problem of cryptography without key exchange was eventually solved in the mid 1970s by two American cryptogeniuses, W. Diffie and M. Hellman (and also R. Merkle, independently). Their idea was to use the principle of one-way functions in modular arithmetic. Here is the first proposed algorithm (now referred to as *Diffie–Hellman* or *Diffie–Hellman–Merkle*):

Step 0: Alice and Bob call each other (or send an e-mail) to agree on a choice of two numbers (m, n), regardless of the possibility that Eve could have wired the line; they can even make this choice openly public;

Step 1: Alice chooses a *large* integer number A and keeps it secret; then she computes

$$a = m^A \bmod n$$

and sends the result (e.g., by phone or e-mail) to Bob;

Step 2: Bob chooses a *large* integer number B and keeps it secret; then he computes

$$b = m^B \bmod n$$

and sends the result (e.g., by phone or e-mail) to Alice;

Step 3: on her side, Alice computes

$$k = b^A \bmod n$$

Step 4: on his side, Bob computes

$$k' = a^B \bmod n$$

As it tuns out, the results of these last two computations are equal, i.e., $k = k' = m^{AB}$ (modulo n). This means that both Alice and Bob have the same number, which they can use as a common key! Eve is not able to get this information, even if she has intercepted the information on the (a, b) values that Alice and Bob communicated to each other. Although Eve also knows m, n, she has no clue of what could be the exponent AB of the key. And since Alice and Bob chose large secret numbers, AB is also very large! For instance, if A and B are of the order of 1,000, the exponent of the key is of the order of 1,000,000, representing a 20-bit key in the binary system. They might as well chose numbers in the order of 10,000, which generate 100,000,000 different key possibilities (26-bit key). Note that from Alice's or Bob's sides, the whole procedure actually involves only two steps (1–3 and 2–4, respectively), which we have decomposed here for clarity.

It is completely remarkable and counter-intuitive that two persons can mutually agree on a secret key by openly exchanging information via a public communication channel (even a wired line) without this secret ever being known. Such a discovery is considered to represent the most important breakthrough in the history of cryptography, following the invention and breaking of the Vigenère cipher, of the Beale papers, and of the Enigma machine.

The Diffie–Hellman–Merkle algorithm of "key generation without key exchange" can even be generalized to parties of three or more. Consider the following sequence with Alice, Bob and Cindy, now grouped into simultaneous steps:

Step 1: Alice chooses a large integer number A and keeps it secret; she computes

$a = m^A$ mod n and sends the result to Bob;

Bob chooses a large integer number B and keeps it secret; he computes

$b = m^B$ mod n and sends the result to Cindy;

Cindy chooses a large integer number C and keeps it secret; she computes

$c = m^C$ mod n and sends the result to Alice;

Step 2: Alice computes $c' = c^A$ mod n and sends the result to Bob;
Bob computes $a' = a^B$ mod n and sends the result to Cindy;
Cindy computes $b' = b^C$ mod n and sends the result to Alice;

Step 4: Alice computes

$$k = (b')^A \text{ mod } n;$$

Bob computes

$$k' = (c')^B \text{ mod } n;$$

Cindy computes

$$k'' = (a')^C \text{ mod } n.$$

As immediately checked, the three results k, k', k'', computed separately by Alice, Bob and Cindy, are all identical to m^{ABC}, which now represents their secret key. Note that the initial secret numbers A, B, C don't need to be as large as in the previous case. Numbers of the order of 1,000 generate as many as 1 billion keys. But there is no reason not to use greater numbers to give a trillion possibilities.

Here is another variant of the Diffie–Hellman–Merkle algorithm (referred to as *Hughes*), where Alice wants the secret key to be some specific number.

Step 1: Alice chooses a large integer number A and keeps it secret; she computes

$$k = m^A \bmod n;$$

Step 2: Bob chooses a large integer number B and keeps it secret; he computes

$$b = m^B \bmod n \text{ and sends the result to Alice;}$$

Step 3: Alice computes

$$a = b^A \bmod n \text{ and sends the result to Bob;}$$

Step 4: Bob computes

$$c = 1/B$$

and

$$k' = a^c \bmod n$$

The result of Bob's operation is $k' = a^{1/B} = (b^A)^{1/B} = (m^{AB})^{1/B} = m^A = k$ (modulo n). Thus Alice has successfully communicated her preferred key choice to Bob. The advantage of Hughes algorithm is that Alice can initiate on her own the communication of crypted messages, not only to Bob but to any group of persons. She can send them the key (step 3) at a later time.

The above approaches concern *secret-key cryptography*, which, except for the absence of key exchange, resort to the same classical approach when it comes to encryption and decryption. Furthermore, the two parties share the same (secret) key, therefore they can both encrypt and decrypt with that key. In the communication channel, their relationship is symmetric, hence the name of *symmetric-key* cryptography. An interesting alternative possibility would be to have different keys for encryption and decryption: Bob can send to Alice crypted messages using a shared key, but only Alice is able to decrypt them, using her own secret key. Bob and Alice's communication being thus asymmetric, they are using a new form of *asymmetric-key* cryptography. In the next subsection, we shall see that modular arithmetic opened the path to asymmetric cryptography, better known as *public-key cryptography* (PKC).

3.2.5 Public-Key Cryptography and RSA

The approach of *public-key cryptography* (PKC) was developed by another trio of American-geniuses: R. Rivest, A. Shamir and L. Adleman, who gave the name to the RSA standard. The RSA principle is based upon the property that it is extremely difficult to factorize large numbers. Given a number n, factorization consists in finding the unique set of prime numbers $(p_1, p_2, \ldots, p_k)$ whose product $p_1 \times p_2 \times \cdots p_k$ is equal to n. Note that these prime factors are not generally all different. For instance, the number 1,000 is factorized into $2 \times 2 \times 2 \times 5 \times 5 \times 5 = 2^3 \times 5^3$. We can then use just two sufficiently large and different prime numbers (p, q) to generate a bigger number $n = pq$, knowing that if someone knows n he would have a hard time figuring out the two primes (p, q). For instance, consider the number $n = 62,615,533$. To find its factorization, we need to divide it by prime numbers, trying them out one after another. In this specific case, the answer is $p = 7,919$ and $q = 7,907$, which are the highest two primes at the top of the first-thousand-prime list. This choice made the factorization easy. But what about $n = 15,773,077$? The answer ($p = 2,383$, $q = 6,619$) is less immediate, since it takes 200 division tests to find q starting from the top of this first-thousand-prime list. Assume next that we select (p, q) from a huge list of known primes, for instance up to 10^9, yielding numbers of the order of 10^{18}. Even at a rate of 10^9 division tests per second (a computing power that only a few states could be able to afford), this would leave about 10^9 seconds or *31.7 years* to check it out!

Consider now how RSA works. As previously stated, PKC is based upon the "asymmetric" principle of using two different cipher keys, one for encryption and one for decryption. Define the following keys:

For *encryption*: a number e, such that it is relatively prime with $\phi(n) = (p-1)(q-1)$;

For *decryption*: the number d which verifies $ed = 1$ mod ϕ, or $d = e^{-1}$ mod ϕ.

We shall call e the *public key* and d the *private key*. The public key is also called *RSA public exponent*), and the private *RSA private exponent*. As its name indicates, the public key should be available to anyone (any Bob) who wants to send a crypted message to Alice. On the other side, Alice keeps her "private" key in absolute secrecy. The number $n = pq$ (also called *RSA modulus*) is known to everyone as being the modulo reference for both encryption and decryption operations.

The operation of *message encryption* with the public key e consists first in decomposing the plaintext into many numerical blocks having a size smaller than n. In the binary system, a 64-bit block represents a maximum number of 1.8×10^{19}, therefore it is an eligible block size if $n \geq 2 \times 10^{19}$. For messages of arbitrary length, it is always possible to pad the blocks with zero nulls on the left, so that they fit in a convenient standard size. Encryption of each of these blocks with number value m is performed by calculating the (cipher) number

$$c = m^e \bmod n$$

For Alice, who is the only one to have the private key d, the operation of block-by-block decryption consists in computing

$$m' = c^d \bmod n$$

Let's now look at the modular-arithmetic value of the m' that Alice gets. It takes a few substitution steps to figure out that the result comes to:

$$\begin{aligned}
m' &= c^d \bmod n \\
&= (m^e)^d \bmod n \\
&= m^{ed} \bmod n \\
&= m^{k(p-1)(q-1)+1} \bmod n \\
&= m \times m^{k(p-1)(q-1)} \bmod n \\
&= m \times [m^{(p-1)(q-1)} \bmod n]^k \bmod n \\
&= m \times 1^k \bmod n \\
&= m
\end{aligned} \qquad (3.17)$$

which shows that (as expected) Alice has retrieved the original plaintext block m. Note that for this demonstration, we used the different properties listed in the modular arithmetic subsection, and in particular the Euler theorem. We also used the preset property of Alice's key, $ed = 1 \bmod \phi$, which means that $ed = k(p-1)(q-1) + 1$. Alice is thus able to decrypt messages that are sent to her from various Bob correspondents who use her public key.

For illustration purposes, let us consider a practical example of RSA encryption and decryption. Here is Bob's personal declaration to Alice:

Plaintext	I	l	o	v	e	y	o	u
PT-ASCII	1001001	1101100	1101111	1110110	1100101	1111001	1101111	1110101

Consistently with the RSA algorithms, we can encrypt the plaintext by blocks, which can represent for instance two letters each. With the above ASCII message, a two-letter block is a number of $2 \times 7 = 14$ bits, corresponding to a maximum size of $2^{14} - 1 = 16{,}383$. Let us assume that this is the standard. It just takes for Alice to pick up two primes (p, q) whose product is greater than or equal to 16,383, for instance $p = 73$ and $q = 227$, which gives $n = pq = 16{,}571$, and $\phi = (p-1)(q-1) = 72 \times 226 = 16{,}272$. The number $n = 16{,}571$ should also be known by everyone. Of course, the numbers p, q, ϕ remain only known to Alice.

Following the standard, Bob must convert his message into the 2-letter blocks and the blocks into decimal numbers, as follows:

Plaintext	I	l	o	v	e	y	o	u
PT-ASCII	1001001	1101100	1101111	1110110	1100101	1111001	1101111	1110101
Block	10 01 00 11 10 11 00		11 01 11 11 11 01 10		11 00 10 11 11 10 01		11 01 11 11 11 01 01	
Decimal	9,452		14,326		13,049		14,325	

Alice has also made a choice for her public key, e. This public key must be relatively prime with $\phi = 16{,}272$. In order to be able to perform the encryption/decryption computations with a pocket calculator (just for the sake of this example!), assume that Alice picked for her public key $e = 5$ which represents the smallest eligible

3.2 Modern Cryptography Algorithms

value (3 and 4 divide 16,272). Going to Alice's web site, Bob, like any other visitor, can read "Alice's encryption instructions: please kindly use two-ASCII block encryption modulo 16,571 as the standard; my private key is number 5."

Bob then immediately proceeds to encrypt his declaration according to the formula $c_i = m_i^e \bmod n$, where m_i is the decimal block number "i" in Bob's message sequence $m_1 m_2 m_3 m_4$. This gives

$$\begin{cases} c_1 = m_1^5 \bmod 16{,}571 = 9{,}452^5 \bmod 16{,}571 = 3{,}704 \\ c_2 = m_2^5 \bmod 16{,}571 = 14{,}326^5 \bmod 16{,}571 = 766 \\ c_3 = m_3^5 \bmod 16{,}571 = 13{,}049^5 \bmod 16{,}571 = 475 \\ c_4 = m_4^5 \bmod 16{,}571 = 14{,}325^5 \bmod 16{,}571 = 372 \end{cases}$$

In the above computations, it is absolutely essential to make no truncation errors. For instance, with a pocket calculator the function $9{,}452^5$ gives $7.544293311 \times 10^{19}$, but we are missing the last ten digits! To avoid truncations from the pocket calculator, the trick is to use the modulus-arithmetic formula $|m^5|_n = ||m^2|_n \times |m^2|_n \times |m|_n|_n$, in which the different terms and their successive products (after reducing each one to their modulo n residue) fit in the calculator display size. To recall, in order to calculate the residue $|s|_n$, one divides s by n and take the integer part of the result, which we have called $[s/n]$. The residue is then $|s|_n = s - n \times [s/n]$. Having finished these computations, Bob e-mails his cipher message

$$c_1 c_2 c_3 c_4 = 3{,}704 - 766 - 475 - 372$$

to Alice (possibly with a few more words to identify himself, like "Bob your classmate").

Alice is the only person able to decrypt this message. Her private key, d, is defined by the formula $ed = 1 \bmod \phi$, or $d = e^{-1} \bmod \phi$, namely $d = 5^{-1} \bmod 16{,}272$. Thus Alice needs to find the inverse of 5 modulo 16,272. Since 16,272 is not the product of two primes ($16{,}272 = 16 \times 9 \times 113$), she can't use Euler's theorem. Instead, she uses the *extended Euclidian algorithm*, which we have previously described in Section 3.2.3 concerning modular arithmetic. It is not a coincidence that we used these numbers to illustrate the algorithm principle, which gives $d = 6{,}509$ (such a big, 13-bit private key, is the price to pay for having chosen the small public key $e = 5$). Alice can thus use her private key to compute the blocks

$$\begin{cases} m_1 = c_1^{6{,}509} \bmod 16{,}571 = 3{,}704^{6{,}509} \bmod 16{,}571 = 9{,}452 \\ m_2 = c_2^{6{,}509} \bmod 16{,}571 = 766^{6{,}509} \bmod 16{,}571 = 14{,}326 \\ m_3 = c_3^{6{,}509} \bmod 16{,}571 = 475^{6{,}509} \bmod 16{,}571 = 13{,}049 \\ m_4 = c_4^{6{,}509} \bmod 16{,}571 = 372^{6{,}509} \bmod 16{,}571 = 14{,}325 \end{cases}$$

Let us take a short break to consider an important technical issue (this paragraph may be skipped on first reading). Indeed, one may wonder whether it is physically possible to *exactly* compute numbers as monstrous as $100^{6{,}509}$ or $1{,}000^{6{,}509}$. If we make the conversion through the formula $a^x = 10^{x \log a / \log 10}$, we get for the first block

$3{,}704^{6{,}509} = 10^{23{,}228.4794} \approx 3.01 \times 10^{23{,}228}$, namely a number which has a whopping 23,229 digits. It looks like Alice would need to have a supercomputer in her basement to be able to retrieve her cryptomail. But it is not the case. The author performed this decryption computation only using a pocket calculator! The point is not that decrypting *requires* one to use a pocket calculator, but that this kind of operation can be performed *without use of computers* and with such rudimentary means. Let us explain how to proceed. Consider for instance the last cipher block, $372^{6{,}509}$. It takes a few minutes to calculate the following power series (modulo 16,571): $372^2 = 5{,}816$, $372^4 = 4{,}44$, $372^8 = 5{,}393$, etc., up to $372^{4{,}096} = 8{,}695$. Thus $372^{6{,}509}$ is the product $372^{4{,}096} \times 372^{1{,}024} \times 372^{1{,}024} \times 372^{256} \times 372^{64} \times 372^{32} \times 372^8 \times 372^4 \times 372^1$. The result of each successive product must be reduced to its residue so that no truncation occurs. The same procedure must be followed with the other blocks $3{,}704^{6{,}509}$, $766^{6{,}509}$, and $475^{6{,}509}$. This point being made, there is no reason not to develop a simple computer program which can perform all these successive reduction tasks in a wink. With the present example, we have chosen a block format ($2^{14} - 1 = 16{,}383$) whose decimal size allows computing any power of two, i.e., $(2^{14} - 1)^2 = 268{,}402{,}689$, and its successive multiples, with a pocket calculator. But with a personal computer, larger coding formats to encrypt four to eight ASCII/EBCDIC characters (i.e., $2^{32} - 1 = 4{,}294{,}967{,}295$ or $2^{64} - 1 \approx 2 \times 10^{19}$) are possible, provided the program is designed to handle numbers up to $(2^{32} - 1)^2$ or $(2^{64} - 1)^2$ without truncation errors.

Alice then converts the decimal numbers $m_1 m_2 m_3 m_4$ in to four 14-bit binary words, then splits each one into two 7-bit groups and finally converts the resulting codes into ASCII, which gives:

Decimal	9,452		14,326		13,049		14,325	
Binary	10 01 00 11	10 11 00	11 01 11 11	11 01 10	11 00 10 11	11 10 01	11 01 11 11	11 01 01
ASCII	1001001	1101100	1101111	1110110	1100101	1111001	1101111	1110101
Plaintext	I	l	o	v	e	y	o	u

Bob and Alice both having computers to perform RSA and a connection to the Internet, the entire operation of encryption, transmission and decryption of Bob's message (possibly completed with a few more statements!) took in fact less than one millisecond. The details of the official RSA standard, called *PKCS#1*, can be downloaded from the web site of RSA Security (see URL in References).

In the preceding example, Eve would have no trouble figuring out the Alice's secret key. Indeed, the small coding size $n = 16{,}571$ is easy for a computer to factorize into the primes (p, q) from which Alice's private key, $e = d^{-1}$ mod $(p-1)(q-1)$, can be instantly retrieved. With larger coding an public/private key sizes (e.g., $n \approx 2^{64} \approx 10^{19}$, $n \approx 2^{128} \approx 10^{38}$), the problem of factorization is no longer within easy reach, and rather requires extended effort and computing facilities. To promote the advancement and security of cryptography, the company *RSA Security* (see related link in Bibliography) offers challenges to the public which take the form of huge modulus numbers to factorize (namely, given n, find the two primes p and q such that $n = pq$). The current challenges concern different numbers with lengths from 576 bits (prize $10,000) to 2048 bits (prize $200,000). Here is the RSA-2048 number, which has 617 decimal digits (exact reproduction in printed form of this book not guaranteed):

```
25195908475657893494027183240048398571429282 12620
40320277771378360436620207075955562640 1 8525880784
4069182906412495 1 5082 1 892985591491761845028084891
200728449926873928072877767359714 18347270261 89637
50 1497 1 82469 1 1 65077613379859095700097330459748808
4284017974291 0064245869 1 81 71 951 187461 2151517 26546
322822 168699875491 824224336372590851418654 6204357
67984233871847744479207399342365482382428 1 198 163
8 1 501067481 0451660377306056201619676256 1 338441436
0383390441495263443219011465754445417842402 092461
65 1 572335077870774981 71 2577246796292638635637 3289
912 154831438 16789988504044536402352738 1951 3786365
643912 120 10397 1 228221 20720357
```

The two previous challenges, named RSA-140 (140 decimal digits, 465 bits) and RSA-155 (155 decimal digits, 512 bits), were successfully met by international teams of researchers from both academia and industry in February and August 1999, respectively. Note that RSA-155 was renamed RSA-512 so that the number defines bit size rather than decimal-digit size.

> Factoring RSA-140 required 125 workstations (175 MHz) and 60 personal computers (300 MHz) to run for about one month, a combined 8.9 years of CPU time. Including the preparation work, the total elapsed time was actually 2.2 months.

> Factoring RSA-512 (ex RSA-155) required 160 workstations (175–400 MHz) and 120 personal computers (300–450 MHz) to run for 3.7 months, representing a combined 35.7 years of CPU time. The total elapsed time was about 7.4 months.

Based upon these results, it is expected that factorization of RSA-576 may not be completed for a few years (starting 1999) while RSA-2048 may remain unchallenged not only over this current decade but possibly over this century.

The general method of "attack" for the factorization problem (and also finding logarithms A from $a = m^A \mod n$) is known as GNFS (*general number field sieve*), or NFS. The word "sieve" is due to *Erathosthenes* (240 BC). This is the same person that estimated the circumference of the Earth (tens of centuries before the Earth was proven to be a sphere), based upon the observation that at the same time of the year the sun casts shadows with different angles for different latitudes. His nonintuitive "sieving algorithm" for sorting out prime numbers can be described as follows: "List all integers lesser than n and remove multiples of all primes less than or equal to the square root of n; the numbers that are left are the primes below "n." The NFS algorithm is implemented into two stages (RSA Bulletin #13, 2000). The first stage, called "sieving," requires moderate computation power (hence the use of multiple personal computers and workstations in parallel), and leads to a set of equations, the "matrix." The second stage requires a supercomputer with massive internal CPU-memory use (e.g., 0.8 Mbytes for RSA-140 and 3.2 Gbyte for RSA-155/ RSA-512). Finding the solution of the matrix might take as much time as the sieving stage, or just a fraction thereof. Once found, this solution leads to instant factorization or logarithm definition. The two parameters of interest for attacking a number n (as defined by its bit length) are the requirements in time (L) and memory space (S). It can be shown that these requirements are defined by

$$L(n) = \exp[C \times (\log n)^{\frac{1}{3}}(\log \log n)^{\frac{2}{3}}] \qquad (3.18)$$

$$S(n) = \sqrt{L(n)} \qquad (3.19)$$

where C is a constant. Knowing these requirements for a smaller number n', we get the relative estimates

$$l_{n/n'} = \frac{L(n)}{L(n')} = \exp\left\{const \times \left[\begin{array}{c}(\log n)^{\frac{1}{3}}(\log \log n)^{\frac{2}{3}}\\-(\log n')^{\frac{1}{3}}(\log \log n')^{\frac{2}{3}}\end{array}\right]\right\} \qquad (3.20)$$

$$s_{s/n'} = \frac{S(n)}{S(n')} = \sqrt{\frac{L(n)}{L(n')}} \qquad (3.21)$$

Table 3.18 provides the numbers ($l_{n/n'}$, $s_{n/n'}$) corresponding to the different values of (n, n'). The first row corresponds to the data provided in the aforementioned reference, i.e., where estimates are relative to $n' = 512$. Analyzing these data, the constant C involved in the definition in Equation 3.18 was observed to follow a heuristic rule $C = 64.11 \times (n/n')^{0.2542}$. We used this rule to compute the other estimates with $n' = 576, 640, 704, 768, 1,024$ and $2,048$. The new data set thus obtained heuristically provides some estimates for the time and memory size increase factors required to reach larger-number sizes from the best previous achievement. We thus see from the table that the time increase from RSA-512 to RSA-576 represents about an 11-fold factor, corresponding to a raw computing time of 11×7.4 months $= 81.4$ months, or 6 years and 9 months! But if about 1,000 workstations and personal computers were used along with (say) 5 supercomputers, this delay could be reduced to a single year. The corresponding increase factor for the CPU memory space is 3.3, corresponding to 3.3×3.2 Gbytes $= 10.5$ Gbytes. Assume now that RSA-576 was solved in 2003. We see from our estimates that to reach the ballpark of RSA-768, the time and memory-space requirements are increased by about 400-fold and 20-fold, respectively. With the conventional approach, this would require some 400,000 workstations and 100 supercomputers. As far as RSA-1,024 and RSA-2,048 are concerned, it is hard to make any sense of the projections in both time and memory space.

Another way to analyze time projections for the factorization/logarithm problem is to consider the progress actually realized over the last 30 years. Remarkably, the progress in key-size and time is very closely linear, like another *Moore's law* (see Chapter 2, Section 2.6). Note, however, that Moore's law is linear but in a logarithmic performance scale, unlike in the present case. From the aforementioned

TABLE 3.18 Relative time and memory space increases for factoring number *n* or solving a logarithm problem of same field size (top row) in reference to the corresponding data for smaller number *n'* (left columns)

Time Increase

	576	640	704	768	1,024	2,048
512	10.9	101	835	6,000	7×10^6	9×10^{15}
576		1.1	8.2	390	3×10^5	2×10^{14}
640			6.5	39	3×10^{10}	3×10^{12}
704				5.5	2,400	1×10^{11}
768					335	9×10^9
1,024						3×10^6

Memory Space Increase

	576	640	704	768	1,024	2,048
512	3.3	10	29	77	2650	9×10^7
576		2.8	7.7	20	550	1×10^7
640			2.5	6	2×10^5	2×10^6
704				2.5	50	4×10^5
768					20	9×10^4
1,024						1,750

The numbers are expressed in bit size.

reference, the law can be expressed under the phenomenological formula

$$\text{size}_{\text{dec}} = 4.23 \times (Y - 1970) + 23 \qquad (3.22)$$

where Y is the current year and the modulus size is expressed in number of decimal digits. This linear law has been verified over the past 30 years. The projection years for solving 1,024-bit keys (size = 309 decimal digits) and 2,048-bit keys (size = 617 decimal digits) are 2037 and 2110 respectively. Other investigators have suggested a model according to which the size should be now dictated by a cubic law

$$\text{size}_{\text{dec}} = \left(\frac{Y - 1928}{13.25}\right)^3 \qquad (3.23)$$

which gives years 2017–2018 for reaching 1,024-bit keys, and 2041 for reaching 2,048-bit keys. We see that both laws predict that 1,024-bit keys won't be solved for 20 years at the very least. Keys as large as 2,048 bits might take forty years (2040) to over a century (2110). It is important to note that such estimates are only based upon publicly available data. Throughout history, it has always been assumed that government agencies have always been ahead of any progress in this field, and the market implications of cryptography do not even exclude the possibility that private, yet unpublicized efforts, have already reached substantially higher performance. As a final remark, there is no mathematical proof that breaking RSA absolutely requires factorization of the modulus n, it is just what is currently conjectured, until new breakthroughs may happen. But even this remark represents a conjecture by itself.

These facts, data and estimates help one to grasp (at the very least) what the problem of factorization of large numbers (or finding logarithms) mean in terms of time, resource and effort. The fact that teams were able to complete the factorization of RSA-140 and RSA-155/512 does not mean that the RSA code is now broken or "cracked." Instead, these successful experiments provided valuable information on the state of the art in factorization, from algorithms and methods to hardware and time requirements. It is still safe for Alice and Bob to use 512-bit RSA keys (even more so with 1,024-bit keys), knowing the costs in personnel, investment and time that Eve would have to support. Another important consideration is the *lifetime* of the data to be protected. This lifetime can range from the scale of a single day (e.g., certified signature, restricted-access broadcast, see Section 3.3) to several years (e.g., business and financial contracts).

With very large scale integration (VLSI), the IC technology has produced chips able to perform RSA encryption and decryption with various modulus sizes (namely 32, 120, 256, 272, 298, 512, 593, 1,024). Compared to DES (see next section) RSA takes 1,000 times longer from the hardware perspective. From the software perspective, RSA takes only 100 times longer. Encryption is always faster than decryption. Using a workstation, (see Schneir, 1996) the encryption/decryption tasks require 30 ms/160 ms for 512 bits and 80 ms/930 ms for 1,024 bits. The process can be speeded up by an appropriate choice of *public key*. Appropriate values are those which have binary words of value $2^N + 1$, which in binary have the form

(100...01) with $N+1$ bits and only two "1" bits. This choice reduces the operation of exponentiation to only $N+1$ successive multiplications. The standard recommendations are $e = 3$ (11), $e = 17$ (10001), and $e = 65{,}537$ (100001). Such a selection avoids $e = 5$ (101) and $e = 9$ (1001) because of the fact that the public key must be relatively prime with ϕ. And as it conveniently turns out, the numbers 3, 17 and 64,537 are all primes, therefore more likely to be relatively prime with ϕ. In order to complicate Eve's task, the recommendation is also that the private key be sufficiently large (even larger than in our illustrative example) and that the choice of modulus be broad. It is not recommended to use the same private key for signature/authentication. It can be shown (see Schneir, 1996) that nefarious Eve can retrieve Alice's private key by sending her a specially prepared document and asking her to candidly sign/certify it with her private key! Several variants and extensions of PKC and asymmetric-key cryptography can be found in this reference.

■ 3.3 COMMUNICATIONS SECURITY AND APPLICATIONS

After the description of RSA/PKC in Section 3.2, we continue here our review of the main *encryption standards*, with emphasis on the *data encryption standard* (DES) and its recent advanced version, AES. We will also consider a representative rather than comprehensive selection of some important cryptosystems, including *RC-4* and *Pretty Good Privacy* (PGP). The different protocols of *digital signature* and message *authentication* are described in another subsection, with the principle of *hash-function* algorithms and standards (MD5, SHA). We will then briefly consider the issues of *network security* (key management and confidentiality, and *Internet security* with the secure-socket layer (SSL/HTTPS). Finally, we will describe a certain number of applications of cryptography, from basic ones such as *electronic tax filing*, *certified e-mail* and *simultaneous secret exchange* to futuristic ones like *electronic voting/polling* and *digital cash*.

3.3.1 Data Encryption Standard (DES)

The previous section has initiated the reader into some aspects of *cryptography standardization*, which concern the encryption/decryption algorithms as well as the formats to be used for keys and ciphers. As we have seen, the story began in the 1960s with the IBM initiative, called *Lucifer*, which set the grounds for the adoption, in the 1970s, of the official *data encryption standard* (DES). The initial DES version had a 56-bit key, a reduction from the proposed 128-bit format which was meant to facilitate cracking the code should government or military authorities need access for national-security considerations. Rumors have circulated about the existence in DES of embedded *trapdoor functions* which could enable direct decryption by these parties, but such rumors did not prevent its eventual standardization and widespread adoption. At the time it was considered that cracking DES by *brute-force* attack (trying all possible keys at random over the entire $2^{56} - 1$ key-space) would require about 1,000 years of CPU time. The rapid progress in computer speed proved this belief completely wrong. In 1997, the company *RSA Security* issued a public challenge to crack DES. The challenge was successfully met that same year with only 96 days of computation, using a network of no less than 14,000 computers. In the two following years, other teams succeeded in cracking DES in 41 days, then in 56 hours, then in even 22h15mn! The machine used for the 56-hour record, nicknamed "Deep Crack," used 27 parallel motherboards each one with 64 processor chips (1,728 total) capable altogether of performing 90 billion key tests per second.

The simple way to improve DES strength and security was to achieve double and triple encryption with two and three independent keys, respectively (see the following details). These *double-DES* and *triple-DES* (3DES) approaches, which bring the effective key size to $2 \times 56 = 112$ bits and $3 \times 56 = 168$ bits, were endorsed in 1999 by NIST (*National Institute of Standards and Technologies*), the new name of the original standardization body which launched DES. Note that DES was also approved by ANSI (*American Standards Institute*) under the name of *data encryption algorithm* (DEA) and by ISO (International Standards Organization) under the name of DEA-1. In spite of the improvements introduced by double and triple encryption, DES further evolved into the *advanced encryption standard* (AES), with the prospect of lasting use in future decades. In this subsection, we shall first describe the DES and AES algorithms in further detail (official documents being available in NIST Web pages, see Web-site reference list in the Bibliography).

In Section 3.2.1, we described the principle of *Lucifer*, which represented the first *block cipher* algorithm. The term "block cipher" refers to an encryption scheme which processed bits by blocks of any arbitrary length, as opposed to groups of bits corresponding to a single character. Block ciphers, which are made possible though binary coding, replace the earlier cryptography approaches based upon mono- or multialphabetic substitution which lasted for centuries. In DES, the unit block length is 64 bits, which can code numbers as high as $2^{64} - 1 = 1.8 \times 10^{19}$. The algorithm consists in a 16-round scrambling process within these individual blocks, after they have been split into 32-bit subblocks. The DES key requires a little bit of initial processing, as we shall explain next.

The DES key, which we know to be 56-bits long, is written in the 64-bit format while overlooking every 8th bit. For instance, the key defined as $K = $ 0123456789ABCDEF in the *hexadecimal* system writes in binary form

$K = $ 0000 000**1** 0010 001**1** 0100 010**1** 0110 011**1** 1000 100**1** 1010 101**1** 1100
      110**1** 1110 111**1**

In this equation, every 8th bit which has been emphasized in bold is overlooked by DES, while the other 56 bits represent the actual key.

The first step is to transform this initial key through a standard permutation called *key permutation*. Table 3.19 lists the 56 bit permutations to be effected on the 56-bit key. The table data must be interpreted according to the following: bit 57 takes the place of bit 1, then 49 the place of 2, then 41 the place of 3, and so on.

3.3 Communications Security and Applications 409

TABLE 3.19 Rules for DES key permutation (to read row by row, from left to right)

57	49	41	33	25	17	9
1	58	50	42	34	26	18
10	2	59	51	43	35	27
19	11	3	60	52	44	36
63	55	47	39	31	23	15
7	62	54	46	38	30	22
14	6	61	53	45	37	29
21	13	5	28	20	12	4

After completing the permutation cycle, and overlooking the 8th bits we obtain the following 56-bit key (K') from the previous example:

$K' = 1_{57} 1_{49} 1_{41} 1_{33} 0_{25} 0_{17} 0_9 0_1 1_{58} 1_{50} 0_{42} 0_{34} 1_{26} 1_{18} 0_{10} 0_2 1_{59} 0_{51} 1_{43} 0_{35} 1_{27} 0_{19} 1_{11}$
$0_3 0_{60} 0_{52} 0_{44} 0_{36} 1_{63} 0_{55} 1_{47} 0_{39} 1_{31} 0_{23} 1_{15} 0_7 1_{62} 1_{54} 0_{46} 0_{38} 1_{30} 1_{22} 0_{14} 0_6 1_{61} 1_{53}$
$1_{45} 1_{37} 0_{29} 0_{21} 0_{13} 0_5 0_{28} 0_{20} 0_{12} 0_4$
$= 1111000\ 0110011\ 0010101\ 0100000\ 1010101\ 0110011\ 0011110\ 0000000$

We then split the key into two 28-bit key halves called C_0 and D_0, respectively:

$$C_0 = 1111000\ 0110011\ 0010101\ 0100000$$
$$D_0 = 1010101\ 0110011\ 0011110\ 0000000$$

The second step of key processing consists in generating 16 different *subkeys*. These subkeys are first obtained by successive circular left-permutations of (C_0, D_0). If N is the number of permutations at a given stage $P = 1-16$, the rule is $N = 1$ for $P = (1, 2, 9, 16)$ and $N = 2$ for all other stages. Table 3.20 lists the 16 different subkeys (C_0, D_0) obtained after these 16 stages of circular permutation. But this is not over. We must now form a set of 16 other keys by selecting 48 bits from each of the subkeys $C_n D_n$. Since the resulting keys have a smaller size, this stage is called key *compression permutation*. The rules to make these compressed keys are defined in Table 3.21. We see that from any subkey $C_n D_n$ the first bit of the compressed key is 14, the second 17, the third 11, and so on. Table 3.22 lists the 16 compressed keys generated from the 16 subkeys $C_n D_n$ shown in the previous table.

As a mathematical oddity, we should note that there are a small number of keys which are less "safe" than the others. These so-called *weak keys*, to be avoided, do not generate 16 different subkeys. The four weak keys 0000000 00000000, 00000000 FFFFFFFF, FFFFFFFF 00000000, and FFFFFFFF FFFFFFFF (hexadecimal representation over the generic 64-bit format) have 16 identical subkeys. There are six pairs of keys, called *semiweak* keys, that generate only 8 different subkeys. In addition, one key of the pair can decrypt ciphers encoded with the other. Finally, there are 48 *possibly weak* keys that generate only four different

TABLE 3.20 Generation of sixteen 28-bit subkeys from the initial subkey example (C_0, D_0)

C_0 = 1111000011001100101010100000 D_0 = 1010101011001100111100000000	
C_1 = 1110000110011001010101000001 D_1 = 0101010110011001111000000001	C_9 = 0101010100001111000011000110 D_9 = 0111100000000101010101100110
C_2 = 1100001100110010101010000011 D_2 = 1010101100110011110000000010	C_{10} = 0101010000011110000110011001 D_{10} = 1110000000010101010110011001
C_3 = 0000110011001010101000001111 D_3 = 1010110011001111000000001010	C_{11} = 0101000001111000011001100101 D_{11} = 1000000001010101011001100111
C_4 = 0011001100101010100000111100 D_4 = 1011001100111100000000101010	C_{12} = 0100000111100001100110010101 D_{12} = 0000000101010101100110011110
C_5 = 1100110010101010000011110000 D_5 = 1100110011110000000010101010	C_{13} = 0000011110000110011001010101 D_{13} = 0000010101010110011001111000
C_6 = 0011001010101000001111000011 D_6 = 0011001111000000001010101011	C_{14} = 0001111000011001100101010100 D_{14} = 0001010101011001100111100000
C_7 = 1100101010100000111100001100 D_7 = 1100111000000001010101010100	C_{15} = 0111100001100110010101010000 D_{15} = 0101010101100110011110000000
C_8 = 0010101010000011110000110011 D_8 = 0011100000000101010101100011	C_{16} = 1111000011001100101010100000 D_{16} = 1010101011001100111100000000

subkeys. For complete lists of semiweak and possibly weak keys, see for instance Schneir (1996).

We consider next the procedure of *data encryption*. The preparatory phase, called initial permutation (*IP*), consists in rearranging the 64-bits of a given message block (*M*) according to the permutation rule shown in Table 3.23. The

TABLE 3.21 Rules for DES subkey compression (to read row by row, from left to right)

14	17	11	24	1	5
3	28	15	6	21	10
23	19	12	4	26	8
16	7	27	20	13	2
41	52	31	37	47	55
30	40	51	45	33	48
44	49	39	56	34	53
46	42	50	36	29	32

3.3 Communications Security and Applications 411

TABLE 3.22 Definition of 16 compressed keys from rules of Table 3.21 and data from Table 3.20

K_1 = 0000 1011 0000 0010 0110 0111 1001 1011 0100 1001 1010 0101
K_2 = 0110 1001 1010 0110 0101 1001 0010 0101 0110 1010 0010 0110
K_3 = 0100 0101 1101 0100 1000 1010 1011 0100 0010 1000 1101 0010
K_4 = 0111 0010 1000 1001 1101 0010 1010 0101 1000 0010 0101 0111
K_5 = 0011 1100 1110 1000 0000 0011 0001 0111 1010 0110 1100 0010
K_6 = 0010 0011 0010 0101 0001 1110 0011 1100 1000 0101 0100 0101
K_7 = 0110 1100 0000 0100 1001 0101 0000 1010 1110 0100 1100 0110
K_8 = 0101 0111 1000 1000 0011 1000 0110 1100 1110 0101 1000 0001
K_9 = 1100 0000 1100 1001 1110 1001 0010 0110 1011 1000 0011 1001
K_{10} = 1001 0001 1110 0011 0000 0111 0110 0011 0001 1101 0111 0010
K_{11} = 0010 0001 0001 1111 1000 0011 0000 1101 1000 1001 0011 1010
K_{12} = 0111 0001 0011 0000 1110 0101 0100 0101 0101 1100 0101 0100
K_{13} = 1001 0001 1100 0100 1101 0000 0100 1001 1000 0000 1111 1100
K_{14} = 0101 0100 0100 0011 1011 0110 1000 0001 1101 1100 1000 1101
K_{15} = 1011 0110 1001 0001 0000 0101 0000 1010 0001 0110 1011 0101
K_{16} = 1100 1010 0011 1101 0000 0011 1011 1000 0111 0000 0011 0010

first and second bits of *IP* are bits 58 and 50 of *M*, respectively, and so on until bit 7. Although we now know well how this works, let us take a simple example in order to illustrate the continuation of the procedure. For instance

$$M = 10000000\ 00000000\ 00000000\ 00000000\ 00000000\ 00000000$$
$$00000000\ 00000001$$

TABLE 3.23 Rules for initial permutation (*IP*) of 64-bit plaintext block (to read row by row, from left to right)

58	50	42	34	26	18	10	2
60	52	44	35	28	20	12	4
62	54	46	38	30	22	14	6
64	56	48	40	32	24	16	8
57	49	41	33	25	17	9	1
59	51	43	35	27	19	11	3
61	53	45	37	29	21	13	5
63	55	47	39	31	23	15	7

gives

$$IP = 00000000\ 00000000\ 00000000\ 10000000\ 00000001\ 00000000$$
$$00000000\ 00000000$$

We then split *IP* into left and right 32-bit blocks, which we call L_0 and R_0, respectively. We note from Table 3.23 that L_0 contains all the even bits and R_0 all the odd bits of *M*. From the previous example:

$$L_0 = 00000000\ 00000000\ 00000000\ 10000000$$
$$R_0 = 00000001\ 00000000\ 00000000\ 00000000$$

The rest of the DES encoding consists in calculating the 16 following operations ($n = 1\ldots 16$)

$$L_n = R_{n-1}$$
$$R_n = L_{n-1} \oplus f(R_{n-1}, K_n)$$

where *f* is a function we shall specify next. Figure 3.11 shows the flowchart of these successive permutations and operations. As the figure indicates, the block obtained after round 16 is $L_{16}R_{16}$. At this stage, the rule is to switch the two halves to give $R_{16}L_{16}$, which is called the *preoutput*. The final step, to obtain the cipher block, con-

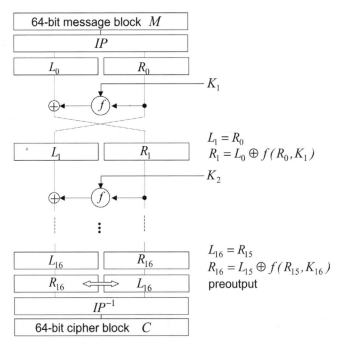

FIGURE 3.11 Coding principle of the data encryption standard (DES) showing the 16 iterations with permutations and transformations of left (L_n) and right (R_n) blocks through the function *f* and the keys $K_1 \ldots K_{16}$ (*IP* = initial permutation, IP^{-1} = inverse *IP*).

sists in realizing the inverse of the initial permutation (IP^{-1}) over $R_{16}L_{16}$. The rules for IP^{-1} are strictly the same as indicated in Table 3.23 for IP, except that the numbers must be interpreted the reverse way. If bit 1 of IP is bit 58 of message M, then bit 58 of IP^{-1} is bit 1 of preoutput $R_{16}L_{16}$. From the IP table, we see that bit 1 of IP^{-1} is bit 40 of $R_{16}L_{16}$, bit 2 of IP^{-1} is bit 8 of $R_{16}L_{16}$, and so on. The resulting IP^{-1} block is the cipher block C.

We focus now on the function $f(R_{n-1}, K_n)$ involved in the above 16 encryption steps. Recall that the 16 subkeys $K_1 \ldots K_{16}$ are 48 bits long, while the blocks R_n are 32 bits long. The procedure then consists in expanding the R_{n-1} into 48-bit blocks, which is referred to as *expansion permutation*. The expansion permutation is defined by an operation $E(X)$ which generates a 48-bit block from the 32 bits of X while repeating 16 selected bits. The resulting operation is defined by the rules of Table 3.24. Expanding our previous R_0 example gives:

$R_0 = 00000001\ 00000000\ 00000000\ 00000000$

$E(R_0) = 000000\ 000010\ 100000\ 000000\ 000000\ 000000\ 000000\ 000000$

The next step of the function $f(R_{n-1}, K_n)$ consists in adding (or "XOR-ing") the key K_n with $E(R_{n-1})$. For $n = 1$ this gives (K_1 defined in Table 3.22):

$K_1 = 000010\ 110000\ 001001\ 100111\ 100110\ 110100\ 100110\ 100101$

$E(R_0) = 000000\ 000010\ 100000\ 000000\ 000000\ 000000\ 000000\ 000000$

$K_1 \oplus E(R_0) = 000010\ 110010\ 101001\ 100111\ 100110\ 110100\ 100110\ 100101$

We have thus obtained $K_1 \oplus E(R_0) = B_1B_2B_3B_4B_5B_6B_7B_8$ which is a block of 48 bits in six groups of 8 bits. But there are two more operations to complete the function $f(R_{n-1}, K_n)$. The first operation consists in transforming the previous result into a 32-bit block. The second operation is a final permutation. We describe next these two operations.

The 48-bit to 32-bit transformation involves the so-called *S-boxes*. This is a somewhat unusual operation, which consists in using the bits in the words B_k as

TABLE 3.24 Rules for bit selection in the expansion-permutation operation $E(X)$ on a 32-bit block X

32	1	2	3	4	5
4	5	6	7	8	9
8	9	10	11	12	13
12	13	14	15	16	17
16	17	18	19	20	21
20	21	22	23	24	25
24	25	26	27	28	29
28	29	30	31	32	1

Repeated bits are shown in bold (to read row by row, from left to right)

addresses to designate a specific element in a matrix, the S-box. Let us write B_k in the form $b_6b_5b_4b_3b_2b_1$, and define $i = b_6b_1$ (the number formed by the two extreme bits) and $j = b_5b_4b_3b_2$ (the number formed by the four intermediate bits). The 2-bit number i has a decimal value between 0 and 3, while the 4-bit number j has a value between 0 and 15. We then use i and j as the row (0–3) and column (0–15) number of the S-box, respectively. Table 3.25 shows the S-box to use for step "1," which is called S_1. We see that to each S-box matrix element (i, j) corresponds to a specific number between 0 and 15, which in binary is a 4-bit word. For instance we have $B_2 = 110010$, which gives $i = 10 \equiv 2$ and $j = 1001 \equiv 9$. The corresponding S-box element in row 2 and column 9 is 12, or 1100 in binary. Thus $S_1(B_2) = 1100$. From the table, we can thus establish the following correspondence between words B_k (k = 1 ... 8) and the S-box substitutes $S_1(B_k)$:

$B_1: i = 0, j = 1$ $S_1(B_1) = 0100$ | $B_5: i = 2, j = 2$ $S_1(B_5) = 1110$
$B_2: i = 2, j = 9$ $S_1(B_2) = 1100$ | $B_6: i = 2, j = 10$ $S_1(B_6) = 1001$
$B_3: i = 3, j = 4$ $S_1(B_3) = 0100$ | $B_7: i = 2, j = 3$ $S_1(B_7) = 1000$
$B_4: i = 3, j = 3$ $S_1(B_4) = 0010$ | $B_8: i = 3, j = 2$ $S_1(B_8) = 1000$

The block resulting from the above operation is defined by $S = S_1(B_1)$ $S_1(B_2)S_1(B_3)S_1(B_4)S_1(B_5)S_1(B_6)S_1(B_7)S_1(B_8)$. In this example, we have

$$S = 0100\ 1100\ 0100\ 0010\ 1110\ 1001\ 1000\ 1000$$

Note that the S-boxes are different for each round (1–16). A complete list of the 16 S-boxes $S_1 \ldots S_{16}$ can be found in the official NIST description of DES, or in Schneir (1996), for instance.

As previously mentioned, there is one more and final operation involved in the function $f(R_{n-1}, K_n)$, which is a bit permutation of S. The rules of this permutation (P) are provided in Table 3.26. We see that the first three bits of the output block are bit 16, 7 and 20 of S, and so on. We finally obtain from our example:

$$f(R_0, K_1) = P(S) = 0_{16}0_{7}0_{20}1_{21}1_{29}0_{12}0_{28}1_{17}0_{1}1_{15}0_{23}0_{26}1_{5}1_{18}0_{31}1_{10}1_{2}0_{8}$$
$$1_{24}0_{14}0_{32}0_{27}0_{3}0_{9}1_{19}0_{13}0_{30}1_{6}0_{22}0_{11}0_{4}1_{25}$$
$$= 00011001\ 01001101\ 10100000\ 10010001$$

Since we now have completed the calculation of $f(R_0, K_1)$, we can complete the first round, using the aforementioned definitions $L_n = R_{n-1}$ and $R_n = L_{n-1} \oplus f(R_{n-1}, K_n)$,

TABLE 3.25 Matrix elements of S-box S_1, as defined by row number i (0–3) and column number j (0–15)

	0	1	2	3	4	5	6	7	8	9	10	11	12	13	14	15
0	14	4	13	1	2	15	11	8	3	10	6	12	5	9	0	7
1	0	15	7	4	14	2	13	1	10	6	12	11	9	5	3	8
2	4	1	14	8	13	6	2	11	15	12	9	7	3	10	5	0
3	15	12	8	2	4	9	1	7	5	11	3	14	10	0	6	13

TABLE 3.26 Rules of permutation (*P*) over the 32-bit block output of function *f(X)*, to read row by row, from left to right

16	7	20	21	29	12	28	17
1	15	23	26	5	18	31	10
2	8	24	14	32	27	3	9
19	13	30	6	22	11	4	25

thus $L_1 = R_0$ and $R_1 = L_0 \oplus f(R_0, K_1)$,

$$R_0 = 00000001\ 00000000\ 00000000\ 00000000$$
$$L_1 = 00000001\ 00000000\ 00000000\ 00000000$$
$$L_0 = 00000000\ 00000000\ 00000000\ 10000000$$
$$f(R_0, K_1) = 00011001\ 01001101\ 10100000\ 10010001$$
$$R_1 = 00011001\ 01001101\ 10100000\ 00010001$$

As illustrated in Figure 3.11, the remaining 15 rounds are conducted according to the same procedure as for round 1, each time using the different inputs $(L_2, R_2) \ldots (L_{16}, R_{16})$, the subkeys $K_2 \ldots K_{16}$ and the S-boxes $S_2 \ldots S_{16}$ and the same permutation *P*. The final block $L_{16}R_{16}$, in then switched into $R_{16}L_{16}$ (preoutput) and fed to the IP^{-1} transformation, which yields the cipher block *C*. Since the encryption is so successful in scrambling the bits into what looks like an alphabet soup, why 16 rounds and not less, or more? The reason for 16 is that crypto-analysts showed that this is a number just under which DES can be more easily cracked with cribs and frequency analysis. In short, 16 rounds achieve the minimum required level of randomization at and above which only brute-force attack remains the only cracking solution. Experts have also questioned the choice of the S-boxes in the DES standard. Because the explanation is officially design-sensitive, some suspect that the S-boxes have embedded patterns, or *trapdoor functions*, which help in breaking DES without the knowledge of the key. In spite of extended analysis the issue has remained so far an open and speculative one.

We have thus completed the description of the procedure for DES encryption. What about decryption? The nice feature of DES decryption is that it is just the *very same* algorithm and program! The only simple requirement is to use the subkeys in the reverse order, namely K_{16} for the first round, K_{15} for the second, and so on with K_1 for the sixteenth round. Note also that these subkeys are not calculated the same way as for the encryption. The circular shift is the same as defined for encryption, but it must be performed from the right.

Our description concerned the encryption and decryption of message (*M*) and cipher (*C*) blocks by independent groups of 64 bits. Such a mode of operation is referred to as *electronic code book* (ECB). The ECB approach is only one out of four main standard ways to encrypt messages (as with any other block cipher algorithms). The others are referred to as *chain block coding/cipher block chaining* (CBC), *cipher feedback* (CFB), and *output-feedback mode* (OFB), respectively. The principle of CBC, CFB and OFB is that each message cipher block is made dependent upon the result of previous cipher-block operations through XOR additions.

- In CBC, each message block is initially XORed with the previous cipher block before encryption. For decryption, the same XOR operation is performed on each recovered message block;
- In CFB, the message block is encrypted by smaller sub-blocks, for instance by 8-bit groups (referred to as *8-bit CFB*) which encrypt one ASCII character at a time. This approach is advantageous for rapid encryption/decryption in communication channels where the transmission speed may be too slow. The concept also applies to a single bit (*1-bit CFB*), although it is clearly not very efficient. One may also save processing power and CPU time by encoding at once the n-bit message block (*N-bit CFB*);
- In OFB, a group of N-bits from each successive cipher block is queued at the right-most positions of the next block, and the reverse is done for decryption.

The advantage of CBC, CFB and OFB is that they conceal with increasing efficiencies any message block patterns, making frequency analysis difficult or intractable. Thus, the strength of single DES can be improved by these different techniques and their many other variants. Their relative merits are measured in term of CPU time, enhanced security and resistance to occasional bit-errors (as affecting a single message block to the entire message).

The VLSI technology used for DES implementation has impressive performance. Messages can be encrypted at over gigabit/s rates (e.g., 200 Mbyte/s), corresponding to processing 25 million DES blocks/s. Some chips also support both ECB/CBC modes or have 3DES CBC or OFB capability. With standard microprocessors (50 MHz) and workstations (125 MHz), the encryption rates range between 80 kbytes/s and 1.5 Mbytes/s.

Since, in its first initiation, DES was rapidly cracked (see facts at the beginning of the subsection and further discussion below), it made sense to further increase its resistance by applying DES twice or three times, using two or three different keys. These approaches are referred to as *double-DES* and *triple DES*, according to the following principles:

- In *double-DES*, the algorithm is implemented just twice in a row with two keys, which brings the effective key length to $56 + 56 = 112$. Note that doubling the key length squares the complexity, namely $(2^{56})^2 = 2^{112} = 10^{33.7} \approx 5 \times 10^{33}$. Thus, *Deep Crack* which is able to check out 10^9 keys per second would now require 5×10^{24} seconds of CPU time, corresponding to 1×10^{17} years or about 10^{15} centuries. Even after an improbable one billion-fold increase of CPU power, this brute-force attack would require 1×10^8 years or one million centuries! But Diffie and Hellman once proved that the effective number of keys is not given by the above figures; the keyspace is just doubled, namely 2^{57}, which does not increase safety so much with respect to standard DES;
- In *triple-DES (3DES)*, three different keys are used; one can just iterate DES with three keys; a different approach, referred to by ANSI as TDEA (*triple*

data encryption algorithm) is to use a combination of encryptions and decryptions with the three keys: if K_1, K_2, K_3 are the three keys and E_K/D_K the encryption and decryption functions with a given key K, the cipher C and message M are defined by $C = E_{K3}\{D_{K2}[E_{K1}(M)]\}$ and $M = D_{K1}\{E_{K2}[D_{K3}(C)]\}$, respectively; with 3DES, the complexity is really 2^{112}.

The number of possible DES keys, $2^{56} = 72{,}057{,}594{,}037{,}927{,}936$, apparently provides a phenomenal level of security against any brute-force attack. Such a number defines a degree of *complexity*, namely the size of the keyspace or the number of checks that must be performed in order to complete a brute-force attack. Yet the progress in VLSI realized since the inception of DES has it that, with only the technology in the mid-1990s, a $1Million machine could theoretically complete the full exploration of this keyspace within 3.5 hours (5,700 billion checks per second!). But it is only a speculation that such machines have ever been constructed. The problem becomes more serious when considering some of the proposed attack strategies which can reduce the keyspace size by orders of magnitude. For instance, a possibility is to check first all possible keys based upon ASCII-printable English words. DES crackers can also have a statistical database of plaintext and ciphertext to make comparisons. The development since 1990 of *differential cryptanalysis* has further improved the efficiency of brute-force attacks. The principle of the analysis consists in looking at certain *ciphertext pairs* corresponding to known plaintext pairs exhibiting particular difference patterns. The occurrences of such pairs are not statistically uniform. The more rare ocurrences provide some information on the subkey structure in the different rounds. Therefore, if one can build a large database of pairs, one can discard a large numbers of keys in the keyspace, whose size is thus reduced. Under differential cryptanalysis, the complexity of a full attack drops to 2^{37}, another venerable monster, but 520,000 times smaller than its 2^{56} sibling. Such a reduction factor represents the difference between one second and six days. It can be shown that DES is fully resistant against differential cryptanalysis if the number of rounds is raised to 19, and is equivalent to brute-force attack if it is raised to 17–18. Another approach for cryptanalytic attack is called *linear cryptanalysis*. The principle is based upon the fact that XORing some of the plaintext bits and some of the ciphertext bits then XORing the result produces a XOR of some of the key bits. Using a known plaintext and ciphertext database, one can take a well-educated guess at the possible key-bit values, thus reducing the keyspace and the search. Combining differential and linear cryptanalysis has also proven effective in recovering some of the key bits with an 80–95% probability of success. The other bits can then be searched with brute-force attack within a keyspace reduced by as many bits. These refined attacks work at least if the S-boxes are known, which is the case with the DES standard. But if the S-boxes are modified according to specific rules, then these attack strategies are of no avail. Nowadays, some commercial VLSI chips offer the capability of loading S-boxes. A risk is however to make a poor S-box choice, which weakens, instead of strengthening, the resulting new DES against differential/linear cryptanalysis. Several DES variants which offer higher resistance against these attacks (i.e., which nullify the complexity reduction) have been proposed as well. But by far, the alternative S-box approach remains the strongest if one knows how to make good S-box choices. Alternatively, *3DES*, which uses the standard S-boxes but has a huge complexity of 2^{112}, may also be recommended.

3.3.2 Advanced Encryption Standard (AES)

The previous considerations concerning DES security against cryptanalysis and brute-force attacks have led NIST to promote even stronger encryption algorithms. This led to the adoption in 2001 of the *advanced encryption standard* (AES), to become effective by mid-2002.

Basically, AES is a symmetric block cipher which uses blocks of 128-bit length (twice the size of DES). The key size is either 128, 192 or 256 bits, representing two to four times that of DES (as defined in 64-bit format). The blocks are represented in the standard *binary* and *hexadecimal* forms, using lower-case characters and expressions in braces. For instance, one writes $a = \{1100\ 0001\} = \{c1\}$. If an additional (higher-weight) bit is required for some operations, the new word will be written $a = \{01\}\{1100\ 0001\} = \{01\}\{c1\}$. Blocks or arrays of 128 bits (16 bytes), are designated by $a_0 a_1 \ldots a_{15}$, where the a_k are the successive bytes and a_0 the byte of highest weight. All operations are performed on a $N \times N$ byte matrix called the *state*. The number N of rows/columns is the block length divided by 32, namely $N = 4$. The state thus contains $4 \times 4 = 16$ elements each one being a byte a_k of the block $a_0 a_1 \ldots a_{15}$. Each of these individual byte elements can be designated by its row/column matrix indices ($r = 0\text{--}3$ and $c = 0\text{--}3$), which are written as s_{rc} or $s[r, c]$ as shown in Table 3.27. The mapping between the input array $a_0 a_1 \ldots a_{15}$ and the state, and between the state and the output array $b_0 b_1 \ldots b_{15}$ is done according to the rules

$$s[r, c] = a_{r+4c} = b_{r+4c} \qquad (3.24)$$

For instance the elements $s[0, 0]$, $s[0, 2]$, $s[1, 3]$ and $s[3, 1]$ correspond to the input block elements a_0, a_8, a_{13} and a_7, respectively, and the same correspondence applies to the output elements b_k. The 4 columns of the state are called w_0, w_1, w_2, w_3. Each of the columns can be conceived as representing a 32-bit word, for instance $w_2 = s[0, 2]s[1, 2]s[2, 2]s[3, 2]$ (see Table 3.27), or $w_2 = a_8 a_9 a_{10} a_{11}$. Similarly, we get $w_0 = a_0 a_1 a_2 a_3$. We thus see that this way to map the input data into the state is

TABLE 3.27 State matrix used in the advanced encryption standard (AES), with designation of individual 1-byte elements $s[r,c]$

$s[0,0]$	$s[0,1]$	$s[0,2]$	$s[0,3]$
$s[1,0]$	$s[1,1]$	$s[1,2]$	$s[1,3]$
$s[2,0]$	$s[2,1]$	$s[2,2]$	$s[2,3]$
$s[3,0]$	$s[3,1]$	$s[3,2]$	$s[3,3]$

equivalent to filling the state matrix by columns rather than by rows. The state can also be viewed as the one-dimensional array $w_0 w_1 w_2 w_3$ made of four 32-bit subblocks.

The exciting novelty introduced by AES ciphering is the use of *multiplication*, which completes the addition operation (XOR) used in the previous DES. In order for multiplied numbers to remain within a 1-byte size, this multiplication is however defined modulo a certain "prime" number. To define how such a multiplication operation works, we need to use another representation for binary numbers, which is the *polynomial representation*. Defining multiplication with polynomials is going to take a small investment in some more arithmetic, but it is worth the effort if we want to understand AES encryption.

A *polynomial* is an element which can be defined according to the expression $m(x) = c_n x^n + c_{n-1} x^{n-1} + \cdots + c_1 x + c_0$, where $c_n, c_{n-1} \ldots c_1, c_0$ are called the polynomial *coefficients* and n the polynomial's *degree*. Since a byte is the 8-bit word $c_7 c_6 c_5 c_4 c_3 c_2 c_1 c_0$ ($c_k = 0$ or 1), its equivalent polynomial representation is $m(x) = c_7 x^7 + c_6 x^6 \cdots + c_1 x + c_0$. Let's look now at the XOR operation between two polynomials. Consider the two numbers $\{1010\ 0010\} = \{a3\}$ and $\{0101\ 1111\} = \{5f\}$. In binary, hexadecimal and polynomial representations, we get

$$\{1010\ 0010\} \oplus \{0101\ 1111\} = \{1111\ 1101\}$$
$$\{a3\} \oplus \{5f\} = \{fd\}$$
$$\{x^7 + x^5 + x\} \oplus \{x^6 + x^4 + x^3 + x^2 + x + 1\} = \{x^7 + x^6 + x^5 + x^4 + x^3 + x^2 + 1\}$$

which is quite simple to grasp.

Polynomial multiplication, modulo a polynomial $m(x)$, is a bit trickier, but no more difficult to grasp. Let's call this strange operation by the symbol $\bullet$. The idea is that the reduction modulo $m(x)$ should give a polynomial whose degree is strictly less than 8, so that the result of the multiplication can be always represented by a 1-byte word. The modulus $m(x) = x^8 + x^4 + x^3 + x + 1$, which is used in AES, meets such a requirement. This specific polynomial is called *irreducible*, because it has no other dividers than 1 and itself. It is the conceptual equivalent of a prime number in the polynomial field. The polynomial dividers of $m(x)$ are any numbers $u(x), v(x)$ which verify $u(x) \times v(x) = m(x)$, where $\times$ is the usual algebraic multiplication. Let us look now at how multiplications "$\times$" and "$\bullet$" work through a simple illustrative example. Take $u(x) = x^3 + x^2 + 1$ and $v(x) = x^2 + x$. Following usual algebra, we first get the product

$$u(x) \times v(x) = (x^3 + x^2 + 1)(x^2 + x) = (x^5 + x^4 + x^2) \oplus (x^4 + x^3 + x)$$
$$= x^5 + x^3 + x^2 + x$$

To obtain this result, we use $\oplus$ to perform the addition between the polynomial coefficients of the same order, since this is the operation used in the corresponding binary representation. The second task is to reduce the above result modulo a given polynomial, say $m(x) = x^4 + 1$. By definition, we have for any polynomial $p(x)$

$$p(x) = q(x) \times m(x) + r(x) \tag{3.25}$$

where $q(x)$ is the *quotient* and $r(x)$ the *remainder*, also noted $p(x)$ mod $m(x)$. Note that the degree of the remainder must be lower than that of $m(x)$, otherwise it could divide it. Taking the previous example, we must find the polynomials $q(x)$ and $r(x)$ which verify

$$x^5 + x^3 + x^2 + x = p(x) \times (x^4 + 1) + q(x) \tag{3.26}$$

From this relation, we know that $p(x) = ax + b$ and $q(x) = cx^3 + dx^2 + ex + f$, where a, b, c, d, e, f are the unknowns. Substituting these two definitions into the RHS yields $a = c = d = 1$, $b = e = f = 0$, which gives $p(x) = x$ and $q(x) = x^3 + x^2$. Thus

$$x^3 + x^2 = x^5 + x^3 + x^2 + x \bmod x^4 + 1 \tag{3.27}$$

which illustrates how modulus reduction is performed.

With the modulus-reduction tool in hand, we can now consider an illustrative example which uses the irreducible polynomial $m(x) = x^8 + x^4 + x^3 + x + 1$. Take for instance the two numbers $\{4A\}$ and $\{18\}$. What is $p(x) = \{4a\} \bullet \{19\}$? The conversion into polynomial representation gives

$$p(x) = \{0100\ 1010\} \bullet \{0001\ 1001\}$$
$$= (x^6 + x^3 + x) \times (x^4 + x^3 + 1) \bmod m(x)$$
$$= x^{10} + x^9 + x^7 + x^5 + x^4 + x^3 + x \bmod m(x) \tag{3.28}$$
$$= x^7 + x^6 + x^5 \bmod m(x)$$
$$= \{1110\ 0000\} = \{e0\}$$

which is the result of the operation. Since we have defined a multiplication law for polynomials (modulo $m(x)$), we must also introduce the corresponding concept of *inverse polynomials*. The inverse polynomial of a non-zero polynomial, $p(x)$, is noted $p^{-1}(x)$ and is defined by the equivalent identities

$$\begin{cases} p(x) \bullet p^{-1}(x) = 1 \\ p^{-1}(x) = p(x) \bmod m(x) \end{cases} \tag{3.29}$$

We also know that the inverse exists because $m(x)$ is irreducible, i.e., $p(x)$ and $m(x)$ do not have a common divisor. As with modular algebra, the search for the inverse can be performed by the *extended Euclidian algorithm*. Referring to the previous section concerning modular algebra, and the implementation of this algorithm to find inverses modulo m, we implement the algorithm strictly the same way, but this time with polynomials and modulo $m(x)$. For instance, let us find the inverse of $p(x) = x^5 + x^3 + 1$. The table below shows the calculation steps (for explanation refer to previous examples in modular algebra):

1	$x^8 + x^4 + x^3 + x + 1$	$x^8 + x^4 + x^3 + x + 1$
2	$x^5 + x^3 + 1$	1
3	$(x^3 + x)(x^5 + x^3 + 1) =$ $x^8 + x^4 + x^3 + x$	$(x^3 + x) \times 1 = x^3 + x$

continued

| 4 | $(x^8 + x^4 + x^3 + x + 1) \oplus$ $(x^8 + x^4 + x^3 + x) = 1$ | $(x^8 + x^4 + x^3 + x + 1) \oplus$ $(x^3 + x) = x^8 + x^4 + 1$ |

Since we obtained "1" in the last line of left column, the result in the right column, $x^8 + x^4 + 1$, is the inverse $p^{-1}(x)$. We can readily check that

$$p(x) \bullet p^{-1}(x) = (x^5 + x^3 + 1) \times (x^8 + x^4 + 1) \bmod m(x)$$
$$= x^{13} + x^{11} + x^9 + x^8 + x^7 + x^5 + x^4 + x^3 + x \bmod m(x)$$
$$= 1 \bmod m(x)$$

which is the proof.

We now have the basic conceptual tools to get into the details of the AES cipher algorithm. This algorithm was conceived by two Belgian cryptoanalysts, J. Daemen and V. Rijmen, which led them to call it *Rijndael* (pronounce "rhine/reign dahl" or "rain doll"). See dedicated Web site in Bibliography. Both encryption and decryption are byte-oriented transformation involving several computation rounds. As previously mentioned, there are three possible key sizes, i.e., 128 bits (4 × 32-bit), 192 bits (6 × 32-bit) and 256 bits (8 × 32-bit), which correspond to increasing levels of security. These options are referred to as *AES-128*, *AES-192* and *AES-256*. The *number of rounds* for AES-128, AES-192 and AES-256 is $N_r = 10$, 12, and 14, respectively. As in DES, each round uses four different 4-byte subkeys derived from the cipher key, which are called *round keys*. Each round involves four computational steps:

1. Byte substitution with S-box (operation SubBytes);
2. Shifting state rows by certain amounts (operation ShiftRows);
3. Mixing data in state columns (operation MixColumns);
4. Adding round key to the state (operation AddRoundKey).

We note as the only exception that the last round does not have the operation MixColumns. We describe now these four operations.

The operation *SubBytes* is a matrix-element substitution similar to that effected with the S-boxes in DES. A first difference are that the S-box transformation in DES has 6-bit words for input and 4-bit words for output, while in AES the word length for input and output is 1 byte. Consistently, the AES S-box size is 16 × 16. Given a hexadecimal byte $a = \{xy\}$, the corresponding output byte is the one that the S-box indicates at the intersection of row "x" and column "y." The design rule for generating this S-box is the following: first take the multiplicative inverse $b = a^{-1}$ (modulo $m(x)$); then compute the output byte b' according to explicit formula

$$b'_i = b_i \oplus b_{i+4} \bmod 8 \oplus b_{i+5} \bmod 8 \oplus b_{i+6} \bmod 8 \oplus b_{i+7} \bmod 8 \oplus c_i \quad (3.30)$$

where b_i is bit i ($i = 0 \ldots 7$) of byte b, and c_i is bit i of byte $\{63\} = \{0110\ 0011\}$. The resulting S-box is shown in Table 3.28.

Before starting the sequence of rounds and proceeding to *SubBytes* of round one, the cipher key must be XORed to the input state. This stage is also referred

TABLE 3.28 S-box used in AES for the SubBytes round, showing the output element (hexadecimal) corresponding to any input byte {xy}, with x = 0-f being the row number and y = 0-f the column number

	0	1	2	3	4	5	6	7	8	9	a	b	c	d	e	f
0	63	7c	77	7b	f2	6b	6f	c5	30	01	67	2b	fe	d7	ab	76
1	ca	82	c9	7d	fa	59	47	f0	ad	d4	a2	af	9c	a4	72	c0
2	b7	fd	93	26	36	3f	f7	cc	34	A5	e5	f1	71	d8	31	15
3	04	c7	23	c3	18	96	05	9a	07	12	80	e2	eb	27	b2	75
4	09	83	2c	1a	1b	6e	5a	a0	52	3b	d6	b3	29	e3	eb	84
5	53	d1	00	ed	20	fc	b1	5b	6a	cb	be	39	4a	4c	58	cf
6	d0	ef	aa	fb	43	4d	33	85	45	f9	02	7f	50	3c	9f	a8
7	51	a3	40	8f	92	9d	38	f5	bc	b6	da	21	10	ff	f3	d2
8	cd	0c	13	ec	5f	97	44	17	c4	a7	7e	3d	64	5d	19	73
9	60	81	4f	dc	22	2a	90	88	46	ee	b8	14	de	5e	0b	db
a	e0	32	3a	0a	49	06	24	5c	c2	d3	ac	62	91	95	e4	79
b	e7	c8	37	6d	8d	d5	4e	a9	6c	56	f4	ea	65	7a	ae	08
c	ba	78	25	2e	1c	a6	b4	c6	e8	dd	74	1f	4b	bd	8b	8a
d	70	3e	b5	66	48	03	f6	0e	61	35	57	b9	86	c1	1d	9a
e	e1	f8	98	11	69	d9	8e	94	9b	1e	87	e9	ce	55	28	df
f	8c	a1	89	0d	bf	e6	42	68	41	99	2d	0f	b0	54	bb	16

to as "round 0." To provide a simplified example (AES-128), assume that the input message and the cipher key (in hexadecimal forms) are:

Input = a0 c1 b2 d3 e4 f5 06 17 28 39 4a 5b 6c 7d 8e 80
Key = b5 32 00 00 00 00 00 00 00 00 00 00 00 00 00 32

Note that this choice of key is very poor by all cryptographic standards, but here the idea is to fasten the byte-to-byte operations and help visual verifications. Figure 3.12 shows the input state matrix, the cipher key and the result of the addition. This is the state to be used as input for round 1.

Input

a0	e4	28	6c
b1	f5	39	7d
c2	06	4a	8e
d3	17	5b	80

⊕

Round key

b5	00	00	00
32	00	00	00
00	00	00	00
00	00	00	32

=

Start of round 1

15	e4	28	6c
83	f5	39	7d
c2	06	4a	8e
d3	17	5b	**b2**

FIGURE 3.12 Initiation of AES ciphering showing example of input state matrix, addition (XOR) of cipher key example, and resulting state matrix to be used at start of round one.

3.3 Communications Security and Applications

Round	Start				SubBytes				ShiftRows				MixColumns					Round key				
1	15	e4	28	6c	59	69	34	50	59	69	34	50	e1	86	da	b0		db	db	db	f7	
	83	f5	39	7d	ec	e6	12	ff	e6	12	ff	ec	e9	22	cd	57	⊕	51	51	51	0f	=
	c2	06	4a	8e	25	6f	d6	19	d6	19	25	6f	b4	7a	c5	b0		23	23	23	fb	
	d3	17	5b	b2	66	f0	39	37	37	66	f0	39	3d	0b	dc	88		63	63	63	83	
2	3a	5d	01	37														af	74	af	58	
	b8	73	9c	58													⊕	5e	0f	5e	51	=
	97	ca	e6	4b														cf	ec	ef	14	
	5e	66	bf	0b														0b	68	3b	b8	

FIGURE 3.13 Completed round one of AES-128, showing (from left to right, top) the input state, the resulting state after SubByte (see S-box Table 3.28), after ShiftRows, and after Mix-Columns. The corresponding round key (top right), which has been calculated from the steps detailed in Table 3.29, is then added (XOR) to provide give the input state of round two.

The next operation, *SubBytes*, then consists in substituting the state bytes by their corresponding value indicated by the S-box (Table 3.28). The result of this operation is shown in the second state-matrix appearing in Figure 3.13.

The next operation, *ShiftRows*, consists in effecting a left-circular permutation of the state rows with the following shifts: 0 for row 0, 1 for row 1, 2 for row 2 and 3 for row 3. The result of this operation is shown in the third state-matrix appearing in Figure 3.13.

The next operation, *MixColumns*, consists in computing, for each column "c" the four new elements defined by the operations:

$$\begin{cases} s'_{0,c} = (\{02\} \bullet s_{0,c}) \oplus (\{03\} \bullet s_{1,c}) \oplus s_{2,c} \oplus s_{3,c} \\ s'_{1,c} = s_{0,c} \oplus (\{02\} \bullet s_{1,c}) \oplus (\{03\} \bullet s_{2,c}) \oplus s_{3,c} \\ s'_{2,c} = s_{0,c} \oplus s_{1,c} \oplus (\{02\} \bullet s_{2,c}) \oplus (\{03\} \bullet s_{3,c}) \\ s'_{3,c} = (\{03\} \bullet s_{0,c}) \oplus s_{1,c} \oplus s_{2,c} \oplus (\{02\} \bullet s_{3,c}) \end{cases} \quad (3.31)$$

with the multiplication being performed modulo $x^4 + 1$, and by definition, $\{02\} = x$ and $\{03\} = x + 1$.

Let us consider a practical example to illustrate these previous operations. Take the four elements $s_{r,0}$ in column "0" of the ShiftRow state in Figure 3.13, namely,

$$s_{0,0} = \{59\} = \{0101\ 1001\} = x^6 + x^4 + x^3 + 1$$

$$s_{1,0} = \{e6\} = \{1110\ 0110\} = x^7 + x^6 + x^5 + x^2 + x$$

$$s_{2,0} = \{d6\} = \{1101\ 0110\} = x^7 + x^6 + x^4 + x^2 + x$$

$$s_{3,0} = \{37\} = \{0011\ 0111\} = x^5 + x^4 + x^2 + x + 1$$

We then obtain:

$$s_{0,0} \bullet x = x^7 + x^5 + x^4 + x \bmod x^4 + 1 = x^3 + 1$$

$$s_{1,0} \bullet (x+1) = x^8 + x^5 + x^3 + x \bmod x^4 + 1 = x^3 + 1$$

Note that this conversion modulo $x^4 + 1$ is effected by using the convenient property

$$x^k \bmod x^4 + 1 = x^{k \bmod 4} \tag{3.32}$$

After going through the computations defined by equation (3.31) while using the above values and property in equation (3.32), we obtain the following values or the elements of column 0 in the MixColumn state.

$$s'_{0,0} = x^7 + x^6 + x^5 + 1 = \{1110\ 0001\} = \{e1\}$$

$$s'_{1,0} = x^7 + x^6 + x^5 + x^3 + x^2 = \{1110\ 1100\} = \{e9\}$$

$$s'_{2,0} = x^7 + x^5 + x^4 + x^2 = \{1011\ 0100\} = \{b4\}$$

$$s'_{3,0} = x^5 + x^4 + x^3 + x^2 + 1 = \{0011\ 1101\} = \{3d\}$$

These values, along with those concerning the other columns 1–3 are shown in Figure 3.13 in the state MixColumns.

The last step of this first encryption round (and at the end of all rounds) consists in adding (XOR) a specific *round key* to the MixColumn state. Recall that for round 0, we XORed the cipher key on a byte-to-byte basis with each of the input-state elements (Figure 3.12). We might as well conceive of the cipher key as the array $w[0]w[1]w[2]w[3]$, made of 32-bit/4-byte elements $w[i]$ ($i = 0$–3), or in our example,

$$w[0] = b5\ 32\ 00\ 00 \quad w[2] = 00\ 00\ 00\ 00$$
$$w[1] = 00\ 00\ 00\ 00 \quad w[3] = 00\ 00\ 00\ 32$$

$$Key = b5\ 32\ 00\ 00\ 00\ 00\ 00\ 00\ 00\ 00\ 00\ 00\ 00\ 00\ 00\ 32$$

For round 1, what we need is a new key subset $w[4]w[5]w[6]w[7]$, for round 1 we need $w[8]w[9]w[10]w[11]$, and so on, until round 10 with $w[40]w[41]w[42]w[43]$. The full set of subkeys is called a *key schedule*. In AES terminology, calculating the key schedule is called *key expansion*. The rules of key expansion are somewhat complex and tricky, but we should have been prepared with the earlier DES. Likewise, the driving idea is to generate a pseudo-random key subset for each round, which is made by using circular permutations and random substitutions. With AES, these substitutions use the S-box and an extra XOR with the corresponding subkeys of order $i - 4$. The way to proceed to compute these subsets, which applies to 128-bit keys (and has generalized versions for 192/256-bit keys) can be conveniently illustrated using our example. Table 3.29 summarizes the procedure up to rank $i = 11$, namely covering the key-schedule computation up to round 2 inclusive. Below is a detailed description of this succession of operations.

Looking at the first column in the table, we see that the variable *temp* corresponds to the word $w[i - 1]$. There are two cases: either i is a multiple of 4 or it is not. In the first case (e.g., $i = 4, 8$, as shown), we must develop the computation further. Consider first the simpler case, where i is not a multiple of 4. We see from the table that the word $w[i]$ is calculated by the simple operation

TABLE 3.29 AES-128 key expansion for 4×4 states using the cipherkey example shown in bold

cipherkey = **b5 32 00 00** w[0]   w[1] 00 00 00 00   w[2] 00 00 00 00   w[3] 00 00 00 32

i	temp = w[i-1]	RotWord	SubBytes	Rcon [i/4]	temp' = SubBytes XOR Rcon[i/4]	w[i-4]	w[i] = temp(') XOR w[i-4]	
4	00 00 00 32	00 00 32 00	63 63 23 63	01 00 00 00	62 63 23 63	b5 32 00 00	**db 51 23 63**	= w[4]
5	**db 51 23 63**	-	-	-	-	00 00 00 00	db 51 23 63	= w[5]
6	db 51 23 63	-	-	-	-	00 00 00 00	db 51 23 63	= w[6]
7	f7 0f fb b1	-	-	-	-	00 00 00 32	f7 0f fb 83	= w[7]
8	f7 0f fb 83	0f fb 83 f7	76 0f ec 68	02 00 00 00	74 0f ec 68	db 51 23 63	af 5e cf 0b	= w[8]
9	af 5e cf 0b	-	-	-	-	db 51 23 63	74 0f ec 68	= w[9]
10	74 0f ec 68	-	-	-	-	db 51 23 63	af 5e ef 3b	= w[10]
11	af 5e ef 3b	-	-	-	-	f7 0f fb 83	58 51 14 b8	= w[11]

Each row indexed by "i" corresponds to the different steps involved in the computation of the round key $w[i]$. The resulting round keys are added (XOR) to the MixColumn states according to the rule shown in Figure 3.13.

$w[i] = w[i-1] \oplus w[i-4]$. In the other case (e.g., $i = 4,8$), we must perform three more operations. The first operation, called *RotWord*, is a left circular rotation by a 1-byte shift. The second operation, called SubBytes, is the substitution of each byte through the S-box (as defined in Table 3.28). Then comes a strange constant called *Rcon[i/4]*. As can be seen, its only non-zero byte is the integer quotient of $i/4$, which is put in the left. The variable *temp* is then transformed into $temp' = $ temp XOR Rcon. The result of the operation is in this case $w[i] = temp' \oplus w[i-4]$. The generation of the key schedule thus consists in filling out this table row by row (up to $i = 43$ for DES-128) and by the same successive steps. We can now use the results for $i = 4-7$ to finalize round 1, namely to XOR the subkeys $w[4]w[5]w[6]w[7]$ to the *MixColumn* state in order to yield the input state of round 2. The data are summarized in Figure 3.13. We may easily imagine the continuation of this procedure up to round 10, except for the absence of any *MixColumn* operation in this last round (as previously stated). The output state obtained at the end is the cipher block version of the input, 128-bit message/plaintext block. The encryption procedure for AES-192 and AES-256 are pretty much similar, except for the number of required rounds (12 and 14, respectively) and details in the associated key schedule computations.

What about AES decryption? As with DES, it is a matter of performing all the operations in the reverse way, using in the first round the last round key of the key schedule all the way down to round 10. The other reverse operations are called *InvShiftRows* (right-circular shift), *InvSubBytes* (inverse S-box) and *InvMix-Columns*. The last operation involves the same type of computation as defined in equation (3.31), except that different polynomial factors, $\{0b\},\{0d\},\{09\},\{0e\}$ must be used, according to the following:

$$\begin{cases} s'_{0,c} = (\{0e\} \bullet s_{0,c}) \oplus (\{0b\} \bullet s_{1,c}) \oplus (\{0d\} \bullet s_{2,c}) \oplus (\{09\} \bullet s_{3,c}) \\ s'_{1,c} = (\{09\} \bullet s_{0,c}) \oplus (\{0e\} \bullet s_{1,c}) \oplus (\{0b\} \bullet s_{2,c}) \oplus (\{0d\} \bullet s_{3,c}) \\ s'_{2,c} = (\{0d\} \bullet s_{0,c}) \oplus (\{09\} \bullet s_{1,c}) \oplus (\{0e\} \bullet s_{2,c}) \oplus (\{0b\} \bullet s_{3,c}) \\ s'_{3,c} = (\{0b\} \bullet s_{0,c}) \oplus (\{0d\} \bullet s_{1,c}) \oplus (\{09\} \bullet s_{2,c}) \oplus (\{0e\} \bullet s_{3,c}) \end{cases} \quad (3.33)$$

While the encryption and decryption algorithms are different in terms of operations definitions, some properties make it possible to bring the two closer together, as described in the NIST reference document.

The Rijndael algorithm was selected by NIST among other potential candidates (Mars, Serpent, Twofish) to become the official AES. The choice seems to have been guided by considerations of algorithm simplicity, S-box size, number of rounds, resistance to linear/differential cryptanalysis techniques and overall, the "complexity" defined by the keyspace (here up to 256-bits). Cryptanalysts are now actively investigating the potential weaknesses of AES, with the goal to reduce its complexity by a significant factor. By the end of 2002 the complexity of AES would have already been reduced from 2^{256} to 2^{100}, which is close to 3DES (2^{112}), within three orders of magnitude. But as we have shown earlier, such a huge number still constitutes a formidable obstacle against brute-force attacks. No doubt that the future of AES will be rich in new developments, partial-cracking announcements, rumors and controversies, which remain some of the most intriguing components of cryptographic science!

3.3.3 Other Encryption Algorithms and Standards

In this chapter, we have extensively described three main standard ciphers that are known as RSA or PKCS (Section 3.2), DES and AES (this section). Predictably, there are many other cipher algorithms, some of which have being standardized, or remain commercial/government property, or are available as downloadable freeware. Some have been specifically developed for protection in mobile telephony, wireless LAN, banking, e-commerce or e-mail, to quote a only few. In this section, we shall provide a selection and brief description of these other cryptosystems and applications, with no pretense to be exhaustive. The idea is both to illustrate their variety and to provide the reader with further reference background.

Here are some of these cryptosystems and features, including the earlier alternatives to DES (see for instance Schneier (1996) for detailed algorithmic descriptions):

- *NewDes* (1985): Despite what its name suggests, this is based upon a completely different algorithm. It works on 1-byte blocks with 17 rounds. While its 120-bit key provides a theoretical complexity of 2^{120}, it was shown that with a 2^{33} plaintext/key database it can be broken in 2^{48} attempts, making it somewhat weaker than simple DES.

- *FEAL* (1987) resembles DES with fewer rounds (e.g., 8). It has been subject to intense cryptanalysis work, which revealed some weaknesses, even with enhanced versions with 128-bit keys.

- *REDOC-II* (1990), works on 80-bit blocks with a 160-bit key and makes use of special masks that change the function at each round, based upon the key schedule. Its unchallenged complexity is 2^{160}. Interestingly enough, the later version (REDOC-III), which uses variable-length keys as big as 20,480 bits (!), proved weak under cryptanalysis attacks.

- *LOKI* (1990) uses a similar structure to DES with improved S-box/expansion/permutation design and key schedule.

- *Khufu* (1990), named after an Egyptian pharao, uses up to 32 rounds with key-dependent S-boxes and a 512-bit key, offering a proven complexity of $2^{512} = 10^{154}$.

- *RC-2/RC-4* (1987), or *Rivest Cipher*: both of proprietary nature, the first claimed higher resistance to cryptanalysis than any other DES variants proposed, but no algorithm details were available unless under special agreements. Yet the full details of *RC-4* were once given away by an anonymous remailer (some entity which forwards e-mail without showing its origin or originator). The RC-4 algorithm is remarkably simple: it consists in XORing plaintext bits with pseudo-random bits, and the reverse for decryption. These unpredictable (but inversely calculable) XOR bits are generated by modulo-256 additions with a 256-element S-box, which comes to the unconceivable number of possibilities of $255! \times 255^2 = 2^{2,700}$. The secret S-box permutations are generated by a variable-length key. As a licensed cryptosystem, RC-4 is used today in a large variety of computer and wireless-LAN products and also for crypted Internet (see Section 3.3.5 on secure-socket layer). Note that U.S. export restrictions apply for key sizes exceeding 40 bits, corresponding to $2^{40} = 1.1 \times 10^{12}$ possibilities, a number within reach of the most basic attack (i.e., comparison with known plaintext/ciphertext/key databases).

IDEA (1992), or *international data encryption algorithm*: initially called IPES (*improved proposed encryption standard*), is considered to be the strongest block cipher publicly available. It is based upon a 128-bit key, 64-bit block, 8-round algorithm which works both ways as in DES. One of its original features is that the rounds are based upon three functions: XOR, addition modulo 2^{16} and multiplication modulo $2^{16} + 1$ (a prime). Its proven complexity is 2^{128}, and the level can be increased by double and triple encryption. Note that due to the key size, even double encryption is more secure than 3DES. Another important feature of IDEA is that it is a constituent of PGP (see further below).

CAST (1993) is a 64-bit block, 64-bit key, 8-round algorithm using six S-boxes. The originality of the design is that the S-boxes can be constructed according to the application. The new version, called CAST5 (or equivalently CAST128) uses a 128-bit key and up to 16 rounds. This algorithm is known to be resistant to any known type of cryptanalysis. It may be adopted as an official Canadian standard.

Blowfish (1994) was specifically designed for simplicity of implementation and is freely available to the public. It is a 64-bit block cipher which works either way, and has a variable-size key. The security can be set to the appropriate level with a maximum key size of 488 bits, representing 2^{488} or 8×10^{146} key combinations. The algorithm uses four 32-bit S-boxes with 256 entries each and 16 rounds.

PGP (1991), which stands for *pretty good privacy*, was developed by P. Zimmermann with the idea of freely offering to the wide public a means for highly secure encryption to protect private life/information and civil liberties (for a riveting account of the story surrounding the birth of PGP, see for instance Singh (1999) and Levy (2001)). The starting concept of PGP is to make combined use of a *symmetric* cipher (such as 3DES, IDEA, CAST5) and an *asymmetric* cipher (namely RSA/PKC). With respect to the second, the drawback of the first type of cipher is that the key must be communicated to the intended recipient. On the other hand, it is however more rapid to implement in both encryption and decryption (namely, some 1,000 times faster), and can encode significantly larger amounts of information. The idea of PGP is thus to combine the two approaches, which is referred to as a *hybrid cryptosystem*. Alice uses IDEA (for instance) to generate the bulk ciphertext, then encrypts the IDEA key with RSA/PKC, and communicates both ciphertext and crypted key to Bob. Bob then uses his private (RSA) key to retrieve the IDEA key, then uses the IDEA key to retrieve the plaintex. PGP includes more features. First, it compresses the plaintext data, which increases resistance to frequency analysis and saves transmission time. Second, it automatically generates a one-time pad, 128-bit *session key*, which is based upon the analysis of Alice's keystrokes and random mouse movements. This key, which is used for the (IDEA) ciphertext, is then encrypted with Bob's public key, which PGP finds in Alice's database, and this crypted key, along with the ciphertext, is forwarded to Bob on the public channel (such as e-mail). But PGP provides two other features which are *authentication* (certification of signature) and *message integrity* (certification of unaltered message). This works by using a one-way *hash function* (such as SHA or MD5) see description of principle in next subsection. This hash function is carried over the entire plaintext, which produces a short data file called a *message digest*, which is unique to the plaintext. This digest is then crypted with Alice's private key and included in the plaintext. Using Alice's public key, PGP on Bob's side can verify that the same hash function produces the same result on the recovered plaintext, which authenticates Alice as the originator and that the text has not been altered. Other applications of PGP concern crypted, real-time voice communications (PGPfone) and secure hard-disk partitions (PGPdisk). A very, very important note: while different freeware versions of PGP are available from some internet sites, their importation (including downloading) and use in several countries, including in Europe, may be subject to interdiction or qualified authorization; also note that PGP is a commercial and licensed product.

We consider next the issue of privacy and user protection in the field of *cellular telephony*. With GSM (*global system for mobile telecommunications*), radio communications are protected by two features: *user authentication* and *two-way voice encryption*. These prevent fraudulent use or cloning (e.g., in case of a stolen or lost handset), or eavesdropping of call/conversation contents. To address the first issue, the cellular handset is equipped with a *SIM card* (*system/subscriber identity module*). It is a "smart" plastic card which enables authentication through a *personal identification number* (PIN). This PIN is a 4–8 decimal digit which must be successfully entered before any tone/connection be provided by the operator. Three successive failures leads to immediate SIM deactivation, like automatic-teller machines swallow up credit cards! The SIM card, which has a 2–3-kByte, non-volatile memory (ROM), contains both the user's PIN and a unique 128-bit key (called K_i) which is registered in a network authentication center (AUC). It is worth mentioning here that the SIM memory has many other spy functions. Indeed, it has records of the full user's identity and home-network location (international phone number and HLR), his/her temporary VLR identity for roaming, his/her personal phone book and received SMS. The SIM ROM also contains encryption algorithms called A3/A8. For initiation purposes (first SIM activation in the home location or during roaming), the network sends a 128-bit pseudo-random word (called RAND) to the SIM, as illustrated in Figure 3.14. With the secret key K_i and A3, the SIM computes, via a certain *hash function*, a 32-bit word (called SRES) and sends it through the line. The authorization center does the same and compares the two SRES. If the test proves successful, the user's SIM is authenticated and is registered in both HLR and temporary VLR. This authentication process can be repeated by the network as many times as required, even during the course of a call. Since the

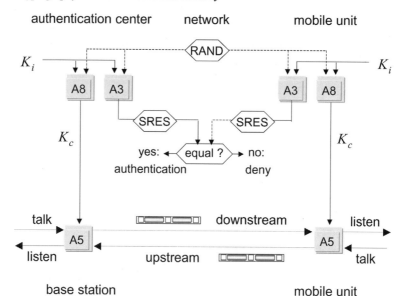

FIGURE 3.14 Authentication and encryption/decryption in the GSM system (see text for description).

RAND is different at each test, the SRES produced by the SIM is also different. A SRES that was eavesdropped cannot be re-used in the future to authenticate a cloned SIM because it would correspond to an outdated RAND.

The second level of GSM protection is provided by two-way *voice encryption*, under the standard referred to as *A5* (A for algorithm). To understand how this encryption works, recall that the upstream/downstream GSM frames have a payload of 114 bits. At the onset of the connection, the SIM first generates a 64-bit "session" key (called K_c), which is generated from both K_i and the network RAND (see Figure 3.14). This key K_c is then used by A8 from the SIM to encrypt the compressed-voice data (or anything else using this phone channel). A second level of encryption, called A5, is built into the mobile's hardware. Using the key K_c, the A5 algorithm encrypts the 114-bit upstream data and decrypts at the same time the 114-bit downstream data broadcast from the base station. Why is the downstream data encrypted the same way? Because the network also has the same key K_c from the knowledge of K_i and RAND.

The first existing A5 standards were called (somewhat illogically) A5/1 for the strongest encryption with 64-bit keys, A5/2 for the weakest with 16-bit keys, and "A5/0" for no encryption. As we have seen, the initial authentication mechanism, which can be repeated at will, is based upon a certain hash function between RAND and K_i, known only to the operator. This secret function, called COMP128, has the weakness of providing the same 32-bit SRES for many different 128-bit RAND inputs (as one could expect from a good hash function). But through the analysis of matching SRES, this weakness can be advantageously exploited to reveal the user's key K_i. Based upon this principle, the first GSM key cracking was proven in 1988 within a 8-hour-only search. Another weakness of A5/1, to have been discovered, was that for some unknown reasons the actual keyspace used only 54 bits, representing 1,000 times fewer encryption possibilities. As in the story of RC-4, the ultra-confidential A5/1 algorithm eventually leaked in 1994 and became openly available as a source code in certain web sites. Note that knowing the encryption algorithm is not a condition for successful cracking, which is proven by the fact that many current standards are publicly available (e.g., DES and AES, as described in this chapter). But anyway A5/1 was eventually cracked in 1999, using a brute-force attack of 2^{40} attempts. The worse was yet to come. By the end of 1999, another crack was performed in *less than one second* using a single PC and 73-Gbyte memory space, based on two minutes of conversation recording. The risk situation for communications under the various 2G/3G protocols GSM/GPRS/HSCSD/EDGE led ETSI (*European Telecommunications Standards Institute*) with other standardization groups to develop, in 2002, a new encryption standard referred to as *A5/3*. Concerning cdmaOne, the encryption standard is called *Oryx* (key of 96 bits in US and of 32 bit outside). For UMTS used in Europe and Japan, the standard is called *Kasumi* (key of 128 bits).

We consider next the issue of cryptography protection for data transport in *wireless local-area network*s (WLAN), in particular the networks associated with the WI-FI or 802.11b standards. The original cryptosystem which was developed for WI-FI is called *wireless equivalent privacy*, or WEP. In its original form, it used a 40-bit key RC-4, which was originally thought relatively secure for public use. One first crack in the 802.11b protocol is that crypto-security is activated only as an option, so most users may think it is just built in. A second problem is that the encryption keys must be shared by all the network users, which means that they are not changed often. A third problem is the weakness of the key management or scheduling used, which allows attackers to eliminate orders of magnitudes in possibilities. This weakness is aggravated by the fact that WLAN data are full of known plaintext. The attack is then based on comparing the ciphertext with known plaintext/ciphertext/key data libraries and recovering the corresponding key when some strong correlation is identified. Another strategy consists in trying "frequently used" password dictionaries which are built into the key management system. The proof of existence for the above problems was provided by the appearance of several WEP-cracking freeware/shareware such as *Airsnort*, *WEPCrack* or *Netstumbler*. For instance, after gathering 5–10 million encrypted packets, Airsnort is reportedly able to recover the encryption password within one second or less. Note that these same tools can be also usefully applied to probe the existence of underground WLANs in the vicinity of a given location, building or campus! An improvement to this embarrassing situation was introduced by the IEEE standard 802.1x, which uses for instance Ethernet-based authentication and generates encryption keys on the fly during each session. This standard is however not addressing the core of the whole encryption issue in WLAN. Because of the large number of users, the entire keyspace may be exhausted within one hour, increasing the probability of key reuse and the excitement of eavesdroppers. Before newly appeared standards such as AES-128 were considered as a radical replacement solution for RC-4, progress was also accomplished in the development of a backward-compatible *temporary-key integrity protocol* (TKIP, formerly called WEP2), which is based upon 128-bit keys. But it appears that this new cryptosystem is just as weak with respect to "dictionary attacks" as WEP was, it may be the matter of extra seconds for attacks to succeed. Before other solutions come from the evolving IEEE security standard, the higher-level answer to increase WLAN security is to place network *firewalls* between the Internet and the WLAN access point. They are however expensive to implement and operate, and slow down the network evolution and connectivity. A higher level of protection is provided by the Internet's *secure-socket layer* (SSL), corresponding to the protocol HTTPS, which are both described in Section 3.3.5.

3.3.4 Digital Signature and Authentication

Finally, we shall consider the concepts and standards concerning *digital signature* and *authentication*, which were evoked earlier in this section when describing PGP. As its name indicates, a digital signature is a way to certify that the author of a message is really the person claimed, or that the signature is *unforgeable*. If Bob sends a love message to Alice with a digital signature, Alice will know that it's really coming from Bob, not Eve! Digital *authentication* is different. It ensures that nothing in the message was deleted, added, or altered, even to a single punctuation mark. Such an authentication is important in all matters pertaining to titles, official records, agreements, contracts, authorizations, and confidential-sensitive orders, for instance. As we shall describe in the following text, these documents can be sent in the open along with a digital signature. The signature may or may not authenticate their contents, according to the options. Alternatively, the documents can be sent in a crypted form with signature included in the plaintext (like a normal message), or with the digital signature as a separate attachment. Obviously, RSA/PKC provides a direct way to authenticate both signature and contents, since the communication channel is perfectly safe. However, asymmetric encryption is 1,000 times slower than with symmetric ciphers and cannot easily handle large-size files. And in many applications, the point is not so much to hide the message as to guarantee that (1) it is really from the sender, and (b) the received contents are absolutely identical to the original. These two conditions can be realized independently or together, representing different possible options, as we shall describe.

We first focus on *digital signature*. Many algorithms have been developed for this purpose, most of which are based on RSA/PKC approaches. Let $E_K^A(X)$ refer to the encryption of block X with key K by person A, and $D_{K'}^B(Y)$ to the decryption of block Y with key K' by person B. We know from PKC principles (see Section 3.2) that encryption by Bob then decryption by Alice is based on the following property:

$$D_{\text{Alice\_secret\_K}}^{\text{Alice}}(E_{\text{Alice\_public\_K}}^{\text{Bob}}(M)) = M \qquad (3.34)$$

where Bob used Alice's public key and Alice uses her private key.

A first possibility for Bob to digitally sign a message is to encrypt it with his *private* key. The only way for Alice to decrypt the message is to use Bob's public key:

$$D^{\text{Alice}}_{\text{Bob\_public\_K}}(E^{\text{Bob}}_{\text{Bob\_private\_K}}(M)) = M \tag{3.35}$$

This way, Alice is positively certain that the message is from Bob. But the message should not be confidential since anybody can decrypt it, unless Bob's public key is kept confidential between a group of trusted persons of which Alice is a member.

Another way to sign a message is to use PKC as a means to communicate a one-time *session key*. The message itself is crypted with any symmetric cipher, say DES. This is the principle of hybrid cryptosystems such as PGP. Alice thus encrypts her message to Bob with DES. She must then communicate the DES key (κ) to Bob. For this, she encrypts κ with PKC using Bob's public key, which gives the crypted key $E^{\text{Alice}}_{\text{Bob\_public\_K}}(\kappa)$. She then forwards both DES crypted message and crypted key to Bob. The first thing Bob does is to recover the DES key through the operation:

$$D^{\text{Bob}}_{\text{Bob\_private\_K}}(E^{\text{Alice}}_{\text{Bob\_public\_K}}(\kappa)) = \kappa \tag{3.36}$$

By itself, this procedure does not certify to Bob that Alice is really the author of the message, unless Bob's public key is kept confidential within a trusted group to which Alice belongs. It is very unlikely that a person from this trusted group would try to impersonate Alice and play a trick on Bob. Since the session key (κ) is a one-time key to be immediately discarded at the end of the session, it is unlikely that Eve will have the time to decrypt it and mingle in the conversation by impersonating Alice. But there is always a risk that Eve finally obtains Bob's private key, in which case she can do all of the above, pretending she is Alice, without Bob suspecting anything.

Another possibility for digital signature is to use a *trusted third party*, also called *arbitrator*. Alice and Bob have both chosen Trent as their arbitrator. Using PKC, they regularly provide Trent with their DES secret keys. Alice encrypts her message to Bob with her DES secret key and sends it to Trent (Bob can't decrypt it). Trent then decrypts the message with Alice's secret key. He does not even read it (but this would probably not matter to Alice) and immediately encrypts it with Bob's DES secret key and forwards the result to Bob. Bob then decrypts the message with his secret key. The whole procedure is summarized by the following:

$$\begin{aligned} D^{\text{Trent}}_{\text{Alice\_secret\_K}}(E^{\text{Alice}}_{\text{Alice\_secret\_K}}(M)) &= M \\ D^{\text{Bob}}_{\text{Bob\_secret\_K}}(E^{\text{Trent}}_{\text{Bob\_secret\_K}}(M)) &= M \end{aligned} \tag{3.37}$$

This way to proceed guarantees to Alice and Bob that their messages are authentically from each other. But can Trent be 100% trusted? This is the crucial issue with arbitrators. Trent may be the most honest person on Earth, but his database containing the secret keys can be regularly visited by Eve's hackers, including from the parking lot. Or at the extreme opposite, it is not that Trent is normally dishonest, but Eve offered him in exchange for the keys the $100,000 that he needed to pay his debts. Or better, Eve knows something about his life that he does not want anyone to know, so she gets the keys for free. But the last two situations are much less likely than the first.

The alternative to these signature options is to use *one-way encryption* through the so-called *hash functions*. One-way encryption also provides a means for message *authentication*, as we shall explain. Unlike conventional encryption, one-way encryption is final and irreversible. It is like reducing a digital color photograph to a pixel histogram. There is a unique histogram to that picture, but the picture cannot be reconstructed from its histogram information. The second point is that it is very unlikely that two pictures will have strictly the same histogram. Most generally, a hash function H reduces a message source X of any length into a hash message h, which is called the *message digest*. The four basic properties of a good hash function are the following:

1. Given X, it is very easy to compute h;
2. Given h, it is extremely difficult to compute X;
3. The probability that two sources X, X' give the same h is extremely small;
4. A very small alteration of X produces a very different h.

In hash-function jargon, the event that two sources give the same digest is called a *collision*. The hash-function algorithm should then be designed to be *collision resistant*. We will see further below how important this condition is.

How are hash functions defined? Basically, the function consists in an algorithm that takes a given block of source message, X_i, to produce a *hash block* $h_i = f(X_i, h_{i-1})$ of smaller length. The function f is a *one-way compression* function. As indicated, the compression function always uses the hash-block results of the previous step, which have been initialized to some value h_0. A possible hash function could consist in taking the value modulo m ($m = 9, 17, 33$) of the sum of all ASCII characters from a 64-bit block and the previous hash block. The hash result would then fit on 8, 16, or 32 bits. The drawback of this approach is that the number of collisions is very high. It would take indeed little effort to find another block having the same hash (attack 1), or two different blocks having the same hash (attack 2). This example was just to illustrate the principle of hashing. Real hash-function algorithms are based upon far more complex block operations and multiple one-way encryption rounds. Also, they are designed to be highly resistant to attacks of types 1 and 2.

One of the main hash-function standards is called MD5 (*message digest* 5), which represents an improved version of MD4. Both produce $4 \times 32 = 128$-bit hash blocks and were designed to make it impossible to find matching hash pairs, unless one uses brute-force attack. The MD5 algorithm takes source messages by $16 \times 32 = 512$-bit blocks. The message is initially *padded* in order for its length to be a multiple of 512 bits minus 64 bits. Padding just consists in putting a "1" bit at the end and filling with "0" the gap in-between. A set of four 32-bit variables (a, b, c, d), which are called *chaining variables*, is first initialized with some constants (A, B, C, D), namely $A = 01234567$, $B = 89abcdef$, $C = fedcba98$ and $D = 76543210$. The computation takes four rounds (three in MD4) each including 16 steps based on nonlinear operations. Using the Boolean notation (see definition in Section 3.2.1), the four basic nonlinear functions used through the process are defined as follows:

$$\begin{cases} F(x,y,z) = (x \wedge y) \vee (\neg x \wedge z) \\ G(x,y,z) = (x \wedge y) \vee (y \wedge \neg z) \\ H(x,y,z) = x \oplus y \oplus z \\ I(x,y,z) = x \oplus (y \vee \neg z) \end{cases} \quad (3.38)$$

These basic functions are then used to form the following four variable-transformation types:

$$\begin{cases} FF(a,b,c,d,M_i,s,t_j) \Longleftrightarrow a = b + [(a + F(b,c,d) + M_i + t_j)_{<<<s}] \\ GG(a,b,c,d,M_i,s,t_j) \Longleftrightarrow a = b + [(a + G(b,c,d) + M_i + t_j)_{<<<s}] \\ HH(a,b,c,d,M_i,s,t_j) \Longleftrightarrow a = b + [(a + H(b,c,d) + M_i + t_j)_{<<<s}] \\ II(a,b,c,d,M_i,s,t_j) \Longleftrightarrow a = b + [(a + I(b,c,d) + M_i + t_j)_{<<<s}] \end{cases} \quad (3.39)$$

where M_i is the sub-block i in the 16-block message sequence ($i = 0 \ldots 15$), and $<<<s$ means a left-circular shift with s bit positions. The constant t_j is given by the integer result of the formula $t_j = 2^{32}|\sin(j)|$, where j corresponds to the step number ($j = 1 \ldots 64$) as expressed in radians (the result fitting in 32 bits). Each round has 16 steps which involve the 16 subblocks M_i. Round 1 uses exclusively FF, round 2 exclusively GG, and so on to round 4. Within each round and at each step j, the functions apply to different permutations of the chaining variables a, b, c, d, different subblocks M_i and different shifts s. The tabulated sequence of functions, variables, subblocks, shifts and values of t_j can be found in RFC 1321 or Schneier (1996), for instance. Here, we shall only reproduce, for clarification and illustration purposes, the first eight steps of round 1:

$$\begin{cases} FF(a,b,c,d,M_0,7,t_1) & FF(a,b,c,d,M_4,7,t_5) \\ FF(d,a,b,c,M_1,12,t_2) & FF(d,a,b,c,M_5,12,t_6) \\ FF(c,d,a,b,M_2,17,t_3) & FF(c,d,a,b,M_6,17,t_7) \\ FF(b,c,d,a,M_3,22,t_4) & FF(b,c,d,a,M_7,22,t_8) \end{cases} \quad (3.40)$$

If we detail the first four steps, we see that they correspond to the following chain-variable substitutions and rotations:

$$\begin{cases} a = b + [(b + F(b,c,d) + M_0 + d76aa478)_{<<<7}] \\ d = a + [(d + F(a,b,c) + M_1 + e8c7b756)_{<<<12}] \\ c = d + [(c + F(d,a,b) + M_2 + 242070db)_{<<<17}] \\ b = c + [(b + F(c,d,a) + M_3 + c1bdceee)_{<<<22}] \end{cases} \quad (3.41)$$

After 4 rounds representing 64 steps, the final result (a, b, c, d) is added to (A, B, C, D). The algorithm is then repeated with the next 512-bit block of message. The 4×32-bit end-result of the operation is the final hash block. This whole description provides an intuitive sense of what a thorough and secure "hashing" algorithm is all about. MD5 is used in the e-mail encryption standard called PEM (see next subsection) together with MD2. This last hash-function is based upon a simpler but slower algorithm. It uses a random permutation of message bytes (as based upon the digits of π) with XOR operations between successive blocks of 16 bytes. From any message size (padded to a multiple of 16-bytes) MD2 yields a 128-bit hash.

The NIST organization has developed another approach called *secure hash algorithm* (SHA, also called SHA-1). It is based upon a principle similar to MD4, while using five chain variables a, b, c, d, e, which produce a $5 \times 32 = 160$-bit hash from a 512-bit block (compared to 128-bit hash in MD5). The hash results from 4 rounds of 20 operations each (80 steps). The increase from 16 to 20 steps in each round is due to the expansion of the 16×32 message block into a 80×32 message block. Such an expansion further mixes together the message bits. The SHA algorithm is even more difficult to attack than MD5, since the hash length is 160 bits, instead of 128 bits. The SHA algorithm is used in NIST's *digital signature algorithm* (DSA) which is at the root of the *digital signature standard* (DSS). Basically, DSA uses a combination of PKC and SHA. The operations involved in the different steps are not easy to justify mathematically. Here is a simplified description whose purpose is only to illustrate how the parties proceed.

A common group of users first agree on three numbers (p, q, g), where (a) p is a prime number whose bit-length L is a multiple of 64 and is comprised between 512 and 1,064, (b) q is a prime factor of $p - 1$, and (c) g is equal to an integer power $(p - 1)/q$, modulo p. Alice's public key (y), is also known to the group. Alice calculated it from her private key (x) according to $y = g^x \mod p$. For Alice to sign a message M to Bob, and for Bob to authenticate both signature and message, here are the steps:

Alice first computes the hash $H(M)$ though SHA. She then computes two numbers (r, s), which involves $H(M)$ and her private key x. She forwards (r, s) to Bob, along with M (here M can be an open e-mail, a text file or a contract, for instance).

Using SHA, Bob first calculates the hash $H(M)$. He then performs four operations which involve $r, s, H(M)$ and Alice's public key y, giving a final result, v. If Bob gets $v = r$, the signature is valid! And in addition, the full-text contents are authenticated!

Any minute modification or alteration of the initial message by Eve during its transfer to Bob would have changed the hash function into some value $H(M') \neq H(M)$. Then Bob could never obtain $v = r$. At least this is true if Eve were unsuccessful in finding a modification for which $H(M') = H(M)$, corresponding to a purposeful collision. If she succeeded in attacking the hash and finding a collision (in such a way that the message modifications would not raise suspicion), Bob would not see anything and would wrongly trust that the received message is the absolute original.

There are many other powerful algorithms for digital signature and authentication, including DSA variants, which we shall overlook here. We shall conclude this subsection by reviewing some interesting features and applications offered by these two principles.

As we have seen, hash functions provide a safe way to both sign and authenticate. We have seen that attacking a hash-based cipher is a case of either finding a message

M' whose hash $H(M')$ exactly matches a known hash $H(M)$, referred to as attack 1, or finding two different messages for which $H(M) = H(M')$, referred to as attack 2. The first type of attack is called "*birthday attack*." The name comes from the problem of finding a person in a group who shares your birthday. How many people should be in the group for the probability of finding at least one such person to be 50%? The answer is 253 since $1 - (364/365)^{253} \approx 0.5$. But the required number (n) is much less if one considers the same probability of finding *two people* sharing the same birthday. The answer is given by $1 - (364/365)^{n(n-1)/2} \approx 0.5$, where the exponent $C_n^2 = n(n-1)/2$ represents the number of possible pairs in the group. Since $253 = 23(23-1)/2$, the answer is $n = 23$, corresponding to a surprisingly ten times smaller group.

The birthday attack is like the first problem. Eve wants to find the message M' that has the same hash as M. The second attack consists in finding two arbitrary messages (M, M') that have a common hash, which is easier, since $H(M) = H(M')$ is not a prerequisite. Let us now see the proof. With a birthday attack, finding a message that gives the same 64-bit hash will require trying out a maximum of $2^{64} = 1.8 \times 10^{19}$ possibilities. At an attack rate of one *billion* tries per second, the search would be completed in 1.8×10^{10} seconds or 570 years! In the second case, finding matching hash pairs reduces the number of tries (for equivalent success) to 2^{32}. This is because the database of pair-candidates grows as the square of the number of tries (explicitly, the number of pairs after n tries is $C_n^2 = n(n-1)/2 \approx n^2/2$; the probability of a pair match after n tries is the same as after 2^{64} tries, provided that $n^2/2 = 2^{64}$ or $n \approx 2^{64/2} = 2^{32}$). At the same attack rate of 10^9 tries per second, the time required to complete the full search of $2^{32} = 4.3 \times 10^9$ tries is about $4\frac{1}{2}$ seconds, a significant difference with the previous birthday-attack case. Even with a lower attack rate of only one *million* times per second, the search would be completed in 1h11mn70s, to compare with 570,000 years!

The interest of the second type of hash-function attack is not immediately apparent. Why would a person be interested in generating two different messages hashing to the same result? Here is a bad story which illustrates why.

Alice and Bob have a common and flourishing business, but don't agree on certain issues of ownership and shares attribution. Alice proposes to make a contract to finalize these issues. She prepares two versions, one that reflects Bob's position and is essentially favorable to him, and another that reflects her position and is essentially favorable to her. The two versions initially give different hash results. Then Alice goes through a process of introducing invisible characters in both documents, like a few spaces before each carriage return or within the tab zones, and many other possibilities. This is for the documents to look normal, but the contents are still different. As we have seen, for a 64-bit hash, this takes Alice one hour or so to find a matching pair. Alice then sends to Bob the version of the contract that is favorable to him. Bob is pleased with it, and returns to Alice his digital signature on the hash. Now Alice has a version that is favorable to her, with a matching hash signed by Bob, which definitely authenticates both Bob's signature and document. She can use this fact in court in the future. She also "forgets" to send to Bob her signature on the hash, which otherwise would put Bob in the same position (contract favorable to Bob with Alice's signature on the hash). That was not very nice of Alice, but it worked. In the case of a court dispute, Bob could bring his version of the contract, and show evidence that it reduces indeed to the hash he signed. But Alice's lawyers may be successful in convincing the court that it is Bob who made the attack and has a phony version. Although we are no specialist to tell how such a situation could conclude, it is clear that Alice has the advantage.

This fictitious story illustrates the importance of *protocols* in all digital-signature and authentication processes. Here are a few (and safer) alternative ways to proceed, which are of indicative value:

> Alice and Bob turn to Trent, the arbitrator. They have agreed on a contract version and send it to Trent. Trent then includes at the bottom of the text a few lines of his own, and asks both Alice and Bob to sign the resulting hash. Upon receiving the two signatures, Trent then sends a confirmation to both, for instance including the original, the two signed hashes and his own signature. Bob or Alice may attack it in the future, but Trent remains the only trusted reference.
>
> Alice and Bob send to Trent their document hash encrypted with their secret keys. Trent has their public keys and can authenticate both signatures. He also has the document and can verify that the two hashes correspond.
>
> Alice makes a hash of the document, for instance using DSA. Using PKC, she encrypts the hash with her private key (which is equivalent to a signature) and sends it to Bob with the document. Using DSA, Bob makes a hash of the document. With Alice's public key he then decrypts the hash that Alice sent. If the two hashes correspond, Alice's signature is valid.
>
> Alice signs the document M with her PKC private key end encrypts the results with Bob's public key. This gives

$$E_{\text{Bob\_public\_K}}^{\text{Alice}}\left(S_{\text{Alice\_private\_key}}^{\text{Alice}}(M)\right)$$

Upon receipt, Bob first decrypts the above with his private key then decrypts the result with Alice's public key to get the document, according to

$$D_{\text{Alice\_public\_K}}^{\text{Bob}}\left\{D_{\text{Bob\_private\_K}}^{\text{Bob}}\left[E_{\text{Bob\_public\_K}}^{\text{Alice}}(S_{\text{Alice\_private\_key}}^{\text{Alice}}(M))\right]\right\} = M$$

Bob possibly includes at the bottom of the message "read and approved," and does the reverse operation:

$$E_{\text{Alice\_public\_K}}^{\text{Bob}}\left(S_{\text{Bob\_private\_key}}^{\text{Bob}}(M')\right)$$

Upon receipt, Alice recovers the message M'

$$D_{\text{Bob\_public\_K}}^{\text{Alice}}\left\{D_{\text{Alice\_private\_K}}^{\text{Alice}}\left[E_{\text{Alice\_public\_K}}^{\text{Bob}}(S_{\text{Bob\_private\_key}}^{\text{Bob}}(M'))\right]\right\} = M'$$

She compares M and M' and sees that the two are identical, except for the optional confirmation "read and approved." The fact that Bob returned the message to Alice with his signature is in any case a sufficient proof that he agreed on the document.

There are many variants and more complex approaches in signature/authentication protocols which are protected against attacks, and can be safely recommended by the authorized services. The above examples only provide a flavor of such a variety, and should not be considered as explicit recommendations or representing best solutions.

Two other important features involved in the complex domain of digital signature and authentication are *timestamping* and *key certification*.

Timestamping is the function of sealing the time information with a given document. This is very important for instance in matters pertaining to patents, copyright or stocks operations. It is also vital in digital checking to prevent people repeating the same cash transfer (see last subsection). In its most simple implementation, Alice would send the document to Trent, who will record the time and date and deliver to Alice a certificate with his own signature. The drawback is that this does not work if Alice does not want Trent to see the document's contents. In another possibility, Alice could put a date on top of her document, and end the hash to Trent. Trent will record the date and store the hash with his digital signature, without any possibility of reading the document. For future proof, both Alice and Trent can provide the hashes. Trent's signature validates the time at which he received the hash. Alice's hash being identical, the date shown in the corresponding original is therefore valid. One drawback of this approach is the possibility that Trent may have agreed to do a "little favor" to Alice by recording a different time, back-dated or forward-dated.

A good timestamping protocol should make it virtually impossible to stamp a time different from that of the current day. A solution which meets this requirement is the *linking timestamp protocol*. The idea is the following. Trent keeps a record of hashes (H_i) that Alice sends to him with the corresponding times (t_i). Any time Alice asks for a timestamp, Trent signs and send to Alice the following:

$$S_{\text{Trent}}(\text{Alice}, i, H_i, t_i; H_{i-1}, t_{i-1}, L_i; I_{i-1})$$

This document certifies that according to records, Trent put under number i (in his timestamped sequence) a document from Alice identified by H_i and timestamped at time t_i. The second part of the information is a recall of Trent's previous timestamp (H_{i-1}, t_{i-1}), and the chained hash link $L_i = H(H_{i-1}, t_{i-1}, L_{i-1}, I_{i-1})$. This hash link combines the information (H_{i-1}, t_{i-1}) with the identity (I_{i-1}) of the person for whom Trent did this previous timestamp. The hash thus seals Alice's timestamp with the sequence of all of Trent's clients. Should there be any question about the validity of Alice's timestamp i, Trent will provide to Alice the names of the two clients I_{i-1}, I_{i+1}. They might easily confirm their times t_{i-1} and t_{i+1} between which Alice's document has been timestamped. While it is technically possible for Trent to reverse-engineer the chain and provide a phony timestamp t'_i to Alice, it becomes more than tricky to generate the new hashes H_{i-1}, H_{i+1}. Another approach which avoids this complex certification from Trent consists in Alice using a *distributed timestamp protocol*. She may use a group of friends and ask them to return the hash with the date of receipt and with their digital signature on both. One or two friends may agree to render a "little service" to Alice, but not all of them. This approach won't prove much to a court, since Alice may just show the phony timestamps, and besides, they come from friends. The best approach for distributed timestamping is to put together a network of N anonymous persons who don't know each other, but need a frequent timestamping service. They all put their contact information in a common database, with identifiers $I_1 \ldots I_N$, which does not betray their identity. To obtain a timestamp from this network, Alice first generates a set of pseudo-random numbers (say 10). She then uses these numbers as the identifiers. Alice sends the hash to these ten identifiers who rapidly return the hash with the date and overall signature. Alice now has 10 documents that show the same timestamp. The network rules could be that if contacted members repeatedly fail to respond (within declared availability periods over the year) their subscription fees would increase or they would be denied the timestamp service. One can imagine that if they need the service, they won't fail. The strength of this approach is that it would take a very corrupt society to get a large number of phony timestamps. The weakness is that Alice may introduce a large number of friends into the network and systematically chose them for the timestamping. The network must then be equipped with a monitoring function that verifies that Alice's chosen identifiers are really random. Finally, some of the network users may identify those members that would be willing to make phony timestamps with the same service in exchange, but we can trust that their number may not be as large as to provide a random-choice database.

Key certification is the responsibility of a *certification Authority* (CA). The CA, presided over by Trent, manages the public-key databases for different network groups. The CA can certify to Alice that Bob's key is really Bob's. It can also manage different keys for the same user with various levels of security. It can also attribute separate keys for encryption and for signature, which enhances the user's security. It can manage key protection by introducing time limits and renewals. The CA can register the group of trusted persons who must have access to the user's private keys on request, and the user can introduce new persons into this circle by informing the CA with authenticated messages. Another issue is to manage *private* keys. Some users may not want to spend their week-ends finding big primes whose products fit within some key format, each time they want another secret key! The trusted CA may accomplish this for them, with optimized choices avoiding frequently used keys. Some users may not want the CA to even keep their private keys for back-up purposes, like putting them into a bank safe where they will always be sure to find them. They may not necessarily give the CA *all* of their secret keys, but only those that they use for business or nonprivate activities, for instance. Finally, the CA can play the role of Trent for any contractual or authentication or timestamping purposes.

In case neither Alice nor Bob would trust a CA (or would be willing to buy his service), they can resort to *distributed key management*. The principle is that Bob can ask two friends to sign his public key. These friends are referred to as *introducers*. When remitting Alice his public key, he also provides the two introducers' signatures. Alice may personally know one of the persons, or have every reason to trust them. Short of complete trust, she may even check that the introducers are really what Bob says they are. The underlying concept is that introducers may form a larger network chain. The longer the chain, the less dependable the system.

Also, it does not solve the other key-management issues previously described, which are best handled by a professional CA.

3.3.5 Network and Internet Security

With the development of networks at LAN, MAN (metro) and WAN scales and that of the global Internet, the issue of security, privacy, confidentiality and protection against different types of malicious abuse and threats have become most relevant. The modern forms of such abuse and threat are quite different from the age-old telephone wiretapping and eavesdropping. *Network abuse* corresponds to a vast and seemingly unbounded list of categories which can be summarized as follows:

Infiltration (accessing networks and their databases);

Intelligence (gathering information from private/confidential communications channels, also called *snooping*);

Tampering (modifying, altering or destroying, databases, computer files or messages);

Impersonation (using someone's identity, also called *spoofing*);

Theft of service or ToS (using a service for free, also called *phreaking* for telephone),

Denial of service or DoS (blocking a Web-site server with an overwhelming flow of data),

Spamming and *bombing* (neutralizing servers and search engines or e-mails with overwhelming data),

Harassment (another form of spamming directed at individuals or discussion groups);

Web-site attacks (cracking servers and changing Web pages);

Network attack (creating outages in network servers, routers or access terminals);

Virus infection and propagation (destroying software or nesting executables such as backdoors and Trojan horses);

Hoaxes (circulating stories, false rumors, phony virus alerts with high forwarding potential or money scams);

Carding (use of computer-generated valid credit card numbers);

Illegal materials/contents dissemination (e.g., pedophilia, racial hatred, bomb and drugs formulae, classified information);

Trafficking, *crime associations*, and *terrorism*;

And many other variants and combinations of all of the previous.

In Desurvire (2004), Chapter 3, Section 3.3, we highlighted some words from the rich "*Internet jargon*," which comes with several new concepts pertaining to both network use and network abuse. One typical profile in the second case is the "*hacker*," also known as "cyberpunk" (not to confuse with highly respectable *cipherpunks*, *crytopunks*, or *cryptocrackers*!), a person whose pastime may be some or most of the top items in the above, and who has a taste for celebrity and challenge. The hacker's psychological profile, villain or hero, is not easily circumscribed, as detailed for instance in Shepard (2002). Another definition of the hacker, which sounds nobler, is that of a professional who searches for "holes" in network systems, with no intent to break, tamper with or violate anything, in contrast with the less worthy "cracker" who just has fun trying passwords and explores forbidden zones until he finds something to play with. The hacker would thus render some service to the community by revealing the loopholes and helping with network security. The cracker would also share the information but with a limited group and preferably under an anonymous hero codename. Here we shall leave the debate, which is obviously rich and controversial.

There is still some similarity between the hacker/cracker and the code breaker, but it is only superficial. Indeed, both help to improve network or channel security by revealing the potential cracks. However, the intent is not the same. In the first case, the gratification may be within reach without a Ph.D. in computer science or crypto-arithmetic. Any normally skilled teenage could write his or her own virus code variant and try it on the Internet. The result can be highly destructive, even if antivirus shields are periodically updated. For those who won't go through the pains of design and programming, new *virus* software can be downloadable from the Net to try out. There must be a sense of exaltation in belonging to a *piracy* network, no matter how far the hacker may go to prove anything. In contrast, code crackers (also called cipherpunks, a title to be merited) have years of hard-gained qualifications and experience and work with the declared purpose of making cryptography (and its security) constantly progress over time. Their underlying ethics may also include the ideal of protecting civil rights and privacy against the state, another endless controversial subject. But they never demonstrate anything by means of network abuse. Cracking the next RSA-N may be sufficient for proof and graduation, although it requires international cooperation. Conceiving the ultimate code that resists attack from the best specialists, or finding cracks in the best specialists' ultimate codes is their definition of a pastime. Code-breaking is not piracy, as long as there is no other reward beyond that point. Otherwise it would be called spying, and this activity entails more serious risks and consequences than hacking and cracking.

One of the most serious threats against networks and the Internet in general remains the aforementioned *denial of service* (DoS). With *distributed* denial of service (DDoS), a popular web site can be attacked by thousands of hijacked home computers. For instance, on May 22nd, 2001, the site *www.whitehouse.com* was unreachable for six hours due to a DDoS attack. With the development of the Internet, the number of attack incidents is increasing exponentially. In 1989, the cumulative number of reported incidents was less than 140, by the end of 1999 it was just under 10,000 and by mid-2003 it was about 260,000 (source *www.cert.org*). The periods 2000–2003 and the first half of 2003 represents 90% and 30% of this total, respectively. Note that some of these reported incidents may concern hundreds or sometimes thousands of sites. The financial loss does not reflect this trend directly (as it is now going downward) and strongly varies with the type of attack. According to the 2003 CSI/FBI Computer Crime and Security Survey (*www.gocsi.com*), the cost of DoS is U.S.$65 millions, is under that incurred by theft of proprietary/trade-secret information (U.S.$70 millions), but over that incurred by viruses (U.S.$27 millions), insider net abuse (U.S.$11 millions), financial fraud (U.S.$10 millions), and data network sabotage (U.S.$5 millions). Interestingly, the survey shows that the majority of the attacks are due to independent hackers and disgruntled employees, representing each $2\times$ to $4\times$ the other types of attacks such as from U.S./foreign competitors and foreign governments. Concerning Web-site attacks, DoS and vandalism remain the most frequent (about 35% each), compared with only 6% for theft of information. Such figures illustrate the diversity of reasons and motivations involved, and the difficulties that ISP and corporations/services that use the Internet are now facing.

Leaving now the issues of cyber attacks and network abuse, what about *communications security* in general? The question is how to *protect* and *authenticate access*, *manage keys*, and ensure *confidentiality* at the user level? Within the scope of this book, we can only briefly address these questions, which are extensively covered by dedicated Web sites (see for instance URLs in the Bibliography), and are also intricately related to evolving products and services. Here, we will not address aspects of network security related to *firewalls*, *antivirus software*, *intrusion detection*, *access control*, *file encryption* and the like. Rather, we shall only consider *communication security with the Internet*, which is consistent with this chapter's focus.

In security matters, one may not confuse the algorithm with its protocol chain. For instance, PGP is a protocol, but it can use indifferently DES, IDEA, CAST for encryption and SHA or MD5 for hashing. At network-management level, the protocol must include a stack of layered policies/procedures and a hierarchy of *certification authorities* (CA). There is a variety of choices and standards in this field. We shall only briefly list some of them:

ISDN: Recall the old *integrated-service digital network* protocol (Desurvire, 2004, Chapter 2)? Because of its slow progress and its lasting promises, it was initially dubbed "Innovation Subscribers Don't Need," or "the technology which took 15 years to become an overnight success." But ISDN has gone a long way since, and now offers 64-kbit/s voice/data services with RSA/PKC key exchange and DES encryption. The telephones have embedded secret keys (like a SIM) which cannot be tampered with. They also contain the network public key (for incoming-call encryption) and the owner's public key (for purposes of authenticating owner's key modifications and roaming). Short-term key pairs, to be regularly updated by the network, are also stored in the unit to increase protection. The DES key, used for the actual encryption, is a session key to be destroyed upon call termination, which surely eliminates any possibility of systematic wire tapping.

CCA (*common cryptographic architecture*): Developed by IBM, this is an all-purpose, inter-operable key-management system. It uses the principle of *control vectors* (CV) whose data structure defines the authorized uses and privileges associated with any *session key* (e.g., encryption, decryption, time limits, machine/terminal unit, restrictions). The user's session key is retrieved by hashing the CV and XORing it with a secret master key, which was given to the user through PKC.

SESAME (*secure European system for applications in multivendor environment*): This is a key-exchange and authentication system based upon DES, PCK and MD5. The system algorithm may have some flaws, being especially exposed to password-dictionary attacks.

X.509: Developed by ISO and recommended by ITU-T (hence the name), this protocol represents more a framework than any specific protocol algorithm for key exchange, encryption hashing and signature. The concept is to provide a common policy reference to the hierarchy levels (CA) in delivering certificates and keys according to the local algorithm standard use. Bob and Alice may communicate while they are locally registered under different CAs. The CAs will search for their common CA root and enable proper key allocation and authentication by means of special *X.509 certificates*. This certificate contains all information concerning the user, the time validity, the encryption algorithm to be used, its parameters and the user public key.

Kerberos: Developed by MIT, and named after the three-headed dog who guarded the entrance of *Hades* (Hell) in Greek mythology. A joke has it that the dog does not guard any exit, so people may be free to escape; but since there is no exit, the smart creature is also able to handle this security issue from the unique access point, probably the reason why he was given three brains for face recognition! Kerberos is a trusted-arbitrator, strong authentication and DES encryption protocol. It is a *user-to-server* rather than a *user-to-user* authentication scheme. Its initiation goes according to the following client/server sequence. Alice and Bob may just be machines, so Trent is actually an *authentication server* (AS):

(a) Alice (the client) sends to Trent (the AS) her identity (A) and that of Bob (B) whom she wants to open a session with.

(b) Trent (the AS which knows both Alice's and Bob's secret keys) generates the two crypted messages

$$E^{Trent}_{Alice\_key}(t, l, K, B) \text{ and } E^{Trent}_{Bob\_key}(t, l, K, A)$$

which are forwarded to Alice. In these messages, t is a timestamp, l the authorized duration of the session and K the one-time DES session key. Alice then sends to Bob the following information:

$$E^{Alice}_{K}(t, A) \text{ and } E^{Trent}_{Bob\_key}(t, l, K, A)$$

(c) Since Bob has the session key (K), he is able to confirm he is there and listening by sending to Alice the acknowledgment:

$$E^{Bob}_{K}(t+1)$$

where the timestamp was incremented by one unit. The session between Alice and Bob is thus authenticated in terms of security (Trent), client initiator (Alice), destination (Bob) with parameters such as the time initiation and maximum duration. The two ends (Alice and Bob) can then communicate through the channel secured by the common and one-time DES session key K. The channel service layer may be anything from HTTP to secure e-commerce. This description is a simplified version of the actual and even current version of Kerberos, but it provides an idea of the underlying principle. Kerberos is the encryption/authentication protocol now used in 802.11 WLAN.

PEM (privacy-enhanced mail): Jointly designed by the different Internet standardization groups, this is a protocol for e-mail encryption (confidentiality) and authentication (message signature and integrity). Encryption is realized by DES in CBC mode (508 to 1,024-bit keys). Authentication uses MD2 or MD5 (128-bit) hash functions, which provides a *message integrity check* (MIC). Key management can be symmetric (keys are exchanged by DES/ECB, or 3DES with two keys) or asymmetric (RSA/PKC up to 1,024-bit key lengths, within X.509). The e-mail messages carry all the information concerning the key-management and encryption protocol, the originator and the issuing CA certificates, the MIC and the ciphertext. Although PEM is highly secure as long as the private key remains a well-kept secret. But attacks may be made against the key-management system (PKC) itself, for instance by tampering with the installed PEM software. Unbeknown to the user, the exchanged keys will then be systematically encrypted with Eve's public key, which enables Eve to decrypt any message, regardless of its encryption scheme. Two derivatives of PEM are TIS/PEM (trusted information systems PEM) and RIPEM (Riordan PEM). The first is an enhanced version supporting a multiple hierarchy of certificates which Internet sites may issue or recognize. The second is a (non-exportable) freeware version of PEM to be used for noncommercial purposes.

We consider next the issue of *cryptography over Internet or TCP/IP data*. The corresponding security protocol, referred to as *IPsec*, was developed by IETF (*Internet engineering task force*). Its general specifications are overviewed in RFC 2401 and RFC 2411. IPsec provides means for message authentication and data integrity, and confidentiality/privacy through encryption. The unidirectional, protected logical connection between two IP endpoints (IP host or server) is called a *security association* (SA). A bi-directional channel requires the establishment of two SA. The following does not assume that the reader is necessarily familiar with the TCP/IP datagram structure and its two addressing versions called *IPv4* and *IPv6*. Here the datagram may just be regarded as a packet with an IP header followed by a TCP/data payload field. Whenever TCP is mentioned, it can be substituted by UDP, which is a simpler protocol variant in the same layer.

The *IPsec* protocol can be viewed as representing another layer above TCP. Referring to Figure 3.15, we recall that IP packets, which are called *datagrams*,

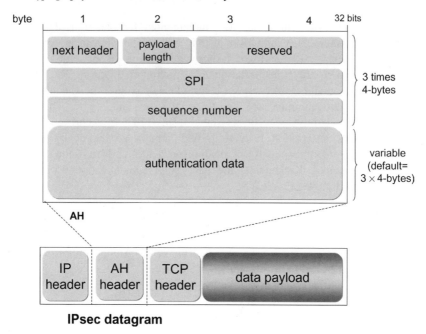

FIGURE 3.15 Field structure of authentication header (AH) in IPsec (SPI = security parameters index). Encapsulation of AH into the TCP/IP datagram (transport mode) is shown at bottom.

have three fields: the IP header, the TCP header (the layer immediately above IP), and the payload data. The payload data can be web pages (HTTP, FTP) or e-mail (SMTP), for instance. The creation of an intermediate security layer above IP but below TCP thus consists in inserting a new header field in between. There are two optional IPsec approaches. The first, used only for authentication purposes, consists in inserting an *authentication header* (AH). The second, which does both authentication and privacy consists in the encryption of the data and the insertion of an *encapsulating security payload* (ESP). Although AH and ESP were developed for IPv6, they are reverse-compatible with IPv4. Thus IPsec is a *mandatory* feature for IPv6 but is only an *optional* feature for IPv4. In either case, the choice between AH or ESP must be indicated by placing the hexadecimal values "51" or "50" (respectively) in the version/protocol field of the IPv4 IP header, or in the next header (NXT HDR) field of the IPv6 header. Next, we shall describe these two approaches.

The detailed AH field structure (RFC 2402) is shown in Figure 3.15. It can be described as follows:

> *Next header field* (1 byte): Designates the type of field immediately following AH;
>
> *Payload length* (1 byte): Designates the full AH length in multiples of 32-bit words, minus two; the default configuration corresponds to the

minimum AH size, namely 6×32-bytes; in this case, the payload-length value is 4;

Reserved (2 bytes): Field assigned for future use, filled with "0";

SPI (*security parameter index*, 4 bytes): Defines with the destination IP address the security association (SA) to which the datagram belongs;

Sequence number (4 bytes): Numbers the datagram in the SA sequence, never allowed to cycle back and allow replay attacks (antireplay mechanism); the maximum is $2^{32}-1$, above which the number is reset to 0 prior to the establishment of a new SA;

Authentication data (variable size with default of 3×4-bytes): Contains the *integrity check value* (ICV); the ICV is an authentication/integrity message which can be a DES cipher or more usually, a MD5 or SHA-1 hash.

Figure 3.15 also shows the AH encapsulation in the so-called "transport mode." Another encapsulation possibility is the "tunnel mode," in which the AH is placed on top of the IP datagram and preceded by a new IP header. In either transport or tunnel modes, the entire IPsec datagram is authenticated, except for "mutable fields" whose values may change during the transit.

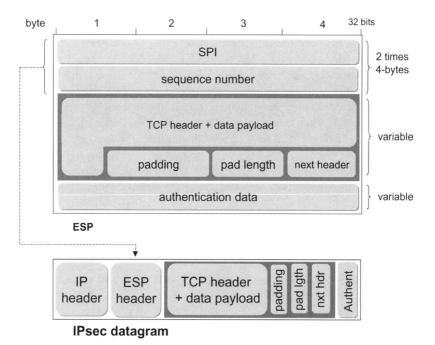

FIGURE 3.16 Field structure of encapsulating security payload (ESP) in IPsec (SPI = security parameters index). Encapsulation of the TCP/IP datagram (transport mode) via ESP is shown at bottom. The dark background defines the encrypted region.

The detailed ESP field structure (RFC 2406) is shown in Figure 3.16. It can be described as follows:

SPI (*security parameter index*, 4 bytes): same as in AH;

sequence number (4 bytes): same as in AH; forms with SPI the ESP header;

TCP header and payload data (variable size): contains the original TCP datagram, as encrypted with DES in CBC mode;

padding (0–255 bytes): use to complete the plaintext block to the length required by the encryption algorithm and to hide the actual size of the payload;

pad length (1 byte): as defined;

next header field (1 byte): identifies the type of data in the previous payload field

authentication (variable size): same as in AH.

The figure shows how the IPsec datagram is constructed by use of these fields (transport mode). All encrypted fields are shown with a dark background. Note that the entire ESP packet is authenticated on top of the encrypted portion. As for AH, the tunnel mode consists in placing the ESP header (SPI + sequence number) before the IP header (instead of after as the figure shows), and topped with a new IP header. Finally, host-to-host SA use either the transfer or the tunnel mode, while router-to-router SA exclusively use the tunnel mode.

Either AH or ESP and transport or tunnel modes are possible to establish a SA. In *tunnel mode SA*, we have seen that the datagram is given a new IP address, like an IP over IP protocol which distinguishes between the IPsec destination (new address) and the final destination of the IP-datagram payload (original IP destination address). As previously mentioned, the authentication generally uses a MD5 or SHA-1 hash. For reference, the coding protocol is called HMAC, for *hash message-authentication code* (RFC 2104). Note that in ESP, authentication is only an option. If one wants to establish an authentication/encryption SA, there are three options: (a) using ESP SA with authentication activated, (b) initiating ESP SA without authentication followed by AH SA, and (c) the reverse of (b).

Key management for AH or ESP in IPsec can be performed under several possible protocols, standards or licenses. These are the following:

ISAKMP (*Internet Security Association key management protocol*, RFC 2407 and 2408): Defines the framework within which keys can be exchanged (regardless of the algorithm); also independently defines the protocol for the initiation and deletion of SA;

Oakley (RFC 2412): Compatible with ISAKMP, is based upon the Diffie–Hellman key-exchange algorithm;

SKEME (Secure key exchange mechanism): A key-exchange technique providing anonymity and quick key renewal;

IKE (*Internet key exchange*, RFC 2409): A hybrid combination of Oakley and SKEME within ISAKMP; an illustrated tutorial of the different IKE steps is provided in Mel and Baker (2001);

Photuris (RFC 2522 and 2523): Generates short-lived session keys without passing them across the Internet;

SKIP (*Simple Key Management Scheme for Internet Protocols*): Licensed by SUN, uses a combination of PKC and symmetric key exchange; provides certificate management, authentication/integrity and automatic encryption/decryption.

Many Internet communications and client-server Web-site applications are concerned with e-commerce, e-business or e-banking. This requires the transmission over the Internet of sensitive data such as access passwords, credit card or bank account numbers. The establishment of IPsec SA channels could be fully adequate for this purpose. But how about a user-transparent, simplified and automatic protocol to be used directly at HTTP level? The answer to this question is the development by IETF of a new protocol called *secure-socket layer* (SSL), with an HTTP-compatible format called *S-HTTP* or *HTTPS* (*HTTP over SSL or secure HTTP*). The SSL protocol, now used by Netscape, features the three functions of client/server authentication, data encryption and message authentication/integrity. In the OSI model, SSL comes as a new intermediate layer above TCP/IP but under HTTP.

Since HTTPS involves time-consuming encryption, only the web applications and pages that are concerned in the acquisition and transmission of sensitive data are placed under this protocol. The key length, which is between 40 bits (U.S.-export version) and 168 bits, defines the security level. Version 2 of SSL only supports RC-2 and RC-4 with 40-bit keys, while version 3 SSL supports DES/RC-4 with 128-bit keys, and triple DES with 168-bit keys. The hash functions are based on MD5 and SHA-1, representing many different cipher/hash combinations options, at least with SSL v3. As an exercise, the unfamiliar reader may check out the SSL V2/3 options from Netscape Navigator, as accessible from the icon "security" in the top menu bar, and see how options may be configured. It might also be the opportunity to set up a PKCS password.

A full SSL session between the initiating client (Bob) and the server (Alice) goes through the following steps:

Bob sends a "hello" message to Alice, which details all possible options he is able to use for key-exchange, cipher and digest;

Alice returns to Bob her choice (e.g., RSA for key exchange, 3DES for encryption and SHA-1 for digest); this choice is called *cipher suite*;

Alice then sends to Bob her certificate.

Bob encrypts a random number using Alice's RSA public key which is in the certificate.

Alice and Bob now need a common secret key, actually three secret keys and one set for each way (see next). Then,

Bob generates a 48-byte random number called *premaster key*, encrypts it with Alice's public key and sends it to Alice;

from this number, Alice first generates three secret keys: one for message encryption (e.g., DES), one for message-integrity digests (HMAC) and a third one for encryption initialization (called IV); these three are for use in Alice-to-Bob messages; Alice also generates three other keys that will be used for the Bob-to-Alice message;

Bob does the same, with three Bob-to-Alice keys and three Alice-to-Bob keys.

Now Bob and Alice have six secret keys in common. The next step consists in Bob authenticating Alice. It does not go the other way, because Alice does not want to make client Bob wait too long and lose him, and most clients like Bob do not have certificates anyway; in these conditions, all Alice needs is Bob's credit card number which she can rapidly validate. Then,

Bob sends to Alice a *finished handshake* message encrypted with his secret keys;

Alice sends Bob the proof that she has the same 6-secret keys, which authenticates her.

Bob can now proceed to type in his credit card number that Alice finally requests, and confirm his order (books, CDs, movie, hotel booking, airplane ticket, stocks buy/sale order, digital check, cash wire-transfer...). Each one-way transaction goes with both cipher encryption and hash authentication, using either Alice-to-Bob keys or Bob-to-Alice keys according to the message direction. The SSL session is then terminated with a mouse click or in the event of any purposeful/accidental Web-site disconnection.

Of course, the entire SSL initiation procedure has taken a few seconds or less, without Bob worrying over anything about key and cipher stuff. This procedure is now routinely used by millions of Internet e-shoppers, although not all e-commerce transactions may be automatically under SSL control. It is noteworthy that a large majority of Internet users are still reluctant to provide their credit card number to a Web site, out of the widespread reputation that the internet is an essentially insecure communications channel. As a matter of fact, the risk of fraudulent transactions with a given credit card number is substantially higher with paper and digital records that the user is constantly leaving behind, from food stores to gas stations to hotels and restaurant bills. Smart cards that have a *personal identification number* (PIN) code, just like GSM cellular phones, are more immune to the abuse. Fraudulent transactions can always be reversed with proof, except if the PIN code has been used for validation. To be sure this point gets across, it would be utterly insane to provide both credit card number and PIN code to a Web site. Interestingly, this is precisely what everyone does with automatic-teller machines! But in this case we trust that the information is not flying over the Internet and instead remains within the private and secured bank network!

In 1997, SSL v3 eventually proved to be breakable. In the meanwhile, IETF developed an enhanced standard called *transport layer security* (TLS), which is described in RFC 2246. The new TLS standard, also called SSL v3.1, includes strong cryptosystems like the Diffie–Hellman key exchanges and supports the *digital signature standard* (DSS).

When SSL was launched, it came into competition with Microsoft's *secure transaction technology* (STT). The STT approach was abandoned in favor of *private communication technology* (PCT), which represents an upper-layer compatible extension of SSL (referred to as *PCT over SSL*), now used in Microsoft browsers and servers. Its claimed advantage, among other additional security features, is to separate the authentication protocol from the encryption protocol, while using the same key. To extend SSL/TLS, Microsoft also developed the *server gated cryptography* (SGC) protocol.

3.3.6 Current and Futuristic Applications of Cryptography

In this subsection, we shall briefly review the most current applications of cryptography, as well as some futuristic ones. As the previous pages have already shown, the scope of modern or digital cryptography is much broader than the age-old, traditional "secret communication" applications, inherited from centuries of spy intrigues and the last two world wars. In the following, we shall first discuss how cryptography has developed well beyond this restricted view in order to meet the concern of telecommunication users in a broad range of new services.

The concerns of privacy protection and confidentiality are definitely not new. But with the Internet and the widespread use of e-mail for private and business communications, progressively replacing traditional "paper mail," it has surely taken new dimensions. With fixed-telephony over public-switched networks (PSTN), the concern is remote, since wiretapping requires some physical means of access to the line or to the local exchange. With cellular/mobile telephony (such as GSM and now UMTS/cdma2000), encryption is a defacto feature, and users are seemingly more worried by (yet unproven) brain damage issues. With future *internet telephony* (VoIP), and *video/multimedia* communications, the situation is different. Business use is not a problem, rather it is highly recommended for expediency and travel-cost savings. Its adoption for private use is much slower. Apart from the issue of terminal and service cost, this could be because personal pictures are more contents-sensitive than voice. This fact is proven by the public's attitude and response, although teenagers and under seemingly love the concept. In broadband/video communications, the need for enhanced security and privacy protection may therefore become more acute. With *wireless LAN* (WI-FI), encryption is built in, but as we have seen, it remains exposed to attack, despite the fact that users or protagonists tend to minimize or deny the risk. Finally, with *e-commerce* (on-line shopping and creditcard mediated transactions) there is increasing trust, thanks to the implementation of SSL. The previous subsection has amply illustrated the variety of cryptosystems and their use for key exchange, message integrity and sender certification/authentication. With IPv6, encryption is a requisite, while it is reverse-compatible with the current IPv4, as an option. These crypted Internet communications represent a formidable progress from the early times. With foreseen convergence between voice, video and data, and Internet as the overarching network protocol, security by cryptography is going to be generalized and embedded in all services. Assuming the same rate of progress in the next ten to twenty years, one might feel at a loss of imagination. We may suggest (as a risqué speculation) that ciphers with 128-bit or even 1,024-bit keys may be cracked some day soon. As the crypto-experts say, considering the 2^N complexities involved, there is not enough silicon in the galaxy to build the number of parallel computers required for such a task! But this applies to *brute-force* attacks. Differential and linear cryptanalysis, and their future developments, remain formidable tools to reduce the task by orders of magnitude. As we have seen even the DES Goliath (and its triple implementation) had to be substituted with AES, with the hope that it will remain unchallenged for the next 20 years. Based on these different observations, we may conclude that cryptosystems are going to take an increasing role in future society, maybe a role so important that they will become unnoticed, just like radio waves in digital cellular telephones.

What could cryptography be useful for, apart from protected e-mails and telecommunications confidentiality, integrity and authentication? The answer to such a question projects us to a different world, even right into the future, to what the twenty-first century may actually resemble in years 2020 to 2050. In the following, we will first discuss the classic applications (although generally little known), then we will review some of the so-called "esoteric" ones. Some of the items covered are inspired by examples in Schneir (1996), yet with our own original descriptions, comments, additions and thoughts.

The course and organization of mature personal life requires one to sign hundreds of paper documents in a lifetime. The unbounded list of required signatures goes from insurance contracts, car loans, rental agreements, house mortgages, employment and mutual-donation contracts to tax declarations. One may expect that the amount of such "paperwork" may not so much decrease in the future, but instead the procedures of access, signature and filing may considerably simplify and increasingly rely upon the Internet.

One highly illustrative example is *electronic tax filing*. Since 2001, the income tax services of several countries such as the United States and France have opened dedicated Web sites to the public. There, not only are all possible forms for individual and business available, but they can be *filed* on line (for free), with the subsequent delivery of an official certificate of receipt, to print or save. The American IRS system, called *e-file*, offers users the option of generating their own PIN with their own choice of five numbers (except 00000). The French system uses the principle of authentication and delivery of a certificate based upon a taxpayer ID appearing in the paper form. In any case, the information is secured and crypted under SSL (e.g., RC-4, 128-bits in the French system). The tax is calculated on-line, and any amendments can be corrected in the future through the same certification/authentication process. *On-line tax payment* is encouraged, directly via creditcard or through other options (such as bank-account withdrawal). In 2002, 47 millions U.S. taxpayers used electronic filing of returns, representing 21% of the 226 millions returns (source: www.irs.gov). Although the process is no substitution for careful filing and tax-advising (there also exists web sites for this), it illustrates a considerable cultural change and probably a no-return evolution. The apparently low (or slow) adoption rate of e-filing might be attributed to residual reluctance to provide personal/corporate income information via the Internet, even when secured in a SSL channel. It may also be related to individual/corporate habits and in the private case, to the statistics of Internet penetration. No doubt that the evolution of tax e-filing represents a reliable and accurate indicator of confidence in Internet security, because of the population stability and the relatively unchanging character of the application. In contrast, the volume of e-commerce is another indicator, but the services are multiple and their user base also growing, which makes the analysis, even in a given specialized sector (e.g., books and music) far more difficult.

According to all expectations, *digital signatures* will increasingly replace the traditional blue-ink-on-paper seal. As we have seen, digital signatures can authenticate both the signing entity and the document's contents. In this respect, an important feature is called *nonrepudiation*. The concept of nonrepudiation is similar to that of certified mail: Neither sender nor destinee can deny having sent or received the message. The fact of using a digital signature may look like built-in nonrepudiation for a sender like Alice. But a loophole is that it is always possible for her to communicate thereafter her private key to the outside, and then claim that the signature is forged (as referred to as *repudiation*). A prevention against such a possibility is *timestamping*, which was described in the previous subsection. As we have seen, timestamping is not free of loopholes either. But it makes it difficult for Alice to claim later that her private key was compromised prior to the time she used it to sign the document! One of the safest implementations of timestamping requires the arbitration of Trent with a *linking protocol* (see an earlier description) which makes it virtually impossible to cheat with dates, even with Alice–Trent collusion.

Next we look at the concept of *blind signature*. Alice wants Trent to sign a document, but does not want him to know its contents, just like a notarization process. To obtain Bob's blind signature, Alice can simply ask him to sign the hash of the document. Bob could certify later that the signature is valid but never know what the document contains, and Alice could authenticate the document because she has a signed matching hash. But it would be better if the crypted document signed by Bob could be reversed in plaintext by Alice. Instead of the hash, Alice could XOR the document to be signed with a large random number called the *blinding factor*. After Bob's signature, Alice then XORs again the document to recover the plaintext. This requires an algorithm for which the digital signature and the XOR are commutative. The drawback of the blind signature is that Bob, having no idea of what he is signing, could be exposed to some risk (plaintext being for instance: "This confirms that I owe 1 million dollars to Alice, signed: Bob"). The smart technique to prevent such a risk is that Alice submits together N blinded messages to Bob, each one having a different blinding factor. Bob picks at random $N - 1$ messages and asks Alice for the corresponding blinding factors. Then Bob can read the $N - 1$ message contents, and make sure there was nothing unsafe to sign. The odds for the remaining message to be unsafe being $1/N$, Bob can blindly sign it with confidence that the risk is minimal. He then signs each message of the set and returns the whole to Alice. See more below on the application of blind signatures to electronic voting and digital cash.

In certain business situations, it may be required that the two parties have to sign a contract *simultaneously*. Apart from the classical signing ritual in Trent's office (or using him as a trusted arbitrator to collect and validate both signatures), it is possible to use specific digital signature algorithms for this purpose. In the "face-to-face" algorithm, Alice and Bob might pass the contract to each other and affix their digital signature a number of times sufficient to prevent repudiation (after 10 exchanges of the sort with Alice, Bob would have a hard time convincing a court that he was not finished and still thinking about it). A more complex approach uses an agreed time limit, and successive negotiations on the probability p each one feels reasonable to be bound by the contract. Say that Alice agrees not to be bound by Bob's signature over 10% probability ($p_A = p_B - 0.1$). With the passing of time, Alice and Bob negotiate by certified e-mail their numbers p_A, p_B towards a common target value $p = 1$, which should normally seal the deal. In case the date is passed with $p < 1$, a judge picks up a number p' at random, which binds both Alice and Bob. If $p' < p$, the two parties are not bound. Simultaneous signing can be done through cryptography as well. Here is a possible algorithm with a simplified description. Both Alice and Bob prepare a set of $2N$ encrypted documents, each one representing half of their full signature and validation, and send these documents to each other. They also give to each other the keys they used for one half of them. Then each side decrypts whatever halves the keys allow, which prove to each other that the procedure is in good standing and no one cheated. Then Bob and Alice proceed to send to each other, one at a time, the remainder of the keys on a bit by bit basis, until the keys are fully recovered. The contract is then considered signed. This protocol is not immune against brute-force attack from either side with a view to completing the deal at the fastest convenience, especially when the number of remaining key bits decreases to some good computational level. Another weakness of this protocol is that any party may stop the process, or gain time to complete the attack by sending phony key bits. But in court this may show evidence of mischief and invalidate the contract.

The bizarre protocol of simultaneous contract-signing can be extended to two applications: *certified e-mail* and *simultaneous secret exchange*. In certified e-mail, the destinee Bob should acknowledge receipt before being acquainted with Alice's message contents, like in any certified post mail. The idea for Alice is to obtain a nonrepudiation guarantee from the message destinee. The previous protocol, where keys are communicated by halves from both sides to initiate the procedure would seem appropriate. But Alice could mischievously send phony keys to Bob while honest Bob sent good keys, which would provide Alice a phony certificate on a message that Bob could not even read! We shall not detail here the cryptographic solution to this problem, which takes no less than 14 steps! Suffice it to say that the approach is similar to simultaneous signing. Bob will only acknowledge receipt of the message by crypted halves, and so will Alice for building up her proof of receipt. Likewise, the process would bring evidence of any cheating from either side. Simultaneous exchange of secrets calls to mind those famous WWII spy exchanges on a certain bridge between West and East Berlin. The two spies to be exchanged walk towards each other, alone on the bridge and meet at the mid-point under the careful watch of both sides. In a more happy situation, imagine that Alice and Bob are *chefs* in 5-star restaurants and use secret cooking recipes, or (if this does not convey the point well enough), they are owners of two competing pharmaceutical companies with unlicensed/ secret molecule synthesis processes. They agree on a deal to exchange their secrets for mutual business benefit, but the problem is who should speak first and what guarantee there is that the other will play the game. Again, the answer is cryptography! We shall skip the details here only providing the spirit of the exchange. It is exactly like certified e-mail, except that the protocol carries over two messages: Alice and Bob complete their key exchange according to similar partial acknowledgment steps, the secrets being revealed to both sides only at the end of the process. If someone has cheated, that will appear and the court will invalidate. So far so good for safety. But we may ask the following "What is the definition of a secret?," and "How to evaluate the value of a secret with respect to another?." In the case of spies, it is clear that either party puts a high value on retrieving them and what they may know or represent. But when it comes to cooking, what if Alice forgot one ingredient? Or a key step in the synthesis of the molecule? And what if Bob's secrets are phony, or are not his? The second question is beyond the issue. It is assumed that Alice really wants *that* recipe or *that* molecule, not anything else, and is willing to trade her own secret of equal perceived value. We may imagine that the final cross-licensing agreement between Alice and Bob will be endorsed on the basis of due verification. The point here is not how the secrets may impact each other's life but how cryptography comes in handy as a process for simultaneous secret exchange.

We leave the petty domain of two-party protocols to the larger picture of global society agreements. Imagine a government that could renew itself (or hand over to the next team) through *electronic elections*. Imagine this government also making important decisions based on *electronic polling* (also called a *referendum*). Here, "polling" does not refer to its usual meaning which consists in asking a limited set of selected/representative people what they think about the top ten winners or the top ten issues. The rumor has it that media-related "polls" are usually wrong for a variety of scientific and non-scientific causes, which we will not elaborate. By polling, we mean here asking *10–100 million* eligible citizens for a yes/no decision on important and serious matters, which is the true referendum concept. Examples range from consulting a province for wishes on extended autonomy or independence, to adopting a common currency system or new constitution, or abolishing capital punishment. Elections and referendum/polling are closely related processes, even if the outcome is radically different. In ordinary circumstances, a government would not resort to polling unless in a situation of crisis, or as inherently constitutional. Parliaments can do a decision-making job, and even use electronic voting, which is secured by the physical presence and authentication of the representatives. The situation is technically different when considering votes from large public groups, society at large and nations.

In public elections, the systems in place oblige people to be physically present for their vote to be endorsed. This requirement makes it difficult to vote twice or on behalf of nonexistent persons. In states having a single official party, this obligation is in fact coercive since individual abstentions, which are systematically recorded, are considered as a negative vote. Note that single-candidate elections with 99.9% participation are either not new or not over with in history. Even with the use of "secret voting booths," dishonest governments may freely modify the ballot counts in their favor. Citizens must have a reasonable trust in their voting system, which is fortunately the case in most advanced countries. Here, we shall not address the issue of what real face value a democratic "vote" has, when the candidates offer identical programs or promises. At the extreme of the democratic system, no matter the number of rounds and stages involved in the voting protocol, national votes could be seen as equivalent to a massive coin-flipping experiment! This is because the two remaining contenders generally have equivalent value and potential. Then, the theoretical outcome of the consultation is invariably a 50.1% majority winning over a 49.9% minority. Half of the people feel that they won a great victory over the adversary, while the other half has the feeling of failure and unfairness, even if the two sides had practically equal chances to win. Here, the matter is not to analyze such a paradox, but to determine a universal voting system which could be protected against any type of fraud by individuals or manipulation by the state, and could at the same time guarantee secrecy and confidentiality.

What if national votes or pollings were all-electronic, or partially electronic as an option (such as, tax e-filing), through an *e-voting* protocol? Let us first recall three important requirements in all voting procedures:

1. *Uniqueness and equitability*: No individual vote can be multiplied, discarded, or altered;
2. *Verifiability and accountability:* Proof of compliance and fault-free implementation, by distributed monitoring and verification from regional to state levels by independent observers;
3. *Confidentiality and security*: Neither state nor any entity is able to know of an individual's vote.

Less known or not immediately intuitive are three other key requirements:

1. *Impossibility of partial tabulation*: No possibility of access to intermediate tabulation results, and use of that information to influence the ongoing poll;
2. *Proof of vote participation*: A voter *must* be able to prove that he has voted;
3. *Impossibility for a voter to prove his choice*: Protects the individual against the possibility that a third party would require a proof of his vote choice.

The first three requirements constitute the golden rule of any voting system. The last point, which is often overlooked, is of equal importance. If all these prerequisites could be met, e-voting/e-polling could become an international standard in the future. Here again, cryptography comes as the enabling solution. As with any cryptosystem, the vote/polling algorithm should prove effective against attacks. The resulting cryptosystem should rise to an adequate level of public trust and recognition, as DES or SSL have. As expected, there is no limit to the number of possible algorithms. Let us consider some possibilities:

> The simplest solution is that on the polling day, individual voters/pollers enter their choice in a web site managed by a central "tabulating Authority." The vote, which consists in clicking a button, takes the form of an encrypted response using the Authority's public key. A problem is that this system does not prevent the possibility of voting several times, through phony identifications or multiple IP addresses.
>
> A second solution is that the vote is first signed with the voter's private key, then encrypted with the Authority's public key. The drawback is that the Authority will know for whom the individual voted.

A variant of the previous solution is that each individual first obtains a unique certificate from the Authority. The system does not keep track of what certificate number was given to the individual, but only of the fact that *a* certificate, and not more than one, has been issued to this individual. The e-vote is endorsed upon using this certificate, which the Authority recognizes as valid. The drawback of the approach is that the algorithm enabling the generation of certificates may leak, and some individuals may vote multiple times under different IDs, including valid ones (some form of "theft of vote").

The general problem is to find an e-voting protocol whereby the Authority can validate and endorse the vote through some individual certification and authentication, but in a way that dissociates the vote from the individual. A solution to this problem is the use of *blind signatures* (see earlier description). Here is the process description:

- The voter receives from the Authority a set of N "ballot groups," each of these N ballot groups has all voting possibilities;
- The voter assigns to each ballot group a large, randomly chosen identification number, then blinds it with a random blinding factor;
- The voter returns the N ballot groups to the Authority, along with all the blinding factors, and his/her identification;
- The Authority checks from a database that the voter has not made the same request previously, and is a legitimate/registered voter; if it is a first and valid request, the Authority records the voter's name in the database;
- The Authority opens all ballot groups, but one, in order to verify that the $N - 1$ are in good standing; it signs the only one it did not open, and returns it to the voter;
- The voter selects his/her one preferred option from the signed ballot group, encrypts it with the Authority's public key and anonymously submits it to the Authority through its Web site;
- The Authority decrypts the ballot with its private key, checks the ballot signature, records the identification number and verifies that it has not been previously used;
- The Authority publishes the list of identification numbers with the matching votes, proving to all voters that they have been properly tabulated.

It is easy to check from the above that the Authority has no way of knowing who is the voter. It can only validate and endorse the vote. It also prevents the possibility of voting more than once with the same ballot, or requesting a ballot signature more than once, or making the request under a different personal identification. This last point requires some secure means of personal identification. It could be providing the name, birth date, tax-payer ID, social-security number and the like. A simple system would be that the state Authority assigns and sends an identification number to all legitimate/registered voters. While confidentiality and safe vote submission/endorsement is now ensured, the remaining issues of security concern the tabulation code. The tabulation code could contain a hidden loop with the instruction "IF percentage of votes p_x for candidate X is less than 50.1% THEN $p_x = p_x + \delta$ and $p_y = p_y - \delta$," or something less trivial. Alternatively, the code could reject ballots past the point when the number p_x has reached a majority threshold. Or it could make use of valid ballots from abstaining voters to increase p_x. The tabulation code could however be submitted to an International Authority (IA) for validation. The executable master would then be timestamped and encrypted with the IA private key. Timestamping would prevent the IA from producing modified versions if it colluded with a preferred state. Encryption would prevent a state from running the vote without IA mediation to decrypt the executable master. In order to rule out any possibility for state fraud (running a different executable master without IA awareness), the results could be tabulated directly by the IA. Following these different procedure possibilities, all the aforementioned requirements are met, except the last one: the voter keeps indeed opportunity to prove his or her choice. Third parties (such as political parties or public organizations) could then require the individual to show a record of such proofs. An ugly possibility, but not unrealistic. For those unconvinced, suppose that a dictatorship (say an authoritarian regime) takes over a democracy. Then only the individuals that could provide this proof of past and present adherence would be granted access to administration careers, business licenses, employment, education, healthcare, private ownership, exit visas, and other privileges. A complete nightmare scenario, but witnessed multiple times in the past.

There are many other possible and more complex protocols, some working without the intermediacy of a central Authority, or distributed between different independent Authorities. For reference, alternative solutions can be based on *homomorphic protocols*. The property is simply defined as $E_K(M + M') = E_K(M) + E_K(M')$, where K is a given public key and M, M' are different messages. Put simply, the approach makes it possible to compute the outcome of the consultation with all authentication guarantees, but without decrypting any of the voters' identities. The drawback is that it is limited to YES/NO votes/polls, unlike in the previous approach.

These situations just convey a flavor of what electronic consultations could be about. The principle is not so far off in the future, since for instance, it is already being studied by the European Union under the project name E-POLL. Note that E-POLL is an "off-line" system where participants must use dedicated voting stations or "kiosks." However they can use them from any location in the EU they may happen to be in on the polling day. See the following conclusion on electronic voting versus electronic cash.

A second futuristic application of cryptography concerns *digital cash*, or more accurately, *anonymous* digital cash. With the use of today's creditcard and checking system, life is not really private. Anything one purchases, or any service one uses through credit card or check payment leaves an "audit trail." Such a trail betrays what foods, drinks, restaurants, entertainment, or specific items, from books to cars to office supplies, have been purchased where, when, how many times and for how much. This may be a reason for people not to trust the Internet for creditcard purchase, but just maybe for tax filing. The citizen who worries about *Big Brother* may just use the creditcard to get cash, which cannot normally be back-traced. But the problem becomes more serious when Alice and Bob want to do some anonymous, but expensive cash transaction. For instance, Alice is a famous singer and wants to give $1,000,000 to the charitable association presided over by Bob, but she does not want either Bob or the press to know she ever made a donation. Bob must also make sure, before he cashes it, that the money is of "private and clean" origin. Here again, cryptography provides an array of safe solutions, of which we can only convey a flavor.

One of the most basic protocols for (anonymous) digital cash, which is based upon the principle of blind signature, is the following:

- Alice generates a set of money orders for a fixed amount (say here $1,000);
- Alice blinds each set with different blinding factors and puts each into one of N envelopes;
- Alice forwards the N envelopes to her bank;
- The bank asks Alice for $N - 1$ blinding factors to verify that the orders are in good standing (e.g., all orders concern $1,000 transactions, not a single cent more);
- Upon verification, the Bank blindly signs the remaining envelope and sends it to Alice; no serial number or other information is attached, other than Alice having been issued a signature; the bank also withdraws $1,000 from Alice's account;
- Alice then uses the signed money order contained in that unique envelope to send it to Bob as an anonymous $1,000 money order.

When the bank receives from Bob's the $1,000 order, Bob is credited by this amount. Neither the bank nor Bob will ever know the origin of the order. All that they both know is that it is coming from one of the bank's clients and that this (anonymous) client has already pre-paid the amount. Alice may try to cheat by making one of the N envelopes a $1,000,000 money order. The odds for the bank *not* to find it are $1/N$, say 1%, for instance. Two things can happen. Case 1: Alice is not this "lucky" (which is the most likely occurrence). The bank either gives Alice a "last-chance" warning before prosecution (Alice is one of the best customers and a major stockholder); or maliciously, the bank signs the spotted $1,000,000 order. Alice does not see the plot and does the transaction. When Bob shows up at the bank, he is held responsible for nothing, but the transaction is denied. Alice is then prosecuted and sent for a while to a remote place where she can learn more on cryptography matters. In any case, Alice would not take such a risk. Case 2: Alice is lucky. Either the bank did not spot the $1,000,000 money order in the $N - 1$ random selection, or the bank employee was coerced into signing it anyway (it being understood that he gets 50%). Since some of the most fortunate customers also use $1,000,000 orders, the bank cannot see any problem in Bob's money order and the bank approves the transaction. When it comes to book-keeping, the bank discovers that there is one money order too many for this specific amount. Different possibilities exist from this point. Ideally, the bank may use a cryptosystem where any signature has a timestamp, and an employee stamp and the detailed outcome of the $N-1$ verifications. But such a verification procedure should only be activated under the circumstances, and for the time period considered. Then it is easy to figure out who cheated: Alice alone, or Alice and the employee? The timestamp should cover a period wide enough to allow for many possible signatures, while the information of timestamp and employee identification should be sufficient to bring the evidence of who may be at the origin of the fraud.

Another requirement for digital cash (whether open or anonymous) is that it should be used only once. Nothing prevents Alice from making multiple transactions with the same order, or Bob and his friends from showing up in different branches of the bank to cash it, the same day or at different times in the future. For open digital cash, the obvious solution is timestamping. If Bob shows up a second time, the bank will recognize that this order was already cashed. The bank may just deny the transaction, or pretend to cash the order and let the FBI do the rest with Bob. For anonymous digital cash, the solution is to attach to each money-order signature a unique hash which does not betray the client's identity. Upon cashing the order, this unique hash is put into the bank records. If Bob shows up with the same money order and associated hash (like a photocopy of the original), the same bad things as in the open system will happen to Bob. Bob may however be of good faith, having just received one of these anonymous donations for his organization! Other protocols exists in which Bob may require the "donor" to provide a one-time identity through a random-number string. Bob can safely cash the order, and the bank records the string to prevent future cashing possibilities with the same order. More complex and safer algorithms are possible. This brings us to the point of a "perfect crime" situation, where a criminal organization could perform anonymous blackmailing or racketeering on private people, associations or states. It could also use the system as a perfect money-laundering machine. The safeguard of anonymous digital cash is that the recipients or "beneficiaries" of any transaction are identified anyway. A person may deny knowing the origin of a $1,000,000 deposit in his account, but cannot hide having approved the transaction and utilized the funds for some purpose. Some day, the law may declare it illegal for private people to accept anonymous transactions above a certain level, or require the banking system to allow some form of source authentication for any audit purposes.

This section on current and futuristic applications of cryptography has illustrated the vastness and complexity of the subject, especially when it comes to the border of individual rights and privacy protection issues. It would be naive to trust that rights and privacy would be better protected if under the full control of computing systems. The two examples of electronic voting and digital cash point to future controversies and new forms of potential abuse. A perfect electronic voting system could be hacked or attacked, causing the nullity of any elections, and putting democracy at an unprecedented risk. Governments might be tempted to overuse their e-polling system in order to consolidate their legitimacy and guide their daily decisions, but to the point of demobilizing citizens in their use of voting rights. Such a trend is already observed in the mistrust of opinion polls during election campaigns. A perfect digital cash system would have as many advantages for practical and convenient use as drawbacks for potential misuse or abuse. The matter is where the control should end in order to fully protect privacy, and where it should begin in order to fully protect society. The same paradox applies to private communications. Cryptography is a formidable tool which we are just learning how to handle properly. But this apprenticeship may still take another century.

3.4 QUANTUM CRYPTOGRAPHY

This last section, dedicated to *quantum cryptography*, comes as a conclusion of this chapter and also this book. At this stage we must have clearer ideas of what cryptography represents for telecommunications in general and for integrity/privacy/authentication in particular. We have seen how powerful standard and nonstandard encryption algorithms can be and how they are already in use on the Internet and wireless access. What ever difference could quantum cryptography make to this picture? The answer to this question is not unique. If quantum cryptography is a constituent of scientific knowledge, and represents a track towards future discoveries and breakthroughs, then it makes a real difference. If it is a complicated but most elegant way to exchange secret keys, it may not be so interesting. A difference that the quantum world introduces to this subject is the concept of *absolute security*. This concept is as absolute as the impossibility of splitting a photon, the elementary quantum of light (see Chapter 2). The reader may view the subject as an important conceptual complement to the field of cryptography, rather than a mere curiosity of physics or a solution looking for a problem (this is how the laser was initially regarded). We also note that commercial products based on its principles have already been released, although the potential for widespread use is still unknown. Here, we will describe it as briefly and practically as possible, while avoiding any academic complexity and formalism.

3.4.1 Photons, Polarization States and Measurements

As previously stated, the main application of quantum cryptography is absolutely secure key exchange. The main implementation protocol, known as "BB84" after a 1984 proposition from C. Bennett and G. Brassard of IBM, consists in using the property that *photons*, the elementary particles of light, can exist in two polarization states. More precisely, photons may be found in four possible polarization states, which, as described in Chapter 2 (Section 2.2), are called *linear* (vertical and horizontal) and *circular* (left and right). Polarization is a property of electromagnetic waves. The case of linear polarization is easy to conceptualize. It is simply defined as the plane within which the electric field of the wave oscillates. With respect to a reference plane, this oscillation plane can be oriented at a $0°$ angle (horizontal polarization) or at a $90°$ angle (vertical polarization). For this reason, these two possible polarization states are said to be orthogonal. If we combine two orthogonally polarized E-field waves being either in phase or out of phase, the polarization of the combined E-field is oriented at $\pm 45°$. It is still a linear polarization. If we redo this experiment with the two waves being in quadrature (phase difference of $\pm \pi/2$), the resulting E-field rotates and is said to be circularly polarized. For the reader who is familiar with trigonometry, the instant coordinates of a circularly polarized E-field are the $\cos \omega t$ and $\sin \omega t$ projections in the trigonometric circle. Each of these projections oscillates in orthogonal directions (ox, oy) and their combination yields a vector which rotates at a angular rate ωt radians per seconds. The rotation direction, clockwise (right circular) or counter-clockwise (left circular), is determined by the sign of the phase difference, $\pm \pi/2$. Thus, a circular polarization is the sum of two orthogonal, linearly polarized waves which oscillate in quadrature ($\pm \sin \omega t = \cos(\omega t \mp \pi/2)$). Conversely, one may conceive of a linear polarization as being the superposition of a right-circular and a left-circular wave. The phase difference between the two determines the polarization direction of the E-field.

Polarization is thus a classical concept associated with classical electromagnetic waves. As we have seen (Section 2.2), actual E-field waves are quantized, meaning that they carry these elementary grains of light called photons. If the number of photons, proportional to the wave's power, E^2, is large, the E-field closely resembles the classical, sinusoidal wave. As the number of photons decreases, the wave becomes more "noisy," which is illustrated in Figure 2.8. Where does this noise come from? The answer is that it reflects the uncertainty associated with the measurement of the E-field. If we could measure the number of photons, $n \propto E^2$, and the associated E-field phase, φ, with equal accuracy, we would obtain a smooth waveform, $\sqrt{n} \sin(\omega t + \varphi)$. But a principle of quantum mechanics, referred to as *Heisenberg's uncertainty principle*, forbids any such possibility. One of the formulations of this principle is $\delta n \times \delta \varphi \geq 1$, where δx means the standard deviation or the uncertainty of the measured quantity x. We see that if we manage to make a very good measurement of the photon number, i.e., obtaining $\delta n \to 0$, we will be left in complete uncertainty about the phase, i.e., $\delta \varphi \geq 1/\delta n \to \infty$. The

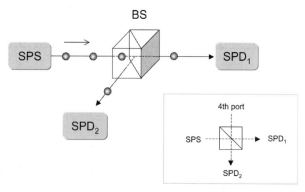

FIGURE 3.17 Experimental apparatus to measure single photons detect single photons passing through a 50/50 beam-splitter. The inset shows a top view with the 4th port of the beam-splitter (SPS = single-photon source, SPD = single-photon detector, BS = polarization beam-splitter).

reverse is true: if we get a good measurement of the phase, the photon number is highly uncertain. Hence the noisy waveform obtained at low photon numbers, as shown in Figure 2.8.

Now that we are familiarized with polarization and photons, we can approach the issue of measurement with a different and nonclassical view. Consider the apparatus shown in Figure 3.17. It includes a single-photon source (SPS), a 50/50 beam-splitter and two single-photon detectors (SPD). Both SPS and SPD do exist in the real world, although we will not elaborate on how they work. Suffice it to know that they can be built. This apparatus splits the incoming light beam into two paths, corresponding to the single-photon detectors labeled "1" or "2." Assume first that the SPS is intense, i.e., the number of photons emitted by the source per unit time is high (e.g., 10^3). According to all expectations, the beam is split into two beams of equal power and both detectors measure $n/2$ photons per second. Next, we decrease the power of the SPS to the point that only one photon is emitted about every second. This leaves us the time to look at what is going to happen. We have also attached to both detectors electronic circuits which makes a sound each time a photon is detected. Detector "1" makes a "ping," and detector "2" makes a "pong," so we can tell which one is hit by a photon and when. As we switch on the electronic circuit, what we hear is quite surprising. Something like (read out loud for a nice effect!):

ping/ping/pong/ping/pong/pong/pong/ping/pong...

This music immediately tells us two things: The photons are split into two paths, which was expected; the photons are split at random over these two paths, which was *not* expected. Our faulty classical intuition would suggest that both detectors should make a simultaneous sound, since the beam-splitter equally divides the input power into two outputs. What we have learned is that we can't cut photons into halves (as a matter of fact, into any fraction). When the photon hits the beam-splitter surface, it must make a tough and quick decision: going straight-through or making a right? The outcome of the decision is completely random. The photon chooses either path at random, like flipping a coin. This explains why we can't hear the sound of two SPD at the same time. It also explains why we hear several pings or pongs in series, with irregular and unpredictable patterns. Over a sufficiently long time, however, we hear as many pings as pongs, that is to say that the two SPDs receive equal average powers. The average SPD powers are nearly equal over one minute (about 60 counts) and perfectly identical over one hour (about 3,600 counts). Then we turn off the loudspeakers, because this music may become annoying, and we try to explain what is the process that makes a single photon "choose" one path or the other.

The only physical explanation to this failed photon-splitting experiment, which has been called *Bernoulli's random partitioning*, is an effect of E-field interference. Some parasitic E-field must be there to mix with the signal at the beam-splitter interface and determine where the E-field wave will be steered to with its unique photon. Looking back at Figure 3.17, we observe that the beam splitter has an "unused" fourth port. Any parasitic E-field wave entering from there would mix and interfere with the signal. This parasitic E-field, which we described in Chapter 2, is a feature of quantum mechanics, and is called *vacuum fluctuations*, or *vacuum noise*. As the name indicates, vacuum, or the absence of any matter, atoms, particle, photon, or radiation of any kind in purely empty space, is not quiet. The nonintuitive characteristic of vacuum is that it contains its own E-field waves in all possible frequencies, and propagation directions, with random polarization and phase. Its infinite energy cannot be extracted or used, however, because it does not have photons to carry it, or more accurately, only the *equivalent* of one-half of a photon. The fourth port of the beam-splitter is thus capturing vacuum noise, exactly like wind or street noise comes in through a window left open. Then what we have observed about random photon partitioning begins to make sense. The input signal E-field mixes and interferes with vacuum noise. What happens at this meeting point is not easy to describe with classical arguments. Because it is a single-photon event, the phase of the signal E-field is also absolutely uncertain! Then the random mixture between the signal and the vacuum field determines a best path where the E-field happens to be maximized, due to coincidental in-phase combinations. Quantum mechanics has it as a postulate that the photon, as a quantum of energy, cannot be divided. So we must accept that the choice of the photon direction is not completely explained by the classical argument of a constructive or destructive interference effect. The photon has to go somewhere, and vacuum noise makes the photon pick one of the two path possibilities. The important lesson learnt is that photons cannot be split. If we place a 50/50 beam-splitter in the photon path, the probability that it will choose one direction over the other is 50%, but we cannot predict the outcome of any single event. It is exactly like the outcome of flipping a coin. This is an important property we can use for a cryptographic communication channel.

Assume Alice and Bob install a unidirectional fiber-optic line. Alice has the source and Bob the receiver. They have agreed on a binary code where one photon is a "1" bit and zero photon is a "0" bit, at a rate of one bit per second. But malicious Eve manages to tap the line by inserting a 10%/90% beam-splitter between two fiber connectors. Eve is a very skilled electrical engineer, but never studied quantum mechanics, and this is the reason why she is going to be disappointed. Indeed, Eve gets 10% of the photons passing through the fiber. Each time she gets a photon, that's definitely one of Alice's "1" bits. But she misses 90% of the other "1" bits sent by Alice. When Eve's detector gets nothing, she can't tell if this corresponds to a real "0" bit or a missed "1" bit. Clearly, Eve cannot make any sense of all this information! There is no possibility for eavesdropping a transmission line with single-photon signals, another property of quantum mechanics. But something even more serious happened. Indeed, suspecting that Eve may be trying to do this, even to no avail, Alice agrees with Bob to do an experiment. On a given day, Alice will be sending a long bit-sequence message, and will record it as well on a piece of paper. Bob will record the result of what he received the same way. Then Alice and Bob will meet somewhere and compare their papers. What they immediately see that *all* the "1" bits received by Bob are correct. But on average, 10% of the original "1" bits are missing. This also means that 10% of the "0" bits received by Bob were "1" bits actually sent by Alice. This can have only one explanation: Eve has placed a 10/90 beam-splitter in the line! Had Eve chosen a $N/(1-N)$ beam-splitter, the error-rate or proportion would be $N\%$. A realistic fiber transmission channel is never without some error-rate, because of the small absorption loss or connector loss. But this error-rate may be as low as 10^{-9}, for instance, so Bob would expect to get one error out of 10^9 bits. But a 10% or even 1% error-rate definitely has Eve's signature on it. Knowing that their fiber has been "wired," a big mistake for Alice and Bob would be to lower the bit error-rate by having Alice repeat her message transmissions. By comparing successive transmission results, Bob could locate where the missing "1" bits are randomly falling and reconstruct the correct information with good accuracy. But then Eve will do the same, although not as fast and as accurately since Bob gets 90% of the "1" bits. Another way to detect that the line could have been wired by Eve is for Bob to use the two detectors SPD_1 and SPD_2. Each time slot must then give one photon count, regardless of the photon polarization and of Bob's mode of measurement (linear or circular). The event of zero count is however possible, due to the effect of the fiber loss (see the following discussion). Since the fiber loss is known and stable, any loss increase caused by Eve's tapping photons somewhere from the line will immediately be noticed.

Being now more familiar with some of these quantum world mysteries, we are in a better position to address the issue of measuring photons with *polarization* as a degree of freedom. Figure 3.18 shows a measurement apparatus similar to the previous one, except that the 50/50 beam-splitter is changed into a *polarization* beam-splitter (PBS). Additionally, we have introduced a polarization controller (PC) in front of the source. This controller makes it possible to generate any state of polarization (SOP), from linear at any angle incidence to circular left or right. As shown in the figure, the principle of this PBS is to transmit 100% signals that are vertically polarized (referred to as TE for *transverse-electric* E-field) and to reflect 100% signals that are horizontally polarized (referred to as TM, for *transverse magnetic*). To verify that the experimental setup works, we can adjust the PC so that the photons incident on the PBS are in the TE mode. The result is that 100% of the photons are being detected in the straight-though path (SPD_1) and

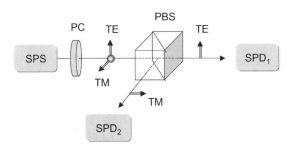

FIGURE 3.18 Experimental apparatus to measure single photons in linearly polarized states. The double arrows indicate TE and TM E-field components (SPS = single-photon source, SPD = single-photon detector, PC = polarization controller, TE = transverse-electric E-field, TM = transverse-magnetic E-field, and PBS = polarization beam-splitter).

0% in the reflected path (SPD$_2$). Readjusting the PC to set the photon in the TM mode gives the opposite result. Next we install a control system which makes the PC randomly switch from TE to TM modes, or the reverse, every second. The system records the PC configuration, but we don't have access to the information. Then we remove the second detector, SPD$_2$. The only information we now have is that photons are emitted in one of two linear polarization states (TE or TM), and at a rate of one photon per second. What we observe is that our straight-through detector makes pings at irregular intervals! That is:

$$ping/\text{-}/\text{-}/\text{-}/ping/\text{-}/ping/\text{-}/\text{-}/ping/ping/\text{-}/\ldots$$

Clearly, each ping represents the detection of a TE photon. The empty intervals must then correspond to TM photons which have taken the reflected path. We conclude with certainty that for this sequence, the photons were emitted according to the pattern:

TE-TM-TM-TM-TE-TM-TE-TM-TM-TE-TE-TM

The lesson learned is that if we use a PBS to measure linearly polarized photons, we know for sure their initial polarization state. A positive measurement (ping) tells us that the photon polarization is aligned with the PBS, and negative measurement (no ping) tells us the opposite.

Consider next photons with circular polarization. As we have seen, a circularly polarized wave is equivalent to the combination of two waves with orthogonal linear polarization (TE/TM) and a phase difference of one quadrature ($\pi/2$). What could then be the photon polarization if there are two orthogonal E-field components? The answer is that the photon has both polarizations with the same 50% probability. If we put a PBS in the photon path, we have equal chances to measure TE or TM. In this case, quantum mechanics says that the photon is in a *superposition of states*. To make a comparison, assume we put a coin inside a box, and shake the box to the point we have no idea on which side the coin is now. If we open the box, we can see that it is either "heads" or "tails." As long as we don't open the box, it can be either. The coin is in a state superposition of "heads" and "tails"! It is only when we make the measurement (open the box) that the coin "chooses" the state it is in. We may argue that it is not the fact that we open the box that flips the coin, since the coin must already be in one position or the other. But since we don't know which one, what is the difference? It can be either, and as long as we don't open the box, the coin is in a superposition of the two possible states. The same is true for the circularly polarized photon. If we try to measure its polarization with a PBS, we get either TE or TM. The photon is initially in a superposition of TE and TM states. Measuring it with a PBS is like opening the box: the output state is one or the other with the same 50% odds. If we pass a stream of photons with random circular polarizations (i.e., randomly flipping between left or right) through a PBS, we obtain the same result as with a 50/50 splitter (the ping-pong music of Bernoulli's random partitioning). On average, 50% of the photons will go to SPD$_1$ and the other 50% to SPD$_2$. If we don't know *a priori* what is the polarization type of the input photons (circular or linear) there is nothing we

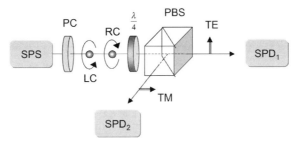

FIGURE 3.19 Experimental apparatus to measure single photons in circularly polarized states. Note the insertion of a quarter-wave plate ($\lambda/4$) to convert the right/left circular polarizations into TE/TM linear polarizations, respectively. See Figure 3.18 for key.

can learn from this measurement. The 50/50 partitioning could be generated by any random combinations of purely circular or purely linear states, or mix of both with equal proportions.

If we know that a photon is circularly polarized, can we then determine whether it is a right or a left? The answer is provided by the measurement apparatus shown in Figure 3.19. It is the same as before, but we have introduced a *quarter-wave plate* in front of the PBS. A quarter-wave plate is a slice of birefringent material. In this material, light travels faster if linearly polarized parallel to some privileged axis direction, and slower if polarized orthogonal to that axis direction. When the input is circularly polarized, its two linearly polarized E-field components travel through the plate at different speeds. The result is that their phase difference increases (or decreases, whatever). The thickness of the quarter-wave plate is such that the accumulated phase difference φ' is exactly one quadrature ($\varphi' = \pi/2$). Thus the resulting phase difference is $\delta\varphi = \pm \pi/2 + \varphi' = 0$ or π. The sign depends upon whether the input signal is right-circular or left-circular. This shows that upon passing through a quarter-wave plate, a circularly polarized signal is transformed into a linearly polarized signal. The linear state is exactly TE or TM, depending upon the input to be exactly right-circular or left-circular. The one-to-one correspondence depends upon the choice of orientation of the plate's birefringent axis (i.e., 45° to the right or to the left with respect to the vertical direction). We conclude from this that the apparatus shown in Figure 3.19 makes it possible to measure the polarization state of circularly polarized photons with no ambiguity. In the configuration chosen here, right-circular photons follow the straight-through path towards SPD_1, while left-circular photons follow the reflected path towards SPD_2. A ping from SPD_1 tells us that the incident photon was a right-circular (RC), and a pong from SPD_2 tells us the opposite (LC). The music ping/-/-/-/ping/-/ping/-/-/ping/ping/-/... now has the meaning **RC**-LC-LC-LC-**RC**-LC-**RC**-LC-LC-**RC**-**RC**-LC. If we were to use linearly polarized input photons, we would get the same result as in the earlier apparatus with circularly polarized photons, namely a random partitioning of the photon stream with 50/50 probabilities.

TABLE 3.30 Probabilities of single-photon count associated with four possible input polarization states

Polarization/Apparatus	$\updownarrow$	$\leftrightarrow$	$\cap$	$\cup$
linear measurement	100/0	0/100	50/50	50/50
circular measurement	50/50	50/50	100/0	0/100

($\updownarrow$ = linear vertical or TE, $\leftrightarrow$ = linear horizontal or TM, $\cap$ = circular right, $\cup$ = circular left) and the two possible measurement apparatus corresponding to Figure 3.18 (linear-polarization measurement) and Figure 3.19 (circular-polarization measurement). The data X/Y correspond to the probabilities of detector SPD_1 (X) and SPD_2 (Y).

This preceding description has highlighted two important properties:

Photons can exist in four different polarization states;

There are two possible apparatus to measure photon polarization: The first unambiguously measures linear-polarization states but does not tell anything about circular-polarization states, while the second does the opposite.

These two statements are summarized in the data shown in Table 3.30. The table provides the probabilities associated with each of the four possible polarization states when either measurement apparatus is used. We can call "+" the linear-polarization measurement apparatus and "O" the circular-polarization measurement apparatus. We see from comparing the two Figures 3.18 and 3.19 that the difference between the two is only the presence (O) or absence (+) of the quarter wave plate. Thus, if we can manage to rapidly insert or remove the plate between two successive photon incidences, we can switch from O to + or from + to O.

We are now ready for quantum cryptography! As we shall see next, quantum cryptography is not normally used to transmit crypted messages, as the name would suggest. Transmitting crypted messages can be done through the Internet, or by post mail, it is as safe as the encryption algorithm may be. Instead, the quantum communications channel is primarily used to distribute or exchange keys. It could be a onetime key, a session key, a confidential public key or a permanent private/secret key, this does not matter. The point is to use the quantum properties of polarized photons to safely distribute or exchange any of these keys.

3.4.2 Quantum Key Distribution (QKD)

Alice and Bob have pulled a long fiber cable between their two garages. The distance cannot be too long because the fiber loss would cause single photons to be randomly lost on the path, resulting in high bit-error-rates. Say that distances of 25–50 km, which represents a loss of 5–10 dB (30–10% transmission) is about a maximum. See the following text on whether optical amplifiers or electronic repeaters can be used to extend this distance. The cable has several fiber pairs which Alice and Bob can freely use to distribute keys to each other: Alice can send a key to Bob and Bob can send a different key to Alice. However, they don't need this

3.4 Quantum Cryptography

sophisticated communication channel for other purposes, like communicating encrypted messages, or even discussing setting up the system, as we shall see. The fiber is only used as a *quantum-key distribution* (QKD) channel.

In their garages, Alice and Bob have installed the same photon-measurement apparatus as described in the previous subsection. Their apparatus also includes the switch that inserts or removes the plate at the same rate as the photons are coming in. Alice and Bob also have single-photon sources with a dynamic polarization controller able to switch, on a photon-by-photon basis, the photon's polarization state into any of the four possibilities ($\updownarrow$, $\leftrightarrow$, $\cap$, $\cup$), the last two symbols standing for RC and LC, respectively. Today, Alice called Bob over the phone to tell him that she is going to send a key through their quantum channel. No problem. Bob goes immediately to his garage. Their agreed protocol is that Alice must send first a group of photons in 16 bits. Bob then programs his receiving apparatus according to the randomly chosen sequence (shown here by groups of four, for clarity):

$$+\bigcirc+\bigcirc \quad \bigcirc+++ \quad \bigcirc+\bigcirc\bigcirc \quad \bigcirc++\bigcirc$$

This choice means that the first photon/bit will be measured with the linear operation mode (+), the second with the circular operation mode (**O**) and so on, the two modes being used with equal probability. Bob is all set, and Alice proceeds to send a randomly chosen message which takes the form:

$$\updownarrow\leftrightarrow\cap\cup \quad \leftrightarrow\leftrightarrow\cap\updownarrow \quad \updownarrow\cup\cap\leftrightarrow \quad \cup\cup\cap\updownarrow$$

The above information and results of Bob's measurements are listed in Table 3.31. We observe that in all outcomes, Bob counted a single photon. If the set up is in the linear measurement mode (+), a count on SPD_1 means the result ($\updownarrow$) and a count on SPD_2 means the result ($\leftrightarrow$). Likewise, in the circular measurement mode (O), a count on SPD_1 means the result ($\cap$) and a count on SPD_2 means the result ($\cup$). Bob's result turns out to be:

$$\updownarrow\cup\updownarrow\cup \quad \cup\leftrightarrow\leftrightarrow\updownarrow \quad \cap\leftrightarrow\cap\cap \quad \cup\updownarrow\leftrightarrow\cap$$

Bob then calls Alice to tell her what (+,O) mode sequence he used (namely +O+OO+++O+OOO++O). Each time Bob used the same measuring mode

TABLE 3.31 Result of Bob's measurements from the receiving end of the quantum cryptography channel, following a 16-bit message sent by Alice and Bob's specific measuring mode

Alice message bit value	1 0 1 0	0 0 1 1	1 0 1 0	0 0 1 1
Alice choice of polarization states	$\updownarrow\leftrightarrow\cap\cup$	$\leftrightarrow\leftrightarrow\cap\updownarrow$	$\updownarrow\cup\cap\leftrightarrow$	$\cup\cup\cap\updownarrow$
Bob's measuring mode	$+\bigcirc+\bigcirc$	$\bigcirc+++$	$\bigcirc+\bigcirc\bigcirc$	$\bigcirc++\bigcirc$
Result of Bob's measurements	$\updownarrow\cup\updownarrow\cup$	$\cup\leftrightarrow\leftrightarrow\updownarrow$	$\cap\leftrightarrow\cap\cap$	$\cup\updownarrow\leftrightarrow\cap$
Time slot validation	**1** × × **0**	× **0** × **1**	× × **1** ×	**0** × × ×

The last line shows in bold the time slots recognized to be valid, with their attributed bit values.

as Alice's transmission mode, Alice tells Bob that corresponding time slot is valid. Alice has no knowledge of Bob's polarization measurement results. She can just tell him which outcomes are correct, and this happens every time they both used the same mode. Eve, who never fails to listen to what Alice and Bob say on a public channel, has absolutely no use for this information. She cannot tell which states of polarization Alice used for each time slot, neither which state of polarization was measured by Bob.

After Bob discarded the wrong measurements, the remaining right ones are:

$$\updownarrow - - \cup \quad - \leftrightarrow - \updownarrow \quad - - \cap - \quad \cup - - -$$

which show that Bob's measurements are right in the six time slots 1, 4, 6, 8, 11 and 13. Bob and Alice now know that they should use these slots for transmitting more bits. Alice just needs to tell Bob what bit value is to be attributed to a given measurement. For instance, "1" bits could be represented by sending/receiving ($\updownarrow$, $\cap$) and "0" bits by sending/receiving ($\leftrightarrow$, $\cup$). With this convention, Bob can then interpret Alice's message according to the bit word:

$$1 - -0 - 0 - 1 - -1 - 0 - -- \equiv 100110$$

which represents the beginning of the key, or the *raw key*.

The whole operation can be repeated as many times as required in order to increase the key size. Alternatively, a very long key can be generated at once, using for instance a 3,000-bit long message for a 1,024-bit key. If Alice wants to use a specific key, she just needs to tell Bob to invalidate certain bits. For instance, if Alice wants the key to start by 011, she tells Bob to drop slots 1, 2, and 5. Obviously, this deletion/selection process requires a message at least twice as long, but this is only a small extra effort considering the potential interest for Alice of choosing a specific key number. On the other hand, the possibility of generating a random key, not initially chosen by Alice, may also be advantageous for security purposes.

In the event that some single photons be lost, due to fiber absorption, Bob's receiving apparatus would record as many anomalous "zero counts." Bob will then mention this to Alice and the time slot will simply be discarded. Note that the exact value of the key is unpredictable (due to the random occurrences of Alice and Bob choosing the same transmission/measurement mode). What matters, is that at the end, both Alice and Bob obtain the same number to use as the encryption/decryption key. Note that they can both utilize this same key for either encryption or decryption purposes, which alleviates the need to duplicate the channel terminals (one transmitter and one receiver being sufficient).

Eve still has the possibility of tapping Alice and Bob's fiber during this QKD process. She can cut the fiber and install a "repeating station." Each photon she receives from Alice is counted and then resent to Bob. The problem is that Eve does not know the transmission mode into which Alice sent the photon. She has 50% chance of choosing the wrong measurement mode. If she sets her apparatus in the linear mode and measures $\updownarrow$, it has a 50% chance to be right (input was indeed $\updownarrow$) and a 50% chance to be wrong (input was $\cap$). Short of any solution,

Eve systematically re-sends what she receives to Bob. In this example, a photon with ↕. But Bob has a 50% chance to be in the matched measurement mode, which represents $0.5 \times 0.5 = 0.25$ or 25% chance for Bob to match Alice's choice altogether. Upon doing their comparison exercise, Alice and Bob will realize that instead of the expected match rate of 50%, the value has significantly dropped, namely to 25%. This oddity would immediately reveal the plot. Alternatively, Eve could only tap a fraction of the photons, for instance using a 10/90 beam-splitter. But as previously discussed, this would give Eve only 10% of the message information, and make the rate of missing bits increase to a high level. In summary, Alice and Bob's key-exchange channel can be "repeated" or "tapped" by Eve. But in the first case, the mode matching rate will drop to 25%. In the second case, the missing photon rate will increase to an abnormally high level. In both case, Alice and Bob know that their line is wired.

Eve (who finally took a crash course on quantum mechanics) understood that wiretapping the line would surely betray her presence. She still has one more possibility yet. It consists in cutting the fiber, impersonating Bob and getting the key (K) directly from Alice. Then what she needs to do is to impersonate Alice and send to Bob a different key of her own (K'). For the whole operation to succeed and not to be suspected, Eve must first intercept Alice's crypted message, $E_K^{\text{Alice}}(M)$, do the decryption/re-encryption operation $E_{K'}^{\text{Eve}}(D_K^{\text{Eve}}(E_K^{\text{Alice}}(M)))$ and send the result to Bob. Yet nothing guarantees that Bob will not also get $E_K^{\text{Alice}}(M)$. In this case, his decryption attempt $D_{K'}^{\text{Bob}}(E_K^{\text{Alice}}(M))$ will fail. Then he will immediately suspect Eve's plot, because Alice never sends wrong keys.

Alice and Bob now have a secure channel to exchange their DES or AES keys. They have set up an automated protocol which performs all these operations and stores the key on their hard-disk drives. For times of dense message exchanges, the system can generate one new key every day or every hour, or even within minutes after request, depending on the maximum emission rate of the single-photon source. A sophisticated version of the link may use the other fiber pairs to convey the encrypted messages, which further protects the channel against brute-force attacks (eavesdropping on fiber links being far more difficult than from the Internet or WLANs).

Can the QKD link distance be increased by the use of *optical amplifiers*? The answer is no for conventional laser and scattering amplifiers such as EDFA and Raman (see Chapter 2). This is because these optical amplifiers generate *amplified spontaneous emission* (ASE). In that chapter, it was shown that ASE is equivalent to a minimum of one photon per unit bandwidth and per polarization state. Note that the two ASE photons are orthogonally polarized, but with random phase relation and polarization orientation (at least in absence of other effects such as polarization-dependent gain or loss). Even by using a time-slot measurement scheme, these two ASE photons would irremediably mix with the single-photon signal and blur the intended photon-state information. The answer to the above question is yes, if certain noise-free optical amplifiers are used. These amplifiers have the capability to leave the photon number uncorrupted (high certainty on the photon-number output, corresponding to negligible ASE), but at the expense of high uncertainty in the photon's phase (according to Heisenberg's principle). Here, we shall not discuss the possibility of implementing such amplifiers as "single-photon repeaters" and see whether or not this phase uncertainty may affect the polarization information. Suffice it to say that their implementation is complex and still at advanced-research level.

Then what about *electronic* repeaters? At least one solution is possible, which to the best of the author's knowledge is an original one.

Each repeater would be a "Bob" receiving station and an "Alice" transmitting station, just as Eve tried to do, but now being official. Each repeater has been encapsulated into a heavy-weight, hard-steel container with an internal security system making it impossible to tamper with or listen to. Each repeater just acts like Eve did in her failed attempt to play the self-appointed intermediary. The difference is now that at the end of the process, Alice can interrogate all repeaters and ask them what measurement mode they used for reception (for photon retransmission, repeaters use the same mode as for the measurement). The repeaters send this information (and exclusively this information) to Alice using the same fiber, but at a different wavelength in a "supervisory" channel. Each repeater passes the information coming from the neighboring repeater, so that Alice can interrogate all the chain through this supervisory channel. At the end of this interrogation process, including calling Bob's terminal, Alice has collected the kind of tabulated information shown in Table 3.32. The table, which corresponds to the example of a 2-repeater link, shows the successive data Alice collected from the repeaters. It is seen that repeater 1 only leaves the following matching time-slots and valid bits:

$$-01 - 00 - 1 - 010 - -0-$$

After this first time-slot selection and validation, it is seen that repeater 2 only leaves the following matching time-slots and valid bits:

$$-0---0-1-01-----$$

After this second time-slot selection and validation, it is seen that the terminal (Bob) only leaves the following matching time-slots and valid bits:

$$-0-----1-0------$$

This makes Alice validate to Bob only the time slots 2, 8, and 10. Because of the cascading repeater process and Bob's own choice, only three valid bits have survived out of 16 possibilities. This is close to the 12.5% ($1/2^3$) expected matching rate.

This preceding demonstration shows that it is possible to extend the reach of the quantum cryptography channel with electronic repeaters. We note that eavesdropping the information provided by the repeaters is no threat to security: Eve can reconstruct all the chain by tapping the information on the supervisory channel. She could tap the line to try getting more information, but the repeaters have been designed smart to detect changes in the matching success rate or zero-count events. Any tampering in the fiber path would automatically trigger a general alarm in the system. We observe that the rate of successful key-bit exchange decreases as $1/2^{N+1}$, where N is the number of repeaters. To exchange a key of some given bit length, the message should be at least 2^{N+1} times longer. With one to nine repeaters, the corresponding enhancement factors are 2 to 1,024.

TABLE 3.32 Result from Alice's interrogation of the electronically repeated quantum channel, showing the results of repeaters measurements and retransmission modes (identical), and the terminal (Bob) measurement mode

Alice message bit value		1 0 1 0	0 0 1 1	1 0 1 0	1 0 0 1
Alice choice of polarization states (1)		↕ ↔ ∩ ∪	↔ ↔ ↕ ↕	↕ ∪ ∩ ↔	∩ ∪ ↔ ↕
electronic repeater 1	measuring mode	O + O +	+ + O +	O O O +	+ + + O
	results (2)	∩ ↔ ∩ ↕	↔ ↔ ∪ ↕	∩ ∪ ∩ ↕	↕ ↕ ↔ ∩
	Alice time-slot validation (1)–(2)	- 0 1 -	0 0 - 1	- 0 1 0	- - 0 -
electronic repeater 2	measuring mode	+ + + O	O + + +	+ O O O	O + O +
	results (3)	↕ ↔ ↔ ∩	∪ ↔ ↕ ↕	↔ ∪ ∩ ∩	∪ ↕ ∪ ↔
	Alice time-slot validation (3)–(2)	- 0 - -	- 0 - 1	- 0 1 -	- - - -
terminal (Bob)	measuring mode	+ + O O	+ O O +	+ O + O	+ + O +
	results (4)	↕ ↔ ↕ ∩	↕ ∪ ∩ ↕	↔ ∪ ↕ ∩	↔ ↕ ∪ ↔
	Alice time-slot validation (3)–(2)	- 0 - -	- - - 1	- 0 - -	- - - -
final Alice-Bob time slot validation		- 0 - -	- - - 1	- 0 - -	- - - -

Consider only the last example. At a source rate of *1 Mphoton/s*, a 1,024-bit key with a 9-repeater system (10 spans of 50 km or 500 km) could be transmitted within 1,024 bits × 1,024/10^6 bits/s = 1.05 seconds. At a source rate of *1 K photon per second*, the same transmission would require 17 mn 04 s. At a source rate of *one photon per second*, it would require 12 days, 3 h 16 mn 16 s. We may imagine a system generating keys on a round the clock basis. Repeaters could encrypt the supervisory-channel with their own keys and an algorithm using the hash of the key generated in the previous stage. To accelerate the process, uni-directional fiber pairs (*space-division multiplexing* or SDM) could be used to double the number of successful matching time-slots, and hence, reduce by two-fold the duration of a full key-generation cycle. The system may also carry different wavelengths in the same fiber (*wavelength-division multiplexing* or WDM), using 16–32 parallel channels and generating valid time-slots at 2^9–2^{10} faster rates. In both SDM and WDM, the probability that two valid time-slots may be independently selected is relatively small. In this event, a protocol may select the bit value corresponding to a priority order (fiber of wavelength rank). In order to further expand the haul of the link, one may conceive of super-repeater stations which would be able to interrogate their repeated links through supervisory channels and iterate the procedure over a scale of (say) 10-repeater, 500-km hauls. Each valid key collected by the super-repeater station from a given 500-km span may be used as the seed to the next stage. New keys may also be formed from mixing the new data and the older data generated in the previous key-exchange session, using a pseudo-random permutation algorithm. In all these possible approaches, the end result is that Alice and Bob have the same number to share as a large, random onetime key.

Field demonstrations of QKD, using 1.3-μm or 1.55-μm signals over single-mode fibers, have been conducted over 30-km distances, also showing potential for Mbit/s rates. Other key-demonstrations concern free-space transmissions (FSO) with 1–10-km distances, which could be applied to secure Earth-to-satellite and intersatellite links.

Quantum-cryptography is not the first application of quantum mechanics in telecommunications. As we saw in Chapter 2 indeed, lasers, optical amplifiers and photodetectors can only be explained through quantum mechanics, or for short, through the quantum nature of light, the photon granularity, and the quantum nature of matter, the atomic and lattice energy levels or bands. Quantum key exchange represents the first application of optical telecommunications that exploits the properties of single photons to transmit information. At low photon numbers, light is no longer "classical" in the sense of the familiar electromagnetic wave. New effects such as *light squeezing* (removing uncertainty from either amplitude or phase) and *entanglement* (generating superposition of photon states) can be produced. One application of such effects concerns *quantum teleportation*. Put simply, it is the possibility for Alice to faithfully and securely communicate to Bob some information from a distance, using a common entangled photon pair and a classical open channel. The name comes from the fact that Alice's information (an arbitrary state) is *recreated* on Bob's side without having been physically transported. Thus quantum teleportation defines a futuristic field of *quantum telecommunications*. Another application field is the *quantum computer*, with its *quantum logic gates* and *quantum memories*. Although experimental demonstrations are still lagging behind the theoretical proofs of existence, quantum computers to us may some day be a reality, just as VLSI would have been to scientists from the early 1900s. Such systems could contain as many as 2^{500} entangled states, equivalent to as many binary words of 500-bit lengths. The quantum computer is able to operate over all these states simultaneously, within a single clock cycle. If it became a reality, the problem of factoring large numbers into primes would be forever solved, and brute-force attacks against cipher complexities of this order could be performed in a matter of seconds. This would surely cause a serious threat, if not put an end, to "classical" cryptography. Quantum communications where Alice and Bob could exchange mail with practical quantum computers and with Eve being absolutely unable to know anything may represent the final answer.

a MY VOCABULARY

Can you briefly explain each of these words or acronyms in the cryptography context?

A3, A5, A8 (GSM)	Bernoulli random	Code cracking	DEA-1
A5/n (GSM)	partitioning	Coding efficiency	Decryption
Abbée Thrithème	Biliteral	Collision	Denial of
cipher	Binumeral	Common	service
Advanced	Birthday attack	cryptographic	(distributed)
encryption	Blind signature	architecture	DES
standard	Blinding factor	Commutative	Differential
Airsnort	Block	COMP128	cryptoanalysis
AES	Blowfish	Compression	Diffie–Hellman–
AES-128/192/256	Bombe (machine)	permutation	Merkle
AH (IPsec)	Book cipher	(DES)	algorithm
Algorithm	Boolean algebra,	Concealment cipher	Digital cash
Alphabet	operator	Cracker	Digital signature
Amplified	CA	Crib	algorithm
spontaneous	Caesar shifts	Cipher	Digital signature
emission	CAST	Cipher block	standard
ANSI	CAST5/CAST128	Ciphertext	Digram
Arbitrator	CBC	Control vector	Distributed key
AS	CCA	(CCA)	management
ASCII	Certification	Cracking (code)	Distributed
ASE	Authority	Crypto-analysis or	timestamp
Asymmetric-key	CFB	cryptanalysis	protocol
(cryptography)	Chain block coding	Cryptography	DoS
Authentication	Chaining variables	CV	Double DES
server	Ciberpunk	Data encryption	Double-key
Authentication	Cipher feedback	standard	encryption
header (IPsec)	Cipher suite	Data encryption	DSA
BB84	Cipherpunk	algorithm	DSS
Beam-splitter	Clock arithmetic	Datagram	E-field
Beale papers	Code	DDoS	e-voting
	Code breaking	DEA	E-POLL

e-polling
ECB
EBCDIC
Electronic code book (DES)
Encapsulating security payload (IPsec)
Encryption
Enigma machine
ESP
Entanglement
ETSI
Expansion permutation (DES)
Extended Euclidian algorithm
Factorization
FEAL
Fermat's theorem
Firewall
Frequency analysis
Frequency spectrum
Hacker
Hash function
Hash message
HMAC
Homophonic substitution
Homomorphic encryption
HTTPS
Hughes algorithm
Hybrid cryptosystem
ICV (IPsec)
IDEA
IETF
IKE
Indecipherable
Initial permutation (DES)
Integrity check value (IPsec)
Internet Engineering Task Force
Introducer
Inverse initial permutation (DES)
IPsec
IPv4, IPv6
Irreducible polynomial
ISAKMP
ISDN
ISO
Jargon code
Kasumi
Kerberos
Key (private, public)
Key certification
Key expansion (AES)
Key schedule (AES)
Keytex
Keyword
Khufu
Knight's tour
Light squeezing
Linking timestamp (protocol)
LOKI
Lord Bacon's cipher
Lucifer
MD2, MD4, MD5
Mersenne
Mersenne prime
Message key
Message digest
Message integrity check
MIC
Modular algebra
Modular reduction
Mono-alphabetic cipher
Morse code
Multialphabetic cipher
Netstumbler
NewDES
Nihilist square
NIST
Nonrepudiation
Null cipher
Nulls
OAKLEY
One-way encryption
Onetime pad cipher
Operand
Operator
Optical amplifier
Oryx
PCT
PEM
Personal identification number
PGP
PGPdisk
PGPfone
Philips cipher
Photon
Photuris
Phreaking
PIN
PKC
PKCS#1
Plaintext
Playfair cipher
Polarization (linear, circular)
Polarizing beam-splitter
Polyalphabetic cipher
Polygram substitution
Possibily weak key (DES)
Premaster key (SSL)
Pretty Good Privacy
Prime number
Privacy-enhanced mail
Private communications technology
Private key
Public key (cryptography, encryption)
QKD
Quantum cryptography
Quantum key distribution
Quantum computer
Quantum teleportation
RAND (GSM)
Raw key
RC-2/RC-4
REDOC-II
Relatively prime (numbers)
Repudiation
RIPEM
Rivest cipher
Rijndael (algorithm)
Round key
RSA
RSA modulus
RSA public or private exponent
S-box
S-HTTP
SA (IPsec)
SCG
Secure European system for applications in multivendor environment
Secure hash algorithm
Secure socket layer
Secure transactions technology
Security association (IPsec)
Security parameter index (IPsec)
Semiweak key (DES)
Server-gated cryptography (protocol)
SESAME
SGC
SHA
Single-photon

Source/detector
SKEME
SKIP
Snooping
SOP
SPD
SPI
Spoofing
SPS
SRES (GSM)
SSL
Substitution
 alphabet
State (AES)
State of polarization
Steganography
STT
Subkey
Symbol
Symmetric-key
 (cryptography)
TCP/IP
TDEA
Theft of service
Timestamping
ITS/PEM
TKIP
TLS
ToS
Transport layer
 security
Transport mode
 (IPsec)
Transposition
 (irregular)
Transverse
 electric/magnetic
 E-field
Trapdoor
 function
Trigram
Trinumeral
Triple DES
Tunnel mode
 (IPsec)
Vacuum
 fluctuations
Vaccum noise
Vigenère cipher
Weak keys
 (DES)
WEP
WEPcrack
Wireless equivalent
 privacy
Wireless LAN
WLAN
X.509
Zanotti
 transposition
1-bit/8-bit/N-bit
 CFB
3DES
802.1x
802.1

Solution to Exercises

Section 2.2

2.2.1 A manned craft is sent to Jupiter's vicinity. What is the minimum and maximum time for exchanging messages with the Earth? Express the result in hour/min/s (Parameters: mean Earth distance to Sun $d_E = 149.6$ million km, mean Jupiter distance to Sun $d_J = 778.3$ million km.) Conclusion and solutions to create a duplex communication channel?

Because of the difference in planet positions on their orbits (assumed circular), the two possible distances between Earth and Jupiter are:

Minimum: $d = d_J - d_E = 778.3 - 149.6 = 628.7$ Mkm, or 6.287×10^{11} m
Maximum: $d' = d_J + d_E = 778.3 + 149.6 = 927.9$ Mkm, or 9.279×10^{11} m

The corresponding round-trip delays are given by

Minimum: $T = 2d/c = 2 \times 6.287 \times 10^{11}$ m$/(3 \times 10^8$ m/s$) = 4{,}191.33$ s $= 1$ h, 9 min, 51 s ≈ 1 h, 10 min

Maximum: $T = 2d'/c = 2 \times 9.279 \times 10^{11}$ m$/(3 \times 10^8$ m/s$) = 6{,}0186$ s $= 1$ h, 43 min, 6 s ≈ 1 h, 45 min

Conclusion: Such a 1–2 h round-trip delay is prohibitive for any interactivity.

Solutions for a duplex communication channel: Both ends could send messages to each other at regular times, every $T/2$ or T. Choosing the first rather than the second solution would double the number of messages, providing more up-to-date information to the other end and making optimum use of time, like a simplex channel. But upstream and downstream messages would not correspond to each other, like two letters crossing in the mail, creating confusing loops. The other solution, where the ends take turns for emission and reception, is less effective in communicating up-to-date information and consumes twice as much time, but is more orderly for feedback and decision purposes, as a true interactive or duplex channel. A third solution would be to mix both approaches through two independent sub-channels, e.g., using two carrier frequencies. One subchannel would be more up to date but less responsive (simplex), while the other would be less up to date but more responsive (duplex). A continuous communications channel could thus be established between Earth and Station, compensating the long response times and delayed interactivity by creating a continuous information flow with different sub-channel contents and simplex/duplex processing functions.

2.2.2 Consider the 45° glass-prism structure shown below. What should the minimum value of the glass' refractive index n_1 be in order for light rays to be totally reflected in the 90° direction? (Assume $n_2 = 1$ for air.)

Wiley Survival Guide in Global Telecommunications: Broadband Access, Optical Components and Networks, and Cryptography, by E. Desurvire
ISBN 0-471-67520-2 © 2004 John Wiley & Sons, Inc.

Inside the prism, the incidence angle of the ray with respect to the glass/air interface is $\theta_1 = 45°$. In order to achieve total internal reflection, this angle must be such that $\theta_1 \geq \theta_1^{\text{crit}} = \sin^{-1}(n_2/n_1)$. The condition for total internal reflection is thus

$$n_2/n_1 \leq \sin\theta_1 \iff 1/n_1 \leq \sin 45° = 1/\sqrt{2} \iff n_1 \geq \sqrt{2} \approx 1.414$$

which provides the minimum value for the glass refractive index n_1.

2.2.3 Obtain the time-independent relation of equation (2.7) between the E-field coordinates defined in equation (2.8). Clue: Use the formula $\cos(a+b) = \cos a \cos b - \sin a \sin b$ and then weight and sum/subtract the resulting equations in such a way to eliminate the time-dependent terms through the property $\cos^2 a + \sin^2 a = 1$; also use $\sin(2a) = 2\sin a \cos a$ and $\cos(2a) = \cos^2 a - \sin^2 a$.

Applying the first formula in equation (2.7) gives

$$\begin{cases} X' = \cos\left(\omega t - \dfrac{\Delta\varphi}{2}\right) = \cos(\omega t)\cos\left(-\dfrac{\Delta\varphi}{2}\right) - \sin(\omega t)\sin\left(-\dfrac{\Delta\varphi}{2}\right) \\ = \cos(\omega t)\cos\left(\dfrac{\Delta\varphi}{2}\right) + \sin(\omega t)\sin\left(\dfrac{\Delta\varphi}{2}\right) \\ Y' = \cos\left(\omega t + \dfrac{\Delta\varphi}{2}\right) = \cos(\omega t)\cos\left(\dfrac{\Delta\varphi}{2}\right) - \sin(\omega t)\sin\left(\dfrac{\Delta\varphi}{2}\right) \end{cases}$$

We first weight the two equations by $\sin(\Delta\varphi/2)$ and take the sum

$$\begin{cases} X'\sin\left(\dfrac{\Delta\varphi}{2}\right) = \cos(\omega t)\cos\left(\dfrac{\Delta\varphi}{2}\right)\sin\left(\dfrac{\Delta\varphi}{2}\right) + \sin(\omega t)\sin^2\left(\dfrac{\Delta\varphi}{2}\right) \\ Y'\sin\left(\dfrac{\Delta\varphi}{2}\right) = \cos(\omega t)\cos\left(\dfrac{\Delta\varphi}{2}\right)\sin\left(\dfrac{\Delta\varphi}{2}\right) - \sin(\omega t)\sin^2\left(\dfrac{\Delta\varphi}{2}\right) \\ U = X'\sin\left(\dfrac{\Delta\varphi}{2}\right) + Y'\sin\left(\dfrac{\Delta\varphi}{2}\right) = 2\cos(\omega t)\cos\left(\dfrac{\Delta\varphi}{2}\right)\sin\left(\dfrac{\Delta\varphi}{2}\right) \\ \equiv \cos(\omega t)\sin(\Delta\varphi) \end{cases}$$

where we used $2\sin a \cos a = \sin(2a)$. We do the same operation with $\cos(\Delta\varphi/2)$, but this time we take the difference:

$$\begin{cases} X'\cos\left(\dfrac{\Delta\varphi}{2}\right) = \cos(\omega t)\cos^2\left(\dfrac{\Delta\varphi}{2}\right) + \sin(\omega t)\sin\left(\dfrac{\Delta\varphi}{2}\right)\cos\left(\dfrac{\Delta\varphi}{2}\right) \\ Y'\cos\left(\dfrac{\Delta\varphi}{2}\right) = \cos(\omega t)\cos^2\left(\dfrac{\Delta\varphi}{2}\right) - \sin(\omega t)\sin\left(\dfrac{\Delta\varphi}{2}\right)\cos\left(\dfrac{\Delta\varphi}{2}\right) \\ V = X'\cos\left(\dfrac{\Delta\varphi}{2}\right) - Y'\cos\left(\dfrac{\Delta\varphi}{2}\right) = 2\sin(\omega t)\cos\left(\dfrac{\Delta\varphi}{2}\right)\sin\left(\dfrac{\Delta\varphi}{2}\right) \\ \equiv \sin(\omega t)\sin(\Delta\varphi) \end{cases}$$

To eliminate the time-dependent factors in U and V, we just need to sum the squares $W = U^2 + V^2$, giving

$$\begin{cases} U^2 = \left[X' \sin\left(\dfrac{\Delta\varphi}{2}\right) + Y' \sin\left(\dfrac{\Delta\varphi}{2}\right)\right]^2 = \cos^2(\omega t) \sin^2(\Delta\varphi) \\ V^2 = \left[X' \cos\left(\dfrac{\Delta\varphi}{2}\right) - Y' \cos\left(\dfrac{\Delta\varphi}{2}\right)\right]^2 = \sin^2(\omega t) \sin^2(\Delta\varphi) \\ W = (X')^2 + (Y')^2 - 2X'Y' \left[\cos^2\left(\dfrac{\Delta\varphi}{2}\right) - \sin^2\left(\dfrac{\Delta\varphi}{2}\right)\right] \\ = \left[\cos^2(\omega t) + \sin^2(\omega t)\right] \sin^2(\Delta\varphi) \end{cases}$$

Using the relations $\cos^2 a - \sin^2 a = \cos(2a)$ and $\cos^2 a + \sin^2 a = 1$, one finally gets

$$W = (X')^2 + (Y')^2 - 2X'Y' \cos(\Delta\varphi) = \sin^2(\Delta\varphi)$$

which is the result of equation (2.7).

2.2.4 Show that elliptical SOP with phase $\Delta\varphi$ comprised between 0 and π corresponds to a right-handed (clockwise) rotation of the E-field polarization, and the opposite rotation direction for $\Delta\varphi$ comprised between π and 2π. (Clue: Evaluate the time change of the angle γ defined by $\tan\gamma = Y'/X'$ from $t = 0$, assuming $\Delta\varphi = \pi/2$ and $\Delta\varphi = 3\pi/2$, then generalize the conclusion to all phase-delays $\Delta\varphi$).

By definition, the E-field coordinates are

$$\begin{cases} X' = \cos(\omega t - u) \\ Y' = \cos(\omega t + u) \end{cases}$$

where $u \equiv \Delta\varphi/2$ is the phase delay, which gives the tangent:

$$\tan\gamma = \frac{Y'}{X'} = \frac{\cos(\omega t - u)}{\cos(\omega t + u)}$$

We should then develop the above definition using the properties $\cos(a+b) = \cos a \cos b - \sin a \sin b$, $\cos(-u) = \cos u$ and $\sin(-u) = -\sin u$, which yields, for

$$\tan\gamma = \frac{\cos(\omega t) \cos u + \sin(\omega t) \sin u}{\cos(\omega t) \cos u - \sin(\omega t) \sin u} = \frac{1 + \tan(\omega t) \tan u}{1 - \tan(\omega t) \tan u}$$

Consider next the two cases:

(a) $\Delta\varphi = \pi/2$, or $\tan u = \tan(\pi/4) = 1$.
(b) $\Delta\varphi = 3\pi/2$ or $\tan u = \tan(3\pi/4) = -1$.

Calling the corresponding angles γ_a and γ_b, we have

$$\begin{cases} \tan\gamma_a = \dfrac{1 + \tan(\omega t)}{1 - \tan(\omega t)} \\ \tan\gamma_b = \dfrac{1 - \tan(\omega t)}{1 + \tan(\omega t)} \equiv \dfrac{1}{\tan\gamma_a} \end{cases}$$

At the origin ($t = 0$), the two functions are both equal to $\tan\gamma_a = \tan\gamma_b = 1$, corresponding to the angle $\gamma_a = \gamma_b = \pi/4$ or 45°. Consider first case (a): If time increases by a small amount t, so does the function $\tan(\omega t)$, thus making $\tan\gamma_a$ to decrease. Therefore, the angle γ_a *decreases* from its initial 45° value, which corresponds to a *clockwise* rotation. The same reasoning applied to case (b) yields the opposite conclusion; i.e., the angle γ_b *increases*, which corresponds to a *counterclockwise* rotation. This is also shown more simply by the fact that $\tan \gamma_b = 1/\tan \gamma_a$.

General case: The same demonstration strictly applies to all elliptical polarizations which are characterized by the tangent function:

$$\tan \gamma = \frac{1 + A \tan(\omega t)}{1 - A \tan(\omega t)}$$

where $A = \tan u$ is positive in the interval $\Delta\varphi = [0, \pi]$ and negative in the interval $\Delta\varphi = [\pi, 2\pi]$.

2.2.5 Show that the Stokes coordinates are uniquely defined by the two ellipse angles (ψ, χ) and the power (S_0) according to equation (2.13), using the definitions of (E_o, E_e), (α, ψ, χ), and (S_0, S_1, S_2, S_3) from text.

We first recall the different definitions:

(1) $E_o = E \cos \alpha$
(2) $E_e = E \sin \alpha$
(3) $S_0 = E = E_o^2 + E_e^2$
(4) $S_1 = E_o^2 - E_e^2$
(5) $S_2 = 2E_o E_1 \cos \Delta\varphi$
(6) $S_3 = 2E_o E_1 \sin \Delta\varphi$
(7) $S_0^2 = S_1^2 + S_2^2 + S_3^2 = E^2$
(8) $\tan(2\psi) = \tan(2\alpha) \cos \Delta\varphi$
(9) $\sin(2\chi) = \sin(2\alpha) \sin \Delta\varphi$

Relation (4) with (1)–(2) yields

(10) $S_1 = E^2(\cos^2 \alpha - \sin^2 \alpha) \equiv E^2 \cos 2\alpha$

Relation (5) with (1)–(2), (8) yields

(11) $S_2 = 2E^2 \cos \alpha \sin \alpha \tan 2\psi / \tan 2\alpha \equiv E^2 \tan 2\psi \cos 2\alpha$

Relation (6) with (1)–(2), (9) yields

(12) $S_3 = 2E^2 \cos \alpha \sin \alpha \sin 2\chi / \sin 2\alpha \equiv S_0 \sin 2\chi$

Relations (10)–(11) yield

(13) $S_2/S_1 = \tan 2\psi$, or
(14) $S_1 \equiv U \cos 2\psi$ and
(15) $S_2 \equiv U \sin 2\psi$

Relation (7) with (14)–(15) yield finally

(16) $E^2 = S_1^2 + S_2^2 + S_3^2 = U^2 \cos^2 2\psi + U^2 \sin^2 2\psi + E^2 \sin^2 2\chi$

$\longleftrightarrow E^2(1 - \sin^2 2\chi) = U^2 \longleftrightarrow E^2(\cos^2 2\chi) = U^2 \longleftrightarrow U = E \cos 2\chi$

which gives, with $E = S_0$,

(17) $S_1 \equiv S_0 \cos 2\chi \cos 2\psi$ and
(18) $S_2 \equiv S_0 \cos 2\chi \sin 2\psi$

Equations (12) and (17)–(18) thus define the three Stokes parameters as functions of the ellipse angles (ψ, χ) and the power (S_0).

2.2.6 What is the photon flux (rate of incident photons per second) in a light beam of power $P = 1$ W, 1 mW, 1 µW, and 1 nW at a wavelength of $\lambda = 1.5$ µm (IR light)? Same question for a wavelength of $\lambda = 100$ nm (UV light). Conclusions?

By definition, the photon energy is $h\nu$, where $h = 6.6262 \times 10^{-34}$ J · s is Planck's constant and ν is the frequency in Hz (s^{-1}). The frequency is related to wavelength through $\lambda = c/\nu$, where $c = 3 \times 10^8$ m/s is the speed of light. This relation gives

$$\nu_{IR} = (3 \times 10^8 \text{ m/s})/(1.5 \times 10^{-6} \text{ m}) = 2 \times 10^{14} \text{ Hz for } \lambda = 1.55 \text{ µm (IR), and}$$

$$\nu_{UV} = (3 \times 10^8 \text{ m/s})/(100 \times 10^{-9} \text{ m}) = 3 \times 10^{15} \text{ Hz for } \lambda = 100 \text{ nm (UV).}$$

We thus get the corresponding photon energies:

$$E_{IR} = h\nu_{IR} = 6.62 \times 10^{-34} \text{ J} \cdot \text{s} \times 2 \times 10^{14} \text{ Hz} = 1.3 \times 10^{-19} \text{ J}$$

$$E_{UV} = h\nu_{IR} = 6.62 \times 10^{-34} \text{ J} \cdot \text{s} \times 3 \times 10^{15} \text{ Hz} = 2.0 \times 10^{-18} \text{ J}$$

The photon flux is then given by $N_x = P/E_x$. The values obtained for the different beam powers $P = 1$ W, 1 mW (10^{-3} W), 1 µW (10^{-6} W), and 1 nW (10^{-9} W) are shown in the table below.

Beam power	1 W	1 mW	1 µW	1 nW
IR flux (photons/s)	7.7×10^{18}	7.7×10^{15}	7.7×10^{12}	7.7×10^9
UV flux (photons/s)	5.0×10^{17}	5.0×10^{14}	5.0×10^{11}	5.0×10^8

Conclusions:

(1) The photon granularity is extremely small: at UV–IR wavelengths, a billionth of a Watt (1 nW) still carries between 0.5 and 8 billion photons per second.

(2) Since the photon energy increases with frequency, at equal power the number of photons increases/decreases with frequency (or increases with wavelength).

2.2.7 An optical amplifier has a gain coefficient of $g = 0.069$ cm^{-1}. What are the required amplifier length L examples to achieve gains of $\times 100$ and $\times 1,000$?

By definition, the gain is $G = \exp(gL)$, or $gL = \log G$ (natural logarithm), corresponding to an amplifier length $L = \log G/g$, as expressed in cm units with g being in cm^{-1}. In order to achieve gains of $G = 100$ or $G = 1,000$, the required lengths are therefore:

$$L_{100} = \log(1,00)/(0.069 \text{ cm}^{-1}) = 66.7 \text{ cm} \approx 0.68 \text{ m}$$

$$L_{1000} = \log(1,000)/(0.069 \text{ cm}^{-1}) = 100.1 \text{ cm} \approx 1 \text{ m}$$

2.2.8 A semiconductor amplifier chip of length $L = 250$ μm provides a gain of $G = +25$ dB. What is its gain coefficient in dB/m and m^{-1}?

By definition, the decibel gain is $G_{sB} = 10 \log_{10}(gL)$. We can either write

$$G_{sB} = 10 \log_{10}(gL) = 10 \log_{10}(g_{m^{-1}}) + 10 \log_{10}(L_m)$$

$$\longleftrightarrow 25 \text{ dB} = g_{dB/m} + 10 \log_{10}(250 \times 10^{-6} \text{ m})$$

$$\longleftrightarrow g_{dB/m} = 25 \text{ dB} - 10 \log_{10}(250 \times 10^{-6} \text{ m}) = 25 \text{ dB} + 36 \text{ dBm}^{-1} = 61 \text{ dB/m}$$

$$\longleftrightarrow g_{m^{-1}} = 10^{g_{dB/m}/10} = 10^{6.1} = 1.25 \times 10^6 \text{m}^{-1}$$

or do the initial conversion in linear units:

$$G_{dB} = 25 = 10 \log_{10}(gL) \longleftrightarrow gL = 10^{G_{dB}/10} = 10^{2.5} = 316.22$$

$$\longleftrightarrow g_{m^{-1}} = 316.22/L = 316.22/(250 \times 10^{-6} \text{ m}) = 1.26 \times 10^6 \text{ m}^{-1}$$

$$\longleftrightarrow g_{dB/m} = 10 \log_{10}(g_{m^{-1}}) = 10 \log_{10}(1.26 \times 10^6 \text{ m}^{-1}) = 61 \text{ dB/m}$$

2.2.9 Prove that the two different gain-coefficients definitions in [dB/m] and [m^{-1}] are related through $g_{m^{-1}} \approx 0.2302 \times g_{dB/m}$.

By definition, the gain is $G = \exp(gL)$, which gives

$$G_{dB} = 10 \log_{10}[e^{gL}] = (gL) \times 10 \log_{10}[e] \approx 4.342 \times (gL)$$

$$\longleftrightarrow g_{dB/m} = G_{dB}/L \approx 4.342 \times (g_{m^{-1}} L_m)/L_m = 4.342 \times g_{m^{-1}}$$

$$\longleftrightarrow g_{m^{-1}} \approx g_{dB/m}/4.342 = 0.2302 g_{dB/m}$$

2.2.10 The amount of output power that can be extracted from a laser cavity is given by the approximation formula $P_{out} \approx P_0(1 - R)/[2\alpha L + \log(1/R)]$, where P_0 is a constant, R is the reflectivity of the output mirror ($R < 1$), and $2\alpha L$ is the cavity's round-trip loss. Determine by trial-and-error (within 0.5% accuracy) the optimum reflectivity for which this output power is maximized (assume a loss $2\alpha L = 0.7$, corresponding to $\exp(-0.7) \approx 0.5$ or 3 dB).

The answer is provided by an estimation of the maximum value of $X = (1 - R)/[0.7 + \log(1/R)]$. We can start the estimation from $R = 0.99$ by steps of 2% until we observe that some maximum has been passed (see table below, left). The result show that the maximum is near $R = 0.89$. Then we can proceed again around this value while using smaller increments of 0.5% (see table below, right) and stop until we have passed another maximum. The answer is $R \approx 0.875$.

R	X	R	X
0.99	0.014	0.910	0.111
0.97	0.041	0.905	0.118
0.95	0.066	0.900	0.124
0.93	0.090	0.895	0.129
0.91	0.111	0.890	0.134
0.89	**0.134**	0.885	0.139
0.87	0.109	0.880	0.144
		0.875	**0.149**
		0.870	0.132

Section 2.3

2.3.1 Determine the maximum number of ray incidence-angles making possible for light waves to propagate inside an index-layer structure of thickness $d = N\lambda$, with $N = 100$, $N = 10$, $N = 1$, and $N = 1/2$. Conclusion for the lowest number?

The allowed angles are given by the relation

$$\theta_m = \sin^{-1}\left(m\frac{\lambda}{2d}\right) \text{ where } m \text{ is an integer. Since } d = N\lambda, \text{ we have}$$

$$\theta_m = \sin^{-1}\left(\frac{m}{2N}\right)$$

The maximum value of the $\sin^{-1}$ argument is unity ($\theta_{max} = \pi/2$), corresponding to $m = 2N$ angle possibilities. Since the incidence $\theta_{max} = \pi/2$ corresponds to a ray perpendicular to the structure, for which there is no ray-propagation effect inside the horizontal structure, and the case must be removed.

Therefore, the maximum number of allowed ray angles is $M = 2N - 1$, or $M = 199$, $M = 19$, $M = 1$, and $M = 0$ for $N = 100$, $N = 10$, $N = 1$, and $N = 1/2$, respectively.

2.3.2 We want to realize a planar dielectric waveguide with a thin slab of glass surrounded by air. What should be the slab thickness in order for the waveguide to have a cutoff wavelength at $\lambda_{cutoff} = 10\,\mu m$? (Glass index $n_1 = 1.450$). Same question if the slab is surrounded by another glass material with index $n_2 = 1.449$? Conclusion for practical realization of the waveguide?

By definition, the cutoff wavelength is $\lambda_{cutoff} = 2d\mathrm{NA}$, where d is the slab thickness and NA is the numerical aperture defined by $\mathrm{NA} = \sqrt{n_1^2 - n_2^2}$. Since the outer waveguide layers are made of air, we have $n_2 = 1$, thus

$$\mathrm{NA} = \sqrt{(1.45)^2 - 1} = 1.05$$

To obtain $\lambda_{cutoff} = 10\,\mu m$, the glass-slab thickness must therefore be

$$d = \lambda_{cutoff}/(2\mathrm{NA}) = 10\,\mu m/(2 \times 1.05) = 4.7\,\mu m.$$

Second question: If we surround the slab by another glass material with index, $n_2 = 1.300$, the numerical aperture becomes

$$\mathrm{NA} = \sqrt{(1.45)^2 - (1.449)^2} = 0.053$$

and the required slab thickness becomes

$$d = 10\,\mu m/(2 \times 0.053) = 94\,\mu m.$$

The conclusion is that the second approach is more practical since the required thickness is easier to realize and control compared to the other approach, provided that one knows how to achieve very small index differences such as $\Delta = 1 - n_2/n_1 = 1 - 1.449/1.45 \approx 0.7 \times 10^{-3}$ (it is actually possible to achieve this by modifying the glass doping composition).

2.3.3 We want to realize a planar dielectric waveguide with a thin slab of glass (index $n_1 = 1.45$) with thickness $d = 10\,\mu m$ surrounded by two identical glass layers with a different index n_2. Give the relative index Δ for which the waveguide has a cutoff wavelength of $\lambda_c = 1.4\,\mu m$.

The cutoff wavelength is related to the waveguide thickness, numerical aperture, and inner/outer indices n_1/n_2 through

$$\lambda_c = 2d\text{NA} = 2d\sqrt{n_1^2 - n_2^2}$$

which gives

$$\sqrt{n_1^2 - n_2^2} = \lambda_c/(2d) = 1.4\,\mu\text{m}/(2 \times 10\,\mu\text{m}) = 0.07$$

$$\longleftrightarrow n_1^2 - n_2^2 = (0.07)^2 = 4.9 \times 10^{-3}$$

$$\longleftrightarrow n_2 = \sqrt{n_1^2 - 4.9 \times 10^{-3}} = \sqrt{(1.45)^2 - 4.9 \times 10^{-3}} = 1.4483$$

The relative index is

$$\Delta = 1 - \frac{n_2}{n_1} = 1 - \frac{1.4483}{1.450} \approx 1.2 \times 10^{-3}$$

2.3.4 Based upon the figure below, show that the condition for an external ray to be captured (totally internally reflected) into a waveguide of numerical aperture NA is that its incidence angle be $\varphi < \varphi_a$, where φ_a is an acceptance angle defined by $\sin \varphi_a = \text{NA}$.

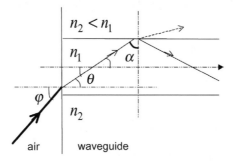

From the angle definitions in the above figure, the condition for total internal reflection is $\alpha > \alpha_{\text{crit}}$ where $\sin \alpha_{\text{crit}} = n_2/n_1$. The corresponding condition for the angle $\theta = \pi/2 - \alpha$ is $\theta < \theta_{\text{crit}}$ with $\sin(\theta_{\text{crit}}) = \sin(\pi/2 - \alpha_{\text{crit}}) = \cos \alpha_{\text{crit}}$. The incidence of the external ray is defined by Descartes/Snell's law, namely $n_0 \sin \varphi = n_1 \sin \theta$ where $n_0 = 1$ is the air index. We thus have:

$$\sin \varphi < n_1 \sin \theta_{\text{crit}} = n_1 \cos \alpha_{\text{crit}} = n_1 \sqrt{1 - \sin^2 \alpha_{\text{crit}}}$$

$$= n_1 \sqrt{1 - (n_2/n_1)^2} = \sqrt{n_1^2 - n_2^2} \equiv \text{NA}$$

2.3.5 Plot the LP_{01} E-field mode envelope $E(r)$ and corresponding power distributions $E^2(r)$ of a step-index optical fiber assuming a numerical aperture of NA $= 0.1$, a core radius of $a = 5.75\,\mu\text{m}$, and three different signal wavelengths $\lambda_1 = 1.3\,\mu\text{m}$, $\lambda_2 = 1.4\,\mu\text{m}$, $\lambda_2 = 1.5\,\mu\text{m}$, $\lambda_4 = 1.6\,\mu\text{m}$, and $\lambda_5 = 1.7\,\mu\text{m}$. You may use a plotting software that includes Bessel and Hankel functions, or alternatively use the approximations

$$J_0(x) \approx 1 - 2.25(x/3)^2 + 1.26(x/3)^4 - 0.31(x/3)^6 \quad \text{for} \quad x \le 3$$

and
$$K_0(x) \approx \sqrt{\pi/(2x)}[1 - 1/(8x)] \exp(-x) \quad \text{for} \quad x > 1$$

(Clue: First make a table with the values of V, U, and W for each wavelength, and then use the above definitions to plot the E-field between $r = 0$ and $r = 3a$. See text for final plots result and conclusions.)

From the fiber parameters (a, NA) we first calculate the cut-off wavelength, i.e.,

$$\lambda_c = \frac{2\pi a \text{NA}}{2.405} = \frac{2\pi \times 5.75 \,\mu\text{m} \times 0.1}{2.405} = 1.502 \,\mu\text{m} \approx 1.5 \,\mu\text{m}$$

A first task is to make a table with the values of V, U, and W for each wavelength. For this, we must use the three definitions

$$V = (2\pi a/\lambda)\text{NA}$$

$$U = \frac{(1 + \sqrt{2})V}{1 + (4 + V^4)^{0.25}}$$

and

$$W = \sqrt{V^2 - U^2}$$

With the above waveguide parameter values, we can fill the following table (below, left):

λ (μm)	V	U	W	$B = J_0(U)/K_0(W)$
1.3	2.779	1.754	2.155	5.970
1.4	2.580	1.712	1.930	4.559
1.5	2.408	1.671	1.733	3.577
1.6	2.258	1.631	1.561	2.877
1.7	2.125	1.591	1.408	2.359

Then we use the approximations for $J_0(x)$, $K_0(x)$ provided to evaluate the amplitude constant $B = J_0(U)/K_0(W)$; see table (above, right). At this point, we can plot the E-field amplitude according to

$$E(r) = \begin{cases} J_0\left(U\dfrac{r}{a}\right), & r \leq a \\ BK_0\left(W\dfrac{r}{a}\right), & r > a \end{cases}$$

with $a = 5.75 \,\mu\text{m}$, while using the $J_0(x)$, $K_0(x)$ approximations provided, for each of the 5 wavelengths and values U, W, B listed in the above table. The corresponding power distributions are given by taking the square of the above expressions.

2.3.6 How much fiber length L can be manufactured from a single glass preform with dimensions $l = 1$ m (length) and od $= 1.5$ cm (diameter)? Same question with od $= 5$ cm.

The principle to use is the conservation of mass and volume for the glass material, from preform to fiber (the overall glass composition is the same before and after melting, and there is no loss of material in the process of melting and re-solidification). The preform's volume in cm³ is

$$v = l \times \pi \mathrm{od}^2/4 = 100\,\mathrm{cm} \times \pi(1.5\,\mathrm{cm})^2/4 = 177\,\mathrm{cm}^3.$$

The fiber volume is $V = L \times S$, where S is the fiber cross-sectional area. Taking OD = 125 μm as the fiber outside diameter, we have

$$S = \pi \mathrm{OD}^2/4 = \pi(125 \times 10^{-4}\,\mathrm{cm})^2/4 = 1.22 \times 10^{-4}\,\mathrm{cm}^2$$

Since $V = v$, the maximum fiber length that can be extracted from the preform is therefore

$$L = v/S = 177\,\mathrm{cm}^3/(1.22 \times 10^{-4}\,\mathrm{cm}^2) = 1.44 \times 10^6\,\mathrm{m} = 14.4\,\mathrm{km}$$

We could also have directly obtained this result by using the relation

$$\frac{L}{l} = \frac{S}{s} = \left(\frac{\mathrm{od}}{\mathrm{OD}}\right)^2$$

If we increase the preform's diameter to od = 5 cm, the fiber length is extended to $L' = 5^2 \times 14.4\,\mathrm{km} = 360\,\mathrm{km}$.

2.3.7 An optical-transmission receiver requires a minimum incident signal power of $P_R = 0.1\,\mu\mathrm{W}$ to guarantee a standard bit error rate of 10^{-9}. What is the maximum possible system length, assuming that the transmitter power is limited to $P_T = 100\,\mu\mathrm{W}$, 1 mW, 10 mW, 100 mW, or 1 W, respectively (fiber attenuation coefficient $\alpha = 0.2\,\mathrm{dB/km}$)? What is your conclusion from the results?

The first two columns in the table below shows the signal power loss corresponding to the different transmitter powers P_T, assuming that in each case the receiver power is $P_R = 0.1\,\mu\mathrm{W}$. For instance, the case $P_T = 100\,\mu\mathrm{W}$ corresponds to a transmission of $T = P_R/P_T = 0.1\,\mu\mathrm{W}/100\,\mu\mathrm{W} = 0.1 \times 10^{-6}\,\mathrm{W}/(100 \times 10^{-6}\,\mathrm{W}) = 10^{-3}$ or a loss of $10\log_{10} T = 10\log_{10} 10^{-3} = -30\,\mathrm{dB}$.

P_T	Loss (dB)	Length L (km)
100 μW	−30	150
1 mW	−40	200
10 mW	−50	250
100 mW	−60	300
1 W	−70	350

The fiber-system lengths corresponding to the above losses are simply given by the ratio

$$L(\mathrm{km}) = \mathrm{Loss(dB)}/\alpha(\mathrm{dB/km})$$

The last column in the table provides the results. It is seen that an increase of one order of magnitude in transmitter power P_T only corresponds to a 50-km increase in transmission distance. Achieving an extra 100-km transmission requires to multiply by one hundred-fold the transmitter power, which is utterly impractical.

2.3.8 Show that the bandwidth-length (BL) limitation of fiber-optics systems:

$$BL \text{ (THz} \times \text{km)} < \frac{1}{|D|_{\text{ps/nm·km}} \Delta\lambda_{\text{nm}}}$$

where $\Delta\lambda \geq (\lambda^2/c)B$ is the signal spectral width can also be put in the B^2L form:

$$B^2L \text{ (THz}^2 \times \text{km)} < \frac{0.3}{|D|_{\text{ps/nm·km}} \lambda_{\mu m}^2}$$

To prove the second relation, we must convert the first relation back into standard units, i.e.,

$$BL \text{ (Hz} \times \text{m)} < \frac{1}{|D|_{\text{s/m·m}} \Delta\lambda_m} = \frac{1}{|D|_{\text{s/m·m}} (\lambda_m^2/c_{\text{m/s}}) B_{\text{Hz}}}$$

which gives

$$B^2L \text{ (Hz}^2 \times \text{m)} < \frac{c_{\text{m/s}}}{|D|_{\text{s/m·m}} \lambda_m^2}$$

Next, we use the relation $|D|_{\text{s/m·m}} = D_{\text{ps/nm/km}} \times 10^{-12}/(10^{-9} \times 10^3) = D_{\text{ps/nm/km}} \times 10^{-6}$, $\lambda_m^2 = \lambda_{\mu m}^2 \times 10^{-12}$, and $c = 3 \times 10^8$ m/s to get

$$B^2L \text{ (Hz}^2 \times \text{m)} < \frac{3 \times 10^8 \text{ m/s}}{|D|_{\text{ps/nm·km}} \times 10^{-6}(\lambda_{\mu m}^2 \times 10^{-12})} = \frac{3 \times 10^{26}}{|D|_{\text{ps/nm·km}} \lambda_{\mu m}^2}$$

and finally, with 1 THz$^2 = 10^{24}$ Hz^2, 1 km $= 10^3$ m:

$$B^2L \text{ (THz}^2 \times \text{km)} < \frac{0.3}{|D|_{\text{ps/nm·km}} \lambda_{\mu m}^2}$$

2.3.9 What should the zero-dispersion wavelengths of NZDSF$^\pm$ fibers be, assuming dispersions of $D = \pm 2.5$ ps/nm-km and identical slopes of $D' = 0.07$ ps/nm^2-km at $\lambda = 1{,}550$-nm wavelength? (Clue: Use linear approximation formula for dispersion.)

The linear-approximation formula yields

$$D(\lambda) = (\lambda - \lambda_0)D'$$

which gives the zero-dispersion wavelength:

$$\lambda_0 = \lambda - \frac{D(\lambda)}{D'}$$

In the case of NZDSF$^-$, we have

$$\lambda_0 = 1{,}550 \text{ nm} - \frac{(-2.5_{\text{ps/nm-km}})}{0.07_{\text{ps/nm}^2\text{-km}}} = 1{,}585.7 \text{ nm}$$

In the case of NZDSF$^+$, we have

$$\lambda_0 = 1{,}550 \text{ nm} - \frac{(+2.5_{\text{ps/nm-km}})}{0.07_{\text{ps/nm}^2\text{-km}}} = 1{,}514.3 \text{ nm}$$

2.3.10 Calculate the fourth-order dispersion coefficient D'' of the SMF ($\lambda_0 = 1{,}300$ nm), assuming its dispersion and slope at $1{,}550$ nm to be $D = +17$ ps/nm-km and $D' = 0.045$ ps/nm^2-km, respectively.

The dispersion is defined according to the expansion formula:

$$D(\lambda) = (\lambda - \lambda_0)D' + \frac{(\lambda - \lambda_0)^2}{2}D''$$

or equivalently

$$D'' = \frac{2}{(\lambda - \lambda_0)^2}[D(\lambda) - (\lambda - \lambda_0)D']$$

Substituting in this expression the values $\lambda_0 = 1{,}300$ nm and $\lambda = 1{,}550$ nm ($\lambda - \lambda_0 = 250$ nm), the dispersion $D(\lambda) = +17$ ps/nm-km, and slope $D' = 0.045$ ps/nm²-km, we obtain

$$D = \frac{2}{(250_{\text{nm}})^2}\left[+17_{\text{ps/nm-km}} - 250_{\text{nm}} \times 0.045_{\text{ps/nm}^2\text{-km}}\right]$$

$$= 1.84 \times 10^{-4} \text{ ps/nm}^3\text{-km}$$

2.3.11 What is the range of maximum allowed PMD (14–30% of the bit period) for bit rates of 2.5 Gbit/s, 10 Gbit/s and 40 Gbit/s? Express the result in picoseconds.

The bit period corresponding to a bit rate B (Gbit/s) is $T = 1/B$ (ns). The table below shows the bit periods in ns and ps, along with the corresponding allowable PMD range:

B (Gbit/s)	T (ns)	T (ps)	PMD (ps)
2.5	0.400	400	56–120
10	0.100	100	14–30
40	0.025	25	3.5–75

2.3.12 Two optical transmission systems A and B are built with fibers with PMD of 0.1 ps/$\sqrt{\text{km}}$ and 0.8 ps/$\sqrt{\text{km}}$, respectively. Make a table showing for systems A and B the maximum achievable transmission distances for bit rates of 2.5 Gbit/s, 10 Gbit/s, and 40 Gbit/s. (Clue: The PMD should not exceed 14% of the bit period.)

The table below shows in the first two columns the conversion between bit rates (Gbit/s) and bit periods (ps). The third column shows the maximum PMD, corresponding to 14% of the bit periods. The maximum system length is given by

$$\sqrt{L \text{ (km)}} = \text{MaxPMD (ps)}/\text{PMD (ps/}\sqrt{\text{km}})$$

where PMD (ps/$\sqrt{\text{km}}$) = 0.1 or 0.8 in systems A or B, respectively. The results for $\sqrt{L}$ and L are shown in the last two columns.

B (Gbit/s)	T (ps)	MaxPMD (ps)	$\sqrt{L}$ (km)		L (km)	
			A	B	A	B
2.5	400	56	560	70	>20,000	4,900
10	100	14	140	17.5	19,600	306
40	0.025	3.5	35	4.4	1,200	19.6

2.3.13 Calculate the full development in single-frequency tones of the third-order nonlinearity

$$P_{\text{NL}} \equiv \chi_3 EEE \equiv \chi_3 \begin{cases} (E_{01}\cos\omega_1 t + E_{02}\cos\omega_2 t + E_{03}\cos\omega_3 t) \\ \times (E_{01}\cos\omega_1 t + E_{02}\cos\omega_2 t + E_{03}\cos\omega_3 t) \\ \times (E_{01}\cos\omega_1 t + E_{02}\cos\omega_2 t + E_{03}\cos\omega_3 t) \end{cases}$$

using the basic rules $\cos^2 x = (1+\cos 2x)/2$ and $2\cos x \cos y = \cos(x+y) + \cos(x-y)$.

We proceed first to develop the triple product into 27 terms while regrouping the repeated ones:

$$P_{\text{NL}} = \chi_3 \begin{cases} E_{01}E_{01}E_{01}\cos^3\omega_1 t + E_{02}E_{02}E_{02}\cos^3\omega_2 t + E_{03}E_{03}E_{03}\cos^3\omega_3 t \\ +3E_{01}E_{01}\cos^2\omega_1 t[E_{02}\cos\omega_2 t + E_{03}\cos\omega_3 t] \\ +3E_{02}E_{02}\cos^2\omega_2 t[E_{01}\cos\omega_1 t + E_{03}\cos\omega_3 t] \\ +3E_{03}E_{03}\cos^2\omega_3 t[E_{01}\cos\omega_1 t + E_{02}\cos\omega_2 t] \\ +6E_{01}E_{02}E_{03}\cos\omega_1 t \cos\omega_2 t \cos\omega_3 t \end{cases}$$

To simplify the calculation we will only consider the terms having for factors (A) $E_{01}E_{01}E_{01}$, (B) $3E_{01}E_{01}$, and (C) $6E_{01}E_{02}E_{03}$. The other terms are of the same type within a straight forward index permutation. For the (A) category, we use the development

$$\cos^3 x = (3\cos x + \cos 3x)/4$$

which is easily obtained using the two rules:

$$\cos^3 x = \cos x(1+\cos 2x)/2 = \cos x/2 + \cos x \cos 2x/2$$

and

$$2\cos x \cos 2x = \cos x + \cos 3x.$$

The total contribution of category (A) terms is thus

$$P_{\text{NL}}(A) = \chi_3 \left\{ E_{01}E_{01}E_{01}\frac{3\cos\omega_1 t + \cos 3\omega_1 t}{4} + E_{02}E_{02}E_{02}\frac{3\cos\omega_2 t + \cos 3\omega_2 t}{4} \right.$$

$$\left. + E_{03}E_{03}E_{03}\frac{3\cos\omega_3 t + \cos 3\omega_3 t}{4} \right\}$$

Consider next category (B). We use the development

$$\cos^2 x \cos y = [2\cos y + \cos(2x+y) + \cos(2x-y)]/4$$

which is easily obtained using the two rules. The total contribution of category (B) terms is thus

$$P_{NL}(B) = \chi_3 \begin{cases} \dfrac{3E_{01}E_{01}E_{02}}{4}[2\cos\omega_2 t + \cos(2\omega_1 + \omega_2)t + \cos(2\omega_1 - \omega_2)t] \\ + \dfrac{3E_{01}E_{01}E_{03}}{4}[2\cos\omega_3 t + \cos(2\omega_1 + \omega_3)t + \cos(2\omega_1 - \omega_3)t] \\ + \dfrac{3E_{02}E_{02}E_{01}}{4}[2\cos\omega_1 t + \cos(2\omega_2 + \omega_1)t + \cos(2\omega_2 - \omega_1)t] \\ + \dfrac{3E_{02}E_{02}E_{03}}{4}[2\cos\omega_3 t + \cos(2\omega_2 + \omega_3)t + \cos(2\omega_2 - \omega_3)t] \\ + \dfrac{3E_{03}E_{03}E_{01}}{4}[2\cos\omega_1 t + \cos(2\omega_3 + \omega_1)t + \cos(2\omega_3 - \omega_1)t] \\ + \dfrac{3E_{03}E_{03}E_{02}}{4}[2\cos\omega_2 t + \cos(2\omega_3 + \omega_2)t + \cos(2\omega_3 - \omega_2)t] \end{cases}$$

Finally, the last category (C) can be developed according to the formula:

$$\cos x \cos y \cos z = [\cos(x+y+z) + \cos(x+y-z) + \cos(x-y+z) + \cos(x-y-z)]/4$$

The contribution of category (C) is thus

$$P_{NL}(C) = \chi_3 \frac{6E_{01}E_{02}E_{03}}{4}[\cos(\omega_1 + \omega_2 + \omega_3)t + \cos(\omega_1 + \omega_2 - \omega_3)t$$
$$+ \cos(\omega_1 - \omega_2 + \omega_3)t + \cos(\omega_1 - \omega_2 - \omega_3)t]$$

We can now write the total development as $P_{NL} \equiv P_{NL}(A) + P_{NL}(B) + P_{NL}(C)$, i.e.,

$$P_{NL} = \frac{\chi_3}{4} \begin{cases} E_{01}E_{01}E_{01}[3\cos\omega_1 t + \cos 3\omega_1 t] \\ +E_{02}E_{02}E_{02}[3\cos\omega_2 t + \cos 3\omega_2 t] \\ +E_{03}E_{03}E_{03}[3\cos\omega_3 t + \cos 3\omega_3 t] \\ +3E_{01}E_{01}E_{02}[2\cos\omega_2 t + \cos(2\omega_1 + \omega_2)t + \cos(2\omega_1 - \omega_2)t] \\ +3E_{01}E_{01}E_{03}[2\cos\omega_3 t + \cos(2\omega_1 + \omega_3)t + \cos(2\omega_1 - \omega_3)t] \\ +3E_{02}E_{02}E_{01}[2\cos\omega_1 t + \cos(2\omega_2 + \omega_1)t + \cos(2\omega_2 - \omega_1)t] \\ +3E_{02}E_{02}E_{03}[2\cos\omega_3 t + \cos(2\omega_2 + \omega_3)t + \cos(2\omega_2 - \omega_3)t] \\ +3E_{03}E_{03}E_{01}[2\cos\omega_1 t + \cos(2\omega_3 + \omega_1)t + \cos(2\omega_3 - \omega_1)t] \\ +3E_{03}E_{03}E_{02}[2\cos\omega_2 t + \cos(2\omega_3 + \omega_2)t + \cos(2\omega_3 - \omega_2)t] \\ +6E_{01}E_{02}E_{03}[\cos(\omega_1 + \omega_2 + \omega_3)t + \cos(\omega_1 + \omega_2 - \omega_3)t \\ + \cos(\omega_1 - \omega_2 + \omega_3)t + \cos(\omega_1 - \omega_2 - \omega_3)t] \end{cases}$$

2.3.14 Determine the SRS and SBS pump thresholds for single-mode optical fibers such as SMF ($A_{\text{eff}} = 80\,\mu\text{m}^2$) and DSF ($A_{\text{eff}} = 50\,\mu\text{m}^2$), using $L_{\text{eff}} \approx 22$ km and the gain coefficients $g_R \approx 6.5 \times 10^{-14}$ m/W and $g_B \approx 5 \times 10^{-11}$ m/W. Express the results in dBm.

By definition, the SRS and SBS pump thresholds verify the respective relations:

$$g_R \frac{P_{\text{th}}}{A_{\text{eff}}} L_{\text{eff}} = 16$$

$$g_B \frac{P_{\text{th}}}{A_{\text{eff}}} L_{\text{eff}} = 21$$

or identically

$$P_{th}(\text{Raman, W}) = 16 \frac{A_{\text{eff}}}{g_R L_{\text{eff}}} = 16 \times \frac{A_{\mu m^2}^{\text{eff}} \times 10^{-12} \text{ m}^2}{6.5 \times 10^{-14} \text{ m/W} \times 22 \times 10^3 \text{ m}}$$

$$= 1.1 \times 10^{-2} \times A_{\mu m^2}^{\text{eff}}$$

$$P_{th}(\text{Brillouin, W}) = 21 \frac{A_{\text{eff}}}{g_R L_{\text{eff}}} = 21 \times \frac{A_{\mu m^2}^{\text{eff}} \times 10^{-12} \text{ m}^2}{5 \times 10^{-11} \text{ m/W} \times 22 \times 10^3 \text{ m}}$$

$$= 1.9 \times 10^{-5} \times A_{\mu m^2}^{\text{eff}}$$

Substituting the values for the effective area $A_{\mu m^2}^{\text{eff}} = 50/80 \ \mu m^2$ for DSF/SMF, we obtain the corresponding pump thresholds ($P_{\text{dBm}} = 10 \log_{10}(P_{\text{mW}})$):

	SMF		DSF	
Raman	880 mW	+29 dBm	550 mW	+27.5 dBm
Brillouin	1.52 mW	+0.18 dBm	0.95 mW	−0.02 dBm

2.3.15 Determine the required pump power to achieve Raman gains of 0 dB, +3 dB, +10 dB and +20 dB at 1.55-μm wavelength, assuming a DSF length of $L = 30$ km with pump/signal attenuation coefficient of $\alpha_p \approx \alpha_s = 0.2$ dB/km (0.046 km^{-1}), effective area $A_{\text{eff}} = 50 \ \mu m^2$ and Raman gain coefficient of $g_R = 6 \times 10^{-14}$ m/W. Same question with SMF ($A_{\text{eff}} = 80 \ \mu m^2$).

By definition, the Raman gain is given by the formula

$$G = \exp\left[g_R \frac{P_p}{A_{\text{eff}}} L_{\text{eff}} - \alpha_s L\right]$$

where $L_{\text{eff}} = (1 - e^{-\alpha_p L})/\alpha_p$ is the effective interaction length. We first get $L_{\text{eff}} \ (\text{km}) = (1 - e^{-0.046 \times 30})/0.046 = 16$ km. The pump power providing the Raman gain G in this fiber is

$$P_p = \frac{A_{\text{eff}}}{g_R L_{\text{eff}}}(\log G + \alpha_s L)$$

or with the fiber parameters

$$P_p(W) = \frac{50 \times 10^{-12} \text{ m}^2}{6 \times 10^{-14} \text{ m/W} \times 16 \times 10^3 \text{ m}}(\log G + 0.046 \text{ km}^{-1} \times 30 \text{ km})$$

$$\longleftrightarrow P_p = 0.052(\log G + 1.38)$$

The four gain values of 0 dB ($G = 1$) +3 dB ($G = 2$), +10 dB ($G = 10$) or +20 dB ($G = 100$) correspond to $\log G = 0$, 0.69, 2.3, or 4.6, respectively, which yields $P_p \approx$ 80 mW, 105 mW, 190 mW, and 310 mW, respectively. In the case of SMF, these powers must be multiplied by the effective-area ratio 80/50, yielding $P_p \approx$ 130 mW, 170 mW, 305 mW, and 495 mW, respectively.

2.3.16 Show that the condition of 1-dB power penalty in a DSF-based WDM system having an effective length L_{eff} and M channels and spaced by $\Delta\lambda$ with individual power P_0 is equivalent to the following:

$$M(M-1)\Delta\lambda_{sm}P_0^{mW}L_{eff}^{km} \leq 42{,}000 \text{ mW} \cdot \text{nm} \cdot \text{km}$$

(Clue: Use the condition $(M(M-1)/2)\Delta\lambda/\Delta\lambda_R g_R^{peak} P_0 \frac{L_{eff}}{A_{eff}} \leq 0.205$ with $\Delta\lambda_R \approx 120$ nm, $g_R^{peak} = 6 \times 10^{-14}$ m/W, and $A_{eff} = 50 \ \mu\text{m}^2$.)

Translating all the relevant parameters into proper units and replacing the numerical values yields

$$\frac{M(M-1)}{2}\left(\frac{\Delta\lambda_{nm}}{\Delta\lambda_R^{nm}}\right) g_R^{m/W} (P_0^{mW} \times 10^{-3}) \frac{L_{eff}^{km} \times 10^3}{A_{eff}^{\mu m^2} \times 10^{-12}} \leq 0.205$$

$$\frac{M(M-1)}{2}\left(\frac{\Delta\lambda_{nm}}{\Delta\lambda_R^{nm}}\right) g_R^{m/W} P_0^{mW} \frac{L_{eff}^{km}}{A_{eff}^{\mu m^2}} \leq 0.205 \times 10^{-12}$$

$$\frac{M(M-1)}{2}\left(\frac{\Delta\lambda_{nm}}{120_{nm}}\right) 6 \times 10^{-14} P_0^{mW} \frac{L_{eff}^{km}}{50_{\mu m^s}} \leq 0.205 \times 10^{-12}$$

$$M(M-1)\Delta\lambda_{nm}P_0^{mW}L_{eff}^{km} \leq 4.2 \times 10^4 \equiv 42{,}000 \text{ mW} \cdot \text{nm} \cdot \text{km}$$

2.3.17 Show that for SMF and DSF transmission fibers at $\lambda = 1.55$-μm wavelength, the nonlinear constant γ associated with the nonlinear phase $\phi_{NL} = \gamma P_i L_{eff}$ is of the order of $\gamma \approx 1 \text{ km}^{-1} \text{W}^{-1}$ (Clue: Use the definition $\phi_{NL} = n_2(2\pi/\lambda)P_i L_{eff}/A_{eff}$ with $n_2 = 2.3 \times 10^{-20} \text{ m}^2/\text{W}$ and $A_{eff} = 50 \ \mu\text{m}^2$ (DSF) or $A_{eff} = 80 \ \mu\text{m}^2$ (SMF).)

From the two definitions of the nonlinear phase, we obtain

$$\gamma = n_2 \frac{2\pi}{\lambda} \frac{1}{A_{eff}} = 2.3 \times 10^{-20} \text{ m}^2/\text{W} \frac{2\pi}{1.55 \times 10^{-6} \text{ m } A_{\mu m^2}^{eff} \times 10^{-12} \text{ m}^2}$$

$$= \frac{0.0932}{A_{\mu m^2}^{eff}} \ (\text{m}^{-1}\text{W}^{-1}) \equiv \frac{93.2}{A_{\mu m^2}^{eff}} \ (\text{km}^{-1}\text{W}^{-1})$$

Thus for DSF and SMF ($A_{eff} = 50 \ \mu\text{m}^2$ and $80 \ \mu\text{m}^2$), the nonlinear constant is $\gamma = 93.2/50 = 1.86 \text{ km}^{-1} \text{W}^{-1}$ and $\gamma = 93.2/80 = 1.16 \text{ km}^{-1} \text{W}^{-1}$, respectively, which corresponds to the (coarse) approximation and $\gamma \approx 1 \text{ km}^{-1} \text{W}^{-1}$.

2.3.18 What is the minimal linewidth of a RZ signal at $\lambda = 1.55$-μm wavelength transmitted at rate $B = 1$ Gbit/s, assuming Gaussian-envelope pulses?

By definition, the linewidth is

$$\Delta\lambda = \frac{\lambda^2}{c} \frac{\text{TBP}}{\Delta\tau}$$

where TBP $= 0.44$ for Gaussian pulses. The rate $B = 1$ Gbit/s correspond to RZ pulse widths of $\Delta\tau < 1$ ns (since there is one pulse per bit). Replacing the different parameter values into this formula, we obtain

$$\Delta\lambda > \frac{(1.55 \ \mu\text{m} \times 10^{-6} \text{ m})^2}{3 \times 10^8 \text{ m/s}} \frac{0.44}{1.0 \times 10^{-9} \text{ s}}$$

$$= 3.5 \times 10^{-12} \text{ m} = 0.035 \text{ nm}$$

2.3.19 What is the fundamental-soliton peak power in a 10-Gbit/s SMF-based 1.55-μm transmission system using soliton pulses of width $\Delta\tau = 25$ ps? (Use for SMF dispersion and effective area are $D = +17$ ps/nm-km and $A_{\text{eff}} = 80$ μm^2, respectively.) Same question for a DSF-based system ($D = +0.2$ ps/nm-km, $A_{\text{eff}} = 50$ μm^2).

We just need to apply the soliton peak-power formula

$$P^{\text{peak}}_{\text{mW}} \approx 8.1 \times \lambda^3_{\mu m} A^{\text{eff}}_{\mu m^2} \frac{D_{\text{ps/nm-km}}}{\Delta \tau^2_{\text{ps}}}$$

or in the SMF case

$$P_{\text{peak}} \approx 8.1(1.55_{\mu m})^3 80_{\mu m^2} \frac{17_{\text{ps/nm-km}}}{(25_{\text{ps}})^2} \equiv 65.5 \text{ mW}$$

and in the DSF case

$$P_{\text{peak}} \approx 65.5 \times \frac{50^2_{\mu m} 0.2_{\text{ps/nm-km}}}{80_{\mu m^2} 17_{\text{ps/nm-km}}} = 0.48 \approx 0.5 \text{ mW}$$

Section 2.4

2.4.1 Calculate the net loss experienced by a signal through a 4 × 4 star coupler according to the two implementations shown in Figure 2.40 (A, based upon 4 × 4 modules; B, based upon perfect shuffle). Same question for 16-port, 32-port, and 64-port star couplers. Application: The 2 × 2 unit couplers are assumed to have a 0.2 dB coupling/excess loss and have a −3-dB splitting loss.

Looking at the figure, implementations A require input signals to cross $m = 4$ crossings of 2 × 2 coupler units, while $m = 3$ in the (perfect-shuffle) configuration B. The net loss is therefore

$$L_{\text{dB}} = 10 \log_{10}(1/4) + m \times (-0.2 \text{ dB})$$

or

$$L_{\text{dB}}(A) = -6 \text{ dB} + 4 \times (-0.2 \text{ dB}) = -6.8 \text{ dB}$$

for implementation A and

$$L_{\text{dB}}(B) = -6 \text{ dB} + 3 \times (-0.2 \text{ dB}) = -6.6 \text{ dB}$$

for implementation B.

The generalization for star couplers with $N = 16$, 32, or 64 is straightforward. We get $m = 8$, 16, or 32 for implementation A and $m = \log_2 N = 4$, 5, or 6 for implementation B. Using

$$L_{\text{dB}}(N \times N) = -10 \log_{10}(1/N) + m \times (-0.2 \text{ dB})$$

we get

$$L_{\text{dB}}(A, 6 \times 16) = -10 \log_{10}(1/16) + 8 \times (-0.2 \text{ dB}) = -13.6 \text{ dB}$$
$$L_{\text{dB}}(B, 16 \times 16) = -10 \log_{10}(1/16) + 4 \times (-0.2 \text{ dB}) = -12.8 \text{ dB}$$

for the 16 × 16 star (corresponding to a loss difference of 0.8 dB), and

$$L_{\text{dB}}(A, 32 \times 32) = -10 \log_{10}(1/32) + 16 \times (-0.2 \text{ dB}) = -18.2 \text{ dB}$$
$$L_{\text{dB}}(B, 32 \times 32) = -10 \log_{10}(1/32) + 5 \times (-0.2 \text{ dB}) = -16.0 \text{ dB}$$

for the 32 × 32 star (corresponding to a loss difference of 2.2 dB), and

$$L_{dB}(A, 64 \times 64) = -10\log_{10}(1/64) + 32 \times (-0.2 \text{ dB}) = -24.4 \text{ dB}$$
$$L_{dB}(B, 64 \times 64) = -10\log_{10}(1/64) + 6 \times (-0.2 \text{ dB}) = -19.2 \text{ dB}$$

for the 64 × 64 star (corresponding to a loss difference of 5.2 dB), which illustrates the benefits of implementation B as the number of ports N increases.

2.4.2 Determine the path imbalance Δl (mm) of Mach-Zehnder interferometers to be used as a MUX/DMUX device for $\lambda_0 = 1.5$-µm WDM systems with $\Delta \nu = 100$-GHz frequency spacing. (Clues: Use $n_{\text{eff}} = 1.5$ for the effective index, $\Delta \omega = 2\pi \Delta \nu$, and $\omega_0 = 2\pi c/\lambda_0$ for the reference frequency.)

At the reference frequency, ω_0, the condition of maximum output transmission is

$$\beta(\omega_0)\Delta l \equiv m\pi \longleftrightarrow n_{\text{eff}} \frac{2\pi}{\lambda_0} \Delta l = m\pi \longleftrightarrow \Delta l = \frac{m\pi\lambda_0}{2\pi n_{\text{eff}}} \equiv \frac{m\lambda_0}{2n_{\text{eff}}}$$

On the other hand, the period of the MZI frequency oscillation is defined as

$$\Omega = 2\frac{\omega_0}{m} = 2\frac{2\pi c}{m\lambda_0} = \frac{4\pi c}{m\lambda_0}$$

The condition for splitting WDM channels by odd/even groups is $\Omega = 2\Delta\omega$, where $\Delta\omega$ is the channel spacing in radians/s. Substituting this condition in the previous definition yields

$$\frac{4\pi c}{m\lambda_0} = 2 \times 2\pi\Delta\nu \longleftrightarrow m = \frac{c}{\lambda_0 \Delta\nu} = \frac{3 \times 10^8 \text{ m/s}}{1.5 \times 10^{-6} \text{ m} \cdot 100 \times 10^9 \text{ Hz}} = 2,000$$

From this result, we find the path imbalance

$$\Delta l = \frac{m\lambda_0}{2n_{\text{eff}}} = \frac{2,000 \times 1.5 \text{ µm}}{2 \times 1.45} = 1,035 \text{ µm} \approx 1.03 \text{ mm}$$

To continue splitting the channels, one requires to pass the signal outputs to a sequence of other MZI having for oscillation frequencies: $\Omega = 4\Delta\omega$, $\Omega = 8\Delta\omega$, etc. In these cases, we find $m = 2,000$, $m = 1,000$, etc., yielding for new path imbalances $\Delta l = 0.515$ mm, $\Delta l = 0.257$ mm, etc.

2.4.3 Determine the fiber Bragg-grating lengths required to achieve a reflectivity of $R = 99\%$, at 1.55-µm wavelength, assuming a power fraction in the fiber core of $\eta = 0.5$ and index changes of $\delta n = 10^{-2}$–10^{-3}.

The result is provided by the strict application of the formula

$$R = \tanh^2\left(\eta \frac{\pi L \delta n}{\lambda_B}\right)$$

or

$$L = \frac{\lambda_B}{\pi \eta \delta n} \text{arctan}(\sqrt{R})$$

with $\sqrt{R} = \sqrt{0.99} = 0.9949$. If we don't have a pocket calculator with the reciprocal arctan(x) function, we can still find rapidly that the argument $x = 2.99$ gives tanh(x) ≈ 0.9949, meaning arctanh(0.9949) ≈ 2.99. Thus, we get

$$L = \frac{1.55 \times 10^{-4}\,\text{cm}}{\pi \times 0.5 \times \delta n} \times 2.99 = \frac{2.95 \times 10^{-4}}{\delta n}$$

or with $\delta n = 10^{-2} - 10^{-3}$, $L \approx 3 \times 10^{-2}$ to $L \approx 3 \times 10^{-1}$ centimeters (300 μm to 3 mm).

2.4.4 Determine the frequencies of the RF drive signals to realize an acousto-optic tunable filter (AOTF) in LiNbO$_3$ capable to select 1.55-μm WDM signals spaced by 10 nm. (Use for the index difference $\delta n = 0.07$ and for the sound velocity $V_a = 3.75$ km/s.) For a given resonant wavelength λ_B, the RF frequency f_a is simply given by the formula $\lambda_B = \delta n V_a / f_a$, or

$$f_a = \frac{\delta n V_a}{\lambda_B} = \frac{0.07 \times 3.75 \times 10^3\,\text{m/s}}{\lambda^B\,\mu\text{m} \times 10^{-6}\,\text{m}} = \frac{0.07 \times 3.75 \times 10^3\,\text{m/s}}{\lambda^B\,\mu\text{m} \times 10^{-6}\,\text{m}}$$

$$= \frac{2.62 \times 10^8\,\text{Hz}}{\lambda^B\,\mu\text{m}} \equiv \frac{262\,\text{MHz}}{\lambda^B\,\mu\text{m}}$$

For a 1.55-μm WDM comb with 10-nm (0.01-μm) spacing, we obtain the successive RF frequencies

$$f_a(1{,}540\,\text{nm}) = \frac{262\,\text{MHz}}{1.54} = 170.12\,\text{MHz}$$

$$f_a(1{,}550\,\text{nm}) = \frac{262\,\text{MHz}}{1.55} = 169.03\,\text{MHz}$$

$$f_a(1{,}560\,\text{nm}) = \frac{262\,\text{MHz}}{1.56} = 167.94\,\text{MHz}$$

$$f_a(1{,}570\,\text{nm}) = \frac{262\,\text{MHz}}{1.57} = 166.87\,\text{MHz, etc.}$$

which differ by approximately 1.1 MHz.

2.4.5 Determine the pump powers required to cancel the link loss at 1.55-μm wavelength, assuming a DSF/SMF lengths of $L = 50$ km and 100 km with attenuation coefficient of $\alpha = 0.2$ dB/km (0.046 km^{-1}), effective area $A_{\text{eff}} = 50/80$ μm^2 and a Raman gain coefficient of $g_R = 6 \times 10^{-14}$ m/W.

For a trunk of length L and attenuation α, the net Raman gain is defined through

$$G = \exp\left(g_R \frac{P_p}{A_{\text{eff}}} L_{\text{eff}} - \alpha_s L\right) \equiv \exp(gL)$$

The condition for loss compensation is therefore $g = 0$, or

$$g_R \frac{P_p}{A_{\text{eff}}} \frac{L_{\text{eff}}}{L} - \alpha_s = 0$$

$$\longleftrightarrow P_p = \frac{\alpha_s L A_{\text{eff}}}{g_R L_{\text{eff}}}$$

Replacing the numerical values into the above expression, we find

$$P_p(W) = \frac{0.046\,\text{km}^{-1}}{6 \times 10^{-14}\,\text{m/W} \times 22 \times 10^3\,\text{m}} L_{\text{km}} A_{\mu\text{m}^2}^{\text{eff}} \times 10^{-12}\,\text{m}^2$$

$$\approx 3.5 \times 10^{-5} L_{\text{km}} A_{\mu\text{m}^2}^{\text{eff}}$$

which gives, in the four cases of interest:

$$P_p(\text{DSF}, 50\,\text{km}) = 3.5 \times 10^{-5} \times 50\,\text{km} \times 50\,\mu\text{m}^2 = 8.7 \times 10^{-2}\,\text{W} = 87\,\text{mW}$$
$$P_p(\text{DSF}, 100\,\text{km}) = 2 \times 87\,\text{mW} = 175\,\text{mW}$$
$$P_p(\text{SMF}, 50\,\text{km}) = 3.5 \times 10^{-5} \times 50\,\text{km} \times 80\,\mu\text{m}^2 = 140\,\text{mW}$$
$$P_p(\text{SMF}, 100\,\text{km}) = 2 \times 140\,\text{mW} = 280\,\text{mW}$$

Section 2.5

2.5.1 Calculate the linewidth of LEDs emitting with a peak wavelength in the range $\lambda' = 0.5\text{--}1\,\mu\text{m}$ at room temperature ($k_B = 1.38 \times 10^{-23}\,\text{J/K}$).

The result is the straightforward application of the formula

$$\Delta\lambda = 1.45(\lambda')^2 k_B T$$

where $k_B T$ is expressed in electron-volts and λ' in microns. At room temperature ($T = 300\,\text{K}$), we first obtain

$$k_B T = 1.38 \times 10^{-23}\,\text{J/K} \times 300\,\text{K} = 4.14 \times 10^{-21}\,\text{J}$$

which in eV units gives

$$(k_B T)_{\text{eV}} = 4.14 \times 10^{-21}\,\text{J}/(1.6 \times 10^{-19}\,\text{C}) = 2.5 \times 10^{-2}\,\text{eV}$$

For peak wavelengths in the range $\lambda' = 0.5\text{--}1\,\mu\text{m}$, the corresponding linewidths are

$$\Delta\lambda = 1.45(0.5)^2 \times 2.5 \times 10^{-2} = 9.1 \times 10^{-3}\,\mu\text{m} \approx 10\,\text{nm}$$
$$\Delta\lambda = 1.45(1)^2 \times 2.5 \times 10^{-2} = 36.2 \times 10^{-3}\,\mu\text{m} \approx 35\,\text{nm}$$

Section 2.6

2.6.1 Assume that the WDM channels form a regular comb with increasing frequencies $f_1, f_2, \ldots, f_N$ such that the frequency spacing $\Delta f = f_{i+1} - f_i$ is uniform or constant, with $N = 32$, $\Delta f = 100\,\text{GHz}$, $\lambda_1 = 1.550\,\mu\text{m}$, and $c' = 3 \times 10^8\,\text{m/s}$. Calculate the wavelength spacing corresponding to the first two (1, 2) and the last two ($N - 1$, N) channels, respectively, and show that there are not strictly identical. What can be concluded?

According to the above assumption, the frequency comb is defined by the rule $f_i = f_1 + (i - 1)\Delta f$, where $i = 1, \ldots, N$. By definition, we also have $f = c'/\lambda$. Thus the first and the last frequencies of the comb are

$$f_1 = c'/\lambda_1 = 3 \times 10^8\,(\text{m/s})/(1.550 \times 10^{-6}\,\text{m}) = 193.54 \times 10^{12}\,\text{Hz} \equiv 193.5483\,\text{THz}$$
$$f_N = f_1 + (N - 1)\Delta f = 193.5483\,\text{THz} + (32 - 1) \times 0.1\,\text{THz} = 196.6483\,\text{THz}$$

Consider then the first two and the last two wavelengths of the comb:

$$\lambda_1 = 1.550\,\mu\text{m}$$
$$\lambda_2 = c'/f_2 = c'/(f_1 + \Delta f) = 3 \times 10^8/(193.5483 \times 10^{12} + 0.1 \times 10^{12}) = 1.54920\,\mu\text{m}$$
$$\lambda_N = c'/f_N = 3 \times 10^8/(196.6483 \times 10^{12}) = 1.52556\,\mu\text{m}$$
$$\lambda_{N-1} = c'/f_{N-1} = c'/(f_N - \Delta f) = 3 \times 10^8/(196.6483 \times 10^{12} - 0.1 \times 10^{12}) = 1.52634\,\mu\text{m}$$

which gives the first and last wavelength spacings:

$$\Delta\lambda_{12} = \lambda_1 - \lambda_2 = 1.5500\,\mu m - 1.5492\,\mu m = 8 \times 10^{-4}\,\mu m = 0.8\,nm$$

$$\Delta\lambda_{N-1,N} = \lambda_{N-1} - \lambda_N = 1.52634\,\mu m - 1.52556\,\mu m = 7.8 \times 10^{-4}\,\mu m = 0.78\,nm$$

We thus observe that the two extreme wavelength spacings are different by 0.02 nm, showing that the wavelengths are not equally spaced in the comb. If we were to use an equal wavelength spacing—for instance, of $\Delta\lambda = 0.8$ nm—the last wavelength of the comb would be defined by $\lambda'_N = \lambda_1 - (N-1)\Delta\lambda$, or $\lambda'_N = 1.550\,\mu m - (32-1) \times 0.8 \times 10^{-3}\,\mu m = 1.52520\,\mu m$. The corresponding frequency would be $f'_N = c'/\lambda'_N = 3 \times 10^8/(1.5252 \times 10^{-6}) = 196.6955$ THz. This frequency value is in excess of 2×10^{-4} with respect to the exact frequency $f_N = 196.6483$. The key conclusion of this exercise is that one can neglect the small error introduced by using an equal-wavelength spacing definition to replace the equal-frequency spacing definition.

2.6.2 Determine the graphs of the physical and virtual topologies associated with the network shown in Figure 2.111.

Consider first the physical topology on the left side of the figure. Let us call V_1, V_2, V_3, V_4, V_5 the nodes (or vertices) A, B, C, D, E. The set of vertices is thus

$$V = (V_1, V_2, V_3, V_4, V_5)$$

The inventory of directed physical links (or directed edges) comes as follows:

$$A \to B \quad B \to A \quad A \to E \quad E \to A$$
$$B \to C \quad C \to D \quad D \to C \quad D \to E$$

corresponding to

$$E_p = (V_1, V_2), (V_2, V_1), (V_1, V_5), (V_5, V_1), (V_2, V_3), (V_3, V_4), (V_4, V_3), (V_4, V_5)$$

The virtual topology is different for each wavelength. Using the left side of the figure, we find the corresponding link lightpaths:

$\lambda_1(k=1)$:	$A \to B \quad B \to C \quad D \to E$
$\lambda_2(k=2)$:	$A \to B \quad B \to C \quad C \to D$
$\lambda_3(k=3)$:	$A \to B \quad C \to D \quad D \to E$
$\lambda_4(k=4)$:	$B \to A \quad A \to E \quad D \to C$

or equivalently:

$\lambda_1(k=1)$:	$(V_1, V_2) \quad (V_2, V_3) \quad (V_4, V_5)$
$\lambda_2(k=2)$:	$(V_1, V_2) \quad (V_2, V_3) \quad (V_3, V_4)$
$\lambda_3(k=3)$:	$(V_1, V_2) \quad (V_3, V_4) \quad (V_4, V_5)$
$\lambda_4(k=4)$:	$(V_2, V_1) \quad (V_1, V_5) \quad (V_4, V_3)$

which form the lightpath (or logical-connection) set noted $E_v = \{(V_i, V_j)_k\}$:

$$E_v = (V_1, V_4)_1, (V_4, V_5)_1, (V_1, V_4)_2, (V_1, V_2)_3, (V_3, V_5)_3, (V_2, V_5)_4, (V_4, V_3)_4$$

The physical and virtual graphs are thus completely defined.

Bibliography

📖 General and Introductory

Desurvire, E., *Wiley Survival Guide in Global Telecommunications: Signaling Principles, Network Protocols, and Wireless Systems*, John Wiley & Sons, Hoboken, NJ, 2004

Dodd, A., *The Essential Guide to Telecommunications*, 3rd edition, Prentice-Hall PTR, Upper Saddle River, NJ, 2002

Dutton, H.J.R., *Understanding Optical Communications*, Prentice-Hall, Englewood Cliffs, NJ, 1999

Killen, H.B., *Fiber Optic Communications*, Prentice-Hall, Englewood Cliffs, NJ, 1991

Levy, S., *Crypto*, Penguin Books, Harmondsworth, 2001

Palais, J.C., *Fiber Optics Communications*, 2nd edition, Prentice-Hall, Englewood Cliffs, NJ, 1988

Shepard, S., *Telecom Crash Course*, McGraw-Hill, New York, 2002

Singh, S., *The Code Book*, Anchor Books, New York, 1999

📖 Advanced and Technical

Aggrawal, G.P., *Fiber-Optics Communications Systems*, John Wiley & Sons, New York, 1992

Aggrawal, G.P., *Nonlinear Fiber Optics*, Academic Press, New York, 1989

Azzam, A. and Ransom, N., *Broadband Access Technologies*, McGraw-Hill, New York, 1999

Bates, B. and Gregory, D., *Voice & Data Communications Handbook*, McGraw-Hill, New York, 1996

Boisseau, M., Demange, M. and Munier, J.-M., *High Speed Networks*, John Wiley & Sons, New York, 1994

Born, M. and Wolf, E., *Principles of Optics*, Pergamon Press, New York, 1980

Chesnoy, J., Editor, *Undersea Fiber Communication Systems*, Elsevier Science, New York, 2002

Clark, M.P., *Networks and Telecommunications, Design and Operation*, 2nd edition, John Wiley & Sons, New York, 1991

Comer, D.E., *Internetworking with TCP/IP*, Volume I, 3rd edition, Prentice-Hall, Englewood Cliffs, NJ, 1995

Wiley Survival Guide in Global Telecommunications: Broadband Access, Optical Components and Networks, and Cryptography, by E. Desurvire
ISBN 0-471-67520-2   © 2004 John Wiley & Sons, Inc.

Desurvire, E., *Erbium-doped Fiber Amplifiers, Principles and Applications*, John Wiley & Sons, New York, 1994

Desurvire, E., Bayart, D., Desthieux, B. and Bigo, S., *Erbium-doped Fiber Amplifiers, Devices and System Developments*, John Wiley & Sons, New York, 2002

EURESCOM, Project P-918-GI Deliverable N.1, "IP over WDFM, transport and routing," EURESCOM, October 1999

Fouché Gaines, H., *Cryptanalysis*, (formerly *Elementary Cryptanalysis*) Dover Publications, New York, 1956

Goralski, W.J., *ADSL & DSL Technolofies*, 2nd edition, Osborne/McGraw-Hill, Berkeley, 2002

Green, P., *Fiber Optic Networks*, Prentice-Hall, Englewood Cliffs, NJ, 1993

Jeunhomme, L.B., *Single-Mode Fiber Optics Principles and Applications*, 2nd edition, Marcel Dekker, New York, 1990

Jukan, A., *QoS-based Wavelength Routing in Multi-Service WDM Networks*, Springer-Verlag/Wien, New York, 2001

Kahn, D., *The Codebreakers, the Story of Secret Writing*, Scribner Publishers, New York, 1967

Kaminov, I.P. and Tingye Li, *Optival Fiber Telecomunications IV (A and B)* Elsevier Science (USA), 2002

Kartalopoulos, S.V., *Understanding SONET/SDH and ATM*, IEEE Press, New York, 1999

Kartalopoulos, S.V., *Introduction to DWDM technology*, IEEE Press, New York, 2000

Keiser, G., *Optical Fiber Communications*, 2nd edition, McGraw-Hill, New York, 1991

Mel, H.X. and Baker, *Cryptography Decripted*, Addison-Wesley, New York, 2001

Mukherjee, B., *Optical Communications Networks*, McGraw-Hill, New York, 1997

Proakis, J.G., *Digital Communications*, 4th edition, McGraw-Hill, New York, 2001

Ramaswami, R. and Sivarajan, K.N., *Optical Networks, a Practical Perspective*, Morgan Kaufmann Publishers, San Francisco, 1998

Saleh, B.E.A. and Teich, M.C., *Fundamentals of photonics*, John Wiley & Sons, New York, 1991

Schneier, B., *Applied Cryptography*, 2nd edition, John Wiley & Sons, New York, 1996

Senior, J.M., *Optical Fiber Communications, Principles and Practice*, 2nd edition, Prentice-Hall, New York, 1992

Sivalingam, K.M. and Subramanian, S. (Editors), *Optical WDM Networks, Principles and Practice*, Kluwer Academic Publishers, Norwell, MA, 2000

Optical telecommunications, E. Desurvire, Editor, Académie des Sciences Dossier (in English), Vol. 4, No. 1, Elsevier, Paris, 2003

Recommended Web-Site Links

A list of recommended Web sites in a paper book can only be indicative and never complete or exhaustive, current or even still valid. It is in the spirit of the internet that people must browse in every direction and generate their own selection of bookmark preferences. Here are a few URLs which may prove useful for a fresh start. In each Web site, the reader may pay close attention to offered *tutorials*, *white papers*, *FAQs*, *search engines* and *related links*. Direct Internet browsing by keywords is also recommended, but using *boolean functions* for faster convergence. Note that URL bookmarks/preferences may be renamed for archival and memorizing purposes, using the right click "properties" command. As a final recommendation and advice, one may subscribe to *e-mail newsletters*, provided the "unsubscribe" options be clearly stated. This is the opportunity to acknowledge the content owners and Webmasters for all the information freely provided. Some web sites are real information treasures with document jewels and can make one's day. We apologize if we missed important ones, which we surely did without intent. Again, it takes the reader's patience and exploration to compile an up to date list. The following list is grouped by topics, but topics may overlap each other and URLs appear only once. The thing is to try them out.

Broadband Access (see also wireless networks and free-space optics)

www.iec.org/online/tutorials

www.fsanet.net

www.vcedmagazine.com

www.cesti.pm.gouv.fr/uk

Cryptography and Security

www.rsasecurity.com/rsalabs/challenges/

www.cdt.org

www.xat.nl/enigma/

www.enigmahistory.org/enigma.html

www.theory.lcs.mit.edu/~rivest/crypto-security.html

www.csrc.nist.gov/encryption/tkencryption.html

www.esat.kuleuven.ac.be/~rijmen/rijndael/

www.counterpane.com

www.pgpi.com

www.cryptography.org/getpgp.htm

www.utm.edu/research/primes/

www.nsa.gov:8080/museum

www.und.nodak.edu/org/crypto/crypto

www.csua.berkeley.edu/cypherpunks/Home.html

www.secnet.com/references.html

www.garykessler.net/library/crypto.html

www.windowsecurity.com/articles_tutorials/

www.secinf.net/

www.cosic.esat.kuleuven.ac.be/sesame/
web.mit.edu/kerberos/www/
www.netscape.com/security/techbriefs/ssl.html
www.e-poll-project.net/
www.notablesoftware.com/evote.html
www.ecst.csuchico.edu/~atman/Crypto/quantum/quantum-index.html
www.cs.dartmouth.edu/~jford/crypto.html
dir.yahoo.com/Science/Physics/Quantum_Cryptography/
www.gocsi.com

Digital Subscriber Line

www.dslforum.org
www.dsllife.com
www.dslreports.com
www.vdslalliance.org
www.vdsl.org
www.davic.org
www.fs-vdsl.net

Free-space Optics

www.bakom.ch/en/funk/forschung/laserkommunikation/
www.fsoalliance.com
www.freespaceoptics.com
www.freespaceoptics.org

Glossaries of Telecom Terms

www.techweb.com/encyclopedia
www.fcc.gov/glossary
www.glossary.its.bldrdoc.gov/fs-1037
www.atis.org/tg2k
www.alcatel.com/atr

Lightwave Systems

www.lightreading.com
www.opticsnotes.com/onindex

www.ponforum.org

Moore's Law

www.news.com.com (CNET)

www.firstmonday.org/issues/issue7_11/tuomi

Standardization Bodies

www.etsi.org

www.itu.int, www.itu.org

www.fcc.org

Wireless and Mobile Networks (see also broadband access)

www.lmdswireless.com

www.watmag.com

www.bbwexchange.com

www.wireless-wolrd-research.org

www.weca.com

www.80211b.weblogger.com

www.isp-planet.com/fixed_wireless/technology/

www.wapforum.org

www.3g-generation.com

www.4gmobile.com

www.palowireless.com

Index

1+1/1:1/1:N/N:N/M:N protection, 317–320
2-PSK/PAM, *see* Modulation formats
2B1Q, 23, 30
3-level/4-level system/pumping, *see* Pumping
3dB or 50/50 coupler, 201–205
3DES, 408, 416–418, 426, 428, 445, 449
2G/3G/4G (mobile systems), 13–14, 51, 82, 296, 431
3R regeneration, *see* Regeneration
2M-1/2M-3 (DSL), 37
24/26 AWG, *see* AWG (DSL)
802.11a/b/g, 7, 87–88, 431
802.1×, 431
16-ary QAM, *see* Modulation formats

A3/A8, 429
A5/A5-0/A5-1/A5-2/A5-3, 430-431
AAL/AAL1/AAL5, 37, 48–49
Abbé Trithème, 365–366
Absorption
  coefficient (EDFA), 238
  length, 183
  rate, *see* Rate
Abuse (network), *see* Network
AC, 87
Access, 11–12, 14, 51, 57, 75
Acoustic wave, 226
Acousto-optic tunable filter, *see* AOTF
Adaptive/equalization, *see* Frequency
Add-drop multiplexing
  electrical (–), *see* ADM
  optical (–), *see* OADM
AddRoundKey (operation), 421
Adleman, 398
ADM (or EADM), 300, 304, 310, 314, 331

ADSL, 12, 14–15, 18, 20, 35, 50, 57, 76, 84–85
Advanced encryption standard, *see* AES
AES, 5, 407–408, 418–427, 431, 452
AES-128/192/256, 421, 431
AEX, 41
Airsnort, 431
Al-Kindi, 353
Alberti, 350
AlGaAs, 266
Algorithm
  Diffie–Hellmann–Merkle (–), *see* Diffie
  encryption (–), *see* Encryption
  Hughes (–), *see* Hughes
  secure hash (–), *see* SHA
  sieving (–), 403
Alice, Bob and Eve, 389
All-optical regeneration, *see* AOR
Alphabet (plain/cipher), 349–350
Alphabetic
  translation, 348–350
  substitution, 350–352, *see also* Cipher
Alternative current, *see* AC
AM, 23, 49, 87
AM-radio interference, 29, 31
American wire gauge, *see* AWG (DSL)
Ampère, 122
Amplified spontaneous
  emission, *see* ASE
  scattering (SRS), 184–185
Amplifier/amplification
  analog/RF, 76, 78
  bands, 247–248, 294, 303
  bidirectional, 241
  distributed (–), 185–186, 246
  fiber (–), 96, 133, 156, 158, 234–248
  front-end (–), 267–268

Wiley Survival Guide in Global Telecommunications: Broadband Access, Optical Components and Networks, and Cryptography, by E. Desurvire
ISBN 0-471-67520-2 Copyright © 2004 John Wiley & Sons, Inc.

Amplifier/amplification (*Continued*)
  gain, *see* Gain
  lumped (−), 185−186
  optical, *see* Optical
  parametric (−), *see* Parametric
  post- (−), 247, 262
  pre- (−), 247, 266−269
  spacing, 296, 298−299
Ancient scripts (lost languages), 348
AND (logical/Boolean), 289, 384−385
Anisotropic (material), anisotropy, 99, 102
Anonymous remailer, 427
ANSI, 32, 47, 408, 416
Antireflection coating, 197
AOTF, 226−227
AOR, 194, 228, 286
APC, 228, 235, 244, 258, 267−268
APD, 263, 265−266
APD-FET, 268
Apodization (grating), *see* Grating
APON, 57, 62, 65−71, 81
APS, 319
AR/HR coating, 254, 265
Arbitrator, 433, 438−439, 444, 453
Arrayed-waveguide grating, *see* AWG
AS
  (security), 444
  subchannel (DSL), 35−43
AS0/AS1/AS2/AS3, 35−36, 39, 40−43
ASAM, 17
ASCII, 364, 382−383, 394, 400, 402, 417, 434
ASE, 118−119, 194, 228, 286, 290, 296, 299, 471
ASON, 334
ASP, 75
Associativity, 392
Asymmetric
  key cryptography, *see* Key
  subscriber line, *see* ADSL
ATM, 14, 16−17, 34, 51−52, 54, 57, 59, 62, 65, 66−68, 71−76, 81, 281, 296, 313−315, 324, 330, 334−335
  adaptation layer, 37
  DSLAM, 17
  over DSL, 34−37, 48−50
  over SDH, 76, 313, 334
  PON, *see* APON
  switch, 16, 272, 314, 330
  voice over (−), *see* VoATM
ATM-ADSL, *see* ATM
Atomic
  cross-section, 114−115, 126, 238
  energy level, 113
  population, 116, 126
  transition, 95

Attack
  birthday (−), 437
  brute-force (−), see Brute
  code (−), *see* Code
  dictionary (−), 432
  hash (−), 434, 436−437
  network (−), 441
  polling (−), 456, 460
  web-site (−), 441−443
Attenuation
  coefficient, 32, 61, 117, 157−159, 165, 167, 228, 247−248
  fiber (−), 63, 95
  light (−) through atmosphere, 94
ATU-R/C, 16, 18, 40
Audit trail, 458
Automatic teller machine, 429, 450
Authentication
  center (network), *see* AUC
  header, 446−448
  message (−), 407, 429, 431−432, 434−439, 444, 447, 450, 452−453, 461
  server, *see* AS
  signature (−), *see* Digital
AUC, 429
Automatic
  power control, *see* APC
  protection switching, *see* APS
Avalanche
  gain, 265−266
  photodiode, *see* APD
AWG
  (DSL), 32
  (waveguide), 202, 206, 210−212, 217, 233, 279, 293

Babbage, 352
Backbone, *see* Network
Backscattering (Brillouin), 183
Backward compatibility, 5
Baker, *see* Mel
Band (conduction/valence), 121−122, 249, 263
Bandgap, 121, 247, 249−250, 264
Bandwidth, 4
Bandwidth
  × length performance, 161−162
  amplifier (−), *see* Gain
  dynamic (−) allocation, *see* DBA
  efficiency, *see* Spectral
  explosion, 301
  fiber (−), 63, 76
  gain (−), *see* Gain
  infinite/unlimited, 63, 96, 159, 162
  fiber (−), 159

FSO (−), 94
optical (−), 91
overhead, 298
receiver (−), *see* Receiver
Baud, 26
BB
access, 51, 57, 86
DSL, 14
(services), *see* Broadband
BB84, 461
BBOR, 287–288, 290
Beacon (maritime), 93
Beale cipher, *see* Cipher
Beale's papers, 376–377, 397
Bearer (DSL), *see* DSL
Beat length, 103, 166, 168
Bennett, 461
BER, 18, 25, 27, 29, 33, 43, 69, 158, 231, 234, 243, 245, 255, 262–263, 269–270, 286, 291–292, 295, 297–299–300, 333
Bernoulli's random partitioning, 463, 466
Bessel/Hankel function, 154
Bias (voltage, forward/reverse), 251–252, 264–265
Big Brother, 458
Bin (frequency), 27–28
Binary
digit, *see* Bit
number, 382
BIP, 69, 74
Birefringence
axes, 102
elliptical, 167–168
fiber natural (−), 99, 168
geometrical, 166
medium/intrinsic (−), 103, 166, 213
stress-induced (−), 100, 166, 168
Birefringent (material), 99
Bit, 1
Bit
error rate, *see* BER
per symbol, 22
scrambling, 26
synchronization (−), 40
BLEC, 54
Blind signature, 453, 457, 459
Blinding factor, 453, 457
Bloch mode, *see* Mode
Block, 388
Blowfish, 428
BLSR, 320
Bluetooth, 87
BOL, 299
Boltzmann's constant, 252, 267
Bombe, 381

Bombing, 441
Boolean logic, 384–385
Bow-tie fiber, *see* Fiber
BPON, 66
BPSR, 319–320
Bragg
condition, 222
fiber (−) grating, *see* IFBG
reflection grating, *see* Grating
wavelength, 222, 225–226, 257
Brassard, 461
BRI, 20
Brillouin
gain coefficient, 183
shift, 183
Broadcasting (TV), 11–12, 91
Broadband
PON, see BPON
services, 14, 54
Brute-force (attack), 363, 408, 416–418, 431, 451, 454, 471, 475
BSHR, 319
Buffer (fast/interleaved data), 39–43
Buffering (packet), 330
Burst (traffic), 11, 13
Byte, 39

C-band, 244–245, 248, 294
C × D performance, 55, 300–303
C + L (EDFA), 245, 248, 294
C-channel (DSL), 37
CA, 440–441, 443–444
Cable
operator, 53
TV, 4, 12, 53, 76, 80
Caesar, *see* Cipher
CAP
(modulation), *see* Modulation formats
(telecom carrier), *see* Competitive
Capacity × distance (performance)
optical fiber, see C × D
upgrade, 298
(×DSL), 21, 55
Carbon dioxide, 120
Carding, 441
Carrier
charge (−), 121
free (−), 121–123, 249
multi/muliple (−), 22, 27
recombination, 122, 249, 251
(telecom operator), *see* Operator, CLEC, ILEC, IXC, LEC, LXC
single (−), 22–23
sine/cosine, 24–25

Carrierless AM, 25
CAST/CAST5/CAST128, 428, 433
CATV, 12–14, 77, 83
Cavity
　Fabry–Pérot (–), see Fabry–Pérot
　laser (–), 124–126
CBC, 415–416, 445, 448
CCA, 443
CD/DVD player, 56
CD-ROM, 21
CDMA, 79–80
cdma2000, 451
cdmaOne, 431
Central office, see CO
Centrex, 17
CEO, 7
Certificate, 452, 449–450, 457
Certification authority, see CA
Certified e-mail, 407, 454–455
CFB, 415–416
CGM, see XGM
Chaining variables, 434–436
Channel
　communication (–), 92–93
　spacing (WDM), 63, 293
Chip/chipping code, 87
Chirp, 224–225
Cindy, 397
Cipher
　Abbé Trithème (–), 365–366
　ADFGVX (–), 377
　Beale (–), 375–377, 397
　biliteral (–), 364
　block (–), 388, 408
　book (–), 375
　Caesar shift (–), 349–350
　concealment (–), 365
　Enigma-machine (–), 378–381, 389, 397
　homophonic substitution (–), 357
　invulnerable/unbreakable/undecipherable (–), 347–348, 350, 377, 384
　Knight's tour (–), 366–367
　Lord Bacon's (–), 364–365
　monoalphabetic (–), 348–350, 352, 355–356–357, 359, 408
　Morse (–), 377
　multialphabetic/polyalphabetic [substitution] (–), 350, 356–357, 360–362, 371, 378, 408
　nihilist square (–), 368–370
　null (–), 346
　onetime pad (–), 92, 352, 368, 384, 429, 433, 468, 474
　Philips (–), 371–372
　playfair (–), 372–375
　Polybius square (–), 370–371
　punk, see Cryptopunk
　Rivest (–), 427–428
　suite, 449
　transposition (–), 366–370, 377
　trinumeral (–), 365
　unbreakable/indecipherable (–), 347, 350, 377
　Vigenère (–), 350–352, 356, 360–362, 372, 384, 388, 397
　Zanotti transposition (–), 367–369
Cipherpunk, see Cryptopunk
Ciphertext, 347
Circuit/channel (voice), 11
Circulator (optical), 109, 214, 216–217, 276–277
Cladding
　fiber (–), see Fiber
　mode, see Mode
　-pumped EDFA, see EDFA
　waveguide (–), 147, 149
Clarke (Arthur C.), 98
Class (DSL transport-), 37
Classical
　light, 111, 462
　nature of light, 110–115, 474
CLEC, 13, 52–55, 57, 75, 80, 335
Clock
　arithmetic, see Modular algebra
　recovery, see CR
CO, 16–17, 20, 46, 48, 50–51, 53, 55
Codopants (index-raising), see Index
COBRA, 274
Code
　attack, 403, 417, see also Brute-force
　binumeral (–), 357
　breaking/cracking, 347, 353–354, 356, 362–364, 366, 368, 380–382, 389, 406–408, 415–417, 427, 431, 442
　cracker/breaker, 347, 352, 356, 362, 368–369, 372, 417, 442
　cracking, 347
　invulnerable/unbreakable/indecipherable, 347–348, 350, 377, 384
　jargon (–), 346–347, 365
　key, 347, 363
　Morse (–), 93, 377
　open (–), 346
　word, 93
Coding efficiency, 357, 359–363
Coherence (temporal/spatial), 95, 132
Coherent
　light, 113, 115
　system, 167, 270
Collision/collision-resistant, 434, 436
Communications security, see Security

Commutativity, 391–392
Competitive access provider (CAP), 54
Complexity, 416–418, 426–428, 451
COMP128, 430
Component (optical)
   active (−), 96, 194
   passive (−), 96, 194–248
Compression permutation, 409
Confidentiality (protection), 451–452, 456
Connector (fiber)
   cost, 55
   types, 87, 196–197
Constellation
   QAM, 23–25, 28–29
   (satellite), *see* Satellite
Control vector, *see* CV
Copper (wire)
   cost, 55
   gauge, 18, 32
   impedance/resistance, 32
   plant, 55, 56–57, 73
   twisted (−) pair, 14–16, 18, 32–34, 56, 73, 86
Core
   fiber (−), *see* Fiber
   waveguide (−), 147, 149
Corning Glass, 157
Coupler (1xN, NxN), 198–208
Coupling length
   (PMD), 170
   two-mode waveguide, 200–201, 273
CP, 15–17, 20, 48–49, 51, 54–55, 73
CPE, 49
CPM, *see* XPM
CR, 288–291
CR-LDP, 331–332
CRC, 25, 40, 74
Credit card (number), 449–450
Crib, 353, 356, 368, 370, 372, 381, 386
Crime, 441, 443
Cross
   -connect (optical), *see* OXC
   -gain modulation, *see* XGM
   -phase modulation, *see* XPM
   -section (atomic), *see* Atomic
Crosstalk, 29, 33, 212, 243, 245, *see also* Polarization
Cryptology, 363, 381
Cryptanalysis, 3, 386, 391, 415, 417–418, 426–427, 452
Cryptography, 92, 347–348, 442
Cryptopunk/cipherpunk/cryptocracker, 441–442
Cryptosystem, 427, 452, 456
CRZ, 299

CSBH, 254
CSI, 443
CSMA/CA, 87
Customer premises, *see* CP
Cut-off wavelength, *see* Wavelength
CV, 443–444
CWDM, 63–65, 294

D-Day, 346–347
Daemen, 421
Dark fiber, *see* Fiber
Data
   encryption standard, *see* DES
   LEC, *see* DLEC
Datagram (Internet), 13, 73, 330, 445
DAVIC, 47
dB, 32, 117
dB/m, 118
dB/km, 32, 61, 157–159
dBm, 118
DBA, 70, 74, 78
DBR, 256, 258, 261–262
DCF, 161–162, 165, 228–230
DCPBH, 254
DCS, *see* DXC
DCS/DACS, 43
DD, 261
DDM, 65
DDoS, 443
DEA/DEA-1, 408
Decibel, *see* dB
Decibel
   milliwatt, see dBm
   per kilometer, *see* dB/km
   per meter, *see* dB/m
Declaration of Independence, 376
Decryption, 347
Deep Crack, 408, 416
Degeneracy (polarization), *see* Polarization
Demultiplexer, *see* DMUX
Demultiplexing, 202, 206–213
Denial of service, *see* DoS
Depletion layer, 264
Deregulation (telecom), 13, 52–55
DES, 5, 389, 406, 407–419, 421, 424, 427–428, 431, 433, 443–445, 447–449, 452, 456, 471
Descarte's law, 101, 213
DFB laser, 129, 256–259, 261–262
DGD, 169–173, 230–231
DGEF, 233
DH, 253
Dielectric material, 99
Diffraction (light), 94

Differential
    cryptanalysis, 417
    group delay, *see* DGD
Diffie, 396, 416
Diffie–Hellmann (–Merkle), 396–398, 448, 450
Digest (message), *see* Message
Digital
    cash/checking, 439, 453, 458–461
    signature, 432–439, 453
    to analog encoding (DSL), 23
    wrapper, *see* DW
Digram, 353, 355, 362, 372, 374, 387
Diode, 252
Dipole, 175, 244
Direct-detection receiver, *see* DD
Dispersion
    and nonlinearity, 174
    chromatic (–), 160
    compensating fiber, *see* DCF
    compensation, 161, 228–230
    compensator, 96
    fiber (–), 63, 95, 150, 159–165
    -flattened fiber, *see* DFF
    group-velocity (–), *see* GVD
    index (–), *see* Index
    -managed solition, 194
    material (–), 160
    modal or inter-modal (-), 153, 159
    reverse (–) fiber, *see* RDF
    -shifted fiber, *see* DSF
    slope, 164–165
    spectral (–), 33
    third/fourth-order (–), 164–65
    waveguide (–), 160
Dispersion-compensating fiber, *see* DCF
Dispersion-flattened fiber, *see* DFF
Dispersion-shifted fiber, *see* DSF
Distance learning, 51
Distributed
    amplifier/amplification, *see* Amplifier
    Bragg reflector laser, *see* DBR
    feedback laser, *see* DFB
    key management, 440
    timestamping, 439–440
Distributivity, 392
DLEC, 54
DMT, 15, 27, 31, 33, 39
DMUX, 59, 202, 206, 294
Doping (material), 120
DoS, 441–443
Double
    DES, 408, 416
    star topology, 60
Double/triple encryption (DES), *see* Encryption

Downlink/downstream
    channel (TV), 12
    ADSL/xDSL, 15, 17–22, 26, 31, 50
    APON, 67
    GPON, 74
    GSM, 430
    HFC, 77
    PON, 59
DS01-DS024, 45
DSA, 436, 438
DSF, 163–166, 168, 183–185, 193, 196, 246, 301
DSL, 14–15, 55, 73, 80, 84–85, 94, 168
DSL
    bearer, 21, 35
    flavors, 17–22
    framing, 21–22, 34–50
    line codes, 22
    Lite, 20
    over fiber, 73
    spectrum (DMT), 29
DSLAM, 15–18, 43, 50–51, 55, 73
DSS, 436, 450
DSSS, 87–88
Duplex
    communications, 55, 276
    (DSL), 35
    (PON), 62
Duplexing (Time), *see* Time
DVD, 56, 66
DW, 314–316, 330
DW frame/superframe, 315–316
DWDM, 63–64, 293–294
DWMT, 31
DXC/DCS, 306, 310, 314, 319, 330
Dynamic bandwidth allocation, *see* DBA
Dysprosium, 234

e-commerce, 427, 445, 449–451, 453
e-banking, 449
e-business, 449
E-field, 97, 100, 102–103, 107, 109, 111, 114, 118, 124–125, 129–134, 138–139, 143–146, 154–156, 166, 169, 175–176, 179–180, 188–190, 199–200, 205, 251, 260, 264, 273, 461–464
E-poll, 458
e-file, 452
e-mail, 13, 79, 427, 429, 441, 445–446, 451–452, 454
E/O conversion, 77, 87, 197, 281, 308–309
E/O modulator, *see* Electro-optic
E/O switching, 273–275, 282–283
E1, 18, 21, 35, 37, 45–46

Index 513

E1 to E4 (container), 295
EA, 259
EADM, *see* ADM
EAM, 259
Earth, 98, 110
Eavesdropping, 65, 167, 389, 429–431, 464, 471–472
EBCDIC, 382, 402
EC-DBR, 256–257
ECB, 415, 445
Echo (DSL), 34
ECL, 255
EDFA, 7, 121, 228, 232, 235–248, 266, 271, 285, 292–294, 301, 471
EDFA
   cladding-pumped (–), 244
   gain, *see* Gain
   time constant, 242–243
EDGE, 431
Effective
   area, 117, 129, 156, 183, 185
   (interaction) length, 183
Efficiency (spectral/bandwidth), *see* Spectral
EFM, 71–72
EH_{mn}, 151–153
Einstein's special relativity, 98
ELC, 333
Electric charge, 264
Electro-absorption, 259
Electro-optic
   coefficient, 260
   effect, 149
   modulator, 167
Electromagnetic, *see* EM
Electromagnetic
   interference, *see* EMI
   nature of light, 95, 97–98
Electron, 110, 113, 121
Electron
   -hole recombination, *see* Carrier
   -Volt, 249–250, 264
Electronic
   (digital) cash, 407
   repeater, *see* Repeater
   tax filing, *see* Tax
   voting/polling, *see* Voting
ELF, 97
ELSR, *see* LER
EM, 33
EM
   spectrum, 97
   wave/field, 97–98, 110–111, 124–125, 128, 133, 135–136, 138–139, 151–153, 160, 175, 234, 253, 461
EMI, 73, 94

Emission
   spontaneous (–), 113, 118, 252
   spontaneous (–) lifetime, *see* Lifetime
   stimulated (–), 113, 122, 124, 253
   stimulated (–) rate, *see* Rate
Encryption,
   algorithm, 347–348
   definition, 70, 81, 347
   double-key (–), *see* Key
   double/triple (–), 408
   mono-alphabetic (–), 348–352
   one-way (–), 434–438
   pluri-alphabetic (–), 348–352
   two–key (–), *see* Key
   voice (–), 429–430
Energy level, *see* Atomic
English language/alphabet/word, 353, 356, 359, 375, 417
Enigma machine, 378–382, 397
Entropy (information), 357, 359, 361–363
EOL, 299
EPON, 57, 62, 65–66, 71–72, 82
Equalization
   electronic (–), 268
   frequency (–), *see* Frequency
Equivalent
   input noise, 118
   step index, see ESI
ER, 261–262, 270
Erathosthenes, 403
Erbium, 234
Erbium-doped
   fiber amplifier, *see* EDFA
   glass, 120–121, 236, 238
Error correction, 18, 93, 296, *see also* CRC, FEC
ESI, 154
ESP, 446–448
Ethernet, 14, 59, 62, 65, 71–75, 82, 85, 281, 296, 305, 313, 315, 330, 334–335, 431
Ethernet
   gigabit (–), see GigE
   PON, *see* EPON
ETSI, 47, 431
Euclidian algorithm (extended), 395–396, 420
Euler (theorem), 393, 394–395, 400
Evanescent wave, 145
EXC, *see* DXC
Expansion permutation, 413
Extended Euclidian algorithm, *see* Euclidian
Extinction ratio, *see* ER

Fabry–Pérot
   cavity/resonator, 124–125, 234, 249, 252, 293
   interferometer, 124–125

514 *Index*

Fabry–Pérot (*Continued*)
  laser, 254–256
  mode, *see* Mode
Factoring/factorization, 398–399, 402–406, 475
Faraday
  effect/rotation, 102, 109, 214
  isolator, *see* Isolator
FAX, 83
FBG, *see* IFBG
FBI, 443, 460
FDD-DMT, 31
FDDI, 314–315
FDM, 91
FDMA, 79, 87
FEAL, 428
FEC, 25, 39–40, 43, 188, 228, 262–263, 269–270, 298–299, 315–316
Feistel, 388
Fermat (theorem), 393–394
FET, 268
FEXT, 33–34
Fiber Bragg grating, *see* IFBG
Field
  -effect transistor, *see* FET
  electric (–), *see* E-field
  magnetic (–), *see* 97
Filter
  acousto-optic tunable (–), *see* AOTF
  gain-flattening (–), *see* GFF
  gain-equalizing (–), *see* GEF *or* GFF
  optical (–), 96, 217–228, 286–287
  thin-film (–), *see* TFF
Finesse, 220
FFT, 27–31
FHSS, 87
Fiber (optical)
  amplifier, *see* Amplifier
  attenuation, *see* Attenuation
  bandwidth, *see* Bandwidth
  birefringence, 99
  bow-tie (–), 167
  Bragg fiber grating, *see* IFBG
  cable cost, 55
  cladding, 100
  coaxial (hybrid), *see* HFC
  connector, *see* Connector
  core, 100
  dark (–), 56, 297–299
  deep (–), 57
  dispersion, *see* Dispersion
  dispersion-compensating (–), *see* DCF
  dispersion-shifted (–), *see* DSF
  HI-BI (–), 166
  holey (–), *see* Photonic bandgap (–)
  in the loop, *see* FITL

  in VDSL, 16, 46
  large effective-area (–), *see* LEA
  laser, 125, 168, 225, 244, 258, 288
  low-loss (–), 96
  microstructured (–), *see* Photonic crystal
  multimode (–), 95, 149, 252
  multi-Stokes Raman (–) laser, 225, 244
  nonlinearities, *see* Nonlinearity
  optical (–), 14, 16, 55–56, 58, 99, 133, 149–156
  pair, 296–297, 300, 303, 305, 321, 474
  PANDA (–), 167
  polarization-maintaining/preserving (–), *see* PMF
  polymer optical (–), *see* POF
  preform, *see* Preform
  protection (–), 318, 320
  rare-earth doped (–), 234, 247–248, 294
  reverse-dispersion (–), *see* RSF
  rich/lean (PON), 64
  single-mode (–), 149, 152, *see also* SMF
  standard (–), *see* SMF
  splice/splicing, 55, 196
  standard single-mode (–), *see* SMF
  to the home/builing/curb/user, *see* FTTH, FTTB, FTTC, FTTU
  to the "x", *see* FTTx
  working (–), 318–319
Firewall (Internet), 17, 432
First mile (the), 71, 73
FITB, 56
FITL, 14, 20, 46, 55–58, 80
Fluorescence, 114
FM (radio), 91
Fouché Gaines, 353, 364
Four-wave mixing, *see* FWM
Fourier transform
  discrete (–), *see* DFT
  fast (–), *see* FFT
  inverse (–), *see* IFFT
  -limited pulse, *see* FTL
FP, 124–125, 218–221, 254
Frame relay, 51, 330, 334–335
Free
  carrier, *see* Carrier
  -space optics, *see* FSO
  -space planar (star) coupler, 204–206, 210
  -spectral range, *see* FSR
Frequency
  analysis, 350, 352–363, 371–372, 374, 416, 428
  chirp, 290
  definition, 97
  equalization, 33

spectrum, 353
vs. wavelength, 63
Fresnel lens, 93
Front-end amplifier, *see* Amplifier
FS-VDSL, 47
FSAN, 47, 66
FSC/LSC/TDM/PSC hierarchy, 332
FSK, *see* Modulation formats
FSO, 55, 93–94, 474
FSR, 124, 129, 218–220, 255
FTL, 192–193
FTP, 446
FTTB, 56, 58, 66, 75
FTTC, 56, 58–59, 66, 75, 84, 86
FTTCa/FTTCab, 56
FTTH, 56, 58–59, 66, 75, 84, 86
FTTN, 56, 58–59, 66
FTTO, 56
FTTP, 56
FTTU, 56
FTTx, 14, 20, 48, 55, 57–58, 72, 76, 85–86, 94, 168, 312
Fusion splicing, *see* Fiber
FWM, 178–182, 191, 194, 261, 297

G.652/G.653/G.655, 164
G.709, 315–316
G.872, 315
G.8080, 334
G.Lite, 20
g.shdsl, 17, 20
GaAs, 99, 250, 268
GaAsP, 252
Gain
 amplifier (–), 62, 117–118, 296
 avalanche (–), *see* Avalanche
 bandwidth, 239, 244–248
 coefficient, 117–118, 238–239, 246
 decibel (–), 117–118
 dynamics, 242–243
 EDFA (–), 238–242
 flattening, 232–233
 -guided laser, 254
 net ( ), 117, 126
 polarization sensitivity (EDFA), 242
 polarization sensitivity (RFA), 183–184, 246
 polarization sensitivity (SOA), 243
 polarization dependent (–), *see* PDG
 saturation, 239, 288
 small-signal (–), 240
 spectrum, 239–242
 target (–), 240–241
GaP, 252
Gateway, *see* Internet

Gaussian
 Desurvire's (–) mode approximation, 156
 distribution, 154
 mode approximation, 155
 pulse, 192
gcd, 394
GEF, 226, 231–233, 236, 239, 242, 244
GEO (satellite), 12, 98
Geostationnary satellite, *see* GEO
GFF, 218, 226, 231–233, 236, 239, 242, 244
GFP, 74
GI-POF, 167
Gigabit PON, *see* GPON
GigE, 71, 76, 330, 334
Glass
 erbium-doped (–), 120–121, 234, 238
 doped (–), 120–121, 234, 238, *see also* Fiber/rare-earth
 fluoride (–), 149, 159, 234, 247–248
 Ge-doped (–), 177
 laser, 234
 material, 99, 157–158, 242
 silica (–), *see* Silica
GMPLS, 331–334
GNFS, 403
GOF, 167–168
GPON, 66–67, 72, 74–75, 296
GPRS, 431
GPS, 93
Graph, 325
Grating
 apodization, 223, 229
 arrayed-waveguide (–), see AWG
 blazed/tilted (–), 225
 chirped (–), 224–225, 228–229
 fiber (–), *see* IFBG
 long-period (–), 226, 232
 reflection or Bragg (–), 206–207, 255, 257–258
GRIN (lens), 197, 202, 207, 214
Grooming (traffic), 310, 312, 314
Group
 delay, 160–161
 differential (–) delay, *see* DGD
 index, *see* Index
 -velocity dispersion, *see* GVD
GSM, 7 , 429–431, 450–451
Guard band (DMT), 29
GVD, 160–161, 163, 167, 174, 180, 192–193, 229

Hacker, 441–443
Half-wave plate, 215

Hamlet, 375
Harassment, 441
Hash function, 429–430, 434
HDSL, 16, 18, 20–21, 34–35, 43
HDSL 2/4, 18, 44, 46, 49–50
HDTV, 64, 75, 79–80
HE_{mn}, 151–153
He–Ne, 120
Heisenberg's uncertainty principle, 462, 471
Helium–neon, *see* He–Ne
Hellman, 396, 416
Hertz, 97
Heterostructure, 250–251, 253
Hexadecimal system, 408–409
HF (DSL), 33–34
HFC, 58, 76–82, 84–85
Hi-BI fiber, *see* Fiber
High-impedance receiver, 267
Hilbert-transform pair, 25
HLR, 429
HMAC, 448–449
Hoax, 441
Home
  computer, see PC
  networking, 56, 58, 82–88
HomePNA, 86
HomeRF, 88
Homodyne/heretoduyne detection, 270
Homojunction, 250, 252
Homomorphic protocol, 458
HSCSD, 431
HTTP, 445–446, 449
HTTPS, 407, 432, 449–450
HTU-C/HTU-R, 43
Hughes algorithm, 398
Hybrid
  cryptosystem, 428, 433
  fiber-coaxial, *see* HFC

I-path, 24
IAD, 17, 51
IBM, 388, 407, 443, 461
IC, 301–302, 406
ICI, 33
IDEA/IPES, 428–429, 443
IDSL, 20, 50
IEEE, 431–432
IETF, 445, 449–450
IFBG, 129, 165, 177, 218, 220–226, 255, 258, 293
IFFT, 27–28, 31
IKE, 448
ILEC, 13, 52–55, 57, 75, 80, 335
ILM, 249, 259, 262

IM, 261, 290
IM-DD, 270, 294
Impairment (transmission), *see* Transmission
Impersonation, 441
Index (refractive)
  difference, 150–151
  dispersion, 101
  effective mode (–), 100, 142
  ellipsoid, 99
  graded (–), 150–151
  group (–), 160
  -guided laser, 254
  -lowering codopants, 150, 163
  nonlinear (–), 189
  ordinary/extraordinary (–), 99, 102
  perturbation/change by acoustic wave, 226
  photosensitivity, 222–223
  profile (fiber), 150–151
  -raising codopants, 149–150, 162–163
  relative (–), 141
  step (–), 150–151, 163
  temperature-induced (–) change, 233, 282
Infiltration, 441
Information, 357
Infrared, *see* IR
InGaAlAs, 288
InGaAs, 263–264
InGaAsP, 247, 250, 252, 260, 263
Inhomogeneous broadening, 115
Innovation, 53
InP, 99, 211, 250, 263, 264, 290
Intel, 302
Intelligence, 441
Intensity modulation, *see* IM
interexchange carrier, *see* IXC
Intermodal dispersion, *see* Dispersion
Interactive video, 37
Intercarrier interference, *see* ICI
Interferometer (FP), *see* Fabry–Pérot
Interleaving (DMT bin), 31
Internet (the), 4–5, 11–14, 50, 52, 54, 59, 83–84, 88, 96, 296, 450–452, 458, 461, 468, 471
  access (high-speed), 14
  browsing, 14, 50, 84
  gateway, 13, 17, 75
  jargon, *see* Jargon
  PON, *see* IP-PON
  protocol, *see* IP
  security, 407, 427, 441–451
  service provider, *see* ISP
  virus, *see* Virus
  voice over (–), see VoIP
  web-site attack, *see* Attack
Interoperability (system), 31, 293, 335

Intersymbol interference, *see* ISI
Intranet, 50, 59
Introducer, 440
InvMixColumns (operation), 426
InvShiftRows (operation), 426
InvSubBytes (operation), 426
Ionization coefficient, 266
IP, 5, 34–36, 50–51, 330, 334–335
  -based appliance, 85
  DSLAM, 17, 50
  over ATM, 34, 76, 313, 330, 334
  over DSL, 35, 49–50
  over Ethernet, 330
  over WDM, 293, 313, 334–335
  PON, 73–74
  router, 16–17, 314, 330, 334
  voice over (–), *see* VoIP
IPsec, 445–449
IPv4/IPv6, 445–446, 451
IR, 93–94, 97, 157–158, 176
IrDA, 88, 252
IRS, 452
ISAKMP, 448
ISDN, 20, 30, 44, 50–51, 443
ISI, 26, 33
ISO, 408, 444
Isolator (optical/Faraday), 96, 109, 214–216, 228–229, 234–235, 277
Isotropic (material), 99, 102
ISP, 13, 50, 54, 75, 331, 443
ISX POP, 75
ITRS, 302
ITU-T, 20, 47, 66, 164, 293–294, 297, 300, 309–310, 315, 334, 444
IXC, 53, 75

Jargon
  internet (–), 13, 84, 441
  (technology), 7
Jitter (timing), *see* Timing
Johnson noise, *see* Noise
Jupiter, 98

Kahn, 348
Kasumi, 431
kBaud, *see* Baud
KDP, 260
Kerberos, 444
Kerr effect, 177–178, 188
Key (cryptographic)
  16/64-bit (–), 430
  32/96-bit (–), 431
  40-bit (–), 431, 449
  56/64/128-bit (–), 389, 407, 428–429, 449
  120/160/168-bit (–), 427, 449
  128/192/256-bit (–), 418, 421, 427–429, 431–432, 449, 451–452
  488-bit (–), 428
  512/1024/2048-bit (–), 406, 427, 445, 451, 474
  20,480-bit (–), 427
  asymmetric (–) cryptography, 398, 428, 432
  code (–), *see* Code
  certification, 440–441
  double (–) encryption, 382, 389–391
  DES (–), 408–410
  exchange, 347, 350, 390, 448, 461, 468
  expansion, 424
  fixed (–), 384
  management, 440
  message (–), 381
  no (–) exchange, 390–391, 396–398
  one-time, *see* Cipher
  permutation (DES), 408
  premaster (–), 449
  public (–) cryptography, *see* PKC
  public/private (–), 399, 432–433, 440, 468
  random (–), 352, 470
  raw (–), 470
  round (–), 421, 424
  schedule, 424, 426
  secret (–), 347, 363, 384, 398, 449, 461, 468
  semiweak (–), 409
  session (–), 433, 443, 445, 468
  SIM (–), 429–431
  size export restrictions (US), 428
  space, *see* Keyspace
  sub, 409, 417
  symmetric (–) cryptography, 398, 428, 432
  text, *see* Keytext
  two (–) encryption, 290
  variable-length/size (–), 427–428
  weak (–), 409
  word, *see* Keyword
Keyspace, 416–417, 426, 431
Keytext, 375
Keyword, 347, 368
Khufu, 428

L-band, 244–245, 294
L-I responsivity, 252, 254
LAN, 11–12, 50–51, 54, 59, 62, 71, 73, 75, 85–87, 167–168, 203, 305, 310, 312–313, 427, 441
Lanthanides, 234
Large effective-area fiber, *see* LEA
Laser, 92, 95–96, 115, 120, 123–132, 234–235

Laser (*Continued*)
  3-level/4-level (−) system, *see* Pumping
  beam, 125, 127, 234
  c3 (−), 255
  cavity, *see* Cavity
  diode, *see* LD
  distributed-feedback (−), *see* DFB
  external-cavity (−), 255–256
  fiber (−), *see* Fiber
  flux, 126–127
  He–Ne (−), *see* He–Ne
  light, 111, 113, 115, 129
  linewidth, 128
  material, 120, 121, 127
  multimode (−), 128
  semiconductor (−), *see* Semiconductor
  single-frequency (−), 255
  speckle, 132
  threshold, *see* Threshold
  tunable (−), 255, 319
LASER, *see* Laser
Layer1/layer2/layer3 network/protocols *see* Network
Last mile (the), 57, 71, 73, 77, 94
Latency (single/dual DSL), 37
LCF, 70
LD, 92, 121, 149, 249–258
LDP, 331
LEA, 156, 188, 229, 299
LEC, 26, 53–54
LED, 251–252
LEPA-HDSL, 44
LER, 331
Levy, 428
LEX, 41
LF (DSL), 33–34
Lifetime
  carrier (−), 122, 252
  fluorescence or spontaneous emission (−), 114–115, 120–121, 124, 126, 242–243
Light
  -matter interaction, 110–115, 121–123
  path, 310, 313, 324–329
  -year, *see* LY
Lightpath, *see* Light
Lightwave
  communications, *see* Optical
  transmission, *see* Transmission
LiNbO3, *see* Lithium
Line broadening, 115
Line of sight, *see* LOS
Linear
  cryptanalysis, 417

B tablets, 348
  link protection, 317–319
Linewidth, 115, 128
Lithium niobate, 167, 176, 226–227, 259–262, 273–275, 283
LMDS, 14
LMP, 333–334
LOKI, 428
Local
  -area network, *see* LAN
  -exchange carrier, *see* LEC, LXC
Longitudinal mode, *see* Mode
Loop
  fiber in the (−), *see* FITL
  (local/subscriber), 12, 54
Lorentzian (lineshape), 115, 118, 122
Lord Bacon, *see* Cipher
LOS, 93
Loss
  coupling loss (input), 238, 269–270
  excess (−), 61, 195, 202
  insertion (−), 61, 195
  microbending (−), 158
  return (−), 195, 202, 234
  splitting (−), 61, 201
  transmission (−), *see* Transmission
Low-loss fiber, *see* Fiber
LP_{01} mode, 153–154, 156, 159–160, 166–167, 169
LP_{mn} mode, 153–154
LS subchannel (DSL), 35–43
LS0/LS1/LS2/LS3, 35, 39–43
LSP/LSR, 330–333
Lucifer, 388–389, 407–408
Luminescence, *see* Fluorescence
LXC, 53
LY, 98

Mach–Zehnder interferometer, *see* MZI
Magic square, 366, 384
Magnetic field
  component, 97, 99, 147, 152
  static (−), 109
MAN, 54, 75, 78, 294, 305, 310, 312–314, 441
Management
  dispersion (−), 96, 299
  of nonlinearities, 96, 299
Mangler/mangling, 388–389
Marcuse's approximation formula, 155
Marketing (telecom), 6
Mars, 98, 426
MASER, 123
Maxwell's
  classical electromagnetism, 111

Index **519**

laws/equations, 97, 110, 129, 132, 135, 151
Maxwellian distribution, 171
MD2, 436, 445
MD4/MD5, 407, 429, 434–436, 443–445, 447, 449
MDU, 16
Mel and Baker, 448
MEMS, 282–283, 331
Merkle, 396
Mersenne number, 393
Message digest, 429, 434, 449
Metro network, *see* Network
Metropolitan–area network, *see* MAN
MFD, 155
MIC, 445
Michelson interferometer, 284–285
Microbending loss, *see* Loss
MILP (VTD), 329
MINITEL, 83
MIT, 444
MixColumns (operation), 421, 423–426
Mixing product, 176, 180
MM (service), 13, 15, 19, 51, 296, 451
MM (fiber), 149
MMS, 51
Mode
   Bloch (−), 205
   cavity (−), 129–132
   cladding (−), 149, 226
   effective refractive index, 100
   EH/HE/TE/TM, 152–153
   field diameter, *see* MFD
   fundamental (−), 131–132, 135, 138, 143, 152
   guided (−), 132–149
   high-order (−), 131
   -hopping, 255
   linearly-polarized (−), *see* LP_{mn}
   longitudinal or FP (−), 124, 127–128, 254
   normal (−), 198–200, 273
   radiation (−), 95
   spatial/transverse, 128, 130–132
   spot size, 155
   transverse-electric (−), *see* TE
   transverse-magnetic (−), *see* TM
   waveguide (−), 95
Modem, 12, 14–15, 50, 76, 84–85
Modular/modulus
   algebra, 382, 391–396
   reduction, 393, 420
Modulation (direct/external), 249, 258–259
Modulation formats
   16/64/128/256 QAM, 23, 29–30, 79
   256-CAP, 25
   2-PSK/PAM, 23
   ASK, 261, 270

   CAP, 23, 25–26, 31, 39
   FSK, 87, 243, 270
   M-ary, 23, 79
   M-ary QAM, 23
   M-PAM, 23, 30
   M-PSK, 23
   M-QAM, 23, 79
   PSK, 261, 270, 299
   QAM, 23, 25–27, 31, 82
   QPSK, 79
Modulator
   electro-absorption (−), *see* EAM
   optic (−), 96
Mono/pluri-alphabetic encryption, *see* Encryption
Monochromatic light, 128
Moon, 98
Moore, 301
Moore's law, 301–303, 404–406
Morse code, *see* Code
MPLS, 330–332
MPλS, 331
MPOF, 168
MQAM, *see* Modulation formats
MQW, 254, 288
MS-DPRING, 321–322
MS-SPRING, 321–322
MTU, 16
Multiquantum well, *see* MQW
Multicarrier modulation, *see* Carrier
Multimedia, *see* MM
Multimedia messaging service, *see* MMS
Multimode
   fiber, *see* Fiber
   laser, *see* Laser
   Multiplexer, *see* MUX
Multiplexing, 2, 202, 206–213
MUX, 59, 64–65, 202, 206, 235, 294
MxU, 16
MZ, see MZI
MZI, 202, 207–210, 218, 227, 233, 259–260, 284–285, 289–290, 293

NA, 143, 150, 153–154
Native Americans, 92, 348
Navajo tongue, 348
Neodymium, 234
Netscape (Navigator), 449
Netstumbler, 431
Network
   abuse, 441–443
   access, 3, 312–314, 335
   attack, *see* Attack
   backbone, 2, 292–293, 312–313

520 *Index*

Network (*Continued*)
  bandwidth, 324
  cloud, 1, 3, 54
  congestion, 272, 328, 330
  convergence, 293, 329–335
  core, 2–3, 312–313, 335
  edge, 3, 75, 312–314, 335
  evolution, 329–335
  failure, 317
  home (–), *see* Home
  intelligence, 3
  interoperability, *see* Interoperability
  layer, 3–4,
  layer-1/-2/-3, 73, 75–75, 272, 296, 309, 313, 330, 335
  legacy, 4, 309, 330
  local-area (–), *see* LAN
  mesh (–), 322
  metro (–), 3, 75, 78, 293–294, 312–313
  metropolitan-area (–), *see* MAN
  opaque (–), 309, 330
  optical (–), 57
  optical (–) layer, *see* Optical
  optimization, 317, 323–325, *see* also VTD
  outage, 323
  packet (–), 271–272, 328
  passive optical (–), *see* PON
  PC (–), 85, 87–88
  protection, 78, 293, 298, 300, 317–324, 332
  restoration, 317
  ring (–), 78, 305–307, 310–312
  roaming, 4
  scalability, 335
  security, 407
  storage-area (–), *see* SAN
  telecom (–), 1
  topology (tree/star/ring/bus), 57
  transparent (–), 281, 286, 308–309, 330
  virtual (–), 324–329
  wavelength routed/switched (–), 305
  WDM (–), 292–335
  wide-area (–), *see* WAN
NewDes, 428
NEXT, 33–34
NF, 237–238, 243, 267
NFS, 403
NIST, 408, 414, 418, 426, 436
NO, 384
Noise
  amplifier (–), 119
  excess (–) factor (APD), 266–267
  figure, *see* NF
  partition (–), 128, 255
  spontaneous emission (–), 118–119
  thermal or Johnson (–), 267, 270

Nonlinear
  function (encryption), 435
  optics, 174
  polarization, 175–178, 188–189
  refraction, 188
Nonlinearity/nonlinearities
  (fiber), 96, 156, 162, 174–194, 228, 245, 270, 297, 299
  second-order (–), 176, 260, 283
  third-order (–), 177, 182, 188, 283
Non
  blocking switch, *see* Switch
  generalized MPLS, *see* MPLS
  priority traffic, 318
  radiative decay/de-excitation, 115, 120, 236–237, 247
  reciprocity, 214
  repudiation, 453–454
  return to zero, *see* NRZ
Normal mode, *see* Mode
Normalized frequency, 148
NRZ, 190, 261, 281, 283, 287–289, 299
NSFNET, 329
NTR, 35
Null, 365, 366, 368–369, 374–375
Numerical aperture, *see* NA
NVoD, 80
Nyquist (sampling rate), 30
NZDSF$^\pm$, 164–165

O/E conversion, 77, 87, 281, 288, 308–309
O/E/O conversion, 270–273, 278, 282, 314
O/O switching, 271, 273
OADM, 110, 217–218, 224, 236, 244, 279, 294, 300, 304–314, 331
Oakley, 448
OAM, 36, 80, 316, 333, 335
OAN, 59
OAS, 309
OC-3 to OC-768, 295
OC/OCH (optical layer), 309, 321, 324, 330, 334
OCC, 334
OCH container, 315
OCH-SPRING, 321
OD, 157
OFB, 415–416
OFDM, 27
OH impurity, 157–159, 163
Ohm, 34, 99
OKG, 123
OLEC, 54
OLT, 58, 60–62, 64, 66, 70, 73
OMS, 309, 321
On-demand (video), *see* VoD

On-line, 13
On-line tax declaration/payment, *see* Tax
One-way function, 393, 396, 434
Onetime pad cipher, *see* Cipher
ONT, 58, 61, 65–66, 71
   ONU, 16, 20, 46, 48, 58, 61, 65–66, 68, 70–71, 73–74
OPA, 266–267
Operator (telecom), 52–55
OPS, 317, 330
Optical
   access, 75–76
   activity, 109
   add-drop multiplexing, *see* OADM
   all (−) regeneration, *see* AOR
   amplifier, 55, 57, 62–63, 65, 92, 96, 155, 158–159, 174, 202, 285–286, 292, 301, 468, 471
   amplification, 111, 115, 116–121
   cavity, *see* Cavity
   circulator, *see* Circulator
   component (active/passive), *see* Component
   coupler, *see* Coupler
   cross-connect, *see* OXC
   early (−) communications, 92–94, 234, 293, 298, 301
   fiber, *see* Fiber
   (fiber) communications, 58, 91, 116, 121, 123, 132, 292
   filter, *see* Filter
   gateway, 312
   isolator, *see* Isolator
   layer, 296, 300, 309–310, 312–313, 333
   maser, *see* MASER
   packet switch, 317
   power, 110, 117
   preamplification, *see* Amplifier
   rectification, 176
   repeater, *see* Repeater
   ring, 78
   splitter, *see* Splitter
   switch/switching, 96
Optics, 4
OPU/ODU/OTU, 315
OR, 384
Orders of magnitude (units), 97
Oryx, 431
OSI network model, 309, 312, 449
OTS, 309
Outside diameter, *see* OD
OXC, 271–285, 304–309, 312, 317, 319–320, 330–332, 334

P2P, 72–73, 81, 263, 296–304, 309, 317, 320

P2MP, 72–73, 75
p–n junction, 250–251, 253–254, 259, 264–265
Packet mode (DSL), 34, 48
Padding, 434
Painvin, 377
Pair (ciphertext/plaintext), 417
PANDA fiber, *see* Fiber
Parametric
   amplification, 178
   gain, 178
   mixing, 176
Passive optical network, *see* PON
Password, 349
Payload (traffic), 295–296, 310, 319, 321
PBS, 213–216, 465–468
PCT, 450–451
PEM, 435
Perec, 356
Perfect shuffle, 203–205
Perfluoro-POF, 168
Personal identification number, *see* PIN
PBS, 213, 216, 235
PBX, 17, 51–52
PC, 66, 83–85, 88
PCD, 173
PCE, 237
PCS, 81
PDF, 269
PDG, 242
PDH, 46, 74
PDL, 172, 195, 203, 211–213, 217, 233, 281–282
Period (symbol), 24–25
Personal
   communication services, *see* PCS
   computer, *see* PC
Petabit, 301
Peta-meter, 98
PGP, 407, 428–429, 432–433
PGPdisk, 429
PGPfone, 429
Phase, 103
   mask, 223
   -matching (condition), 124, 180, 188, 226, 273–274
   -locked loop, *see* PLL
   nonlinear (−), 189
   velocity, 160
Philips, *see* Cipher
Phonon, 120, 182–183, 185, 236–237
Photo-electric effect, 110, 123
Photocurrent, 123, 263–264, 266–268
Photodetector/photoreceiver, 123, 262–271
Photodiode, 92, 121
Photoelectron, 263

Photon, 92, 110–113, 237, 461–468
Photon
  absorption/capture, 113
  emission/release, 113
  entanglement, 474
  flux, 110, 117, 237, 251
  flux density, 114–117
  splitting, 461–468
Photon/bit (sensitivity), 158, 269
Photonic
  bandgap/crystal fiber, 168, 230
  switching, 271–285, 317
Photonics, 4, 92, 96–97, 317, 330
Photoreceiver, 96
Photosensitivity, 222
Photuris, 449
Pilot tone (DMT), 30
PIN or p-i-n, 263–268
PIN (security), 429, 450, 452
PIN-FET, 268
Piracy, 442
PKC, 348, 382, 391, 396, 398–407, 427–428,
    432–433, 436, 438, 443–444, 449
PKCS/PKCS#1, 402, 449
Plaintext, 347
Planck's
  constant, 110
  theory of quanta, 110
PLL, 269
PLOAM, 67–70, 74
Pluri/polyalphabetic encryption, *see* Encryption
PM, 23, 290
PMD, 109, 166–167, 168–174, 228–231, 299
PMD
  coefficient, 171
  higher-order/second-order, 173
  mitigation, 172, 228, 230–231
  vector, 173
PMF, 100, 108, 165–167, 169–171, 184, 195,
    231, 233, 235, 258, 261
PMMA, 167
Probability-density function, *see* PDF
Pockels
  coefficient, *see* Electro-optic
  effect, 176, 260, 273–274
POF, 86–87, 149, 165, 167–168
Poincaré sphere, 107–109, 166
Point
  of presence, *see* POP
  to point (link), *see* P2P
Polarization
  beamsplitter, *see* PBS
  circular (–), 105, 461
  controller, 465
  crosstalk, 166

degeneracy, 124, 131, 273
-dependent loss, *see* PDL
diversity, 211, 215, 217, 228, 233
ellipse, 103
elliptical (–), 105
-independence/insensitivity, 195, 215–216,
    242, 246, 286
light (–), 97
linear (–), 102, 105, 461
maintaining fiber, *see* PMF
medium (–), 97, 175–177
mode dispersion, *see* PMD
nonlinear (–), *see* Nonlinear
photon (–) states, 461–462
principal states of (–), *see* PSP
rotation, 109
scrambling, 100
spontaneous noise (–), 118–119
state of (–), *see* SOP
TE/TM, *see* TE, TM
Polyalphabetic cipher, *see* Cipher
Polybius, *see* Cipher
Polygram substitution, 372
Polymer
  jacket (fiber), 149
  optical fiber, *see* POF
  types, 167
  waveguide, *see* Waveguide
Polynomial
  coefficient, 419
  degree, 419
  divider, 419
  inverse (–), 420
  irreducible (–), 419
  quotient, 419
  remainder, 419
  representation, 419
PON, 46, 57–75, 80, 84, 203, 293, 312
POP, 53–55, 75, 331
Population
  (atomic), *see* Atomic
  inversion, 117, 119, 237–238, 240, 242
Port (ATM-DSL), 36
Potential barrier, 251
POTS, 4, 11–12, 50–53, 80–81, 85
Power
  budget, 62, 299
  conversion efficiency, *see* PCE
  equalization, 231–234, 286–287
  excursion, 186, 246
  peak (–), 191
  penalty, 187
Praseodymium, 234
PRBS, 299
Preform (fiber), 157

Index 523

Pretty good privacy, *see* PGP
Preoutput, 412, 415
Prime number/polynomial, 393–394, 419, 428, 436, 475
Privacy (protection), 451, 428, 460–461
Private
 key, *see* Key
 virtual network, *see* PVN
Propagation constant, 136, 142, 168, 180, 189, 273
Protection
 line/link/span (–), 322
 path (–), 322
 ring (–), 319–322
Protocols, 4–5, 13–14
Provisioning (traffic/service), 272, 298, 300, 316, 335
Proxima Centauri, 98
PSP, 173
PSTN, 11–12, 16, 30, 43, 49–51, 53, 55, 62, 76, 85, 310, 451
PTT, 53
Public key, *see* Key
Public-key cryptography, *see* PKC
Pump, 119, 202, 235
Pump
 infinite (–) power, 240
 module, 236
 redundancy, 236
Pumping
 1480nm (–), 235–240, 243–244, 247
 3-level (–), 119–120
 4-level (–), 119–120
 980nm (–), 235–238, 240, 243
 efficiency, 236
 electrical (–), 122–123
 forward/backward/bi-directional (–), 184–186, 235–236, 241
 hybrid (–), 236, 243
 laser (–), 124–127
 multi-stage (–), 236
 optical (–), 121–123
 rate, 119, 126–127
 solar (–), 168
 threshold, *see* Threshold
PVN, 54

Q-factor, 262
Q-path, 24
QAM, *see* Modulation formats
QED, 111
QKD, 468–475
QoS, 51, 72, 81–82, 314, 329, 331, 335
Quadrature, 23

Quality of service, *see* QoS
Quantum/quanta, 110, 120, 123
Quantum
 computer, 474–475
 cryptography, 461–475
 efficiency, 251–252, 264
 electrodynamics, *see* QED
 key distribution, *see* QKD
 logic gate, 475
 limit (NF), 238
 memory, 475
 nature of light, 95–96, 110–115, 474
 optics, 111
 telecommunications, 474
 teleportation, 474
Quarter-wave plate, 467–468
Quat, 23
Queuing (packet), 328, 330

Radiative decay, 120
Radio frequency, *see* RF
Radiofrequency interference, *see* RFI
RADSL, 20, 50
Rainbow, 97, 102
Raman
 amplification, 96, 184–186
 bands, 247–248
 fiber amplifier, *see* RFA
 multistokes (–) fiber laser, *see* Fiber
 shift, 182–183
 stimulated (–) scattering, *see* SRS
RAN, 59
RAND, 429–430
Rare-earth (elements), *see* RE
Rate
 (absorption/emission), 114–116, 122
 pumping (–), *see* pumping
 spontaneous decay (–), 116, 119
Ray
 light (–), 133–134
 skewed (–), 141
Rayleigh scattering, 157–158
RC-2/RC-4, 7, 407, 428–429, 431, 449, 452
RDF, 165
RE, 234, 244, 247–248
Receiver
 bandwidth, 267–268
 CAP, 25–26
 coherent (–), 270
 direct-detection (–), 262–271, 270
 (QAM), 24–25
 sensitivity, 263, 269–270
REDOC-II/III, 428
Reed–Solomon code, *see* RS

Referendum, 455
Reflection (light)
  in waveguide, 133–137
  on dielectric interface, 100–101, 249
  total internal (–), 95, 101, 134
Refraction
  double (–), 213
  light ray (–), 94, 100–101, 133
  nonlinear (–), see Nonlinear
Refractive index, see Index
Regeneration/regenerator
  1R, 271, 285–286, 291–292
  2R (–), 281, 286–288, 291, 299, 308
  3R (–), 270–271, 281, 285–292, 294, 299, 308–309
  all-optical (–), see AOR
  black-box (–), see BBOR
  in-line (–), 291–292
  O/E and O/E/O (–), see O/E/O
Rejewski, 380
Relatively prime numbers, 394
Repeater
  electronic (–), 165, 234, 255, 263, 285, 291–292, 297, 468, 472–474
  O/E/O (–), 270–271, 319
  optical (–), 174, 229, 243, 271, 298
  spacing, 188, 255, 296, 298
Repeaterless system, 185
Repudiation, 453–454
Residential access network, see RAN
Residue, 391–393
Responsivity
  LD (–), see L-I
  photodetector (–), 263, 266
Restoration (static/dynamic), 322
Return to zero, see RZ
Reverse-dispersion fiber, see RDF
RF, 24, 77, 87, 167
RFA, 96, 121, 156, 184–186, 228, 245–247, 271, 285, 471
RFC-1321, 435
RFC-2401, 445
RFC-2402, 446
RFC-2104, 448
RFC-2246, 450
RFC-2406, 448
RFC-2407, 448
RFC-2408, 448
RFC-2409, 448
RFC-2411, 445
RFC-2412, 448
RFC-2522, 449
RFC-2523, 449
RFI, 34, 73

Rijmen, 421
Rijndael, 421
Ring network protection, 318–324
RIPEM, 445
Rivest, 398
Rivest cipher, see Cipher
Rivest–Shamir–Adleman (algorithm/standard), see RSA
Rosetta stone, 348
Rotatory power, 109
Round
  DES/AES (–), 388, 408, 412, 415, 417, 421–424, 426–428
  hashing (–), 434–436
Router, 272–273
Routing, 272, see also Wavelength
RS, 25
RS(255,231), 7
RSA, 398–407, 427–428, 432–443, 449
RSA-140/155/512/576/1024/2048, 402–404
RSA-N, 442
RSA Security, 402, 408
RSVP-TE, 331–332
RXCF, 70
RZ, 192–194, 261, 281, 283, 287–289, 299

S-box, 413–415, 417–418, 421–422, 424, 426–428
S-CDMA, 80
S-HTTP, see HTTPS
SA
  (optics), 286–287
  (security), 445, 447–448
SAN, 75
SAR, 72–73
Satellite
  constellation, 12, 99
  geostationnary (–), see GEO
  systems, 12, 14
  TV, 12, 77, 83
Saturable absorber, see SA
SBS, 183–184
Scherbius, 380
Schneier, 395, 406–407, 410, 414, 427, 435, 452
Scrambling (bit), see Bit
SDH, see SONET/SDH
SDM, 474
SDSL, 15, 20, 50
Second-harmonic generation, see SHG
Secure
  HTTP, see HTTPS,
  socket layer, see SSL
Security
  absolute (–), 461

Index 525

association, *see* SA
communications (−), 443–445
Internet (−), *see* Internet
network (−), *see* Network
RSA (−), see RSA
SELD, 258
Self
 -healing ring, 319–324
 -induced SRS, see SI-SRS
 -phase modulation, *see* SPM
 -pulsing LD, 288
Semaphore, 93
Semiconductor
 intrinsic (−), 250, 264
 laser (diode), 95, 125, 235
 material , 99, 121, 250
 n-type/p-type (−), 250, 264
 optical amplifier, *see* SOA
 photodiode, *see* Photodiode
Serpent, 426
Service(s), 4
 access (−), *see* Access
 broadband (−), *see* Broadband
 five-nines, 85
 HFC (−), 80–82
 multimedia messaging (−), *see* MMS
 on-demand (−), 12
 quality of (−), *see* QoS
 TV (−), 12
 voice/telephony (−), 11, 52–55
SESAME, 444
SGC, 451
SH, 319
SHA/SHA-1, 407, 429, 436, 443, 445, 449
Shamir, 398
Shawlow–Townes formula, 128
SHDSL, 16, 20, 50
Shepard, 442
SHG, 176–177
ShiftRows (operation), 421, 423
Shuffling, 388–389
Si, 266, 268, 451
SI-POF, 167
SI-SRS, 186–188, 193, 245
Signal
 crosstalk, *see* Crosstalk
 pre-emphasis, 233–234
 -to-noise ratio, *see* SNR
Signature, *see* Digital
Silica (fused), 99, 149, 157, 163, 174, 182–183, 211, 222, 234, 245, 247
Silicon, *see* Si
Silicon oxide (SiO2), *see* Silica
SIM (card), 429–430, 443

Simplex
 (DSL), 35
 (PON), 62
Simultaneous
 contract signature, 453–454
 secret exchange, 407, 454–455
Singh, 348, 356, 364, 428
Single-photon
 detector, *see* SPD
 source, *see* SPS
Single-mode waveguide, *see* Waveguide
SKEME, 448
SKIP, 449
SLD, 249
SM
 (fiber), 149
 regeneration, 288–292
SME, 75
SMF, 162–166, 168, 183–185, 193, 196, 229–230, 246, 301
SMF plant (upgrade), 165
SMTP, 446
SNCP, 319
Snell's law, 101, 213
SNMP, 244
SNR, 18, 27, 29–30, 228, 231, 233, 266, 286, 290, 296, 308, 333
SOA, 96, 121–122, 234, 243, 249, 266, 283
SOA gate, 282–285, 288
SOHO, 51
Soliton
 propagation/pulse-characteristics, 96, 174, 191–194, 227–228, 286–288, 290–291
 self-frequency shift, see SSFS
SONET/SDH, 5, 14, 18, 34, 39, 45, 65, 67, 74–75, 281, 292, 295–296, 299–300, 309–316, 321, 324, 331–335
SONET/SDH sublayers (path, line, section, physical), *see* Sublayer
SOP, 105–110, 166, 168, 173, 195, 202, 212–213, 230–231, 238, 465
Spamming, 441
SPD, 462
Spectral
 dispersion, *see* Dispersion
 efficiency, 22–23, 79, 82, 159, 220, 294
Speed of light, 63, 98–100
Splice/splicing (fiber), *see* Fiber
Splitter
 1 × N optical (−), 199–201
 (DSL), 16, 50
 (PON), 59
Splitterless DSL, 20
SPM, 177–178, 188–194, 261, 287

Spontaneous
  emission, *see* Spontaneous
  emission factor, 118–119, 184–185, 238, 242
Spot size (mode), *see* Mode
Spread-spectrum, 87
SPS, 462
Squeezing (light), 474
SRES, 429–430
SRS, 121, 182–188, 244–246, 258, 294
SSFS, 194
SSL, 407, 428, 432, 449–452, 456
SSLR, 322
SSP, 75
SSPR, 322
Standard fiber, *see* SMF
Standards, 4–5
Star topology, *see* Network
Stark
  effect, 115
  level, 238
State (AES), 418
State of polarization, *see* SOP
Steganography, 346
Stimulated
  Brillouin scattering, *see* SBS
  emisssion, *see* Emission
  Raman
    gain coefficient, 183, 187
    scattering, *see* SRS
STM mode (DSL), 34, 37, 48, 81
Stokes
  parameters, 106–108
  wave (SRS), 182–183
STT, 450
SubBytes (operation), 421–423
Subcarrier (DMT), 27
Subkey, *see* Key
Sublayer (SDH/SONET), 309–310
Submarine (cable) systems, 165, 195, 295, 298, 300, 312, 317
SUN, 449
Superposition of states, 466
Susceptibility (linear/nonlinear), 175
Switch/switching
  ATM (–), *see* ATM
  blocking/nonblocking, 273–274, 281, 305, 317
  connection-oriented (–), 271
  crossbar (–), 274–275
  packet (–), 317, 328, 330
  photonic (–), *see* Photonic
  photonic vs. electronic (–), 272
Spanke (–), 274
STM-1 to STM-256, 295, 300, 310, 321
SWP, 213

Symbol
  (QAM), 23–24
  rate, 26
Symmetric-key cryptography, *see* Key
Synchronous modulation, *see* SM
System design (WDM), 298–301

T1, 17–18, 21, 35, 45–46,
T1/J1 to T3/J3 (container), 295, 310, 312
T1 TDM/ATM, 17
Tabulation, 456–458
Tampering, 441
TAT-8, 255
Tax filing, 407, 452
TBP, 192–193, 227
TCP/IP (protocol), 13, 40, 74, 76 , 244, 281, 296, 313–315, 324, 330, 334, 445–449
TDD-DMT, 31
TDEA, 416
TDM, 17, 74, 262, 296, 332
TDM (PON), 59–62, 70–71, 75
TDMA, 79, 87
TE, 136, 138–140, 143–148, 213–215, 226–227, 233, 465–467
TE_{mn}, TM_{mn}, 151–153
Technology cycle, 301–303
Telecommuting, 4, 50, 59, 85
Teleconferencing, 51
Telephone, 4
  cellular (–), 12
  circuit/channel, *see* Circuit
  services, *see* Services
Teleportation, *see* Quantum
Teleworking, 4, 50, 59, 83, 85
TEM_{00}, TEM_{lm} cavity modes, 131–132, 138, 153, 249
Temperature insensitivity, 195
TeraKIPS/petaKIPS law, 303
Terrestrial systems, 165, 195, 298–299, 312, 314
Terrorism, 441
TFF, 206–207, 212, 218
Theft of
  service, *see* ToS
  vote, 457
Thermal noise, *see* Noise
Thermo-electric/Pelletier cooler, 235, 262
THG, 178, 180
Thin-film
  filter, *see* TFF
  heater, 233
Third-harmonic generation, *see* THG
Threshold
  current (–), 254
  laser oscillation (–), 126–127
  Raman/Brillouin scattering (–)

Index 527

THP, 26
Thulium, 234
Time-division multiplexing, see TDM
Time
  bandwidth product, see TBP
  duplexing, 31
Timestamp protocol
  distributed (−), 439−440
  linking (−), 439, 453
Timestamping, 439−440, 444−445, 453, 460
Timing jitter, 288−290
TIS/PEM, 445
TKIP, 432
TLS, 450−451
TM, 138−139, 148, 213−215, 226−227, 233, 465−467
Tomlinson–Harashima (coding), see THP
ToS, 81, 441
Traffic
  aggregation, 5
  deaggregation, 5
  grooming, see Grooming
  IP-based (−), 335
  non priority (−), see Nonpriority
  optimization, 324−325, see also VTD
  payload, see Payload
  provisioning, see Provisioning
Transceiver/transponder, 263, 296−298
Transimpedance receiver, 267−268
Transmission
  impairment (DSL), 31−34
  loss (DSL), 32
  loss (fiber), 61, 247−248
  optical/lightwave (−), 96
  window, 159, 248, 293
Transmitter
  (DMT), 27−28
  (QAM), 24
Transverse
  electric (mode), see TE
  magnetic (mode), see TM
Trapdoor function, 407, 415
Trent, 433, 438−440, 444−445, 453
Trigram, 353, 362, 372, 387
Triple-DES, see 3DES
Trusted
  group, 433, 436, 440
  third party, see Arbitrator
TV, see Satellite, Services
Twofish, 426

UDP, 334, 445
ULSR, 320, 322
UMTS, 7, 451
Uncertainty principle, see Heisenberg
Undersea system, see Submarine

Universal DSL, 20
Unpolarized light, 97, 132
Uplink/upstream
  ADSL/xDSL, 15, 17−22, 26, 31
  APON, 68
  channel (TV), 12
  GPON, 74
  GSM, 430
  HFC, 77
  PON, 59
UPSR, 319−320
U.S. Army, 348, 352
USHR, 319
UV, 97, 157, 168, 177, 222−223

V-number, 148, 152−154
Vacuum
  fluctuations, 111, 464
  impedance, 99
  noise, 111, 464
VC-4/VC-n, 321, 332
VCO, 269
VCSEL, 258
VDSL, 16−18, 21, 35, 46−50, 56, 73, 80, 84
VDSL Alliance/Coalition, 47
Velocity
  (EM wave), see Speed of light
  of sound, 183
Verdet constant, 214
Verlaine, 346
Vertex, 325
Video on demand, see VoD
Videoconferencing, 51
Videotelephony, 82
Vigenère, 350−352, see also Cipher
Vigenère
  cipher, see Cipher
  square, 350−351
Virtual topology design, see VTD
Virus infection/propagation, 441−443
Visible light, 97, 176
VLR, 429
VLSI, 406, 416−417, 475
VoATM, 17, 50−51, 81
VoD, 12−14, 50−51, 65, 75−76, 79−80, 85
VoDSL, 51−52, 81
Voice
  circuit, 295
  compression, 295
  enncryption, see Encryption
  over ATM, see VoATM
  over cable, 80
  over DSL, see VoDSL
  over frame relay, see VoFR
  over multiservice data networks, see VoMSDN

Voice (*Continued*)
  over packet, *see* VoP
VoFR, 51
VoIP, 5, 50, 51, 54, 75, 80–81, 451
Voltage-controlled oscillator, *see* VCO
VoMSDN, 51
VoP, 51, 81
Voting/polling (electronic), 407, 453, 455–458
VT-n, 332
VTD, 317, 325–329

WAN, 54, 305, 312–313, 441
Wave-particle duality, 111
Waveguide
  attenuation, 158
  buried (−), 148
  cylindrical (−), 95
  dielectric (−), 138–149
  light (−), 95
  mirror (−), 137–138
  mode, 101
  planar (−), 95, 133, 147–148, 253–254
  polymer (−), 168, 282
  ridge (−), 148
  single-mode (−), 95, 143, 152
  slab (−), 148
  strip (−), 147
Wavelength, 63
Wavelength
  blocking, 308
  Bragg (−), *see* Bragg
  comb, 63, 241–242, 293–294
  conversion, 279–280, 283–285, 288, 308
  cross-connect, *see* WXC
  cut-off (−), 135, 143, 148, 153–154, 156
  definition, 97
  multiplexer/demultiplexer, 57, 96
  routed transport layer, 324
  routing/router, 64, 211–212, 271, 278–280, 304–308, 316, 330
  -selective coupler, *see* WSC
  static (−)router, 211
  zero-dispersion (−), 163–164
Wavelength-division multiplexing, *see* WDM
Wavelet, 31, 82
Wavevector, 135–137
WDM
  bandwidth, 187
  coarse (−), *see* CWDM
  dense (−), *see* DWDM
  bandwidth management, 76
  (HFC), 78
  layer, 78
  networking, 96

(PON), 59–60, 63–66, 72, 75
ring, 75, 305–307, 310–312
swiching, 278–281
systems/techniques/technologies/
  transmission, 91, 96, 102, 162, 164, 168, 178–181, 186–191, 194–195, 206, 210–211, 218, 224, 229–232, 240–244, 248, 257, 262–263, 271, 273, 279, 285–286, 291–292, 292–335, 474
wide (−), *see* WWDM
Weakly guiding fiber, 152
Web page, 13
WECA, 88
WEP, 431–432
WEP2, 432
WEPCrack, 431
WI-FI, 51, 81, 87, 431, 451
WIC, 279–281, 304, 308, 316
Wide-area network, *see* WAN
Wireless
  access networks, 13–14
  cable, 12
  LAN, see WLAN
  link (home), 83, 85, 87–88
  vs. wireline optics, 95
Wireline
  (access), 14
  optics, 94–96
WLAN, 12, 87, 427, 431–432, 451, 471
World Wide Web, *see* WWW
WSC, 202, 278, 304, 308, 316
WWI, 352, 377, 380, 451
WWII, 346, 363, 365, 380, 451, 454
WWDM, 71
WWW, 4, 13
WXC, 278–285, 320

X.509, 444–445
X-junction (coupler), 198–202
xDSL, 14–15, 17–23, 35, 49, 54, 56–57, 71, 73, 75–76, 81, 312
XGM, 283–284
XOR, 384–385, 390, 416–418, 421, 424, 427–428, 444, 453
XPM, 178, 188–194, 261, 283–284, 289
XT, 33
Xtalk, *see* Crosstalk

Y-branch/junction (coupler), 198–201, 203, 276, 284
YIG, 214
Ytterbium, 234, 244

Zanotti, *see* Cipher ZBLAN, 159
Zimmermann, 428